Atlas of Diagnostic Pathology in Nonhuman Primates

Ivanela Kondova-Perseng • Keith G. Mansfield •
Andrew D. Miller
Editors

Atlas of Diagnostic Pathology in Nonhuman Primates

Springer

Editors
Ivanela Kondova-Perseng
Division of Pathology and Microbiology
Animal Science Department
Biomedical Primate Research Centre
Rijswijk, The Netherlands

Keith G. Mansfield
Investigative Pathology
Novartis Biomedical Research
Cambridge, MA, USA

Andrew D. Miller
Department of Population Medicine and
Diagnostic Sciences, Section of Anatomic Pathology
Cornell University College of Veterinary Medicine
Ithaca, NY, USA

ISBN 978-3-031-41279-0 ISBN 978-3-031-41280-6 (eBook)
https://doi.org/10.1007/978-3-031-41280-6

Cover illustration: Author of the selected cover draft (picture of cynomolgus macaque): F. van Hassel at BPRC, The Netherlands.

This Springer imprint is published by the registered company Springer Nature Switzerland AG
The registered company address is: Gewerbestrasse 11, 6330 Cham, Switzerland

We would like to dedicate the atlas to our mentors and teachers who gave us the knowledge and confidence to work in this challenging and important field of research.
"Scientia potentia est."

Preface

This carefully presented collection is the first atlas focusing on the diagnostic pathology in nonhuman primates (NHPs). Compiled by an international team of expert contributors, this work offers extensive image material describing classic and unusual lesions in NHP diseases and pathological conditions.

The chapters are arranged around infectious and non-infectious diseases. Viral, bacterial, fungal, and parasitic diseases as well as nutritional, toxic, and metabolic causes are comprehensively covered. Contributions on genetic, age-related, neoplastic, and non-infectious inflammatory conditions complete this atlas.

Recent scientific evidence supports that NHP remain an indispensable resource for efficacy and safety evaluation of novel therapeutic strategies that target clinically important human diseases. Research with monkeys is critical to preventing and treating emerging infectious diseases such as Zika, Ebola, Middle East respiratory syndrome (MERS), SARS, pandemic flu, and many more.

This book will guide professionals in the diagnostic process toward optimal clinical diagnoses. Not only veterinary practitioners in university facilities, zoos, and biotechnological and pharmaceutical companies but also clinicians, researchers, and students engaged in NHP research will find this atlas of considerable value.

Rijswijk, The Netherlands — Ivanela I. Kondova-Perseng
Cambridge, MA — Keith G. Mansfield
Ithaca, NY — Andrew D. Miller

Acknowledgments

We wish to express our warm gratitude to the editors of Springer, Lee Klein and Annette Klaus, who were very helpful during the preparation of the atlas and encouraged us to overcome all the challenges we faced during the COVID-19 pandemic.

Contents

Editors and Contributors

About the Editors

Ivanela I. Kondova-Perseng Division of Pathology and Microbiology, Animal Science Department, Biomedical Primate Research Centre, Rijswijk, The Netherlands

Keith G. Mansfield Investigative Pathology, Novartis Institute for Biomedical Research, Cambridge, MA, USA

Andrew D. Miller Department of Population Medicine and Diagnostic Sciences, Section of Anatomic Pathology, Cornell University College of Veterinary Medicine, Ithaca, NY, USA

Contributors

Sarah E. Beck Charles River Laboratories, Veterinary Pathologist II, Columbia, MD, USA

Martina Bleyer German Primate Center, Pathology Unit, Göttingen, Germany

Andrew N. Cartoceti Zoetis Reference Laboratories, Parsippany-Troy Hills, NJ, USA

Peter J. Didier Division of Comparative Pathology, Tulane National Primate Research Center (Tulane University School of Medicine), Covington, LA, USA

Amanda L. Johnson Inotiv – Boulder, Boulder, CO, USA

Shannon G. M. Kirejczyk Division of Pathology, Emory National Primate Research Center, Atlanta, GA, USA

StageBio, Mount Jackson, VA, USA

Ivanela I. Kondova-Perseng Department of Animal Science, Division of Pathology and Microbiology, Biomedical Primate Research Centre, Rijswijk, The Netherlands

Anne D. Lewis Pathology Services Unit, Oregon National Primate Research Center, Oregon Health and Science University, Beaverton, OR, USA

Keith G. Mansfield Investigative Pathology, Novartis Biomedical Research, Cambridge, MA, USA

Kerstin Mätz-Rensing Deutsches Primatenzentrum GmbH, Leibniz-Institut für Primatenforschung, Göttingen, Germany

Lars Mecklenburg Labcorp Early Development Services GmbH, Münster, Germany

Andrew D. Miller Department of Population Medicine and Diagnostic Sciences, Section of Anatomic Pathology, Cornell University College of Veterinary Medicine, Ithaca, NY, USA

Katherine J. Olstad Department of Pathology, Microbiology and Immunology, California National Primate Research Center, University of California, Davis, Davis, CA, USA

Heather A. Simmons Pathology Services Unit, Wisconsin National Primate Research Center, University of Wisconsin-Madison, Madison, WI, USA

Viral Diseases in Nonhuman Primates

1

Kerstin Mätz-Rensing and Keith G. Mansfield

Abstract

Viral infections remain an important threat to nonhuman primate colonies and range in severity from unrecognized subclinical infections that may impact research to severe epizootics associated with significant morbidity and mortality. Many may cause serious zoonotic infections. Histological evaluation and molecular localization studies utilizing immunohistochemistry and in situ hybridization are important tools that may assist in rapid diagnosis of such infections. This chapter will review the pathology and diagnosis of common viral infections of nonhuman primates.

Keywords

Nonhuman primate · Viral infections · Epizootic · Zoonosis · Pathology · Diagnosis · Immunohistochemistry

Viral infections remain an important threat to nonhuman primate colonies and range in severity from unrecognized subclinical infections that may impact research to severe epizootics associated with significant morbidity and mortality. Many may cause serious zoonotic infections. Histological evaluation and molecular localization studies utilizing immunohistochemistry and in situ hybridization are important tools that may assist in rapid diagnosis of such infections. This chapter will review the pathology and diagnosis of common viral infections of nonhuman primates.

K. Mätz-Rensing
Deutsches Primatenzentrum GmbH, Leibniz-Institut für Primatenforschung, Göttingen, Germany
e-mail: kmaetz@dpz.eu

K. G. Mansfield (✉)
Investigative Pathology, Novartis Biomedical Research, Cambridge, MA, USA
e-mail: keith.mansfield@novartis.com

1.1 Enveloped DNA Viruses

1.1.1 Poxviridae

Poxviruses are large brick- to ovoid-shaped enveloped DNA viruses and have a characteristic morphology on ultrastructural examination. Their genomes are complex often encoding more than 100 polypeptides that play a role in viral pathogenesis and virulence. The family includes the orthopoxvirus, Yatapoxvirus, and molluscipoxvirus genera, all of which contain viruses that infect nonhuman primates. Several species are also susceptible to variola and vaccinia infection and serve as important animal models for human infection [1].

1.1.2 Monkeypox

Monkeypox is an orthopoxvirus with natural rodent reservoirs restricted to the Congo Basin and West Africa. Infection has been recognized in a variety of New World Primate (NWP) and Old World Primate (OWP) species [2, 3]. It is an enveloped DNA virus 200–250 nm in diameter with a brick-like shape. Two clades based on geography are recognized with the Congo basin (CB) more virulent than the West African (WA) clade. The virus is thought to infect a number of small African rodent species including Funisciurus sp. (rope squirrel), Cricetomys sp. (Gambian-pouched rat), and Graphiurus sp. (African dormouse) which have also served as vectors in a North American outbreak. While African in origin, the first cases were diagnosed in Asian-origin *M. fascicularis* housed in an animal facility located in Copenhagen, Denmark, shortly after importation in 1958 [4]. A second outbreak occurred in macaques housed in Philadelphia, United States in 1959 in both *M. fascicularis* and *M. mulatta* [5, 6]. Two forms of the disease were reported with an acute form characterized by facial edema resulting in dyspnea and asphyxiation and accompanied by a papular exanthema on haired skin, oral mucosa ulcerations, and

I. Kondova - Perseng et al. (eds.), *Atlas of Diagnostic Pathology in Nonhuman Primates*,
https://doi.org/10.1007/978-3-031-41280-6_1

lymphadenopathy. In this severe form, lesions were often hemorrhagic. A second form presented with skin lesions but no other clinical signs and took a more prolonged course.

Histologically, the skin lesions progress through an initial phase followed by proliferative acanthosis and hydropic degeneration (Fig. 1.1). This leads to vesiculation and either erosion/ulceration or a region of central necrosis surrounded by proliferative epithelium termed umbilication or "pock lesion." Eosinophilic intracytoplasmic viral inclusions are evident in infected cells (Fig. 1.2). Lesions often first involve the hands, feet, or face and spread centripetally to the trunk. Experimental inoculation of *M. fascicularis* in which the disease is more severe has been used to further our understanding of orthopoxvirus pathogenesis, prevention, and treatment [7–11].

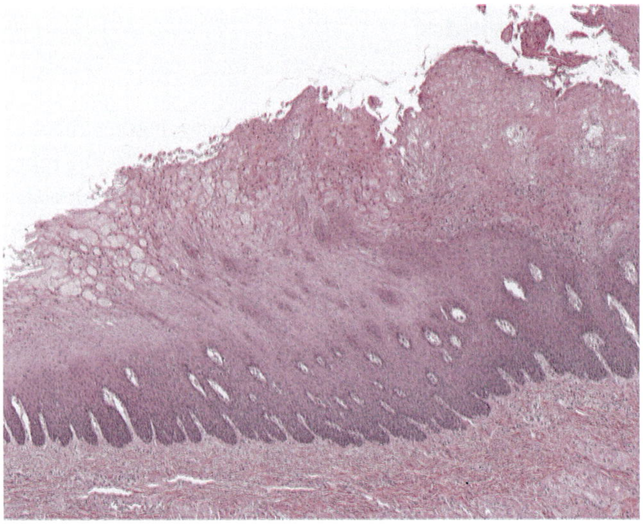

Fig. 1.1 Monkeypox virus (orthopoxvirus genus). Proliferation and hydropic degeneration of mucosal epidermal cells (*Macaca mulatta*, oral mucosa, H&E)

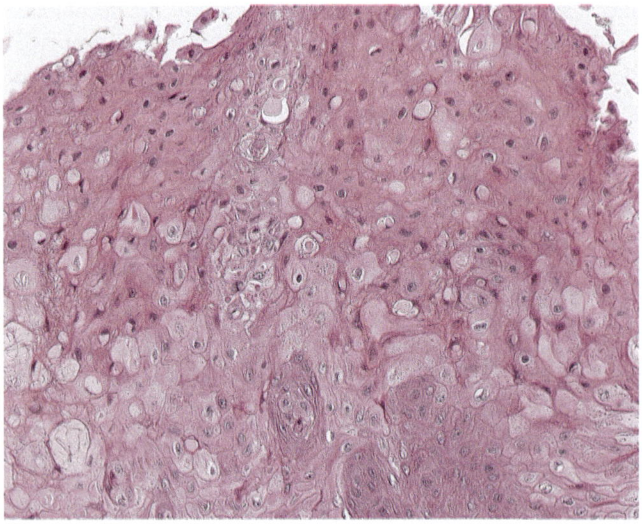

Fig. 1.2 Monkeypox virus (orthopoxvirus genus). Hydropic degeneration and pale eosinophilic intracytoplasmic inclusion bodies (*Macaca mulatta*, oral mucosa, H&E). This proliferative phase precedes vesiculation and umbilication

1.1.3 Cowpox

Cowpox is orthopoxvirus which is carried by rodent vectors and may be inadvertently transmitted to nonhuman primates (NHPs) including callitrichids and macaques. NWPs develop typical pox-like lesions on the face, scrotum, and soles and palms [12–14]. The animals succumb to the viral infection and secondary bacterial pathogens. Typical histologic changes are observed in skin. The condition has been termed "Calpox" due to the susceptibility of callitrichids. A severe epizootic has also been described in a colony of Tonkean macaques (*Macaca tonkeana*). Severe pulmonary involvement was observed in animals that died within 48 h of disease onset and animals that survived longer developed a vesicular exanthema on the face, oral cavity, tongue [15], and inguinal region. Multifocal hepatic necrosis and interstitial pneumonitis were also noted. A modified Vaccinia virus Ankara (MVA) vaccine was used to protect remaining animals. Histologically, the skin lesions with cowpox infection are typical of orthopoxvirus and progress through proliferation, vesiculation, and umbilication phases, and large, brightly eosinophilic intracytoplasmic inclusion bodies were evident (Fig. 1.3).

Cowpox can infect humans and an asymptomatic occupational transmission has been reported in one epizootic [16]. The virus must be distinguished from other orthopoxviruses and molecular techniques may be required for definitive diagnosis.

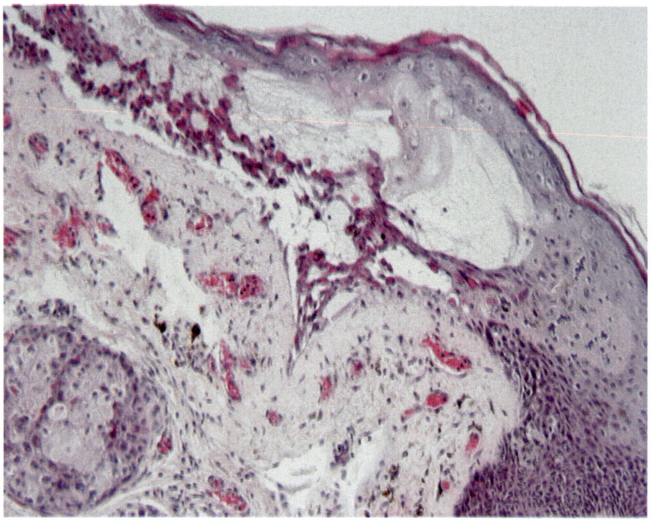

Fig. 1.3 Cowpox virus infection (orthopoxvirus genus). Vesiculation of epidermis with brightly eosinophilic intracytoplasmic inclusions along remaining basal layer and within adjacent sebaceous gland (*Callithrix jacchus*, skin, H&E)

1.1.4 Yatapoxvirus

The Yatapoxviruses contain two virus groupings: the Yaba monkey virus and the Tanapox-like viruses. Yaba monkey virus was first recognized as a cause of subcutaneous tumors in a group of captive macaques in 1957. Natural outbreaks are characterized by the appearance of multiple subcutaneous masses varying in size from small papules to nodules. These masses often are observed on the face and extremities. Larger masses may ulcerate and masses regress after 6–8 weeks. New lesions may develop as older lesions regress. Few naturally occurring infections of African species have been described but captive-born animals are susceptible suggesting pre-existing immunity and widespread infection of wild-caught animals with Yaba or a closely related virus. The characteristic lesion consists of large pleomorphic histiocytes forming a nonencapsulated and infiltrative mass [17]. Mitotic figures may be frequent and large eosinophilic intracytoplasmic inclusions may be evident (Fig. 1.4). Regression is associated with erosion, ulceration, and formation of multinucleated giant cells. Natural infection of humans has previously been described [3, 18].

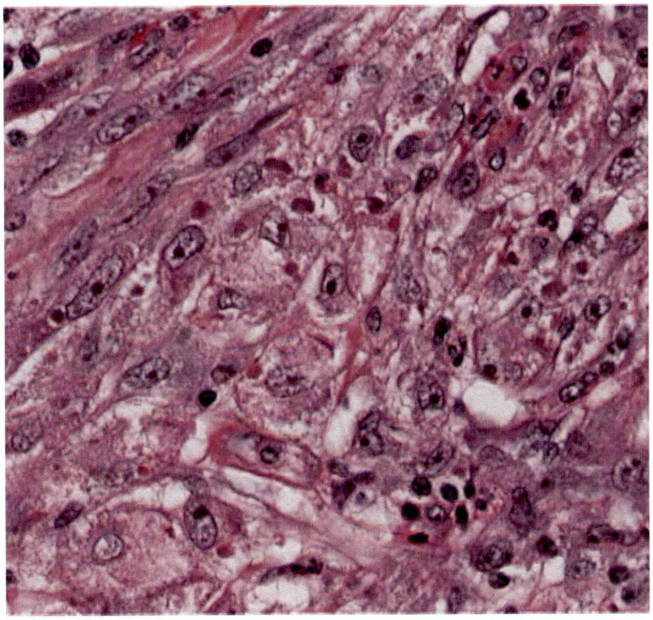

Fig. 1.4 Yaba virus (Yatapoxvirus genus). Neoplastic spindleoid cells within the dermis, some with intracytoplasmic inclusion bodies. Small numbers if infiltrating mononuclear cells are evident (*Macaca mulatta*, dermis, H&E)

Tanapox was first isolated in 1957 and 1962 from humans living along the Tana River Valley in Kenya. A second closely related agent named Yaba-like disease (YLD) virus, Or-Te-Ca poxvirus, or benign epidermal monkeypox (BEMP) was isolated from animal care takers following an outbreak recognized in 1965–67 in Oregon, Texas, and California. Initially thought to represent the same virus as tanapox it is now recognized as a distinct agent. The diseases caused by tanapox and YLD viruses are biologically identical. YLD virus is a relatively benign infection which progresses from small red papules to raised circumscribed nodules up to 1 cm in diameter. Characteristically, these areas are flat and round and often appear on the face or extremities. Focally extensive regions of epidermal proliferation, ballooning degeneration, and eosinophilic intracytoplasmic inclusions are observed. Lesions regress and become umbilicated in 3–4 weeks. The agent may be transmitted by fomites such as tattoo needles or insects. Infected humans developed typical pox-like lesions and constitutional symptoms. The natural reservoir of these viruses is unknown.

1.2 Herpesviridae

Herpesviridae are spread worldwide and can be identified in many species. In their natural hosts, they generally do not cause a fatal disease course. However, cross-species infection can cause acute severe illness and death. Based on biological and genetic sequence characteristics, the Herpesviridae family contains the alphaherpesvirinae, betaherpesvirinae, and gammaherpesvirinae subfamilies, all of which contain viral agents known to infect nonhuman primates.

1.2.1 Alphaherpesvirinae

Human Herpesvirus 1,2 (HHV1,2) (Herpes Simplex 1, 2 (HSV-1,2)

HHV1 causes in its natural host (human) only mild facial lesions like gingivostomatitis or keratitis or inapparent infection. Only in neonates or immune-compromised individuals, can fatal encephalitis or systemic disease occur [19]. In humans, infection with HHV1 is lifelong because of viral latency in sensory neurons. When immunity is compromised through stress or disease, it becomes activated through suppression of the latency-associated transcript genes. It can be shed even in the absence of visible lesions.

Several nonhuman primate species are susceptible to infection with human herpesviruses. Old World Primates like gorillas (*Gorilla gorilla*), bonobos (*Pan paniscus*), white-handed gibbons (*Hylobates lar*), and chimpanzees (Pan troglodytes) can be naturally infected with HHV1 [20–25]. Although single fatal disease outcomes have been described, their virus–host relationship seems to be similar with that in humans, as they usually show only mild symptoms (lesions in mucocutaneous tissue, skin, conjunctiva). However, New World monkey species and prosimians are highly susceptible to HHV 1 infections [26–29]. Typical lesions are erosions and ulcers of oral mucous membranes, especially on the tongue, and mucocutaneous junctions of the lips. Severe necrotizing hepatitis (Figs. 1.5 and 1.6) and multifocal nonsuppurative meningoencephalitis often occur. In the brain, lymphocytic perivascular cuffing and intranuclear viral inclusions are observed (Fig. 1.7).

As a group callitrichids are highly susceptible to severe HHV1 infections. The contact of callitrichids with latently or symptomatically infected humans should be reduced and prophylactic strategies like face masks are strongly recommended to reduce the risk of infection.

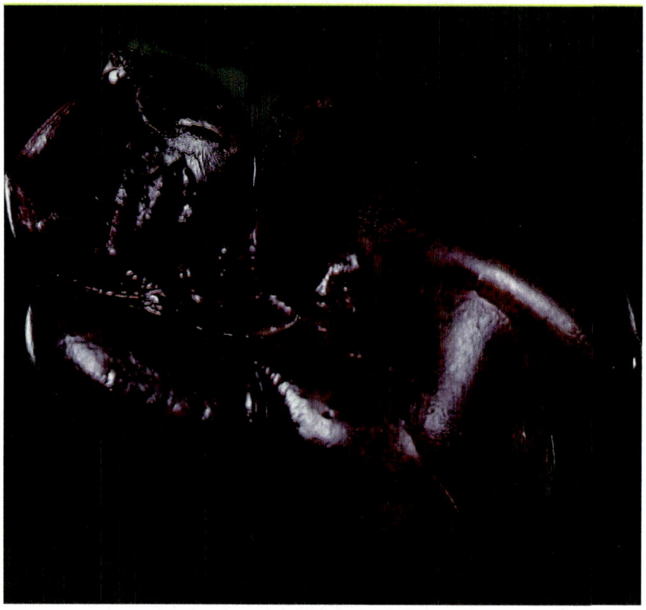

Fig. 1.5 Human herpesvirus 1 (Herpes simplex virus 1; simplexvirus genus). Multifocal to coalescing hepatic necrosis (*Aotus trivirgatus*)

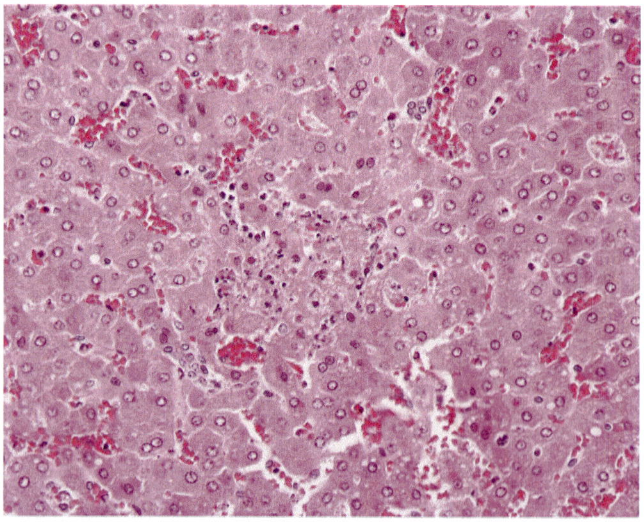

Fig. 1.6 Human herpesvirus 1 (Herpes simplex virus 1; simplexvirus genus). Focal hepatocellular necrosis with pyknotic debris and eosinophilic intranuclear inclusion bodies present within adjacent hepatocytes (*Callithrix jacchus*, liver, H&E)

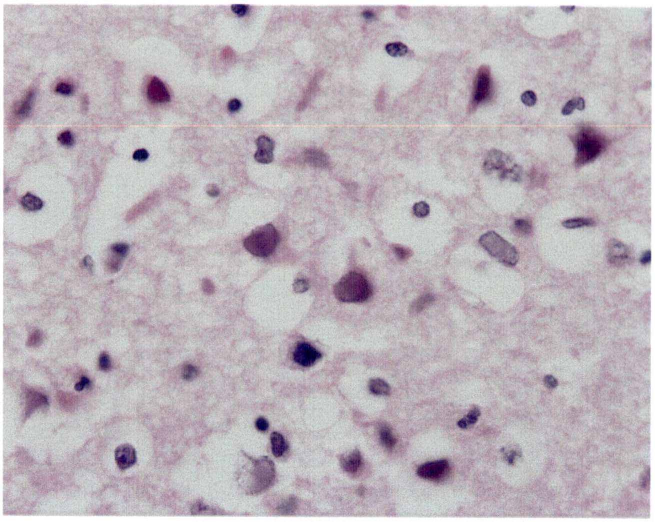

Fig. 1.7 Human herpesvirus 1 (Herpes simplex virus 1; simplexvirus genus). Viral dissemination is common to the central nervous system and may be accompanied by a mixed inflammatory cell infiltration, necrosis, and intranuclear inclusion bodies (*Callithrix jacchus*, brain, H&E)

Macacine Herpesvirus 1, Herpesvirus B, Herpes Simiae, B Virus (BV)

BV is an alphaherpesvirinae within the simplexvirus genus found enzootically in all species of macaques kept in captivity and results in lifelong infection [17]. Specific strains of BV have been identified associated with individual macaque species and may potentially differ in their virulence [30]. Disease in macaques is usually mild or asymptomatic and in conventional colonies seropositivity may reach 70–90% in adults. The virus is transmitted in saliva and other mucosal secretions. Primary infection and seroconversion in breeding colonies occur in juvenile animals and at the time of sexual maturity. Following primary infection, there may be conjunctivitis, vaginitis, or oral vesicles/ulcerations. Lesions heal rapidly and latent infection is established in sensory ganglion [31]. Reactivation and shedding may occur during periods of stress such as breeding, parturition, or concurrent illness. These may or may not be associated with any visible clinical signs of disease.

Rarely in young animals or animals with concurrent illness, severe primary or reactivation disease may be observed. This may be characterized by more severe involvement of mucosal surfaces (Fig. 1.8), involvement of haired skin, and dissemination to multiple internal organs including lung, liver (Figs. 1.9 and 1.10), kidney, adrenal gland (Figs. 1.11 and 1.12), and central nervous system. In the liver, multifocal hepatic necrosis is observed and inclusions can be visualized in remaining hepatocytes at the leading edge of the expanding lesion (Fig. 1.13). In these cases, the clinical presentation may not obviously suggest BV and a high degree of clinical suspicion should be maintained for BV in conventionally reared animals. Immunohistochemistry can be used to confirm the presence of viral antigen (Fig. 1.14).

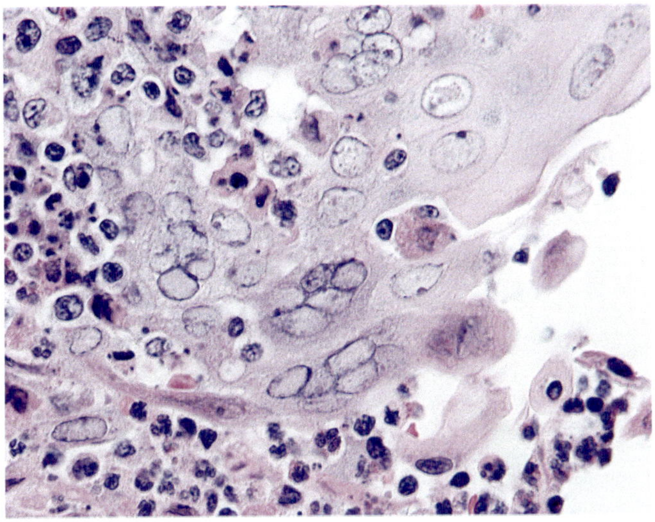

Fig. 1.8 Macacine herpesvirus 1 (B virus; simplexvirus genus). Syncytial cells with intranuclear viral inclusions accompanied by neutrophilic infiltrate (*Macaca mulatta*, tonsil, H&E)

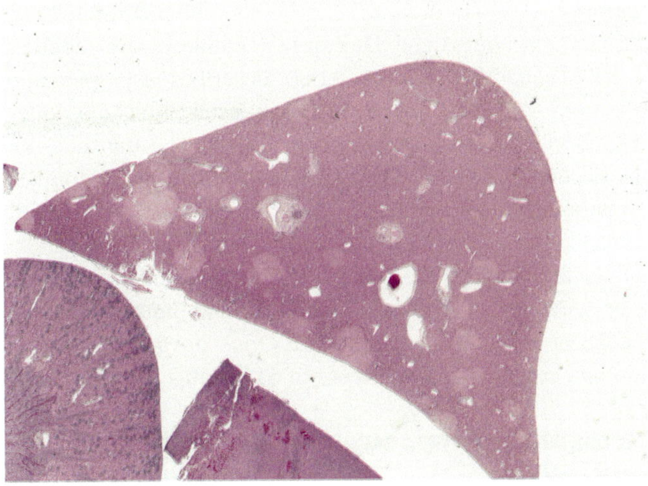

Fig. 1.9 Macacine herpesvirus 1 (B virus; simplexvirus genus). Multifocal to coalescing hepatic necrosis evident as pale foci in liver of neonatal rhesus macaque (*Macaca mulatta*, subgross liver, H&E)

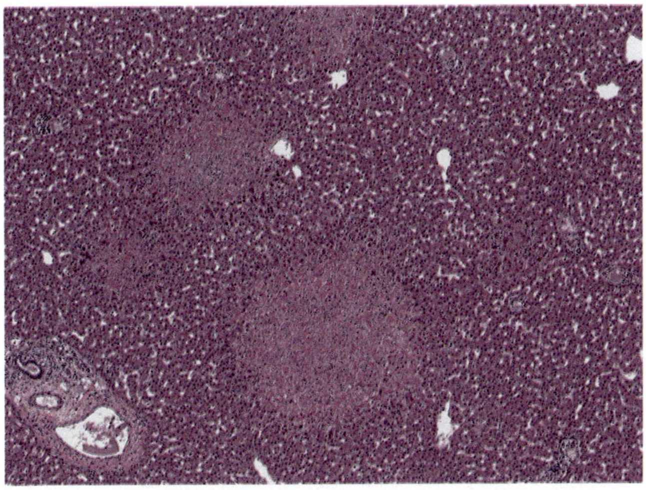

Fig. 1.10 Macacine herpesvirus 1 (B virus; simplexvirus genus). Foci of coagulative hepatocellular necrosis mixed with pyknotic debris (*Macaca mulatta*, liver, H&E)

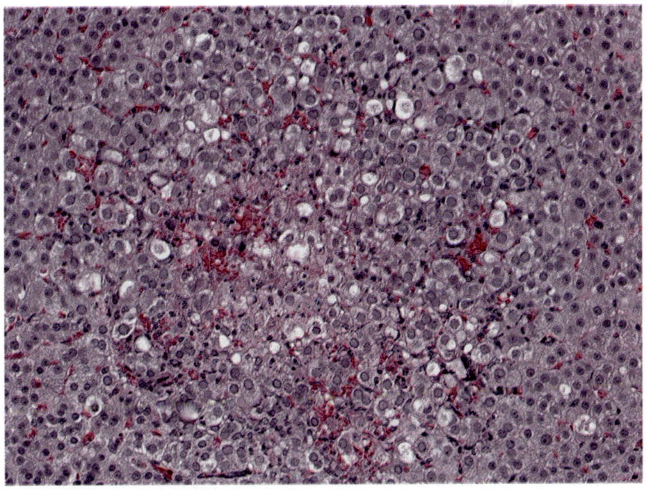

Fig. 1.11 Macacine herpesvirus 1 (B virus; simplexvirus genus). Focus of adrenal necrosis and hemorrhage in neonatal macaque with disseminated B virus infection (*Macaca mulatta*, adrenal gland, H&E)

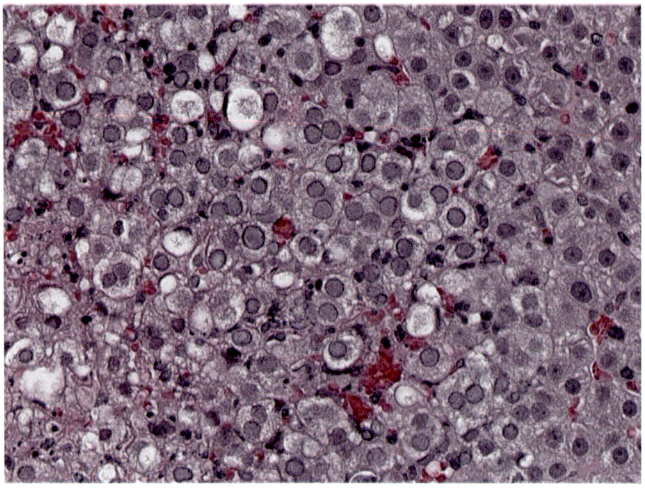

Fig. 1.12 Macacine herpesvirus 1 (B virus; simplexvirus genus). Abundant lightly basophilic intranuclear inclusions evident within adrenocytes (*Macaca mulatta*, adrenal gland, H&E)

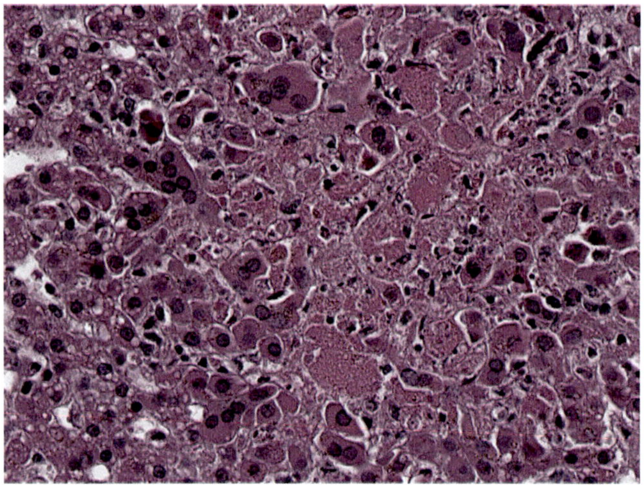

Fig. 1.13 Macacine herpesvirus 1 (B virus; simplexvirus genus). Foci of coagulative hepatocellular necrosis mixed with pyknotic debris. Intranuclear inclusions and rare syncytia are evident at the interface of necrosis and remaining hepatocytes (*Macaca mulatta*, liver, H&E)

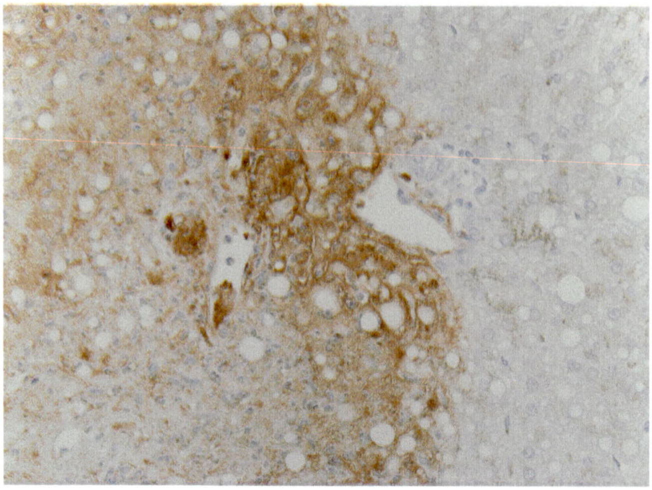

Fig. 1.14 Macacine herpesvirus 1 (B virus; simplexvirus genus). Immunohistochemistry demonstrates simplex viral antigen at interface of viable and necrotic hepatocytes (*Macaca mulatta*, liver, avidin biotin immunostain, 3,3'-diaminobenzidine (DAB) chromogen)

Transmission to other NHP species may result in a more severe disease process and resemble HSV or HVT infection described in callitrichids. Transmission to other primate species can result in fatal disease. African green monkeys, marmosets, Owl monkeys, gibbons, bonnet monkeys, Barbary macaques, and DeBrazza's monkeys are reported to be susceptible [32]. A report describes asymptomatic infections of capuchin monkeys housed with macaques [33]. B virus is a serious zoonotic disease and transmission has occurred through bites, splashes, scratches, or fomites contaminated with monkey saliva. Disease in affected humans is characterized by ascending paralysis, encephalitis, and death. Specific pathogen free (SPF) breeding colonies are now widespread; however, facilities that house macaque species or work their unfixed tissues must have a BV prevention and control plan regardless of SPF status [34, 35].

Cercopithecine Herpesvirus 9, Simian Varicella Virus (SVV)

Simian varicella viruses (SVVs) are a group of alphaherpesviruses including the closely related Liverpool vervet virus, Medical lake virus, and Delta herpesvirus first recognized in 1967–1968. These viruses are antigenically related and cause a similar exanthematous disease [17]. Disease has been recognized in Vervet and Patas monkeys and a variety of macaque species. In African OWPs, the disease may be severe and initially recognized as a cutaneous exanthema progressing to disseminated disease and death within 48–72 h. Following experimental inoculation, animals may have a neutrophilic leukocytosis and elevations in liver enzymes. Mortality may vary and may be dependent on stress and other co-factors.

Vescicular dermatitis appears 6–8 days following experimental inoculation and appears to be the primary lesion. The rash often appears in the inguinal region and spreads centripetally to the face and extremities (Fig. 1.15). In contrast to pox virus infections usually spare the soles and palms [26]. Cutaneous lesions are characterized by multiple vesicles that contain debris, erythrocytes, acantholytic keratinocytes, and less commonly syncytial cells (Fig. 1.16). Intranuclear inclusions can be found in adjacent dermis (Fig. 1.17). Syncytia are occasionally seen but not prominent (Figs. 1.1–1.18). Viral antigen can be found widely disseminated by 8 days including within lung, liver, adrenal gland, skin, and trigeminal ganglion. In severe cases, vesicles may appear blood filled and indicate disseminated infection with involvement of the vasculature (Fig. 1.19). During periods of immunosuppression, latent virus may become reactivated and the infection can disseminate and impact various organs including the gastrointestinal tract, adrenal glands, lungs, and liver (Figs. 1.20 and Fig. 1.21) [36]. This can happen following whole body irradiation or use of solid organ transplantation regimes. In disseminated cases, extensive lymphoid necrosis may be evident (Fig. 1.22). Differentiation from disseminated BV should be considered: (1) BV does not normally cause epizootics in macaques; (2) BV does not normally produce extensive cutaneous involvement; (3) vesiculation with SVV generally occurs higher in the epidermis; and (4) SVV does not cross react with simplexvirus immunohistochemistry. Definitive diagnosis may require viral isolation and/or molecular techniques [37]. Transmission of SVV to humans has not been documented.

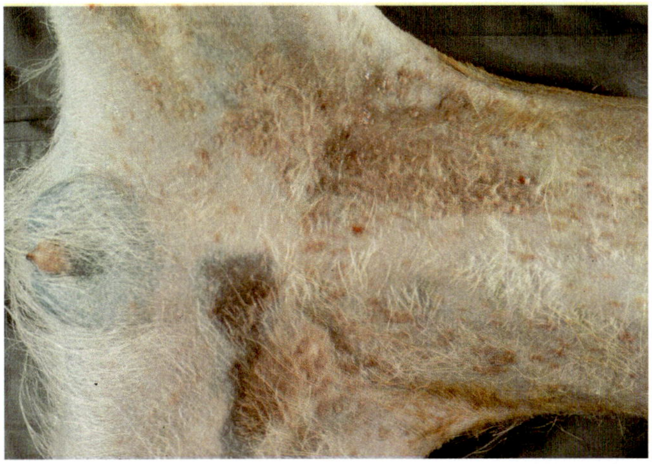

Fig. 1.15 Cercopithecine herpesvirus 9 (Simian varicella virus; varicellovirus genus). Viral exanthema on abdomen due to simian varicella virus infection. In severe cases, lesions may appear as blood filled vesicles (Patas monkey)

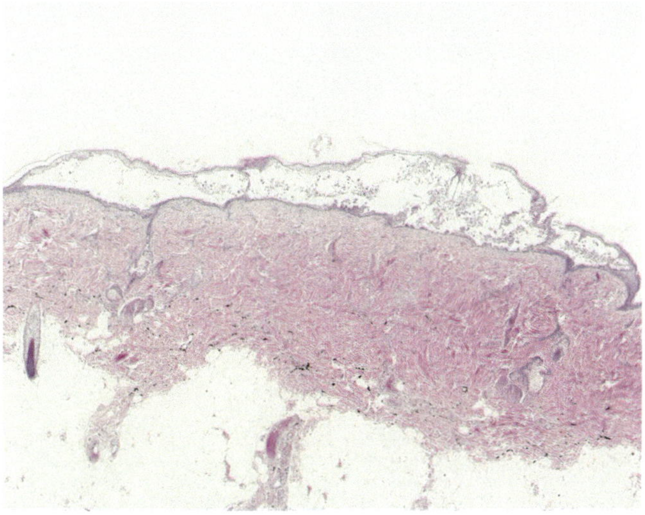

Fig. 1.16 Cercopithecine herpesvirus 9 (Simian varicella virus; varicellovirus genus). Large intraepidermal vesicle containing exudate and acantholytic keratinocytes (*Macaca mulatta*, skin, H&E)

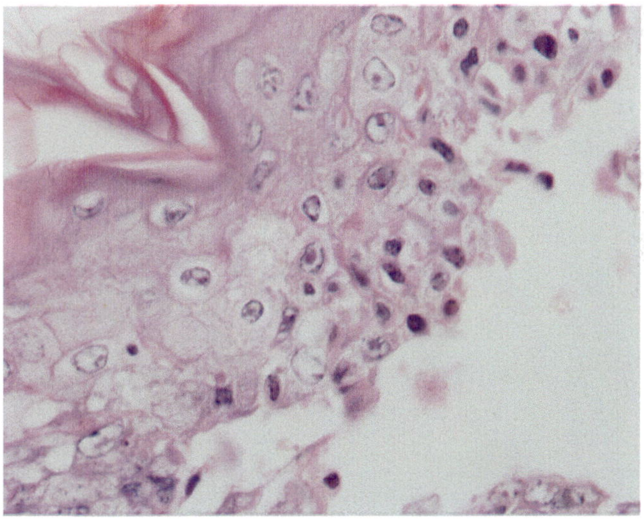

Fig. 1.17 Cercopithecine herpesvirus 9 (Simian varicella virus; varicellovirus genus). Small intraepidermal vesicle containing acantholytic keratinocytes (*Macaca mulatta*, skin, H&E)

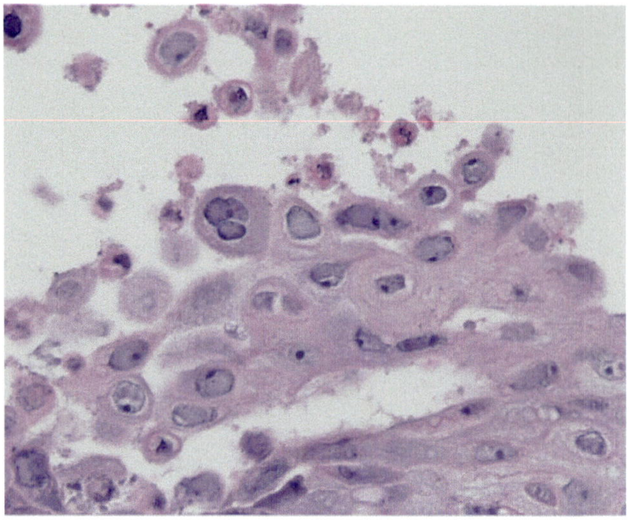

Fig. 1.18 Cercopithecine herpesvirus 9 (Simian varicella virus; varicellovirus genus). Edge of large intraepidermal vesicle demonstrating viral syncytial cells and poorly defined intranuclear inclusions (*Macaca mulatta*, skin, H&E)

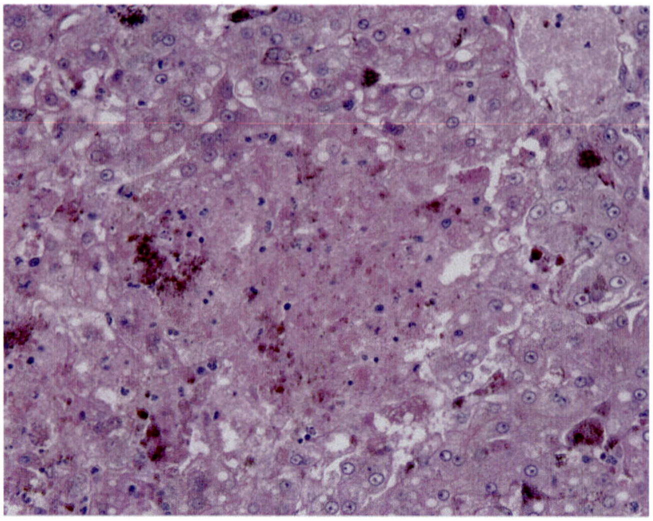

Fig. 1.20 Cercopithecine herpesvirus 9 (Simian varicella virus; varicellovirus genus). Focal hepatic necrosis with swelling and disorganization of surrounding hepatic cords (*Macaca mulatta*, liver, H&E)

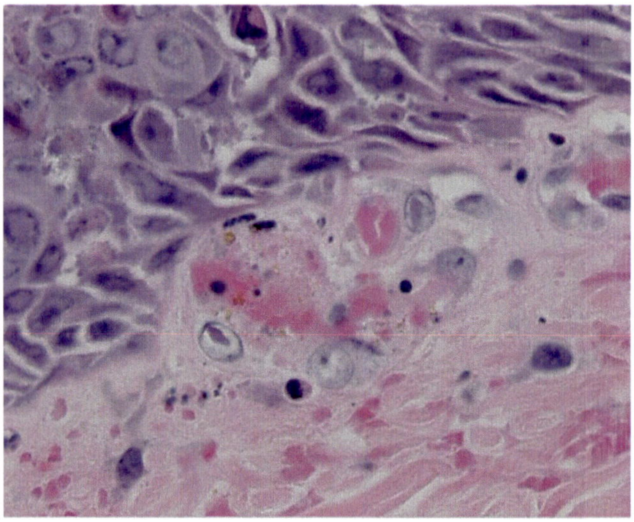

Fig. 1.19 Cercopithecine herpesvirus 9 (Simian varicella virus; varicellovirus genus). Endothelial necrosis with intranuclear viral inclusions in a small dermal capillary. Vascular involvement can be seen with disseminated disease and result in blood filled vesicles observed clinically (*Macaca mulatta*, skin, H&E)

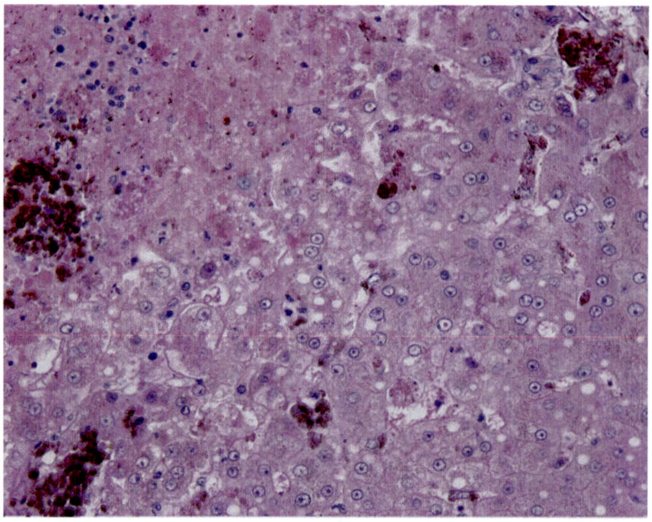

Fig. 1.21 Cercopithecine herpesvirus 9 (Simian varicella virus; varicellovirus genus). Intranuclear viral inclusions are often evident in hepatocytes directly adjacent to areas of necrosis (*Macaca mulatta*, liver, H&E)

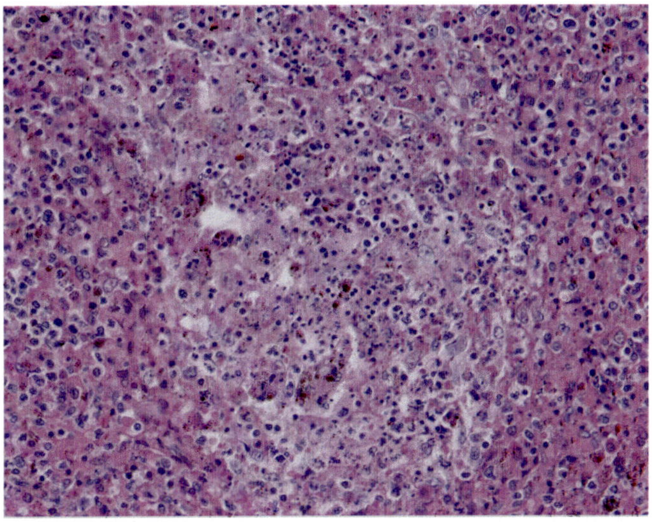

Fig. 1.22 Cercopithecine herpesvirus 9 (Simian varicella virus; varicellovirus genus). Lymphoid necrosis is observed in advanced disease. Intranuclear inclusions are more difficult to recognize but are usually present (*Macaca mulatta*, spleen, H&E)

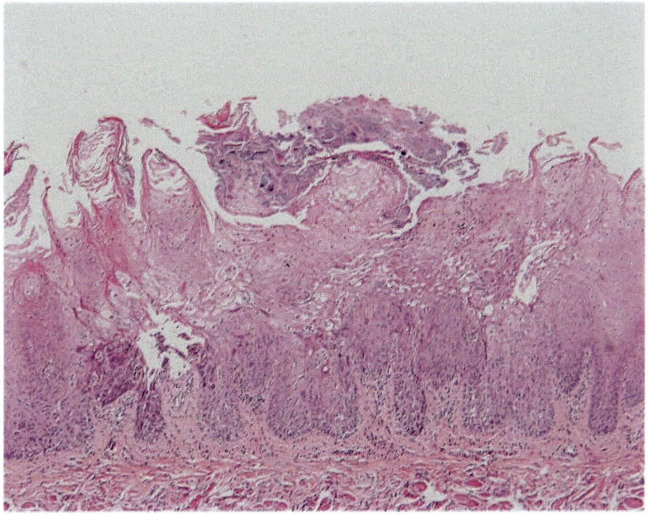

Fig. 1.23 Saimirine herpesvirus 1 (Herpesvirus tamarinus; simplexvirus genus). Early intraepidermal vesicles may be accompanied by minimal inflammatory response due to the rapidly progressive nature of the disease (*Aotus trivirgatus*, oral mucosa, H&E)

Saimirine Herpesvirus 1 (Also Known as Herpesvirus Tamarinus; HVT)

Saimirine herpesvirus 1 is an alphaherpesvirus in the simplexvirus genus found as an enzootic infection of squirrel monkeys [12, 17, 38]. In the natural host, few clinical signs are recognized. Greater than 95% of adult squirrel monkeys are seropositive. However, it is only rarely associated with oral vesicles and ulcerations during primary infection in this species.

Disease is more severe in Owl monkeys, tamarins, and marmosets. Animals may present with conjunctivitis and oral ulcerations progressing to more severe clinical disease; death may be the presenting clinical sign. The virus may be transmitted through bites or fomites. Once established within a Callitrichid or Owl monkey colony, spread may be rapid leading to an epizootic of high morbidity and mortality. In Owl monkeys, tamarins, and marmosets, HVT causes a disseminated cytolytic viral infection initially involving the mucosa and later multiple internal organs including the lymphoid tissue (Fig. 1.23), central nervous system, lungs, liver, and kidney. The lesion is characterized by multifocal to coalescing areas of necrosis and neutrophilic infiltration with syncytia and intranuclear viral inclusions (Figs. 1.24 and 1.25). In peracute disease, the inflammatory reaction to infection may be minimal. Strict separation of NHP species is recommended. There is no known zoonotic potential.

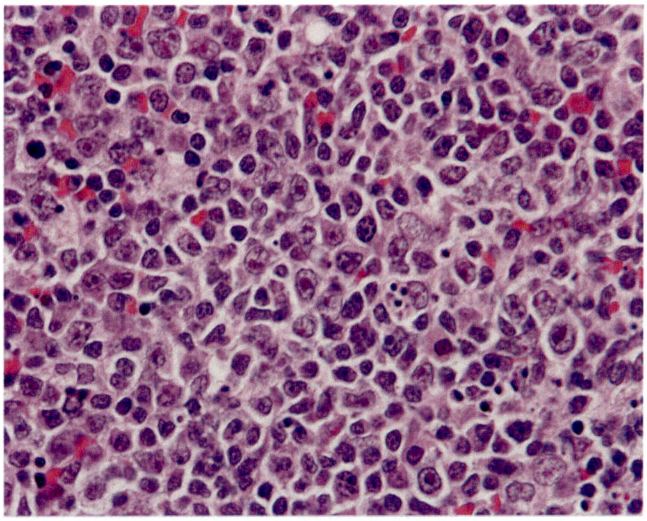

Fig. 1.24 Saimirine herpesvirus 1 (Herpesvirus tamarinus; simplexvirus genus). Lymphoid necrosis with rare intranuclear viral inclusion (*Aotus trivirgatus*, lymph node, H&E)

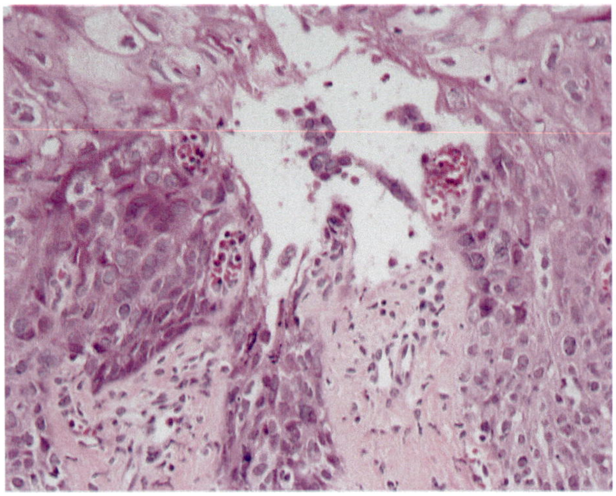

Fig. 1.25 Saimirine herpesvirus 1 (Herpesvirus tamarinus; simplexvirus genus). Forming intraepidermal vesicle containing viral syncytial cells and acantholytic keratinocytes. Numerous intranuclear viral inclusions are evident adjacent to the vesicle and in basal epithelial cells (*Aotus trivirgatus*, oral mucosa, H&E)

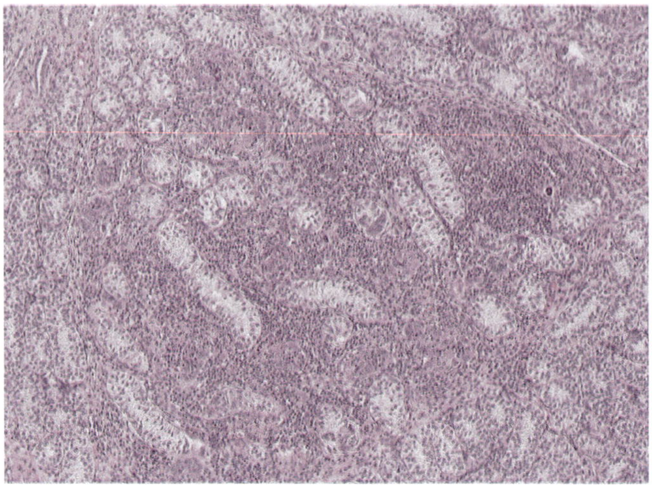

Fig. 1.26 Macacine herpesvirus 3 (rhesus cytomegalovirus; betaherpesvirinae subfamily). Necrotizing neutrophilic orchitis can be observed in primary CMV infection and with reactivation of latent infection with immunomodulation (*Macaca mulatta*, testes, H&E)

1.2.2 Betaherpesvirinae

Cytomegalovirus

CMVs are described in a variety of NWP and OWP species [17]. Infection of the normal mature host is usually asymptomatic and serosurveys reveal that greater than 95% of adult animals may be infected. Disease becomes apparent during periods of immunosuppression or immunodeficiency following reactivation and dissemination. Clinical signs relate to organ system affected and may include diarrhea, respiratory compromise, paralysis, or CNS signs. CMV is a common opportunistic infection in macaques infected with simian immunodeficiency virus (SIV) or simian retrovirus (SRV) and may also be seen in measles virus infected animals [39]. After reactivation, a generalized infection may occur with necrotizing meningitis, neuritis, vasculitis, enterocolitis, and interstitial pneumonia. The histopathologic hallmarks are karyomegaly and large basophilic intranuclear inclusions (Figs. 1.26 and 1.27) and eosinophilic granular intracytoplasmatic inclusion bodies (Fig. 1.28) [39]. The intranuclear inclusions are often said to have an "owl's eye" appearance. The inflammatory infiltrate is often neutrophilic and may produce discrete plaque-like lesions in the gastrointestinal tract when accompanied by epithelial proliferation [40]. Ulceration of the tongue (Fig. 1.29) may be associated with neuritis of branches of the hypoglossal nerve (Fig. 1.30).

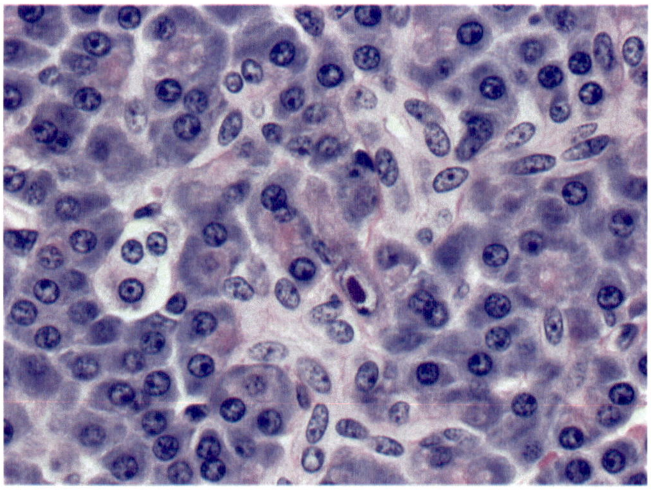

Fig. 1.27 Macacine herpesvirus 3 (rhesus cytomegalovirus; betaherpesvirinae subfamily). Typical "Owl eye" type intranuclear inclusion in a pancreatic acinar cell (*Macaca mulatta*, pancreas, H&E)

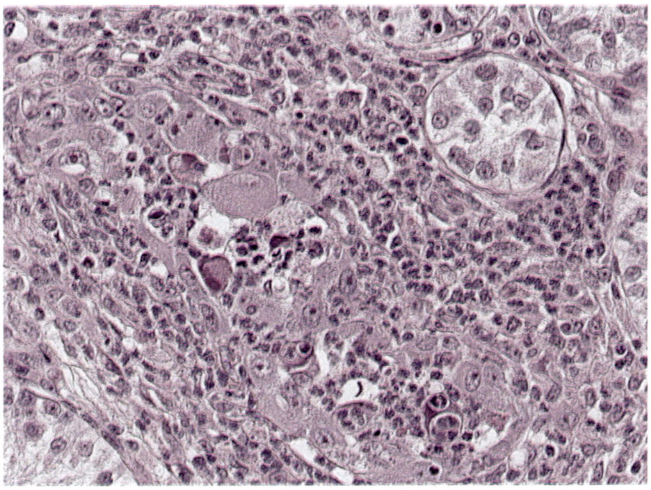

Fig. 1.28 Macacine herpesvirus 3 (rhesus cytomegalovirus; betaherpesvirinae subfamily). Poorly defined intracytoplasmic basophilic inclusions. These inclusions can help differentiate CMV infection from other viral infections associated with karyomegaly, intranuclear inclusions, and neutrophilic infiltrates such as adenovirus and SV40 (*Macaca mulatta*, testes, H&E)

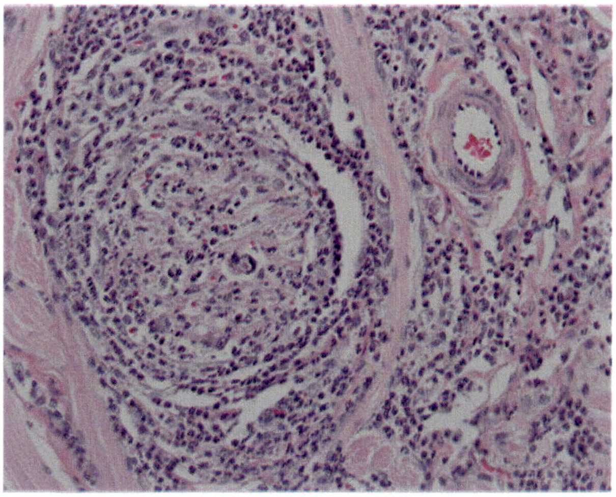

Fig. 1.30 Macacine herpesvirus 3 (rhesus cytomegalovirus; betaherpesvirinae subfamily). Neuritis of the hypoglossal nerve with karyomegaly, "Owl eye" intranuclear viral inclusions, cytoplasmic viral inclusions, and neutrophilic inflammatory response. This combination of findings would be pathognomonic for CMV infection (*Macaca mulatta*, tongue, H&E)

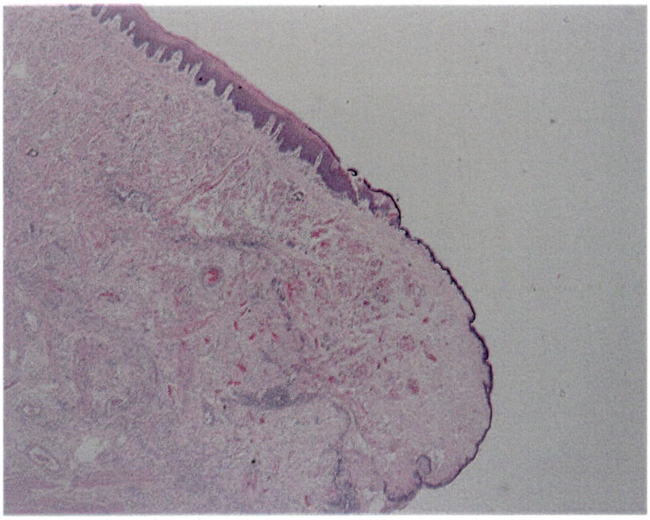

Fig. 1.29 Macacine herpesvirus 3 (rhesus cytomegalovirus; betaherpesvirinae subfamily). Ulceration of the lingual surface associated with CMV neuritis of the hypoglossal nerve. Clinically, this may appear similar to a late B virus infection; however, the CMV lesion does not progress through a vesicular stage (*Macaca mulatta*, tongue, H&E)

This clinical presentation may be similar to BV infection; however, syncytia are absent and both intracytoplasmic and intranuclear inclusion are present. CMV neuritis has been associated with self-injurious behavior [41]. Immunohistochemistry or in situ hybridization may be used to confirm the diagnosis [37]. Often molecular localization techniques detect a greater number of infected cells than recognized on H&E evaluation (Figs. 1.31 and 1.32). CMV myocarditis can be observed in immunomodulated macaques and can be difficult to diagnose due to the absence of typical neutrophils and inclusion bodies (Fig. 1.33). It can be confirmed with molecular localization techniques (Figs. 1.34 and 1.35) [37, 42]. CMV infection should be differentiated from adenovirus and SV40 infections which can also be observed in immunocompromised animals and can produce karyomegaly and intranuclear inclusions. There is no known zoonotic potential.

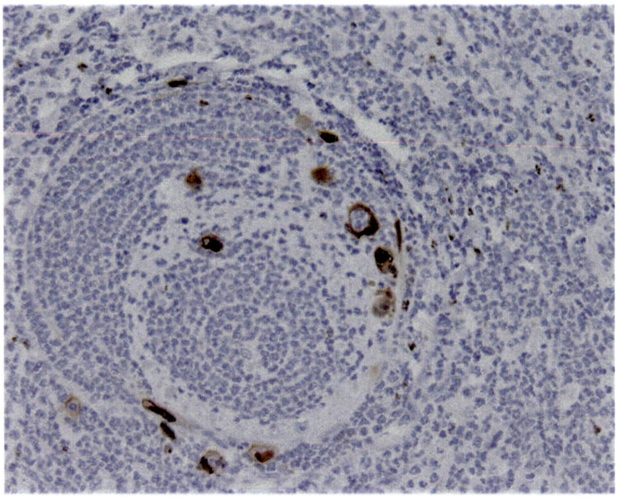

Fig. 1.31 Macacine herpesvirus 3 (rhesus cytomegalovirus; betaherpesvirinae subfamily). Immunohistochemistry for CMV immediate-early protein demonstrates infected cells within the hypoglossal nerve (*Macaca mulatta*, tongue, ABC immunostain; DAB chromogen)

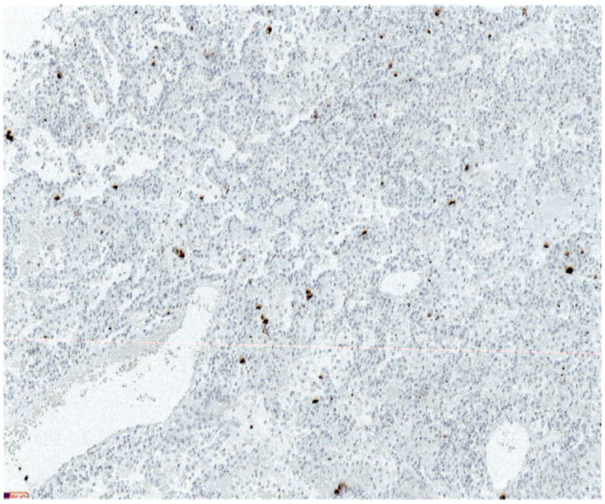

Fig. 1.32 Macacine herpesvirus 3 (rhesus cytomegalovirus; betaherpesvirinae subfamily). CMV may cause interstitial pneumonitis in immune modulated primates. In situ hybridization demonstrates a large number of virally infected cells when only small numbers of cells with viral inclusions were recognized (*Macaca Mulatta*, lung, in situ hybridization, DAB chromogen)

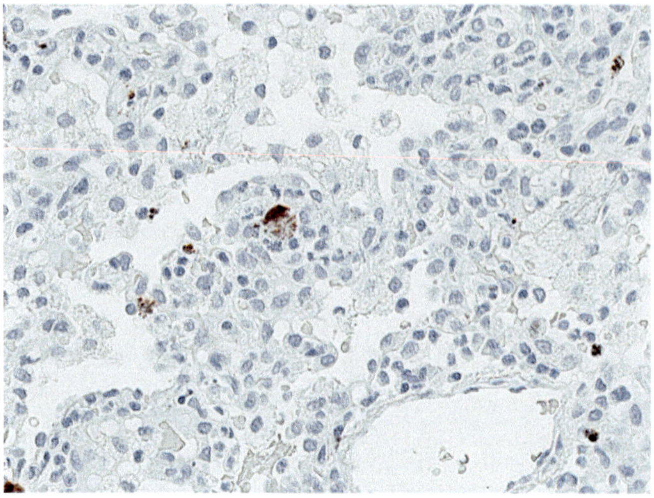

Fig. 1.33 Macacine herpesvirus 3 (rhesus cytomegalovirus; betaherpesvirinae subfamily). Interstitial pneumonitis with in situ hybridization positive cells and accompanying neutrophilic infiltration (*Macaca Mulatta*, lung, in situ hybridization, DAB chromogen)

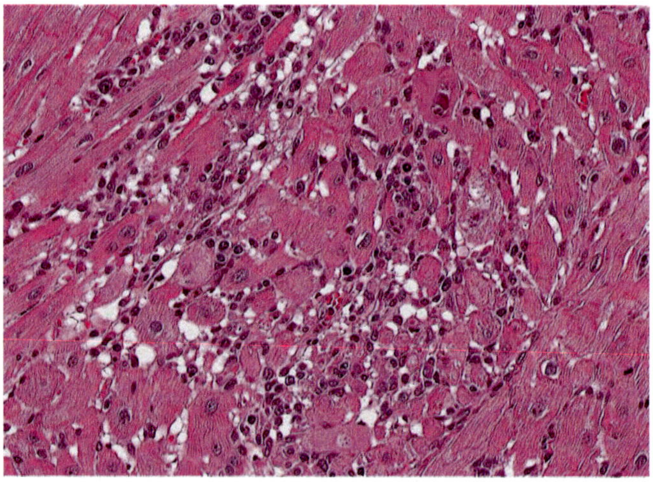

Fig. 1.34 Macacine herpesvirus 3 (rhesus cytomegalovirus; betaherpesvirinae subfamily). Lymphocytic myocarditis can result from CMV reactivation in the myocardium with immune modulation. While karyomegaly and heterochromatic nuclei are observed, definitive identification of viral inclusions can be difficult (*Macaca fascicularis*, heart, H&E)

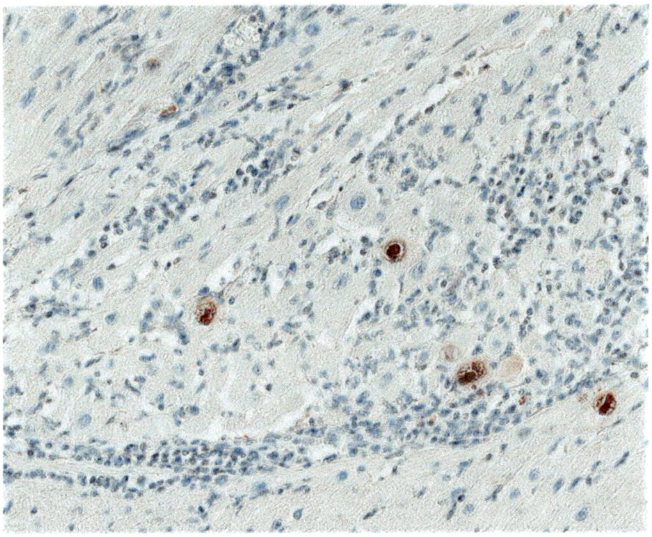

Fig. 1.35 Macacine herpesvirus 3 (rhesus cytomegalovirus; betaherpesvirinae subfamily). Immunohistochemistry for immediate-early protein can positively identify cells in the myocardium when these are difficult to visualize on routine examination (*Macaca fascicularis*, heart, ABC immunostain technique, DAB chromogen)

1.2.3 Gammaherpesvirinae (Lymphocryptoviruses and Rhadinoviruses)

The gammaherpesvirinae subfamily is divided into two distinct lineages: the gamma-1 (lymphocryptoviruses) and gamma-2 (rhadinovirus) groupings. Both contain important viruses that infect NHPs.

Macacine Herpesvirus 4 (Lymphocryptovirus)

A number of simian lymphocryptoviruses (LCV) (gamma-1 herpesviruses) have been isolated from OWPs and NWPs and infect in a species-specific fashion [43]. All are related genetically to the type strain human herpesvirus 4, the Epstein–Barr virus (EBV). There are two distinct lineages of rhesus lymphocryptovirus (RhLCV) which differ in the sequence of their EBNA genes [44]. These viruses are readily transmitted in NHP colonies resulting in high seroprevalence rates. While a common, infection is generally asymptomatic and goes unrecognized. Pathology develops when animals become immunocompromised through concurrent disease or experimental manipulation. In SIV-infected rhesus and cynomolgus macaques, LCV-infections may cause lymphoproliferative disease and malignant lymphoma similar to the development of non-Hodgkin lymphoma in HIV-infected AIDS patients [45]. These are typically B cell lymphomas that develop at extranodal locations like the gastrointestinal tract, kidney, central nervous system, nasal cavity, skeletal muscle, or in the periorbital tissue (Fig. 1.36) [46–49]. These Burkitt-like lymphomas are often classified as centroblastic, immunoblastic, or large cell phenotypes (Fig. 1.37) [50].

Post-transplant lymphoproliferative disease (PTLD) may arise following solid organ transplantation regimes and also commonly affects extranodal sites such as liver (Fig. 1.38) and kidney (Fig. 1.39) [51].

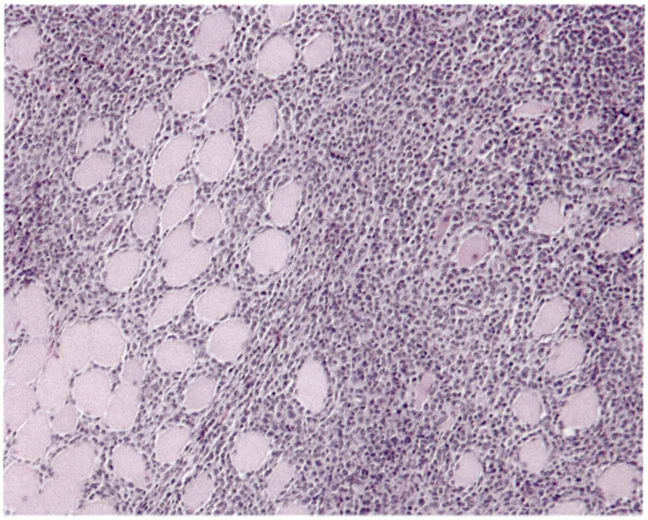

Fig. 1.36 Macacine herpesvirus 4 (rhesus lymphocryptovirus; gammaherpesvirinae subfamily). Lymphoma in diaphragmatic skeletal muscle. Reactivation of LCV during immune modulation can result in extranodal B cell lymphomas and is common finding in SIV-infected macaques (*Macaca mulatta*, diaphragm, H&E)

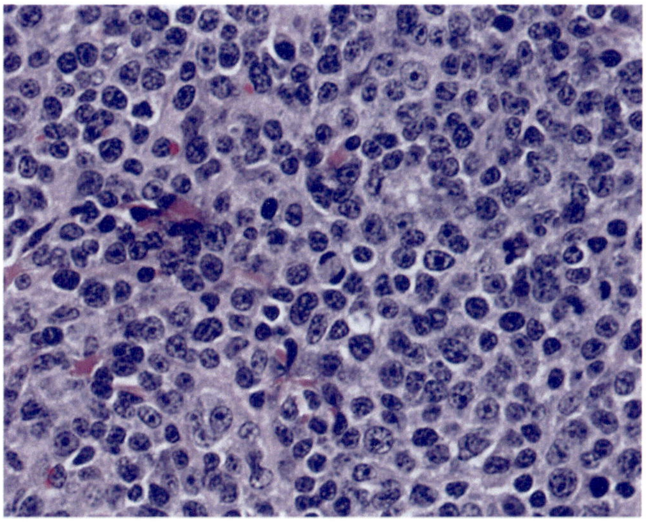

Fig. 1.37 Macacine herpesvirus 4 (rhesus lymphocryptovirus; gammaherpesvirinae subfamily). SIV-associated lymphomas share features with AIDS associated non-Hodgkin's lymphomas in human patients and may demonstrate different phenotypes such as an immunoblastic form here on H&E evaluation. These are invariably CD19 positive B cell lymphomas but can contain a considerable CD8 T cell infiltration (*Macaca mulatta*, lymph node, H&E)

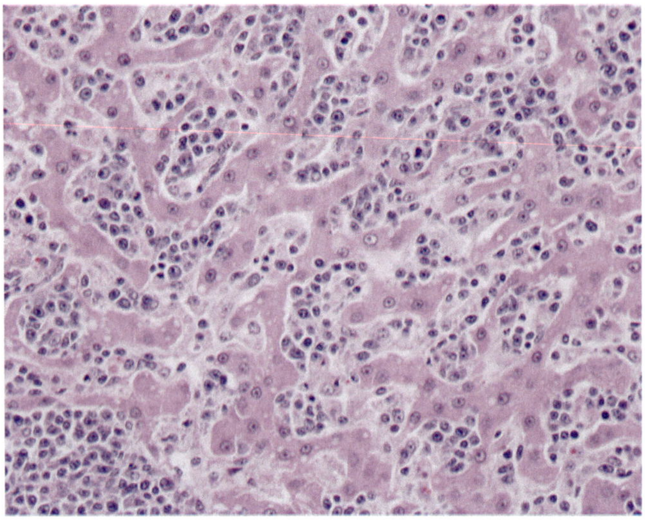

Fig. 1.38 Macacine herpesvirus 4 (cynomolgus lymphocryptovirus; gammaherpesvirinae subfamily). Reactivation of LCV in cynomolgus macaques can occur with immunosuppression induced on solid organ transplantation studies and result in post-transplant lymphoproliferative disease (PTLD). These are often extranodal and may involve the liver (*Macaca fascicularis*, liver, H&E)

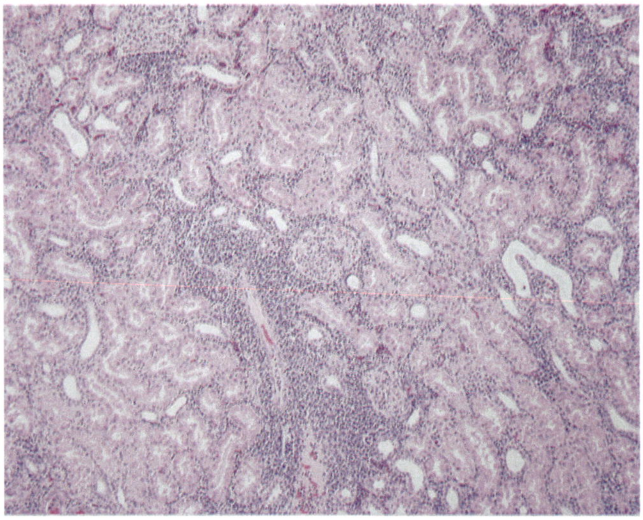

Fig. 1.39 Macacine herpesvirus 4 (cynomolgus lymphocryptovirus; gammaherpesvirinae subfamily). Early PTLD in kidney (*Macaca fascicularis*, kidney, H&E)

An epidermal proliferative lesion termed "oral hairy leukoplakia" has been described in human AIDS patients and SIV-infected rhesus macaques. Leukoplakia is characterized by proliferative pale swollen acanthocytes causing a raised plaque often on mucosal surfaces and less commonly haired skin (Fig. 1.40) [52, 53]. Intranuclear viral inclusions are often recognized in middle and superficial epithelial layers (Fig. 1.41). Infection can be confirmed by immunohistochemistry for latent membrane protein 1 (LMP1) or Epstein–Barr nuclear antigen 2 (EBNA2) and by in situ hybridization for EBER1 and 2 (Fig. 1.42) [54, 55].

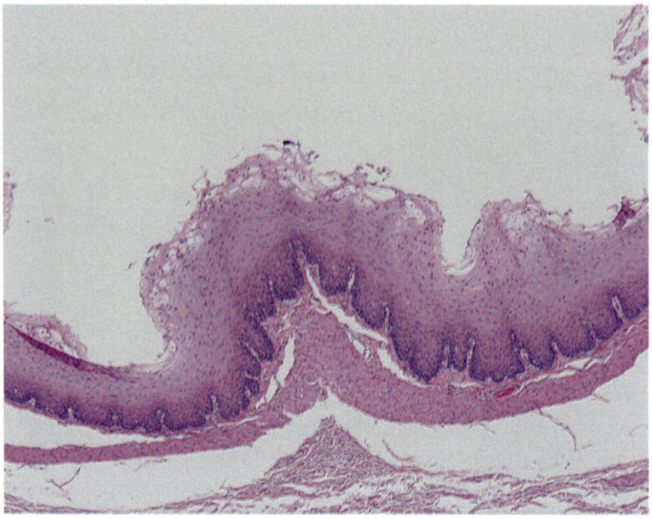

Fig. 1.40 Macacine herpesvirus 4 (rhesus lymphocryptovirus; gammaherpesvirinae subfamily). LCV may also cause epithelial proliferations during immunodeficiency. These often appear on mucosal surfaces but can appear on adjacent haired skin and are termed leukoplakias. Raised plaques are visualized due to proliferation and hydropic degeneration of keratinocytes (*Macaca mulatta*, esophagus, H&E)

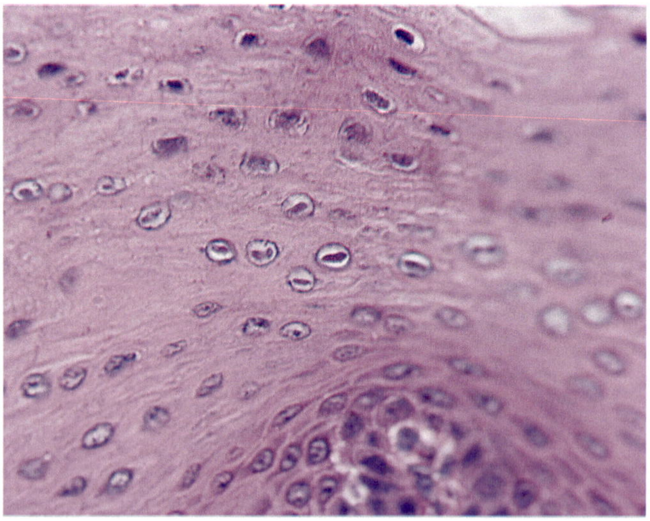

Fig. 1.41 Macacine herpesvirus 4 (rhesus lymphocryptovirus; gammaherpesvirinae subfamily). LCV intranuclear inclusion in mucosal epithelial cell of leukoplakia present in the esophagus (*Macaca mulatta*, esophagus, H&E)

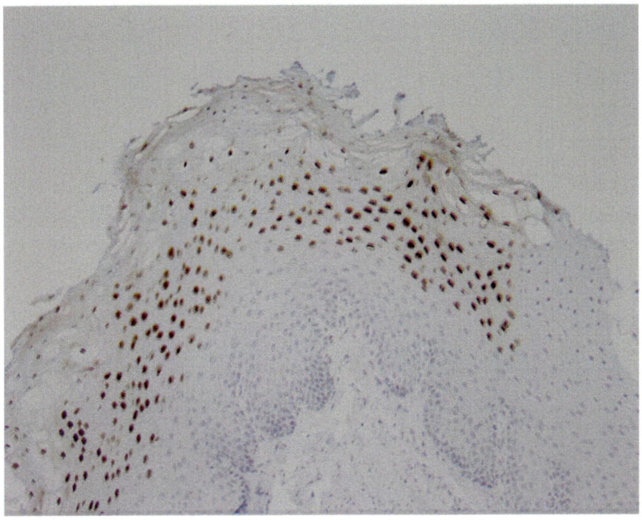

Fig. 1.42 Macacine herpesvirus 4 (rhesus lymphocryptovirus; gammaherpesvirinae subfamily). Immunohistochemistry for the LCV protein EBNA2 demonstrating infected epithelial cells in esophagus (*Macaca mulatta*, esophagus, ABC immunostain, DAB chromogen)

Macacine Herpesvirus 5 (Rhadinovirus) and Retroperitoneal Fibromatosis Herpesvirus

Rhadinoviruses (gamma-2 herpesviruses) include the Human herpesvirus 8 (HHV8) also known as the Kaposi's sarcoma herpesvirus (KSHV) and several important viruses of NHPs including retroperitoneal fibromatosis-associated herpesvirus (RFHV), rhesus rhadinovirus (RRV) and Herpesvirus saimiri. RRV and RFHV are closely related to Human herpesvirus 8 (HHV-8) the causative agent of Kaposi's sarcoma (KS) [56, 57]. RRV was initially recognized in rhesus macaques and subsequently related viruses have been identified in additional macaque species. Infection occurs commonly early in life but is unrecognized and not associated with clinical signs. Seroprevalence in surveys of infected colonies indicates that more than 90% of animals have antibodies and are infected for life. RHFV is commonly found in conjunction with RRV and is also asymptomatic in immunocompetent animals. It was first identified in tissues from animals with retroperitoneal fibromatosis (RF) and subcutaneous fibromatosis (SF). Histologically, these lesions show parallels to KS in human AIDS patients [58–61]. RF generally arises from the mesentery at the ileocecal junction and progressively infiltrates the mesentery, mesentery lymph nodes and gastrointestinal tract resulting in their encapsulation by dense fibrous tissue. SF produces multiple small expansile and ulcerating subcutaneous masses. Both RF and SF are associated with RFHV infection secondary to SRV type 2 immunosuppression (see below). Microscopically, these lesions characterized by spindle-shaped cells are arranged in intersecting fascicles often with a mononuclear inflammatory cell infiltrate and abundant collagen deposition (Fig. 1.43). The tumor cells are desmin, vimentin, and smooth muscle actin positive and express ORF73 (latency-associated nuclear antigen (LANA)) [59, 62, 63].

RF and SF are largely historical diseases as SRV type 2 has been eliminated from most macaque breeding colonies. However, RFHV continues to circulate widely in macaque colonies in conjunction with RRV [56] and has been identified in intestinal stromal tumors. These are well-circumscribed tumors consisting of benign appearing stromal cells which arise in the context of immunodeficiency (Figs. 1.44 and 1.45). Infection may be confirmed using an antibody against the LANA protein of HHV-8 which cross reacts with the homologous RFHV protein and produces a characteristic punctate intranuclear pattern (Fig. 1.46) [63].

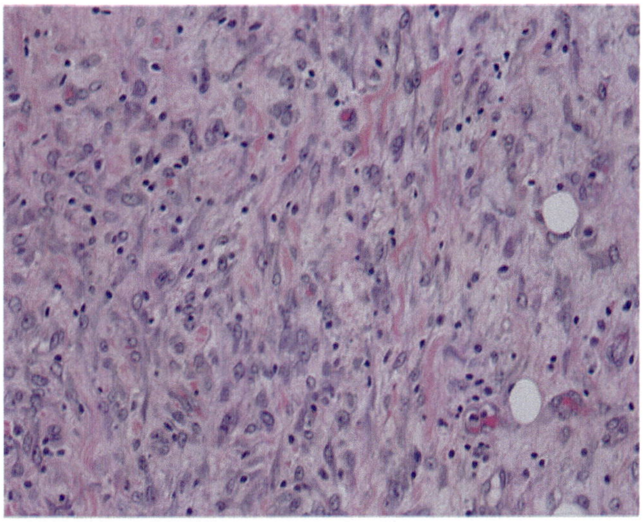

Fig. 1.43 Retroperitoneal fibromatosis virus (rhadinovirus genus; gammaherpesvirus subfamily). Retroperitoneal fibromatosis consisting of immature spindleoid cells admixed with a variable mononuclear cell infiltration. This process often originates in the mesentery and is associated with simian retrovirus type 2 induced immunosuppression and RFHV infection (*Macaca mulatta*, mesentery, H&E)

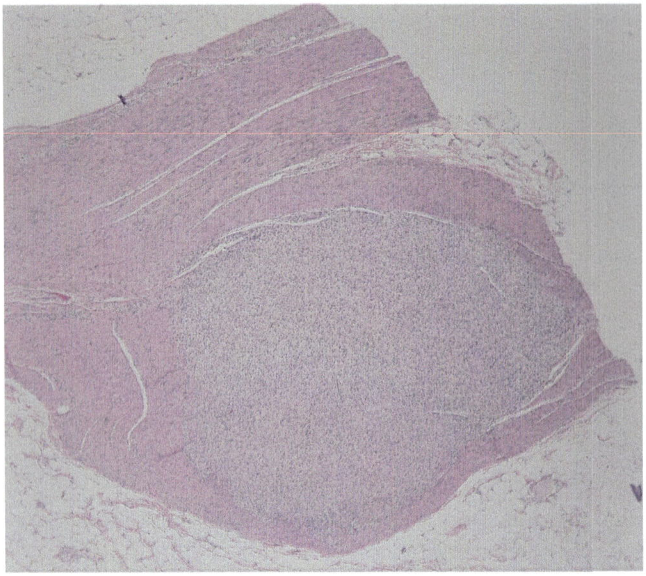

Fig. 1.44 Retroperitoneal fibromatosis virus (rhadinovirus genus; gammaherpesvirus subfamily). Gastrointestinal stromal tumor in muscularis layer of the intestine. This mesenchymal proliferative disease is seen in immunodeficient animals and associated with RFHV infection (*Macaca mulatta*, intestine, H&E)

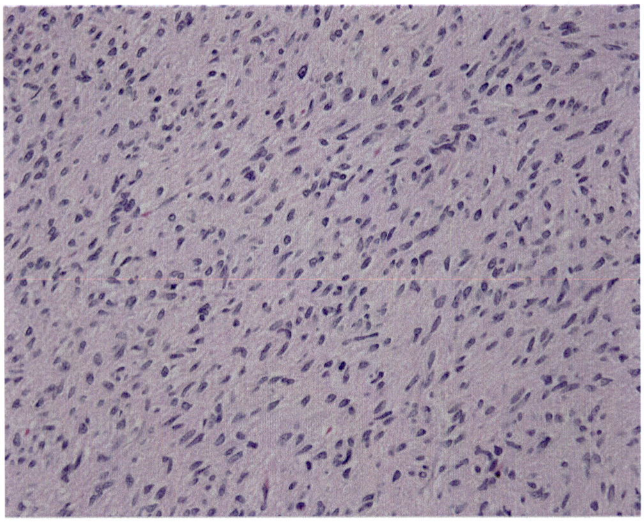

Fig. 1.45 Retroperitoneal fibromatosis virus (rhadinovirus genus; gammaherpesvirus subfamily). Gastrointestinal stromal tumor in muscularis layer of the intestine demonstrating benign spindleoid cells arranged in loose bundles and fascicles (*Macaca mulatta*, intestine, H&E)

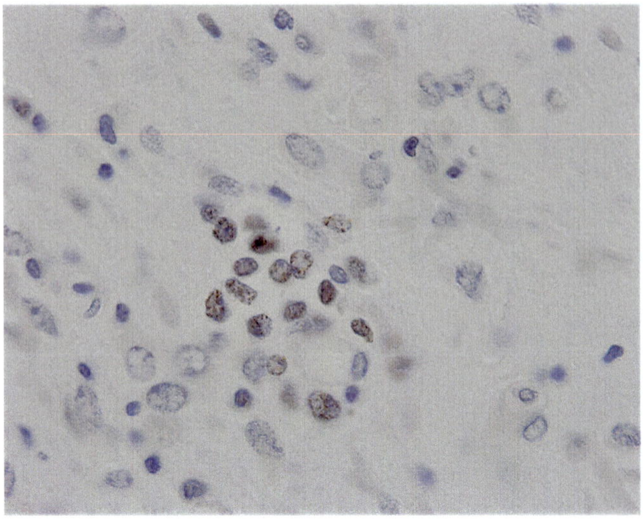

Fig. 1.46 Retroperitoneal fibromatosis virus (rhadinovirus genus; gammaherpesvirus subfamily). Immunohistochemistry for the viral LANA protein demonstrating characteristic punctate intranuclear pattern indicating RFHV infection (*Macaca mulatta*, intestine, ABC immunostain, DAB chromogen)

Herpesvirus Saimiri (HVS), Saimirine Herpesvirus 2 or Herpes Saimiri

HVS is a gamma-2 herpesvirus in the rhadinovirus genus related to HHV-8/KSHV. It is a common enzootic infection of squirrel monkeys in which it causes no recognized disease [64–66]. Serological surveys reveal that greater than 95% of adult squirrel monkeys are infected. However, transmission of HVS to susceptible tamarins, owl monkeys, or marmosets causes an acute lymphoproliferative disorder that resembles malignant lymphoma and has been used as an animal model of herpesvirus acute oncogenesis. Animals infected with HVS subtype C develop clinical signs of anorexia and weight loss within 2–3 weeks of infection. Hepatosplenomegaly may be identified as well as enlargement of peripheral lymph nodes. Leukemia may be seen in blood smears.

A distinct characteristic of gamma herpesviruses is their ability to establish latent infections in lymphoid cells and their close association with abnormal lymphoid proliferation and cancer in a number of primate species. The first open reading frame of the primate gammaherpesviruses has been shown to directly contribute to virus associated pathogenesis. All of these gene products are capable of eliciting cellular signal transduction events resulting in cell growth transformation. The marmoset model has been used to evaluate the in vivo significance of oncogene deletion or substitution. Following infection infiltrates of neoplastic lymphocytes are observed within multiple organs including GI tract, spleen, liver, and kidney [67–69]. These infiltrates are primarily CD3CD8 positive T lymphocytes and often appear to initially infiltrate along blood vessels. A related rhadinovirus, herpesvirus ateles, may induce similar pathology with cross-species transmission [61].

1.3 Hepadnaviridae

1.3.1 Hepatitis B Virus (HBV)

HBV belongs to the Orthohepadnavirus grouping and is a major cause of human hepatitis worldwide. Other important viruses in this group include the woodchuck hepatitis virus, duck hepatitis virus, and ground squirrel hepatitis virus. It is a small enveloped DNA virus with a genome consisting of a single strand of circular DNA. HBV infection has been primarily described in anthropoids including chimpanzees and gorillas [70]. A single report suggests infection of cynomolgus macaques in association with nonsuppurative hepatitis, and more recent work has suggested that infection of Mauritian origin animals may be common [71, 72]. The pathology in anthropoids is similar to that described in humans with the exception that progression to cirrhosis and hepatocellular carcinoma is not observed. In these species findings vary from essentially normal hepatic parenchyma to multifocal nonsuppurative inflammation involving the portal tracts and extending into the hepatic parenchyma. The typical ground glass hepatocytes containing viral nucleic acid may be evident. Several reports suggest macaque species may be susceptible to experimental inoculation but pathology has not been adequately described.

HBV has been studied experimentally in chimpanzees and has been transmitted to anthropoids inadvertently. It represents a zoonotic risk when working with these species or their tissues. Species-specific strains have been identified in chimpanzees and gibbons suggesting the existence of unique variants or strains that circulate within primate populations. While such infections of gibbons have recently been recognized, hepatic pathology has not been described. The zoonotic potential of these strains is not known.

1.4 Nonenveloped DNA Viruses

1.4.1 Adenoviridae

More than 30 serotypes of adenoviruses have been isolated from a variety of nonhuman primates including both OWP and NWP species [73, 74]. Although adenoviruses have been confirmed as causing mild to severe respiratory and gastrointestinal disease in monkeys, many of the isolates have been made from otherwise healthy animals. More severe disease has been described in chemically or virally immunosuppressed animals as well as neonatal animals. The virus is readily transmitted by the aerosolized or fecal–oral route. In most cases, infection of normal animals is asymptomatic but may occasionally be associated with respiratory or gastrointestinal symptoms especially in young animals. In immunodeficient animals, more severe signs may be noted and related to hepatic, gastrointestinal, or pancreatic involvement [75]. The disease course in immunodeficient animals may be chronic active, associated with significant fibrosis in some tissues and prolonged.

Infection in very young or immunodeficient animals may cause severe disease [76]. The respiratory airways and lungs are common sites of adenovirus infection. Grossly the affected portions of the lungs have patchy areas of firmness and gray-white to red areas of discoloration. Microscopically, epithelial cells of the trachea, bronchi, and alveoli are variably necrotic and contain basophilic intranuclear inclusions (Figs. 1.47 and 1.48) which can be confirmed as adenoviral by immunohistochemistry (Fig. 1.49). A severe epizootic with pulmonary involvement and transmission to human contacts has been reported in Titi monkeys (*Callicebus cupreus*) (Figs. 1.50 and 1.51) [77, 78].

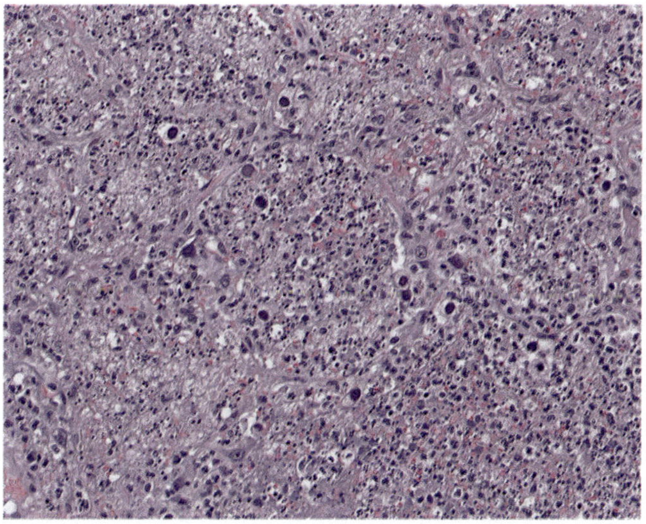

Fig. 1.47 Adenovirus (adenoviridae family). Severe bronchointerstitial pneumonia with alveoli filled with pyknotic and karyorrhectic debris and numerous cells with intranuclear adenoviral inclusions (*Macaca mulatta*, lung, H&E)

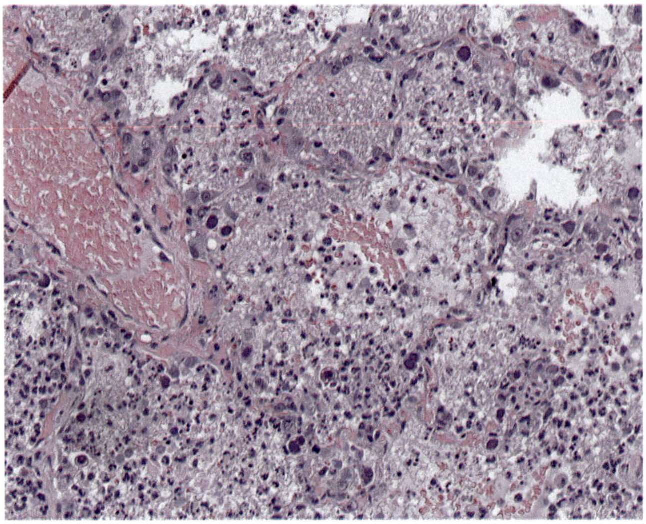

Fig. 1.48 Adenovirus (adenoviridae family). Severe bronchointerstitial pneumonia with intranuclear adenoviral inclusions. Pale eosinophilic foamy material represents concurrent pneumocystis infection in immunocompromised animal (*Macaca mulatta*, lung, H&E)

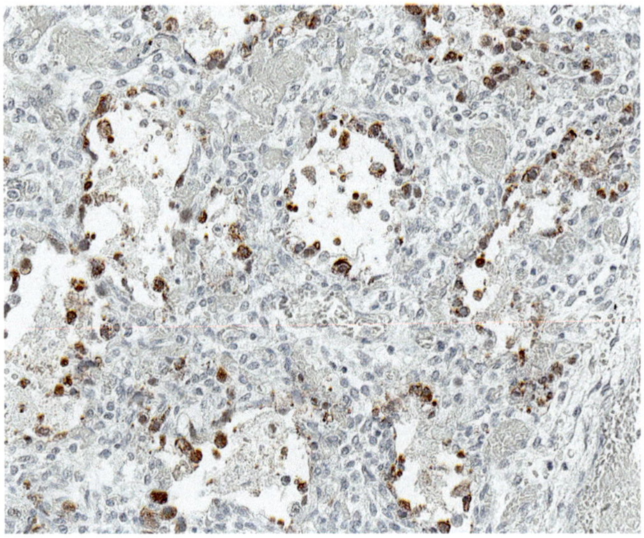

Fig. 1.49 Adenovirus (adenoviridae family). Immunohistochemistry for adenovirus demonstrating infection of type 1 and 2 pneumocytes and extensive septal thickening (*Macaca mulatta*, lung, ABC immunostain technique)

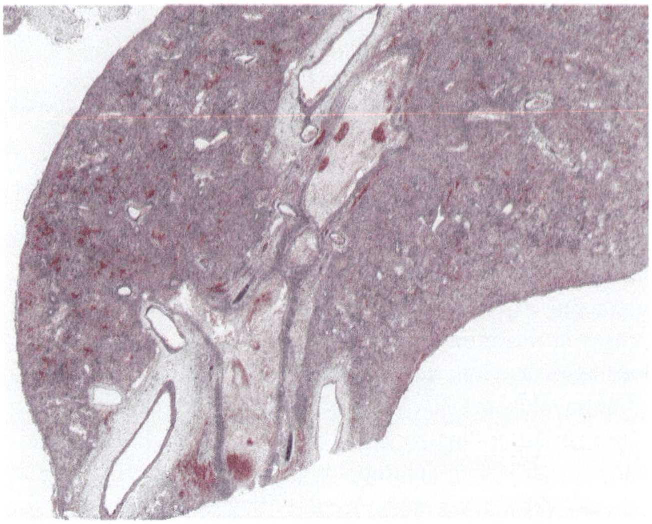

Fig. 1.50 Adenovirus (adenoviridae family). Severe diffuse necrohemorrhagic pneumonitis (Titi monkey, lung, H&E)

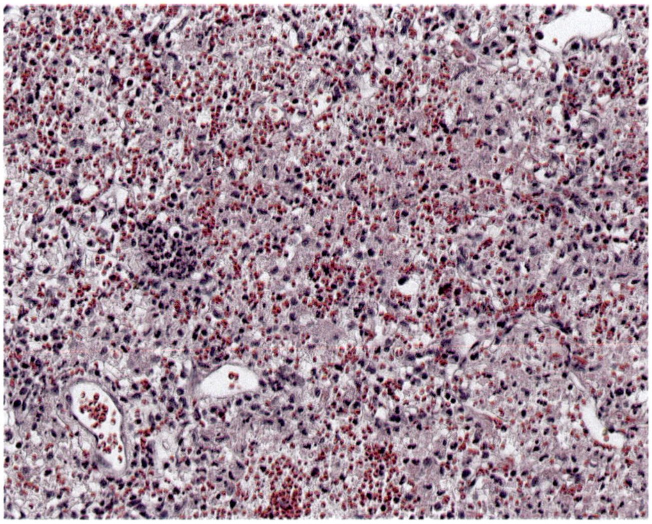

Fig. 1.51 Adenovirus (adenoviridae family). Severe diffuse necrohemorrhagic pneumonitis with scattered intranuclear adenovirus inclusions and neutrophilic foci (Titi monkey, lung, H&E)

Hepatic involvement characterized by multifocal to coalescing hepatic necrosis can be seen in juvenile animals. Intranuclear inclusion bodies are evident along the leading margin of necrosis (Fig. 1.52) and the findings must be distinguished from BV infection which may appear similar. Involvement of the gastrointestinal tract is also frequent and may or may not be found in association with epithelial cell necrosis and loss [75, 76]. Increased numbers of tingible body macrophages in the lamina propria suggest increased epithelial cell turn over (Fig. 1.53). In the absence of necrosis or epithelial cell hyperplasia the role adenovirus infection in clinical disease may be unclear. Histologically, intranuclear inclusions (Fig. 1.54) are often seen in tissues and must be distinguished from herpesviruses including CMV and BV. This sometimes can be difficult and may require ultrastructural examination or immunohistochemistry (Fig. 1.55). Occasionally, adenovirus infection can produce karyomegaly (Fig. 1.56) and can be difficult to distinguish form CMV. Adenovirus inclusions are often preferentially found in epithelial cells of the villus tips of the small intestine and glandular epithelial cells of the colon, whereas CMV will target stromal cells.

A well-characterized form of chronic active pancreatitis is frequently observed in SIV-infected macaques [79, 80]. The condition develops during periods of immunodeficiency and is recognized as multifocal areas of exocrine pancreatic necrosis in which adenovirus inclusions can be visualized in acinar cells (Figs. 1.57 and 1.58). Pancreatic necrosis may be noted grossly. Examination of the pancreas may reveal areas in different stages of disease varying from acute necrosis to lobular fibrosis and chronic active inflammation.

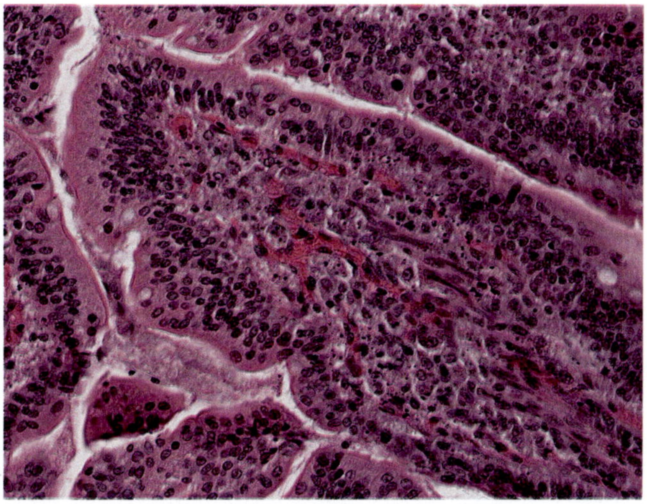

Fig. 1.53 Adenovirus (adenoviridae family). Intestinal villous tip with increased tingible body macrophages, pyknotic debris, and rare intranuclear inclusions typical of adenoviral enteritis (*Macaca mulatta*, small intestine, H&E)

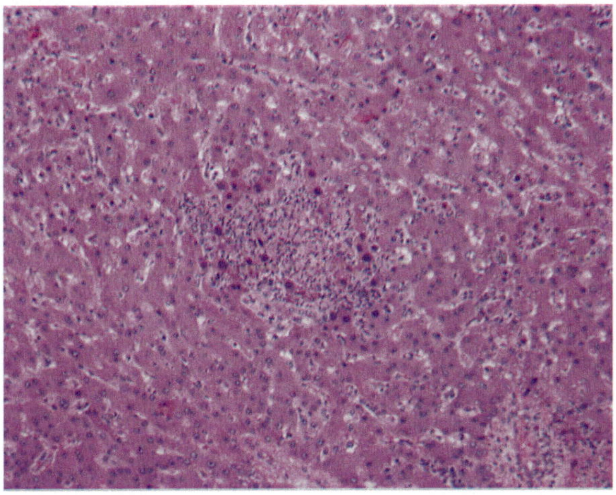

Fig. 1.52 Adenovirus (adenoviridae family). A focus of hepatocellular necrosis with neutrophilic infiltration and large adenoviral intranuclear inclusions (*Macaca mulatta*, liver, H&E)

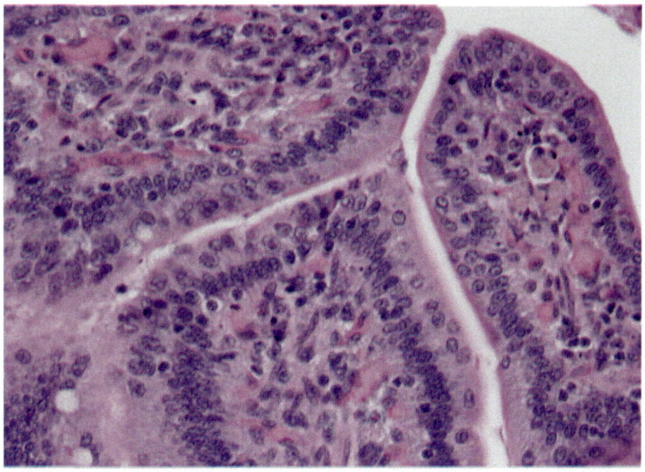

Fig. 1.54 Adenovirus (adenoviridae family). Intestinal villous tip with adenoviral intranuclear inclusions (*Macaca mulatta*, small intestine, H&E)

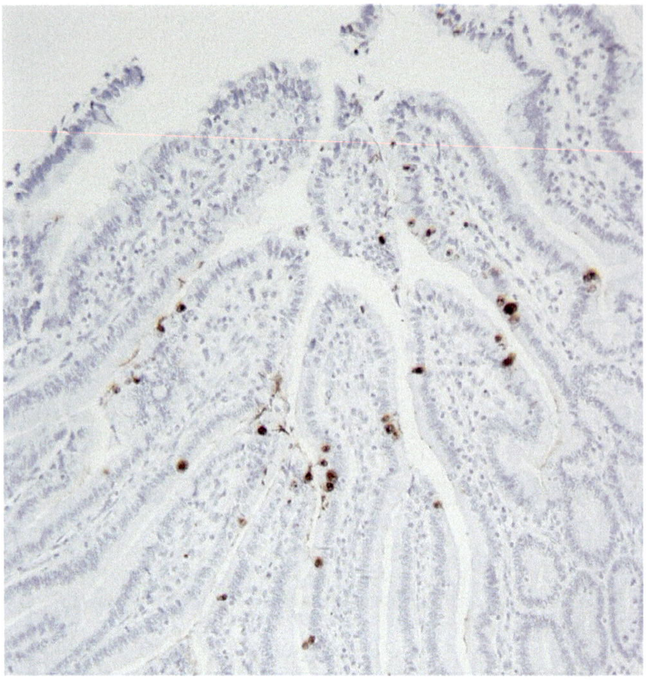

Fig. 1.55 Adenovirus (adenoviridae family). Immunohistochemistry for adenovirus protein demonstrating infection of enterocytes on intestinal villous (*Macaca mulatta*, small intestine, ABC immunostain technique)

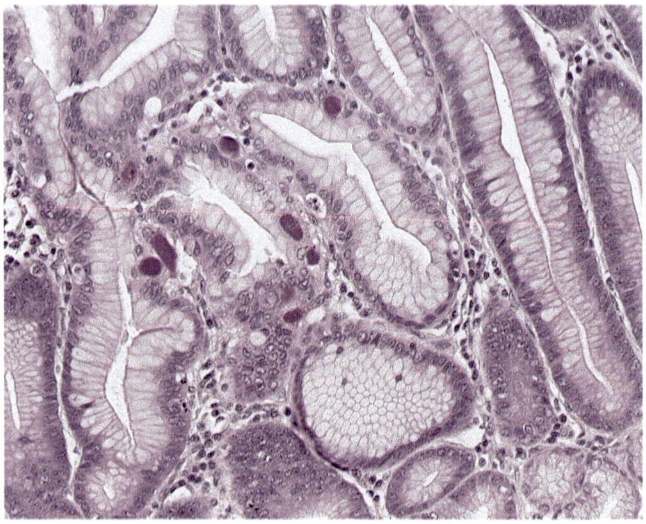

Fig. 1.56 Adenovirus (adenoviridae family). Large intranuclear inclusions present within gastric epithelial cells (*Macaca mulatta*, stomach, H&E)

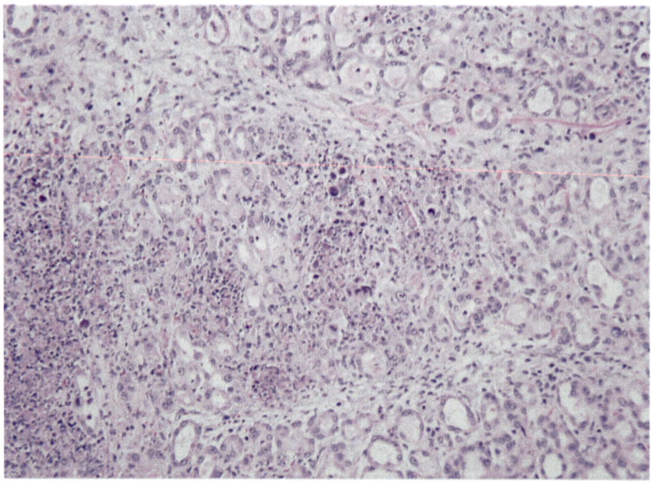

Fig. 1.57 Adenovirus (adenoviridae family). Chronic active pancreatitis with extensive acinar cell necrosis and adenoviral intranuclear inclusions (*Macaca mulatta*, pancreas, H&E)

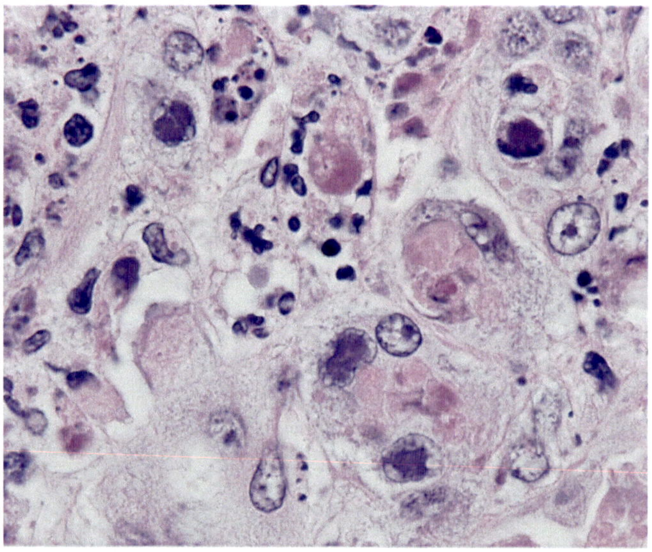

Fig. 1.58 Adenovirus (adenoviridae family). Chronic active pancreatitis with karyomegaly of acinar cells and adenoviral intranuclear inclusions (*Macaca mulatta*, pancreas, H&E)

1.5 Polyomaviridae

1.5.1 SV40

SV40 is a common latent infection of feral and captive Asian macaques and was first isolated from primary macaque kidney cell lines used in polio vaccine production. Because of the oncogenic potential in suckling hamsters, SV40 has been extensively studied and is now a common laboratory contaminant. As a result, identification of SV40 sequences by molecular techniques in clinical samples should always be viewed critically. While SV40 has been linked with human disease, this association is controversial. A closely related SV40-like virus has been recognized in *M. fascicularis* and termed Cynomolgus papovavirus (CPV).

SV40 is a common infection of virtually all macaque species kept in captivity. It appears to be readily transmitted and most adult animals will be seropositive unless colonies are developed to be free of the agent. It does not appear to be associated with disease frequently in normal animals but may cause a number of clinical conditions in immunosuppressed animals. Clinical signs may relate to CNS, pulmonary, or renal involvement. They are generally slowly progressive.

Pathology is usually not observed in normal animals and the occurrence of disease should prompt a search for underlying immunosuppressive disease. Concurrent natural or experimental infection with SIV has been demonstrated in multiple cases and should be considered in others. Animals receiving drugs for solid organ transplantation may be susceptible [81]. Lesions are generally confined to the brain, lungs, and kidney.

Progressive Multifocal Leukoencephalopathy (PML) is a condition observed in human patients infected with JC virus [82]. A similar condition has been described in immunodeficient rhesus macaques. The lesion is characterized by multifocal to coalescing areas of demyelination and gliosis throughout the white matter and subependymal regions (Fig. 1.59). Demyelination results from direct viral infection and destruction of oligodendrocytes. Large basophilic intranuclear inclusions can be found to enlarge and fill nuclei of oligodendrocytes and astrocytes and less frequently neurons (Fig. 1.60) [83, 84]. A distinct form of SV40-induced meningoencephalitis has been described in SIV-infected macaques in which infection of astrocytes and not oligodendrocytes appears to predominate [84]. Sequence analysis suggested infection with a distinct SV40 variant. SV40 immunohistochemistry for the large T antigen and viral isolation may be used diagnostically.

Renal lesions are found primarily within the inner cortex and medulla (Fig. 1.61). Affected medullary tubules are lined by hypertrophied and hyperplastic epithelial cells (Fig. 1.62) that contain intranuclear basophilic inclusions. Inclusions are often very large and visible under lower magnification. Occasionally, affected renal tubular epithelial cells form fronds which project into tubular lumens and appear dysplastic. Renal involvement is often associated with chronic tubulointerstitial nephritis. In mild cases, the number of recognizable inclusions on H&E stains (Fig. 1.63) may be scant and immunohistochemistry will reveal many more infection cells (Fig. 1.64).

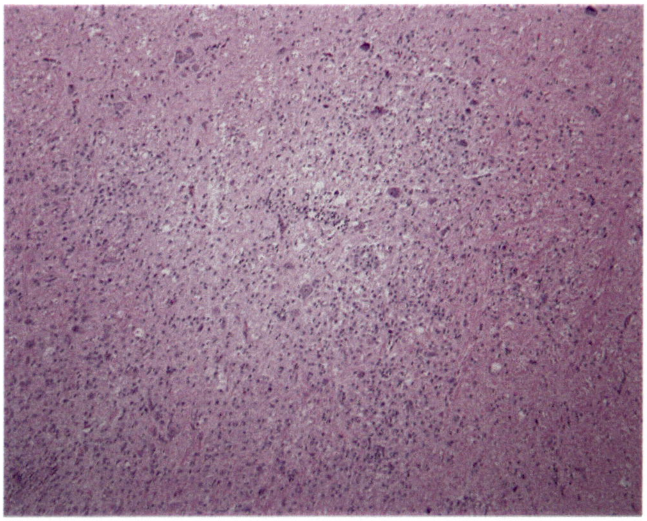

Fig. 1.59 SV40 (Polyomaviridae family). Progressive multifocal leukoencephalopathy characterized by demyelination of white matter and gliosis (*Macaca mulatta*, brain, H&E)

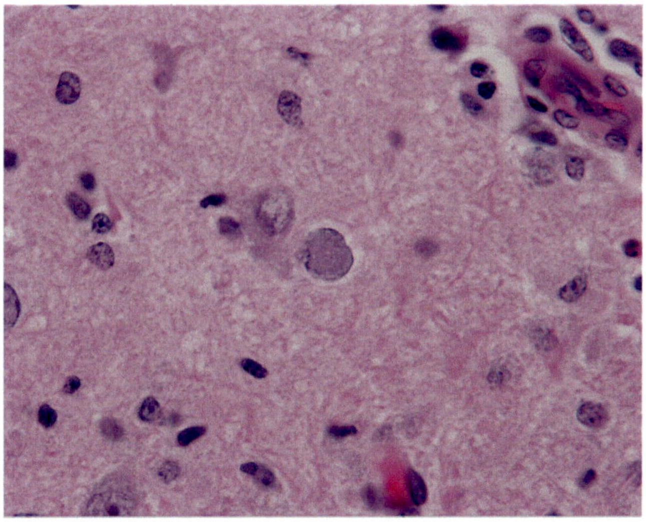

Fig. 1.60 SV40 (Polyomaviridae family). Progressive multifocal leukoencephalopathy with intranuclear viral inclusion (*Macaca mulatta*, brain, H&E)

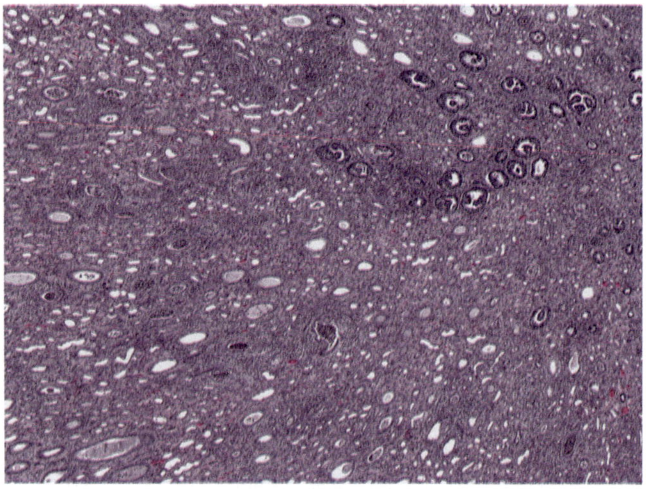

Fig. 1.61 SV40 (Polyomaviridae family). SV40-induced chronic interstitial nephritis (*Macaca mulatta*, kidney, H&E)

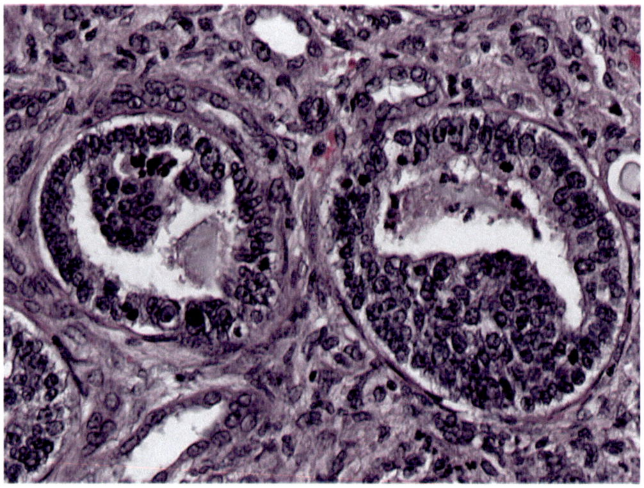

Fig. 1.62 SV40 (Polyomaviridae family). SV40-induced chronic interstitial nephritis with proliferative renal tubular epithelial cells (*Macaca mulatta*, kidney, H&E)

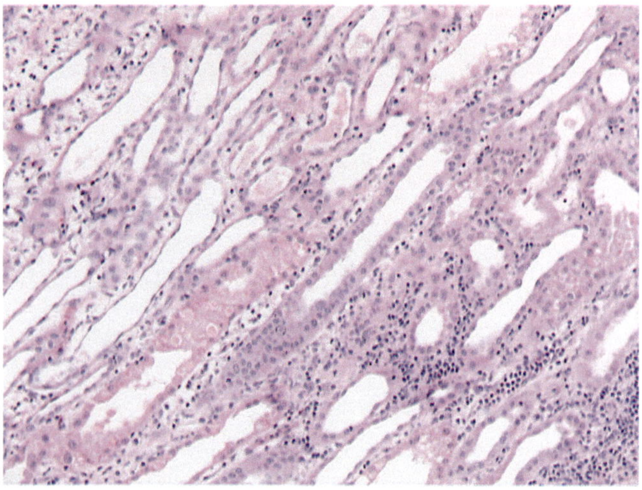

Fig. 1.63 SV40 (Polyomaviridae family). SV40-induced interstitial nephritis with rare intranuclear inclusion bodies (*Macaca mulatta*, kidney, H&E)

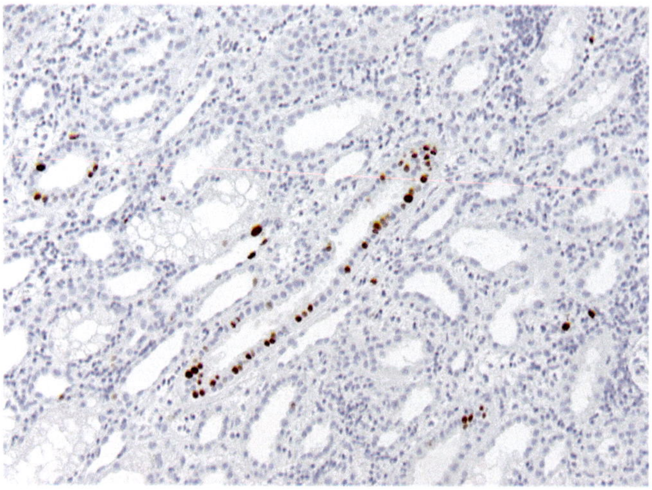

Fig. 1.64 SV40 (Polyomaviridae family). Immunohistochemistry for SV40 large T antigen demonstrating abundant infected renal tubular epithelial cells (*Macaca mulatta*, kidney serial section adjacent to Fig. 1.63, H&E)

Pulmonary involvement is seen less frequently and consists of a proliferative interstitial pneumonitis with inclusions present within hypertrophied type 2 pneumocytes (Figs. 1.65 and 1.66). Hepatic involvement has also been recognized in immunodeficient macaques. Poorly defined brightly eosinophilic inclusions were identified in periportal hepatocytes (Fig. 1.67) and were positive for large T antigen by immunohistochemistry (Fig. 1.68).

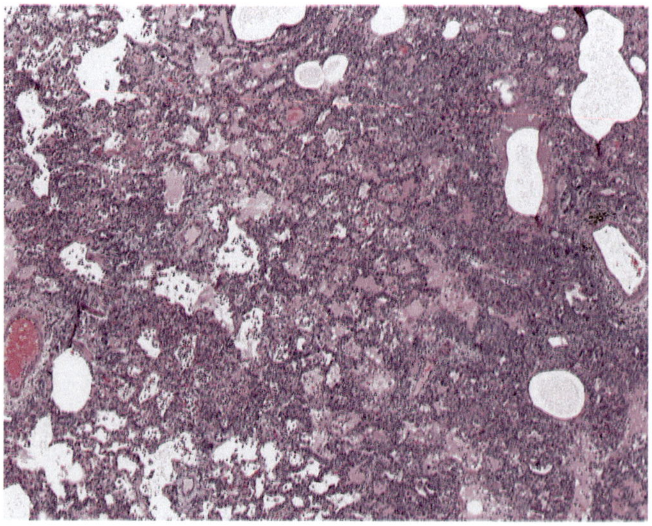

Fig. 1.65 SV40 (Polyomaviridae family). SV40-induced interstitial pneumonitis with mixed inflammatory cell infiltrate (*Macaca mulatta*, lung, H&E)

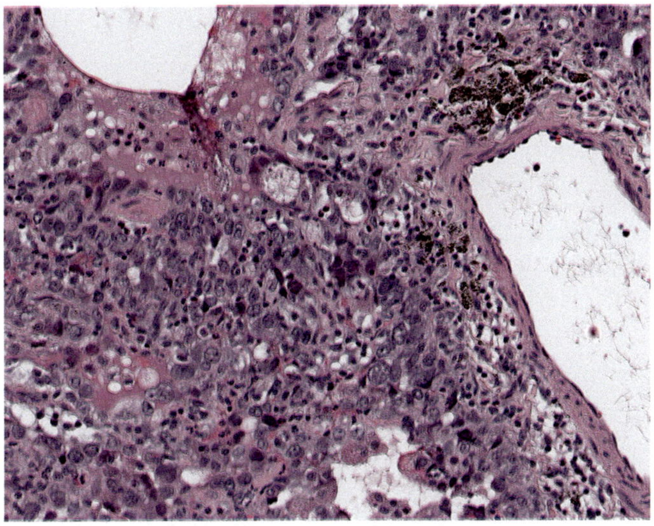

Fig. 1.66 SV40 (Polyomaviridae family). SV40-induced interstitial pneumonitis with intranuclear inclusions (*Macaca mulatta*, lung, H&E)

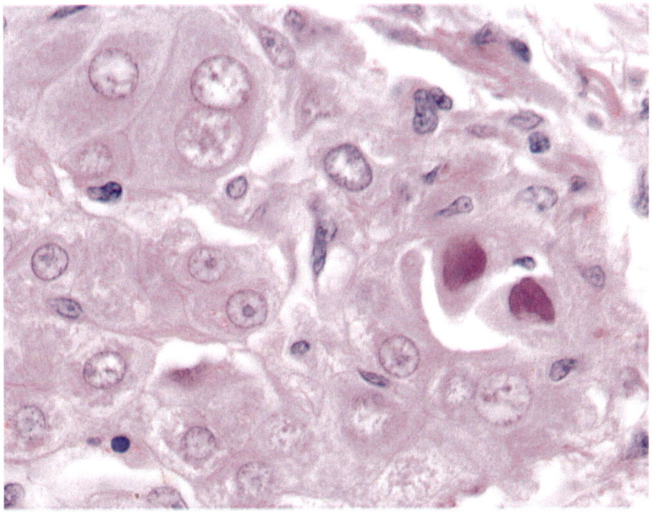

Fig. 1.67 SV40 (Polyomaviridae family). Large amphophilic intranuclear inclusions within periportal hepatocytes (*Macaca mulatta*, liver, H&E)

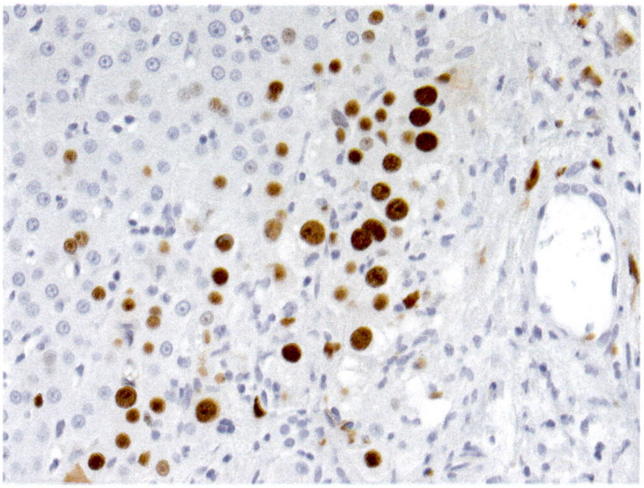

Fig. 1.68 SV40 (Polyomaviridae family). Immunohistochemistry for SV40 large T antigen demonstrates infection of a large number of periportal hepatocytes (*Macaca mulatta*, liver, ABC immunostain technique, DAB chromogen)

1.6 Papillomaviridae

Papillomaviruses are spherical double-stranded DNA viruses with capsids composed of 72 capsomeres. Infection of rhesus and cynomolgus macaques, chimpanzees, and Colobus monkeys has been demonstrated and infection has been suspected in many other species [85–88]. Infection of rhesus may or may not be associated with papillomas. Often there is serologic evidence of infection without clinical signs. Serosurveys indicate that viral transmission occurs often at sexual maturity. When lesions are present, they are observed on mucosal surfaces or haired skin. They have the appearance of a proliferative exophytic mass that is described to have a cauliflower-like appearance. Histologically, there is massive hyperplasia of the stratum spinosum and corneum. Intranuclear basophilic inclusions may be observed but are often absent. Infection in macaques has been associated with penile carcinoma and with cervical vaginal dysplasia/invasive carcinoma in macaques (Fig. 1.69) [89, 90]. In these lesions, viral inclusions are generally absent and infection can be confirmed by immunohistochemistry.

In chimpanzees, focal epithelial hyperplasia is a recognized lesion and characterized by multiple well-circumscribed proliferative structures at the oral mucosa [91–93]. These may persist for extended periods before undergoing spontaneous regression. Typical basophilic intranuclear inclusions may be observed.

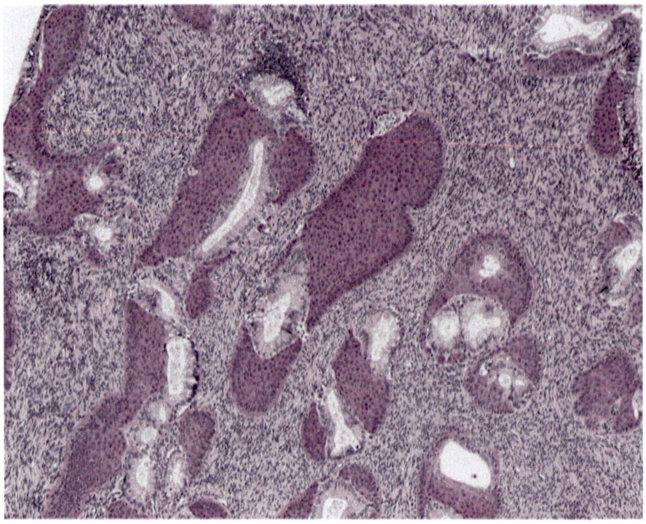

Fig. 1.69 Papilloma virus (Papillomaviridae family). Cervical carcinoma induced by papilloma virus infection (*Macaca mulatta*, cervix, H&E)

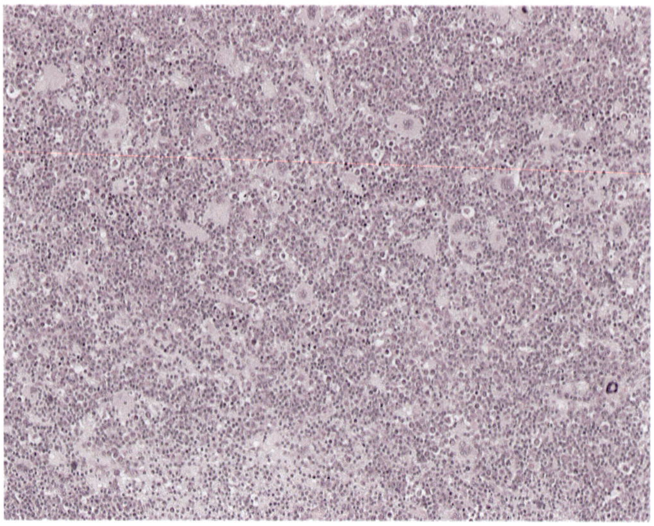

Fig. 1.70 Simian parvovirus (Erythrovirus genus). Marked dyserythropoiesis and loss of mature erythroid precursors in bone marrow (*Macaca mulatta*, bone marrow, H&E)

1.7 Parvoviridae

1.7.1 Simian Parvovirus (SPV)

Parvoviruses are some of the smallest known vertebrate viruses. They are nonenveloped and contain a genome of a single strand of DNA. SPV has recently been described in cynomolgus and rhesus macaques and shares 65% homology with human B19 virus within the major capsid protein [94, 95]. The virus is difficult to culture in vitro.

SPV has been recognized in association with anemia in cynomolgus macaques coinfected with SRV-D and less commonly SIV. Affected animals had clinical signs of anemia, diarrhea, weight loss, and dehydration [96]. It is unclear which clinical signs can be directly attributed to SPV infection as these are frequently observed in SRV-D infected animals. Similar findings have also been recognized in cynomolgus macaques receiving post-transplantation chemotherapy and SIV infection [97, 98].

Examination of bone marrow has revealed marked dyserythropoiesis with a loss of mature erythroid precursors (Fig. 1.70). Large intranuclear inclusion bodies are evident in some animals (Fig. 1.71). Ultrastructural examination and in situ hybridization may be used for diagnosis. Seroreactivity to SPV has been detected in individuals with and without primate contact [99]. The significance of this findings is not known.

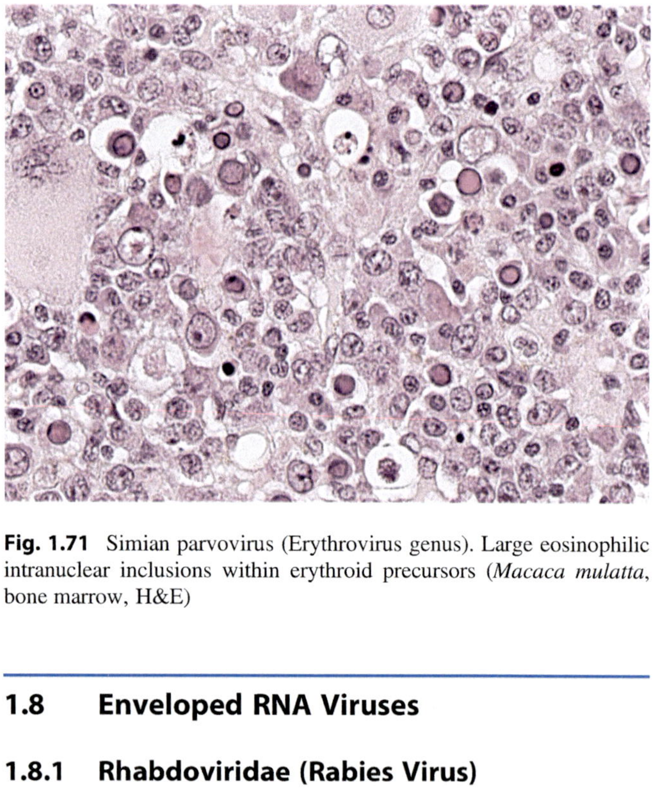

Fig. 1.71 Simian parvovirus (Erythrovirus genus). Large eosinophilic intranuclear inclusions within erythroid precursors (*Macaca mulatta*, bone marrow, H&E)

1.8 Enveloped RNA Viruses

1.8.1 Rhabdoviridae (Rabies Virus)

Rabies is a well-known enveloped RNA virus within the lyssavirus genus. Infection of NWPs and OWPs has been recognized. The occurrence of rabies in domestic captive-bred NHP is extremely rare; however, exposures may occur but it should be considered in sporadic cases of encephalitis particularly in outdoor animals raised in regions where rabies is enzootic. In these regions, rabies virus infection of NHP species may occur and these species may represent vectors for transmission to humans. The disease has also been studied experimentally in NHPs. Recent reports have documented transmission of rabies from common marmosets to human. Limited sequence data suggests the establishment of an enzootic marmoset cycle [100–102].

Both furious and paralytic forms of rabies have been described in NHPs. Following experimental inoculation, the furious clinical signs developed in 15–35 days, whereas the paralytic form was delayed and developed up to 105 days postinfection [17]. Histopathology is similar to that recognized in other species and may vary from few recognized changes to formation of glial nodules, perivascular nonsuppurative encephalitis, and neuronal degeneration. Negri bodies may be evident and immunofluorescence assays may be used for diagnosis. NHPs may be vaccinated with killed vaccine but the efficacy is unknown. Rabies is a serious zoonotic risk and NHP to human transmission has been recognized. Due to the long incubation period, typical quarantine requirements may not be adequate to prevent importation from geographic regions in which rabies virus is enzootic.

1.8.2 Filoviridae

Marburg Virus and Ebola Viruses

Marburg virus is the prototypic filovirus. It was first recognized as a cause of hemorrhagic fever in laboratory workers preparing cell lines derived from African Green monkey tissue in Marburg and Frankfurt, Germany in 1967 [103, 104]. During the initial outbreak, 25 primary cases and a number of secondary and tertiary cases were recognized. Filoviruses are anti-sense single-stranded RNA viruses with a relative small genome of 12.7Kb. They have a characteristic long filamentous appearance in tissue or cell culture.

A second group of closely related viruses were recognized in 1976 as the etiology of human hemorrhagic fever in Zaire and Sudan [105, 106]. The agent named after the Ebola river in northwest Zaire caused a clinical syndrome in humans similar to Marburg and has been identified periodically in sub-Saharan African. A distinct filovirus subtype named Ebola Reston was identified in 1989 and 1990 during an outbreak of hemorrhagic fever in newly imported Asian macaques from the Philippines [107, 108]. The source of infection in these animals is unknown but concurrent infection with Simian Hemorrhagic Fever Virus (SHFV) initially suggested exposure and transmission from an African source. Additional infected macaques were identified in mid- and late-1990s and 2015, all originating from the same geographic location in the Philippines [109, 110]. Subsequent work has demonstrated infection of pigs and potential reservoir in bats [100, 101].

The severity of clinical signs varies depending on the filovirus strain and host. No clinical signs were recognized in African green monkeys during the initial Marburg outbreak and Ebola Reston did not produce illness in humans. In contrast, significant die-off events have been observed in wild chimpanzee populations due to Ebola virus infection in association with human cases [111, 112].

Following experimental inoculation with Ebola virus, there is an incubation period of 7–14 days. Once clinical signs develop, progression to death is rapid and usually occurs within 24 h. Cardiovascular collapse occurs followed by severe depression, coma, and death. Petechiae are noted on the face, chest, and medial aspects of the arms and thighs.

Pathologic findings are similar to Simian hemorrhagic fever from which it must be distinguished. There is extensive lymphoid necrosis and deposition of fibrin within splenic white and red pulp (Fig. 1.72). A characteristic lesion within the liver is a multifocal random hepatic necrosis accompanied by a mild mononuclear inflammatory cell infiltrate (Fig. 1.73). Large amorphic intracytoplasmic inclusions are observed in hepatocytes (Fig. 1.74). Similar necrosis is seen within the zona glomerulosa of the adrenal gland. Adrenal and hepatic lesions help distinguish filovirus infection from SHFV. In addition to these findings, an interstitial pneumonitis has been described and seen in conjunction with disseminated intravascular coagulation (Fig. 1.75). Serology can be used to determine previous exposure of animals to filoviruses and may be required by some facilities prior to importation and quarantine of animals.

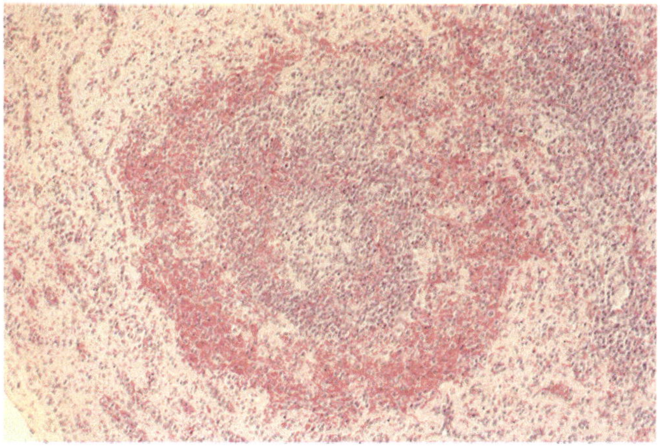

Fig. 1.72 Ebolavirus (Filoviridae family). Lymphoid necrosis and extensive fibrin deposition in red pulp of spleen (*Macaca mulatta*, spleen, H&E)

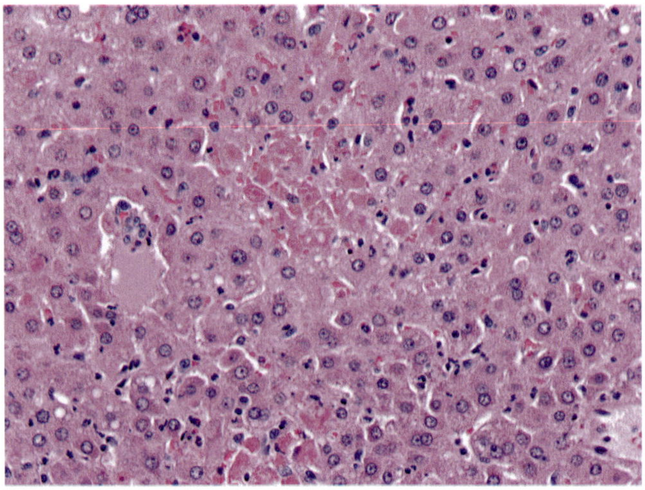

Fig. 1.73 Marburg virus (Filoviridae family). Multifocal random hepatic necrosis (*Callithrix jacchus*, liver, H&E)

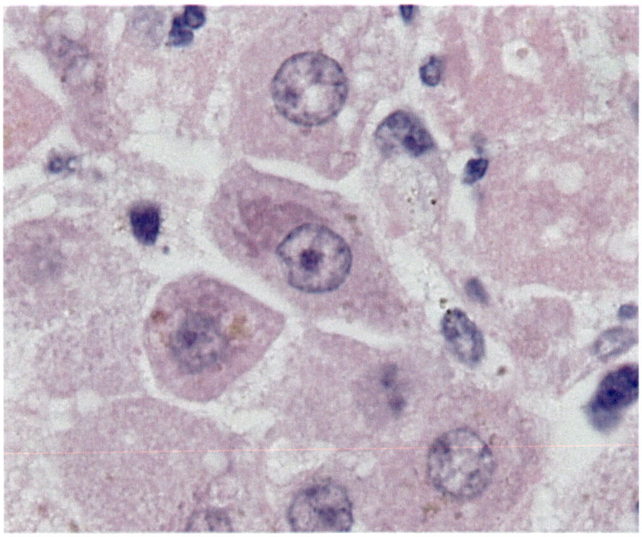

Fig. 1.74 Marburg virus (Filoviridae family). Large amphophilic intracytoplasmic viral inclusion within hepatocyte (*Macaca mulatta*, liver, H&E)

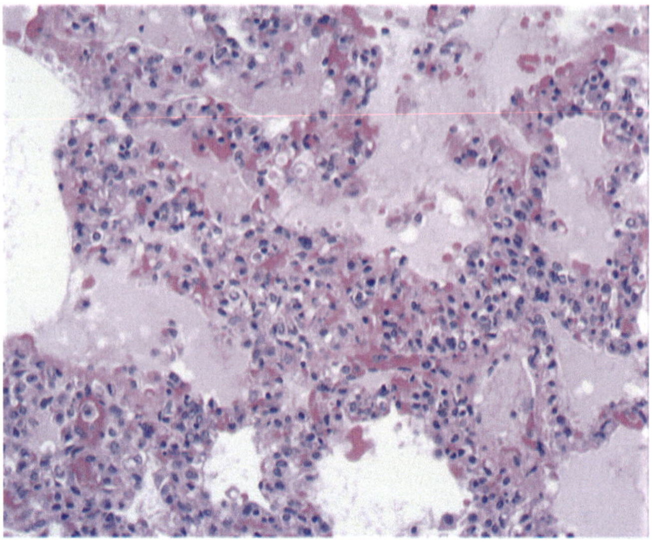

Fig. 1.75 Ebolavirus (Filoviridae family). Necrotizing interstitial pneumonitis (*Macaca mulatta*, lung, H&E)

1.9 Arteriviridae

1.9.1 Simian Hemorrhagic Fever Virus (SHFV)

SHFV belongs to the Simartevirus genus in the arteriviridae family, and represents a group of closely related viruses that may infect a variety of African primates subclinically [113]. It is a positive-stranded RNA virus and is related to equine arteritis virus, lactate dehydrogenase elevating virus of mice and porcine infertility and respiratory syndrome virus.

African OWPs including patas monkeys, African Green Monkeys, and baboons are the natural hosts and the disease is generally mild in these species. Transmission to Asian species of macaques results in an epizootic of high morbidity and mortality [114–117]. Strains differ in their ability to cause persistent infection and disease in the African host species.

In macaques, experimental infection is uniformly fatal following inoculation in 5–7 days. Animals develop hemorrhagic diarrhea and respiratory distress followed by death. Petechiae may be observed especially on mucosal or serosal surfaces. Initial infection of macaques appears to require parental exposure to infected blood or tissue from infected carrier species. Once established in macaque colonies the disease may spread rapidly through aerosolization, direct contact, or fomites resulting in an epizootic of high morbidity and mortality. Infection of African species is usually asymptomatic. Different viral strains have been identified that vary in their ability to produce antibody responses and occasionally disease in patas monkeys. Strains that produce a strong antibody response may be more likely associated with disease (anorexia, lethargy, petechiation) and viral clearance.

In macaques, SHFV infection causes clinical signs compatible with a hemorrhagic fever and the virus may spread rapidly through a colony. At necropsy, there is often extensive hemorrhage and congestion in the duodenum and similar foci may be found throughout the GI tract. Small hemorrhages may also be found in the liver, renal capsule, retrobulbar tissue, subcutis, and lungs. Microscopically, characteristic lesions are found in the lymphoid tissue and consist of extensive lymphoid necrosis and congestion. Perifollicular hemorrhages may be evident and the sinuses may be distended with fibrin (Fig. 1.76).

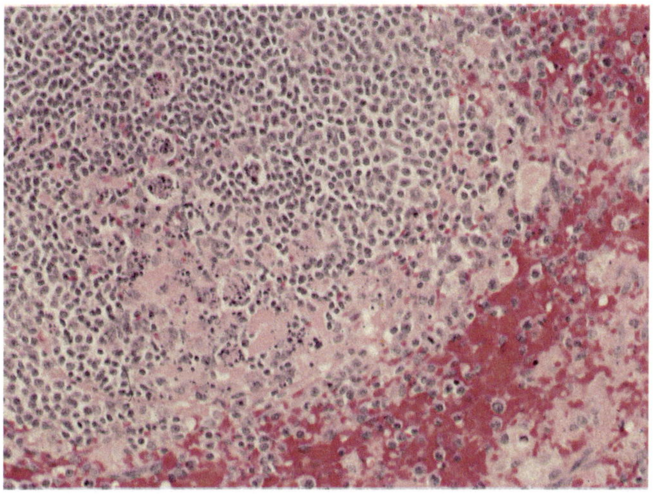

Fig. 1.76 Simian hemorrhagic fever virus (arterivirus genus). Lymphoid necrosis and fibrin deposition in red pulp (*Macaca mulatta*, spleen, H&E)

Overt necrosis with germinal centers is prominent. Lesions elsewhere relate to the development of disseminated intravascular coagulation.

Differential diagnosis in macaque species includes hemorrhagic fever viruses (Ebola, Marburg, Kyasanur forest disease virus) and the absence of adrenal necrosis and absence of viral inclusions in SHFV may be helpful in ruling out these diseases. Filoviruses produce intracytoplasmic inclusion bodies. Strict separation of African and Asian OWPs should be maintained. Potential fomites such as tattoo needles, clippers, and other instruments should be thoroughly disinfected.

1.10 Flaviviridae

1.10.1 Yellow Fever Virus (YFV)

YFV is a member of the Flaviviridae family and may naturally infect and cause disease in humans and nonhuman primates. Mosquitoes of the genera Aedes and Haemagogus serve as vectors with *A. aegypti* the most important, as the virus may persist in this species through the dry season by transovarian transmission.

YFV may produce a severe epizootic hemorrhagic fever syndromes in both NWP and OWP. Infection on OWP has been studied as an experimental infection but the disease may cause epizootics in NWP in their natural setting [118]. In these instances, disease is usually first recognized as die-offs in NHP populations occasionally in association with human cases. The disease may be severe in Howler (*Alouatta* spp.), spider (*Ateles* spp.), and squirrel (*Saimiri sciureus*) monkeys [119, 120].

As a naturally occurring disease, YFV infection is seen in NHP in their normal environment. However, the pathogenesis has been studied extensively in macaque species as an animal model of disease in man. Following experimental inoculation, there is an initial round of replication in regional lymph nodes with subsequent dissemination to multiple organs. In the liver, virus is first identified in Kupffer cells and then hepatocytes. Histologically, liver lesions are characterized by multifocal hepatic necrosis with formation of Councilman and Torres bodies [121]. These changes are often accompanied by fatty degeneration of the remaining hepatocytes. Depletion of lymphoid tissue and periarterial sheaths may be prominent. In humans, three diagnostic hallmarks are said to be characteristic and are most pronounced 3–8 days postinfection: (1) midzonal hepatic necrosis, (2) eosinophilic degeneration, and (3) fatty metamorphosis.

1.10.2 West Nile Virus

West Nile virus is a single-stranded RNA virus in the family Flaviviridae which has recently been introduced into the USA and associated with high mortality in wild and captive bird populations and the occurrence of encephalitis in human patients. Serologic evidence indicates that infection of captive primates is common but asymptomatic in animals housed in outdoor corrals during epizootics in wild bird populations [122–124]. While clinical signs were not observed in these spontaneous cases, fever and encephalitis were described following experimental inoculation. Following experimental inoculation, animals remained persistently infected for prolonged periods and developed morphologic evidence of nonsuppurative encephalitis [125].

1.11 Arenaviridae

1.11.1 Lymphocytic Choriomeningitis Virus (LCMV)

Callitrichid hepatitis is caused by LCMV, a member of the arenaviridae family which are pleomorphic enveloped viruses containing two segments of RNA that encode at least three viral proteins. A characteristic feature of arenaviruses is that they cause lifelong persistent infection of the definitive rodent host. Such hosts show no clinical signs and may shed virus in body secretions leading to contamination of environment. In addition to LCMV, a number of arenaviruses have been identified as important exotic pathogens of man including Lassa fever virus, Mopeia virus, Junin virus, and Machupo virus.

LCMV has been recognized as a cause of rapidly progressive viral hepatitis in NWP species including tamarins and marmosets [126]. Clinical signs include dyspnea, anorexia, weakness, and lethargy. Animals may become jaundiced and evidence of a coagulopathy may develop. In some cases, sudden death with no clinical signs may be seen. Epizootics of LCMV hepatitis were associated with the feeding of infected neonatal mice to primates. If this practice is used, colonies should be screened for the presence of LCMV. In addition, contact of NWP species with wild rodents should be prevented and if biologics of murine origin are to be used, they should be tested for the presence of LCMV.

Hepatosplenomegaly, pleural and pericardial effusions, jaundice, and subcutaneous and intramuscular hemorrhages are characteristic findings on necropsy. Histologically, multifocal hepatic necrosis with infiltration by lymphocytes and neutrophils may be seen (Figs. 1.77 and 1.78) [127, 128]. Acidophilic apoptotic hepatocytes termed Councilman bodies may be evident within sinusoids. Necrosis may also be evident within other tissues including mesenteric lymph nodes and GI tract. LCMV is a zoonotic agent and seroconversion of veterinarians in contact with infected marmosets has been described.

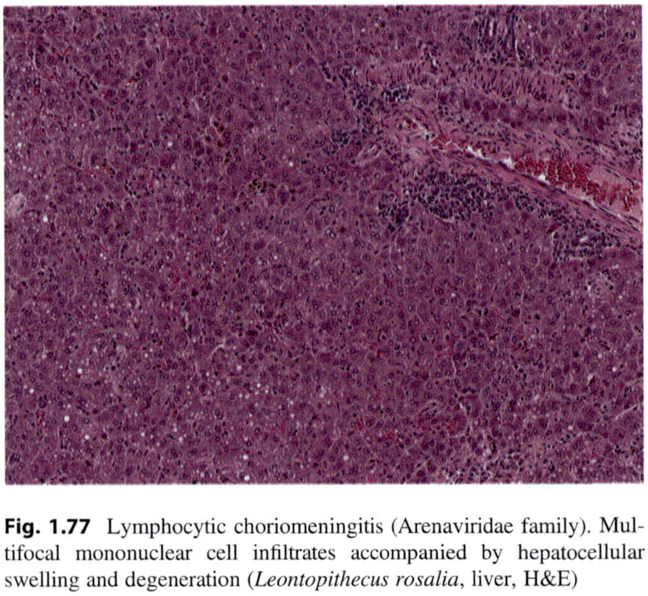

Fig. 1.77 Lymphocytic choriomeningitis (Arenaviridae family). Multifocal mononuclear cell infiltrates accompanied by hepatocellular swelling and degeneration (*Leontopithecus rosalia*, liver, H&E)

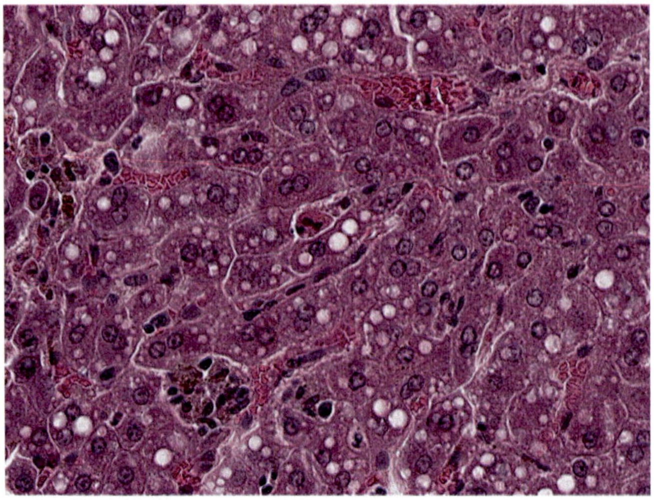

Fig. 1.78 Lymphocytic choriomeningitis (Arenaviridae family). Hepatocellular swelling and degeneration with biliary stasis (*Leontopithecus rosalia*, liver, H&E)

1.12 Paramyxoviridae

1.12.1 Measles Virus

Measles virus is a member of the Paramyxoviridae family and is an enveloped RNA virus. Measles virus infection was once a common and devastating disease of recently captured NHPs. It has been described in many OWP and NWP species and is less frequent now with the control of measles in the human population particularly in the USA [129].

In OWP species, the disease may be mild to asymptomatic unless animals are stressed by concurrent illness or are immunodeficient. The incubation period varies from 6 to 10 days and is followed by a fever and maculopapular exanthema (Fig. 1.79).

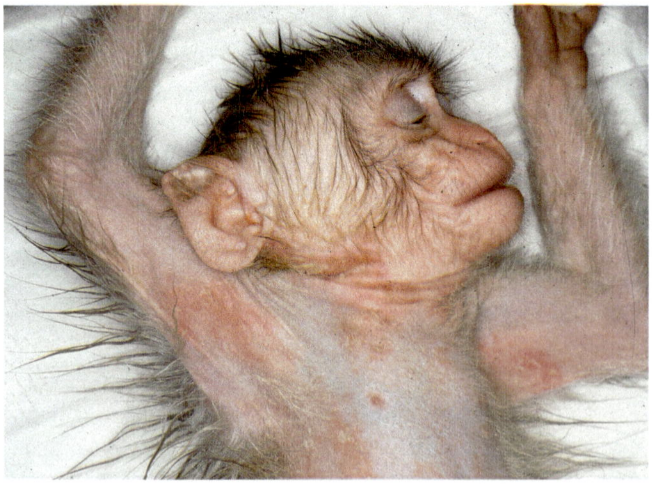

Fig. 1.79 Measles virus (Paramyxoviridae family). Desquamative viral exanthema involve face, thorax, and axilla in juvenile macaque (*Macaca mulatta*)

The rash is most pronounced on the ventral body surface and generally spares the plantar and palmar surfaces of the feet and hands. The rash progresses to a dry or scaly desquamative dermatitis and may continue for 2–3 weeks. In severe cases, respiratory illness may develop and produce a conjunctivitis and cough. Abortion and neurologic signs may occur in some individuals. In NWP, the disease may be more severe producing an epizootic of high morbidity and mortality. The characteristic exanthema is lacking and gastrointestinal signs predominate. Edema of the periorbital region may be observed and mortality may approach 100%.

Initial transmission is from human handlers to animals. The virus is transmitted by the aerosolized route and there is an initial round of viral replication within regional lymph nodes followed by hematogenous spread. In OWPs, it is during the terminal phases of this viremia that the characteristic cutaneous exanthema is observed and appearance of the rash coincides with the appearance of neutralizing antibodies. Histologically, there is mild erythema and parakeratotic hyperkeratosis. Multinucleated syncytial cells with typical inclusions may be found within the epidermis. Epithelial necrosis and such inclusions may be most prominent within hair follicles.

Eosinophilic intracytoplasmic and intranuclear inclusions are found within epithelial cells of affected organs. In OWP, a necrotizing bronchointerstitial pneumonia may be evident (Fig. 1.80) [130–132].

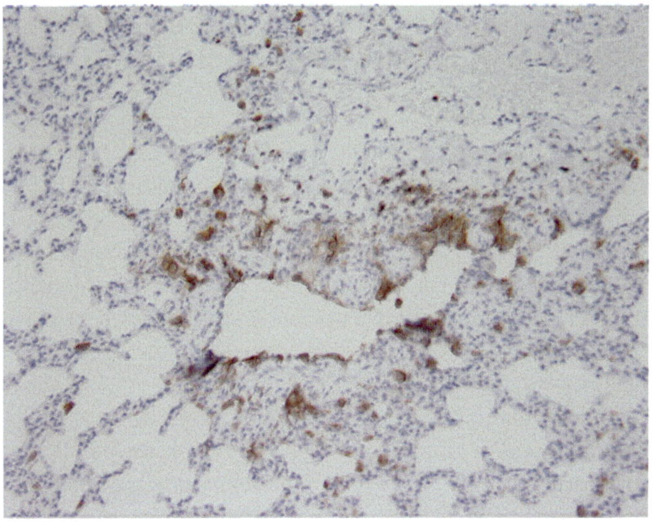

Fig. 1.80 Measles virus (Paramyxoviridae family). Viral bronchointerstitial pneumonia with syncytial cells demonstrated by immunohistochemistry for measles virus (*Macaca mulatta*, lung, ABC immunostain technique, DAB chromogen)

Abortions and still births occur [133]. In NWP, the disease is often centered on the GI tract resulting in a necrotizing gastroenteritis. In these species, the disease may be particularly severe resulting in destruction of virtually all epithelial cells. Viral isolation, immunohistochemistry, and ultrastructural examination may be useful in diagnosis. Grossly, small white foci within the oral mucosa rimmed by a thin raised red border (termed Koplik spots) are pathognomonic of measles virus infection but are infrequently observed. Giant cells (Warthin–Finkeldey cells) lacking viral inclusions may be found within lymphoid tissue.

In the later stages of infection, viral-induced immunosuppression may develop due to the viral effect on thymic function and viral inhibition of interferongamma-gamma upregulation of major histocompatibility complex II antigen and widespread necrosis of lymphocytes. This dysfunction has been implicated in false negative intradermal tuberculin tests and in the reactivation of latent infections such as CMV. A modified live vaccine is available and can be used in NWPs and OWPs. Although the source of infection in NHPs is often an infected human, once established in primate colonies, the virus may be transmitted back to human handlers.

1.12.2 Canine Distemper Virus

An outbreak of canine distemper in multiple Chinese macaque colonies has recently been described and resulted in epizootics of morbidity and mortality [134–136]. Cutaneous involvement characterized by hyperkeratosis of the face and hands/feet was common (Fig. 1.81) similar to hard pad disease described in dogs. Vaccination for measles virus may offer protection [137].

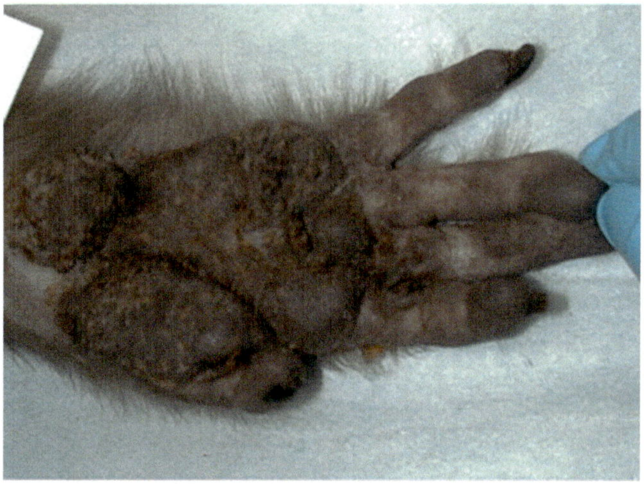

Fig. 1.81 Canine distemper virus (Paramyxoviridae family). Hyperkeratosis of palm resembling hard pad disease of dogs (*Macaca mulatta*)

1.13 Retroviridae

1.13.1 Orthoretrovirinae

Simian Retrovirus

Type D retroviruses are unique to nonhuman primates and may exist as endogenous and exogenous forms. The original type D retrovirus was isolated in 1970 from a mammary neoplasia of a female rhesus macaque and designated the Mason-Pfizer monkey virus (MPMV). Disease conditions are variable and antibody-negative viremic, and antibody-positive nonviremic and viremic states are observed [17]. Clinically affected animals are usually viremic and often show signs of chronic diarrhea and weight loss accompanied by anemia or pancytopenia. Such animals may acquire a number of opportunistic infections including CMV, Rhodococcus, candidiasis, cryptosporidiosis, and noma. SRV-D infected animals appear to be more susceptible to infection with pyogenic bacteria and less susceptible to *M. avium*, Pneumocystis, SV40, and adenovirus infection than SIV-infected animals [138, 139]. This may be related to the wider viral tropism of SRV-D and the occurrence of pancytopenia rather than the selective loss of CD4 T cells. Retroperitoneal fibromatosis is uniquely associated with SRV-2 infection and is a fibroproliferative disorder most frequently observed originating at the ileal–cecal junction and root of the mesentery. Histologically, the lesion shows some similarities to Kaposi's sarcoma of human AIDS patients and there is evidence that a macaque rhadinovirus (RFHV) may play a causative role in the face of immunosuppression [59, 61]. A subcutaneous form may also be noted.

A case definition of SRV-D retrovirus-induced simian AIDS has been developed and includes generalized lymphadenopathy and/or splenomegaly accompanied by at least four of the following clinical and laboratory findings: (1) weight loss >10%, (2) fever (>103F), (3) persistent refractory diarrhea, chronic or opportunistic infections unresponsive to treatment, (4) retroperitoneal or subcutaneous fibromatosis, and (5) hematological abnormalities (including anemia <30, neutropenia <1700, lymphopenia <1600, thrombocytopenia <50,000) [8].

The virus is readily transmitted within macaque colonies and the seropositivity rate may vary depending on husbandry and housing practices. SRV-D may be isolated from the saliva of diseased or carrier animals and is horizontally transmitted through fighting and as a sexually transmitted disease. Vertical transmission may also occur from infected dam to the neonate. SRV-D has a wider cellular tropism than SIV and may infect a variety of cell types including B and T (CD4 and CD8) cells, macrophages, hematopoietic progenitor cells, and epithelial cells (e.g., salivary glands and gastrointestinal tract) [138]. The mechanism by which the virus produces severe immunosuppression is unknown. Animals may harbor the virus for extended periods with few or no clinical signs. Only a subset of animals will progress to clinical disease initially recognized by a lymphadenopathy characterized by varying degrees of follicular and parafollicular hyperplasia progressing to marked lymphoid atrophy and complete effacement of lymph node architecture.

Effective vaccines have been developed experimentally but are not available commercially. SPF colonies have been developed through a test and remove strategy. The antibody-negative viremic animal may represent an important reservoir for transmission and use of antibody detection and viral isolation or PCR are required to detect all infected animals within a colony and eliminate the virus. The zoonotic potential of SRV-D is unclear. A single report claims identification of SRV-D in an AIDS patient with skin disease by PCR. However, the individual had no contact with NHP, and it is unclear how this infection may have occurred. Occupational exposure has resulted in seroconversion but viral sequences could not be detected [140].

Simian T-Lymphotropic Virus (STLV)

STLV or Primate T-lymphotropic viruses (PTLV) are commonly found in numerous OWP species and show close homology to the human pathogen HTLV-1. Strict species specificity is not observed and cross-species transmission may be associated with disease. Three groups of STLV agents are recognized: STLV-1, STLV-2, and STLV-L. A single report of an STLV-2 virus in spider monkeys has not been confirmed and infection appears to be limited to OWP species. SIV-infected rhesus macaques are susceptible to HTLV-1 inoculation and develop an atypical lymphocytosis. There is evidence that HTLVs are derived from PTLVs and that multiple cross-species transmissions have occurred through consumption of bushmeat [141, 142].

In most instances, infection is asymptomatic and not associated with clinical signs [143, 144]. While STLV-1 infection of macaques is common, it has not been associated with overt clinical disease in normal or immunosuppressed animals. Infection in baboons has been associated with malignant lymphoma and leukemia and is believed to have resulted from cross-species transmission of the agent with resulting epizootic [145, 146].

In baboons, infection has been characterized by depression, anorexia, regional or generalized lymph node enlargement, and hepatosplenomegaly. Pulmonary involvement may be frequent. Cutaneous involvement, hypercalcemia, and pleural effusions are observed less frequently. Multilobulated neoplastic cells are observed within peripheral blood of the majority of cases.

Serologic assays indicate a high seropositive rates within many wild and captive populations of African green monkeys, macaques, and baboons. The virus is likely readily transmitted by the parenteral and sexual routes. In most species, infection is not associated with pathology. Epizootics of malignant lymphoma have been observed in baboons and believed to be caused by cross-species transmission of the virus. In these epizootics, involvement of the lymph nodes, spleen, liver, skin, and lungs is common [145, 146]. Overt leukemia may be observed and pulmonary involvement is characteristic. Most cases have been CD3CD4 T cell infiltrates consisting of neoplastic lymphocytes and accompanied by multinucleated giant cells, necrosis, and inflammatory cells. A subset of cases express CD30 a marker consistent with large cell anaplastic lymphoma of man.

Simian Immunodeficiency Virus (SIV)

Primate lentiviruses have a genome of about 10 kb encoding 9 open reading frames that produce 15 distinct proteins. The prototypic primate lentivirus HIV was first recognized in the early 1980s and isolated in 1983. The first isolation of a related nonhuman primate virus was made a year later. Like other primate lentiviruses, SIV is a diploid virus having two positive-sense RNA genomes which are contained within the conical core of a single virion. The virus may infect terminally differentiated cells like tissue macrophages but its principal target during acute infection is the replicating CCR5 positive CD4 T lymphocytes within the GI tract. The virus uses a trimeric env or envelop protein to bind to CD4 and chemokine coreceptors to infect cells. Two important coreceptors are CCR5 and CXCR4, both chemokine receptors involved in immune regulation.

SIVs were first identified as a cause of naturally occurring immunosuppressive disease in macaques at several National Primate Research Centers NPRCs in the 1970s and 1980s. Inadvertent transmission of SIVsm from sooty mangabeys to macaques likely initiated these epizootics.

Pathogenesis and associated pathology differ between African and Asian species:

African species SIV variants infect more than 35 species of African nonhuman primates throughout sub-Saharan Africa [147–149]. Following infection viruses replicate to high titer targeting lymphoid tissue and gut. The viruses are readily transmissible leading to high seroprevalence in some populations. The founder effect has produced populations that are free of viral infection. While infection is common, disease is rare. Following infection, high viral loads are observed and are accompanied by modest CD4 T cell depletion. Animals develop antibody and cellular immune responses. CD4 T cell numbers stabilize and there is no correlation between viral load and CD4 T cell number. Lower expression of activation and proliferation markers on CD4+ T-cells suggests decreased activation-induced cell death (AICD). Understanding how these species co-exist with chronic SIV infection without disease progression may give insight to the disease process in man.

Animals have peak viremia of 10^7 to 10^8 copies/ml of plasma at 2 weeks and an immune response is established leading to a reduction in viral load to a set point of 10^6 o 10^7. Following inoculation, there is a modest dip in CD4 T cell number but these then stabilize. During this acute period, there is a rapid increase in the number of proliferating CD4 T cells that is likely able to compensate for CD4 T cell loss. Despite high viral loads, most animals do not develop progressive loss of CD4 T cells. The reason African species can support vigorous viral replication but do not develop AIDS is an area of intense research as this may lead to insights to the mechanism of AIDS in humans. Pathology is most frequently lacking. Rarely disseminated giant cell disease and opportunistic infections have been described.

Asian macaques SIV was first identified at several NPRCs in the early 1980s as a cause of spontaneous immunodeficiency in animals that had developed opportunistic infections including disseminated *Mycobacterium avium* complex dMAC, pneumocystis, and PML [150]. The first virus was isolated from rhesus macaques and named SIVmac. It was later shown to have a high degree of similarity to SIVsm, a virus of sooty mangabeys. In fact, SIVmac was probably established in rhesus colonies through the inadvertent exposure of macaques to sooty mangabeys or their tissue. This is supported by the fact that SIV is not identified as an indigenous infection of macaques in their natural habitat. Infection of these animals causes progressive loss of CD4 T cells and eventually AIDS mimicking the clinical course of disease in humans [79, 80].

In African species, SIV agents are readily transmitted at about the time of sexual maturity leading to high seropositivity rates. In Asian species, transmission by the parenteral and mucosal routes is observed. Initiation of naturally occurring epizootics in macaques is believed to have developed due to exposure of macaques to sooty mangabeys or their tissues in the late 1960s and early 1970s.

While infection of CD4 positive cells is a hallmark of HIV infection, the mechanism of CD4 T depletion and the development of immunodeficiency are more complex and not fully understood. In all likelihood, a combination of the following mechanisms may be responsible: (1) Direct viral cytolytic infection of CD4 T cells; (2) Destruction of CD4 cells by viral specific cytotoxic T cells; (3) Antibody-dependent cell-mediated cytotoxicity (ADCC); (4) Activation-induced cell death (AICD); and (5) Destruction of normal lymph node and thymic architecture leading to dysfunction in regenerative capacity.

Following inoculation of Asian species of macaques, three infection profiles may be observed: normal progression profile, rapid progression profile, and elite controller profile. Normal progression profile is seen in 75% of animals and is characterized by peak viral loads of 10^7 at 2 weeks. A humoral and cellular immune response develops that is partially effect in controlling infection with viral set-points of 10^5. Progressive loss of CD4 T cells is observed with acquisition of OIs when absolute counts drop below 300–400 cells/mm^3. Mean survival is 18 months. Rapid progression profile is seen in 20% of animals and is characterized by peak viral loads of 10^8. Viral set point is not established and loads remain high throughout the course of infection. An aborted or truncated antibody response is observed and survival is only 3–6 months. Animals may die with normal CD4 counts with no OIs. Giant cell disease is often evident in many organs. Elite controller profile is seen in 5% of animals with peak viral loads of only 10^5 RNA copies/ml plasma. A vigorous immune response develops consisting of humoral and cellular immune elements and viral replication is eventually controlled below the limit of detection. Genetic basis of control may be related to MHC type or polymorphisms in other innate immune system factors.

When absolute CD4 T cell count drops below 300–400 cells/mm^3, animals develop opportunistic infections through reactivation of latent infections or acquisition from the environment. The spectrum and histologic picture of these changes show remarkable similarities to the diseases recognized in human HIV patients. In addition to opportunistic infections, there are a number of viral specific diseases such as AIDS enteropathy, lymphocytic interstitial pneumonitis, giant cell pneumonia, and SIV arteriopathy that the virus may induce in-and-of itself (Fig. 1.82), lungs (Fig. 1.83). SIV encephalitis is a frequent disease entity in macaque species showing similarities to HIV encephalitis in human AIDS patients (Fig. 1.84) [151–158]. These are viral syncytial cells containing large amounts of viral antigen (Fig. 1.85).

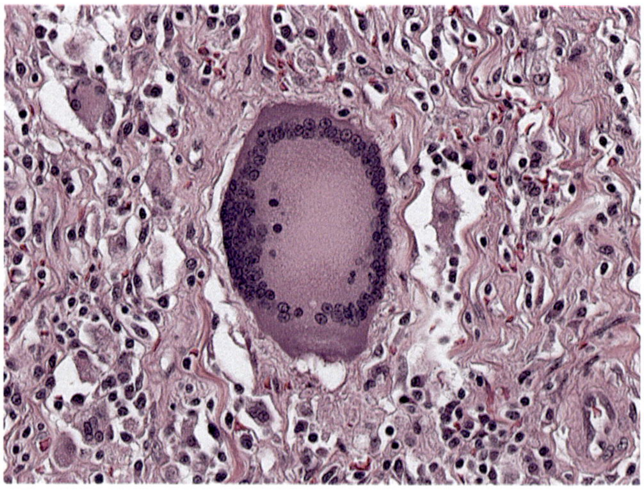

Fig. 1.82 Simian immunodeficiency virus (Lentivirus genus). Large multinucleated syncytial cell in lymphoid of SIV-infected macaque (*Macaca mulatta*, lymph node, H&E)

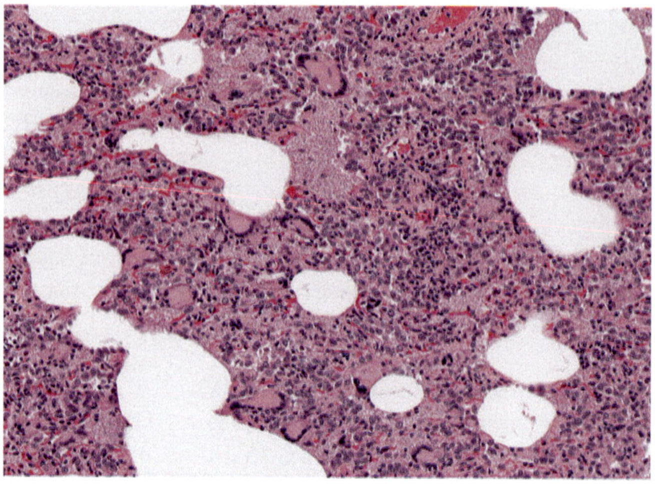

Fig. 1.83 Simian immunodeficiency virus (Lentivirus genus). Giant cell pneumonia in an infected macaque. Eosinophilic foamy material represents concurrent pneumocystis infection (*Macaca mulatta*, lung, H&E)

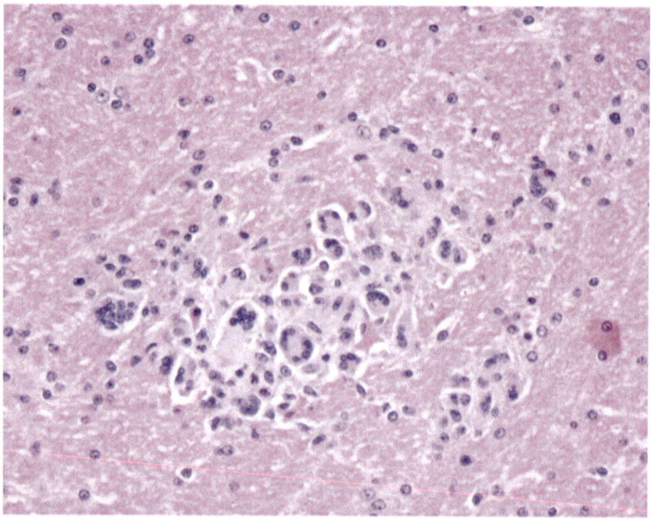

Fig. 1.84 Simian immunodeficiency virus (Lentivirus genus). SIV encephalopathy with multinucleated viral syncytial (giant) cells (*Macaca mulatta*, brain, H&E)

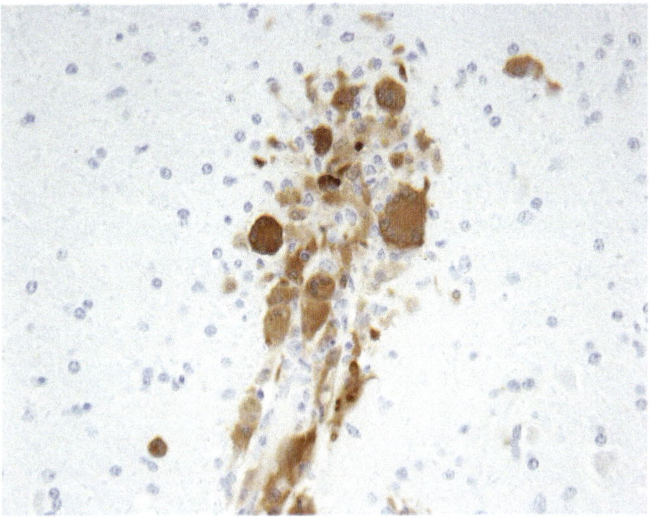

Fig. 1.85 Simian immunodeficiency virus (Lentivirus genus). Immunohistochemistry for SIV p27 antigen demonstrating viral infected syncytial cells (*Macaca mulatta*, brain, ABC immunostain, DAB chromogen)

In humans, this causes HIV dementia and motor cognitive impairment. The pathogenesis involves the recruitment and activation of macrophage/microglia to the CNS. These cells then elaborate mediators of neuroinflammation that then cause neuronal apoptosis and loss. Histologically, the lesion is characterized by the accumulation of perivascular macrophages, formation of giant cells, gliosis, and demyelinization in deep white matter. Perivascular cells often express p28 antigen and IHC will reveal large numbers of CD68 positive macrophages/microglia. CD163 a marker of microglia hyper-ramification is observed in affected animals. A number of diagnostic tools are available and include (1) Antibody response by ELISA and Western blot; (2) Antigen detection by Gag antigen capture; (3) Viral isolation (Co-cultivation with CEMx174 cells); (4) PCR or Q RT PCR sensitive to <30 RNA copies/ml plasma; and (5) ISH/IHC.

Infection of laboratory workers has occurred on several occasions [159]. An anonymous serosurvey performed in the early 1990s revealed 3 of 472 (0.6%) researchers had antibodies to HIV2 or SIV. A SIV prevention and control plan should be developed by facilities housing potentially infected animals and follow the OSHA bloodborne pathogen standard to include: (1) Universal (standard) precautions; (2) Medical surveillance, (3) Personal protective equipment, (4) Appropriate training, (5) Sharps injury prevention plan (hierarchal control), and (5) a post-exposure plan to include risk assessment and stratification, assessment and treatment by physician experienced in HIV/SIV exposure and post-exposure prophylaxis [160].

1.13.2 Spumaretrovirinae

Simian foamy viruses (SFV) have been recognized in many OWP species but have not yet been definitively identified in NWPs. Serosurveys reveal greater than 95% of adult macaques are chronically infected but there is no known disease association. Zoonotic transmission to laboratory and zoo workers has been identified [161, 162]. The virus does cause cytopathic effect and cytolysis in long-term primary cell cultures such as bronchoalveolar macrophages and the veterinary pathologist may be called upon to identify the etiologic agent in cultures where such cytopathic effect is noted. Well-characterized viruses include SFV-1 in rhesus macaques, SFV-3 in African green monkeys, and SFV-6 in chimpanzees [163–165].

1.14 Nonenveloped RNA Viruses

1.14.1 Picornaviridae

1.14.2 Hepatitis A Virus (HAV)

HAV is a common cause of human hepatitis and human isolates may experimentally infect a variety of NHP species. In addition, serologic evidence suggests natural infection of a number of NWP and OWPs [166]. The relationship between human HAV isolates and those associated with naturally occurring disease in NHPs remains unclear. Isolates from Aotus, *M. fascicularis*, and *C. aethiops* have been shown to differ substantially from human clinical isolates suggesting that despite antigenic similarities HAV represents a heterogenous group of viruses [167–169].

Clinical signs are uncommon in natural infections. Elevations in ALT and AST may be observed as well as mild increases in bilirubin [170]. Inadvertent infection may confound experimental work. Transmission is thought to occur through the fecal–oral route. Most work suggests that animals become infected and shed virus for short periods. Exposure to conspecifics and stress during transportation may initiate epizootics. A single report suggests a carrier state may exist.

Following infection, a prolonged incubation period of 20–50 days is observed followed by an increase in liver enzymes. Virus may be shed in the feces for an additional 10–30 days before resolution. This incubation and shedding may surpass the standard quarantine period used at most facilities. Histologically, there is a multifocal nonsuppurative hepatitis mediated by CD8 T cells (Figs. 1.86 and 1.87) and abundant HAV nucleic acid detected by in situ hybridization in hepatocytes and sinusoidal cells (Fig. 1.88). There may be mild hyperplasia of biliary epithelium. The zoonotic potential of simian HAV isolates is unknown; however, HAV variants appear to lack species specificity. The efficacy of the human HAV vaccine against simian variants is unknown.

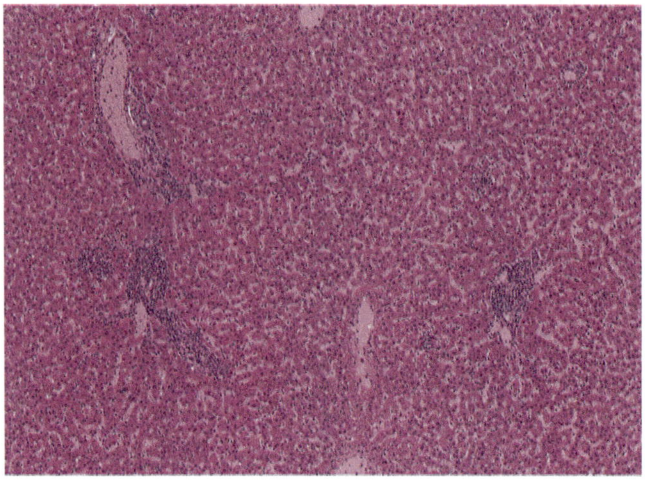

Fig. 1.86 Hepatitis A virus (hepatovirus genus). Multifocal portal and periportal mononuclear cell infiltrate (*Macaca mulatta*, liver, H&E)

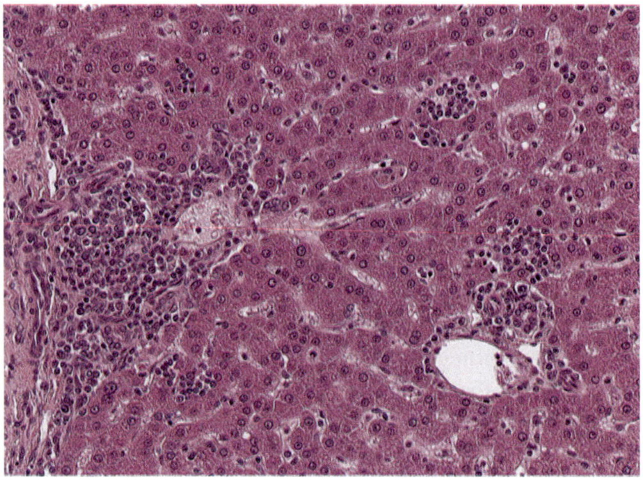

Fig. 1.87 Hepatitis A virus (hepatovirus genus). Multifocal portal mononuclear cell infiltrates extending past the limiting plate into the adjacent hepatic parenchyma (*Macaca mulatta*, liver, H&E)

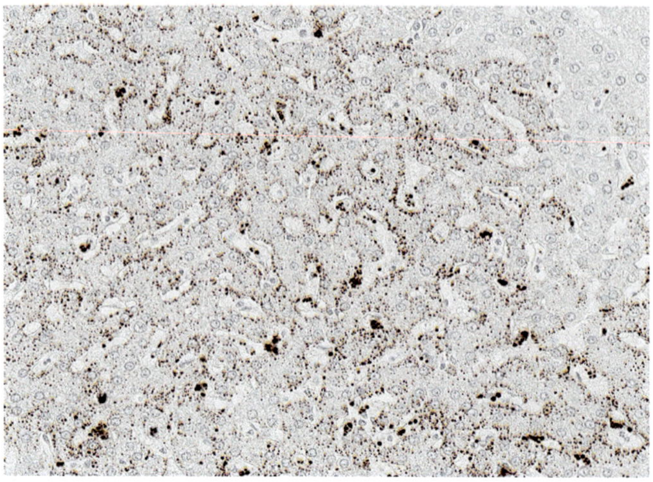

Fig. 1.88 Hepatitis A virus (hepatovirus genus). In situ hybridization demonstrating abundant viral RNA in hepatocytes and sinusoidal cells (*Macaca mulatta*, liver, in situ hybridization, DAB chromogen)

1.14.3 Cardiovirus (Encephalomyocarditis Virus; EMCV)

EMCV is a nonenveloped RNA containing virus within the picornavirus family and has been recognized to cause myocarditis and abortion in several nonhuman primate species [171–175]. Five strains are identified by hemagglutinin assays including Mengo-, MM-, Columbia-SK, ME-, and encephalomyocarditis viruses. As with other picornaviruses, they may be highly resistant to environmental influences. Affected primates are usually found dead with no premonitory clinical signs. In less acute cases, tachypnea, dyspnea, and frothing from the nostrils have been recognized and fetal death has been suspected. In experimental cases, the incubation period has been variable from 4 to 41 days.

The viruses are carried by a variety of rodent species and shed in bodily fluids that may then contaminate the environment. Epizootics have been recognized in both OWP and NWP species. Several outbreaks have been noted in baboon species suggesting a unique sensitivity [173]. While the virus has been identified in mice, epizootics have been most frequently linked to rats. Once established in a species, it may be propagated through intraspecies transmission. Disease is due to the primary effect of viral replication and inflammation within the myocardium and central nervous system. Viral particles may be found within cardiomyocytes and endothelium. At necropsy, pulmonary congestion, pericardial effusion, and mottling of the myocardium may be noted. Histologically, there is a necrotizing multifocal myocarditis (Figs. 1.89 and 1.90) and may be accompanied by mineralization (Fig. 1.91). Karyomegaly of cardiomyocytes may be evident but intranuclear and intracytoplasmic inclusions, as seen with CMV, are absent.

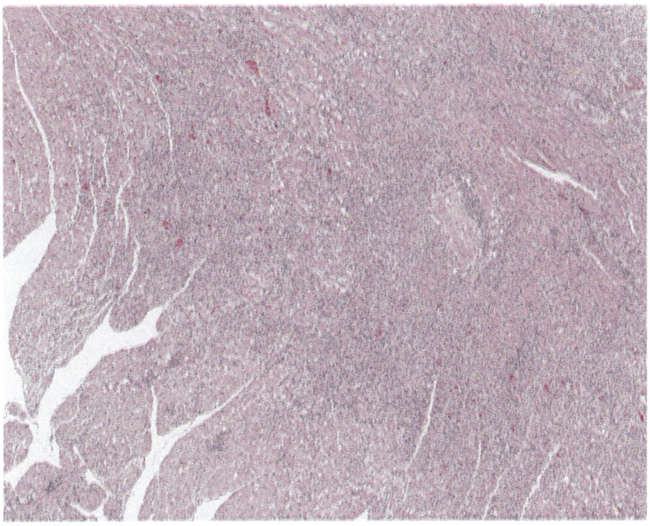

Fig. 1.89 Encephalomyocarditis virus (cardiovirus genus). Severe locally extensive myocarditis (*Macaca mulatta*, heart, H&E)

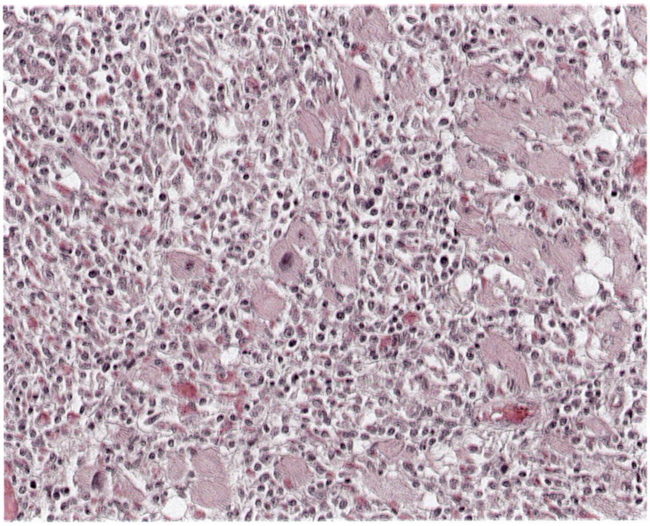

Fig. 1.90 Encephalomyocarditis virus (cardiovirus genus). Nonsuppurative myocarditis with cardiomyocyte degeneration (*Macaca mulatta*, heart, H&E)

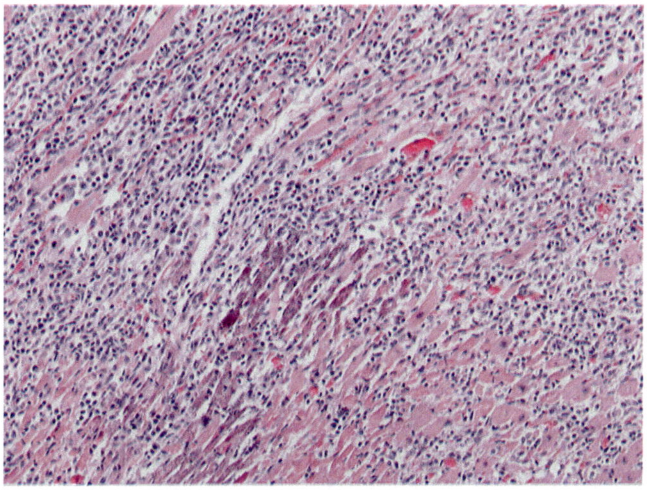

Fig. 1.91 Encephalomyocarditis virus (cardiovirus genus). Necrotizing myocarditis with dystrophic mineralization of cardiomyocytes (*Macaca mulatta*, heart, H&E)

References

1. Johnson RF, Keith LA, Cooper TK, et al. Acute late-stage myocarditis in the crab-eating macaque model of hemorrhagic smallpox. Viruses. MDPI AG. 2021;13(8)
2. Arita I, Jezek Z, Khodakevich L, Ruti K. Human monkeypox: a newly emerged orthopoxvirus zoonosis in the tropical rain forests of Africa. Am J Trop Med Hyg. 1985;34(4):781–9.
3. Downie AW. Serological evidence of infection with Tana and Yaba pox viruses among several species of monkey. J Hyg. 1974;72(2):245–50.
4. Magnus P, Andersen E, ... KP-AP, 1959 undefined. A pox-like disease in cynomolgus monkeys. Wiley Online Library [Internet]. [cited 2022 Aug 3]. Available from: https://onlinelibrary.wiley.com/doi/abs/10.1111/j.1699-0463.1959.tb00328.x
5. Prier JE, Sauer RM. A pox disease of monkeys.
6. Prier JE, Sauer RM, Malsberger RG, Sillaman JM. Studies on a pox disease of monkeys. II. Isolation of the etiologic agent. Am J Vet Res [Internet]. 1960 [cited 2022 Aug 3]; 21:381–384. Available from: https://www.cabdirect.org/cabdirect/abstract/19602203236
7. Nagata N, Saijo M, Kataoka M, et al. Pathogenesis of fulminant monkeypox with bacterial sepsis after experimental infection with West African monkeypox virus in a cynomolgus monkey. Int J Clin Exp Pathol. 2014;7(7):4359–70.
8. Tree JA, Hall G, Pearson G, et al. Sequence of pathogenic events in cynomolgus macaques infected with aerosolized monkeypox virus. J Virol. 2015;89(8):4335–44.
9. Dyall J, Johnson RF, Chen DY, et al. Evaluation of monkeypox disease progression by molecular imaging. J Infect Dis. 2011;204(12):1902–11.
10. Cann JA, Jahrling PB, Hensley LE, Wahl-Jensen V. Comparative pathology of smallpox and monkeypox in man and macaques. J Comp Pathol. 2013;148(1):6–21.
11. Iizuka I, Ami Y, Suzaki Y, et al. A single vaccination of nonhuman primates with highly attenuated smallpox vaccine, LC16m8, provides long-term protection against monkeypox. Jpn J Infect Dis. 2017;70(4):408–15.
12. Mätz-Rensing K, Bleyer M. Viral diseases of common marmosets. In: Marini R, Wachtman L, Tardif S, Mansfield K,

Fox J, editors. The common marmoset in captivity and biomedical research. Elsevier; 2018. p. 251–64.
13. Mätz-Rensing K, Stahl-Hennig C, Kramski M, Pauli G, Ellerbrok H, Kaup FJ. The pathology of experimental poxvirus infection in common marmosets (Callithrix jacchus): further characterization of a new primate model for orthopoxvirus infections. J Comp Pathol. 2012;146(2–3):230–42.
14. Schmitt A, Gan LL, Wahed AA, et al. Dynamics of pathological and virological findings during experimental calpox virus infection of common marmosets (Callithrix jacchus). Viruses. 2017;9(12)
15. Cardeti G, Gruber CEM, Eleni C, et al. Fatal outbreak in Tonkean macaques caused by possibly novel orthopoxvirus, Italy, January 2015 1. Emerg Infect Dis. 2017;23(12):1941–9.
16. Puro V, Fusco FM, Castilletti C, et al. Occupational transmission of an Orthopoxvirus infection during an outbreak in a colony of *Macaca tonkeana* in Lazio Region, Italy, 2015. Zoonoses Public Health. 2018;65(5):578–83.
17. Wachtman L, Mansfield K. Viral diseases of nonhuman primates. Nonhuman primates in biomedical research. Elsevier Inc.; 2012; 1–104.
18. Whittaker D, Glaister JR. A Yaba-like condition in a young baboon (Papio anubis). Lab Anim. 1985;19(3):177–9.
19. Zhu S, Viejo-Borbolla A. Pathogenesis and virulence of herpes simplex virus. Virulence; 2021; 12(1):2670–2702.
20. Sakulwira K, Theamboonlers A, Charoonrut P, Ratanakorn P, Poovorawan Y. Serological evidence of herpesvirus infection in gibbons. BMC Microbiol. 2002;2:1–5.
21. Landolfi JA, Wellehan JFX, Johnson AJ, Kinsel MJ. Fatal human herpesvirus type 1 infection in a white-handed gibbon (Hylobates lar). J Vet Diagn Invest. 2005;17(4):369–71.
22. Heldstab A, Ruedi D, Sonnabend W, Deinhardt F. Spontaneous generalized Herpesvirus hominis infection of a lowland gorilla (Gorilla gorilla gorilla). J Med Primatol. 1981;10(2–3):129–35.
23. Gilardi KVK, Oxford KL, Gardner-Roberts D, et al. Human herpes simplex virus type 1 in confiscated gorilla. Emerg Infect Dis. 2014;20(11):1883–6.
24. McClure HM, Swenson RB, Kalter SS, Lester TL. Natural genital Herpesvirus hominis infection in chimpanzees (Pan troglodytes and Pan paniscus). Lab Anim Sci. 1980;30(5):895–901.
25. Landolfi JA, Wellehan JFX, Johnson AJ, Kinsel MJ. Fatal human herpesvirus type 1 infection in a white-handed gibbon (Hylobates lar). J Vet Diagn Investig. 2005;17(4):369–71.
26. Imura K, Chambers JK, Uchida K, et al. Herpes simplex virus type 1 infection in two pet marmosets in Japan. J Vet Med Sci. 2014;76 (12):1667–70.
27. Mätz-Rensing K, Jentsch KD, Rensing S, et al. Fatal Herpes simplex infection in a group of common marmosets (Callithrix jacchus). Vet Pathol. 2003;40(4):405–11.
28. Costa EA, Luppi MM, Malta MC, et al. Outbreak of human herpesvirus type 1 infection in nonhuman primates (Callithrix penincillata). J Wildl Dis. 2011;47(3):690–3.
29. Bruno SF, Liebhold MM, Mätz-Rensing K, et al. Herpesvirus infection in free-living black-tufted-ear marmoset (Callithrix penicillata E. Geoffroyi 1812) at the State Park of Serra da Tiririca, Niterói, Rio de Janeiro, Brasilien. Berliner und Munchener Tierarztliche Wochenschrift. 1997;110(11–12):427–30.
30. Eberle R, Maxwell LK, Nicholson S, Black D, Jones-Engel L. Genome sequence variation among isolates of monkey B virus (Macacine alphaherpesvirus 1) from captive macaques, vol. 508. Virology; 2017. p. 26–35.
31. Oya C, Ochiai Y, Taniuchi Y, et al. Prevalence of herpes B virus genome in the trigeminal ganglia of seropositive cynomolgus macaques. Lab Anim. 2008;42(1):99–103.
32. Retrospective analysis of an outbreak of B virus infection in a colony of DeBrazza's monkeys (Cercopithecus neglectus). [cited 2022 Aug 1].
33. Coulibaly C, Hack R, Seidl J, Chudy M, Itter G, Plesker R. A natural asymptomatic herpes B virus infection in a colony of laboratory brown capuchin monkeys (Cebus apella). Lab Anim. 2004;38(4):432–8.
34. Ward JA, Hilliard JK. Herpes B virus-specific pathogen-free breeding colonies of macaques: serologic test results and the B-virus status of the macaque. Contemp Top Lab Anim Sci. 2002;41(4): 36–41.
35. Holmes GP, Chapman LE, Stewart JA, et al. Guidelines for the prevention and treatment of B-virus infections in exposed persons. Clin Infect Dis. 1995;20(2):421–39.
36. Gulani J, Koch A, Chappell MG, et al. Cercopithecine Herpesvirus 9 (simian varicella virus) infection after total-body irradiation in a rhesus macaque (Macaca mulatta). Comp Med; 2016; 66(2): 150–153.
37. Mansfield KG, Sasseville VG, Westmoreland S, v. Molecular localization techniques in the diagnosis and characterization of nonhuman primate infectious diseases. Vet Pathol. 2014;51(1): 110–26.
38. Desrosiers RC, Falk LA. Herpesvirus tamarinus and its relation to herpes simplex virus. J Gen Virol. 1981;56(Pt 1):119–30.
39. Baskin GB. Disseminated cytomegalovirus infection in immunodeficient rhesus monkeys. Am J Pathol. 1987;129(2):345–52.
40. Hutto EH, Anderson DC, Mansfield KG. Cytomegalovirus-associated discrete gastrointestinal masses in macaques infected with the simian immunodeficiency virus. Vet Pathol. 2004;41(6): 691–5.
41. Clemmons EA, Gumber S, Strobert E, Bloomsmith MA, Jean SM. Self-injurious behavior secondary to cytomegalovirus-induced neuropathy in an SIV-infected rhesus macaque (Macaca mulatta). Comp Med. 2015;65(3):266–70.
42. Kuhn EM, Stolte N, Mätz-Rensing K, et al. Immunohistochemical studies of productive rhesus cytomegalovirus infection in rhesus monkeys (Macaca mulatta) infected with simian immunodeficiency virus. Vet Pathol. 1999;36(1):51–6.
43. Carville A, Mansfield KG. Comparative pathobiology of macaque lymphocryptoviruses. Comp Med. 2008;58(1):57–67.
44. Cho Y-G, Gordadze A, v., Ling PD, Wang F. Evolution of two types of Rhesus Lymphocryptovirus similar to type 1 and type 2 Epstein-Barr virus. J Virol. 1999;73(11):9206–12.
45. Habis A, Baskin G, Simpson L, Fortgang I, Murphey-Corb M, Levy LS. Rhesus lymphocryptovirus infection during the progression of SAIDS and SAIDS-associated lymphoma in the rhesus macaque. AIDS Res Hum Retrovir. 2000;16(2):163–71.
46. Marr-Belvin AK, Carville AK, Fahey MA, et al. Rhesus lymphocryptovirus type 1-associated B-cell nasal lymphoma in SIV-infected rhesus macaques. Vet Pathol. 2008;45(6):914–21.
47. Feichtinger H, Kaaya E, Putkonen P, et al. Malignant lymphoma associated with human AIDS and with SIV-induced immunodeficiency in macaques. AIDS Res Hum Retrovir. 1992;8(3):339–48.
48. Habis A, Baskin GB, Murphey-Corb M, Levy LS. Simian AIDS-associated lymphoma in rhesus and cynomolgus monkeys recapitulates the primary pathobiological features of AIDS-associated non- Hodgkin's lymphoma. AIDS Res Hum Retrovir. 1999;15(15):1389–98.
49. Baskin GB, Cremer KJ, Levy LS. Comparative pathobiology of HIV- and SIV-associated lymphoma. AIDS Res Hum Retrovir. 2001;17(8):745–51.
50. Kahnt K, Mätz-Rensing K, Hofmann P, Stahl-Hennig C, Kaup FJ. SIV-associated lymphomas in rhesus monkeys (Macaca mulatta) in comparison with HIV-associated lymphomas. Vet Pathol. 2002;39(1):42–55.
51. Page EK, Courtney CL, Sharma P, et al. Post-transplant lymphoproliferative disorder associated with immunosuppressive therapy for renal transplantation in rhesus macaques (Macaca mulatta). Exp Toxicol Pathol. 2013;65(7–8):1019–24.

52. Baskin GB, Roberts ED, Kuebler D, et al. Squamous epithelial proliferative lesions associated with rhesus Epstein-Barr virus in simian immunodeficiency virus-infected rhesus monkeys. J Infect Dis. 1995;172(2):535–9.
53. Kutok JL, Klumpp S, Simon M, et al. Molecular evidence for rhesus lymphocryptovirus infection of epithelial cells in immunosuppressed rhesus macaques. J Virol. 2004;78(7): 3455–61.
54. Hadlock KG, Gish R, Rowe J, et al. Expression of the simian Epstein-Barr virus-encoded latent membrane protein-1 in malignant lymphomas of SIV-infected rhesus macaques. J Med Virol. 2001;65(1):114–20.
55. Pingel S, Hannig H, Mätz-Rensing K, Kaup FJ, Hunsmann G, Bodemer W. Detection of Epstein-Barr virus small RNAs EBER1 and EBER2 in lymphomas of SIV-infected rhesus monkeys by in situ hybridization. Int J Cancer. 1997;72(1):160–5.
56. White JA, Yang X, Todd PA, Lerche NW. Longitudinal patterns of viremia and oral shedding of rhesus rhadinovirus and retroperitoneal fibromatosis herpesviruses in age-structured captive breeding populations of rhesus macaques (Macaca mulatta). Comp Med. 2011;61(1):60–70.
57. Desrosiers RC, Sasseville VG, Czajak SC, et al. A herpesvirus of rhesus monkeys related to the human Kaposi's sarcoma-associated herpesvirus. J Virol. 1997;71(12):9764–9.
58. Tsai CC, Warner TFCS, Uno H. Subcutaneous fibromatosis associated with an acquired immune deficiency syndrome in pig-tail Macaques. Am J Pathol. 1985;120(1):30–7.
59. Giddens WE, Tsai CC, Morton WR, Ochs HD, Knitter GH, Blakley GA. Retroperitoneal fibromatosis and acquired immunodeficiency syndrome in macaques. Pathologic observations and transmission studies. Am J Pathol. 1985;119(2):253–63.
60. Bruce AG, Ryan JT, Thomas MJ, et al. Next-generation sequence analysis of the genome of RFHVMn, the macaque homolog of Kaposi's sarcoma (KS)-associated herpesvirus, from a KS-like tumor of a pig-tailed macaque. J Virol. 2013;87(24):13676–93.
61. Tsai CC, Giddens WE, Morton WR, Rosenkranz SL, Ochs HD, Benveniste RE. Retroperitoneal fibromatosis and acquired immunodeficiency syndrome in macaques: epidemiologic studies. Lab Anim Sci. 1985;35(5):460–4.
62. Burnside KL, Ryan JT, Bielefeldt-Ohmann H, et al. RFHVMn ORF73 is structurally related to the KSHV ORF73 latency-associated nuclear antigen (LANA) and is expressed in retroperitoneal fibromatosis (RF) tumor cells. Virology. 2006;354(1):103–15.
63. Bruce AG, Bakke AM, Bielefeldt-Ohmann H, et al. High levels of retroperitoneal fibromatosis (RF)-associated herpesvirus in RF lesions in macaques are associated with ORF73 LANA expression in spindleoid tumour cells. J Gen Virol. 2006;87(Pt 12):3529–38.
64. Desrosiers RC, Bakker A, Kamine J, Falk LA, Hunt RD, King NW. A region of the Herpesvirus saimiri genome required for oncogenicity. Science. 1985;228(4696):184–7.
65. Laufs R, Fleckenstein B. Susceptibility to Herpesvirus saimiri and antibody development in old and new world monkeys. Med Microbiol Immunol. 1973;158(3):227–36.
66. Laufs R, Fleckenstein B. Malignant lymphoma induced by partially purified Herpesvirus saimiri and recovery of infectious virus from tumorous lymph nodes. Med Microbiol Immunol. 1972;158 (2):135–46.
67. Hunt RD, Garcia FG, Barahona HH, King NW, Fraser CEO, Melendez LY. Spontaneous herpesvirus saimiri lymphoma in an owl monkey. J Infect Dis. 1973;127(6):723–5.
68. Hunt RD, Melendez LV, King NW, et al. Morphology of a disease with features of malignant lymphoma in marmosets and owl monkeys inoculated with herpesvirus saimiri. J Natl Cancer Inst. 1970;44(2):447–65.
69. Meléndez L, v., Daniel MD, Hunt RD. Herpesvirus Saimiri induced malignant lymphoma: recovery of the viral agent from the fatally affected animals. Bibl Haematol. 1970;36:751–3.
70. Utsumi T, Wahyuni RM, Lusida MI, et al. Full genome characterization and phylogenetic analysis of hepatitis B virus in gibbons and a caretaker in Central Kalimantan, Indonesia. Arch Virol; 2015; 160(3):685–692.
71. Dupinay T, Gheit T, Roques P, et al. Discovery of naturally occurring transmissible chronic hepatitis B virus infection among Macaca fascicularis from Mauritius Island. Hepatology. 2013;58 (5):1610–20.
72. Bukh J, Lanford RE, Purcell RH. Persistent human hepatitis B virus infection in cynomolgus monkeys: a novel animal model in the search for a cure? Hepatology. 2013;58(5):1533–6.
73. Vasileva VA, Ivanov MT, Rumel NB, D'Yachenko AG, Kakubava V, v., Danelyan GA. Isolation and biological characterization of an adenovirus of rhesus macaques. Acta Biol Med Ger. 1978;37(8):1281–7.
74. Rogers DL, Ruiz JC, Baze WB, et al. Epidemiological and molecular characterization of a novel adenovirus of squirrel monkeys after fatal infection during immunosuppression. Microb Genom. 2020;6(9):1–13.
75. Baskin GB, Soike KF. Adenovirus enteritis in SIV-infected rhesus monkeys. J Infect Dis. 1989;160(5):905–6.
76. Ochs HD, Morton WR, Tsai C, et al. Maternal–fetal transmission of SIV in macaques: disseminated adenovirus infection in an offspring with congenital SIV infection. J Med Primatol. 1991;20 (4):193–200.
77. Yu G, Yagi S, Carrion R, et al. Experimental cross-species infection of common marmosets by titi monkey adenovirus. PLoS One. 2013;8(7):e68558.
78. Chen EC, Yagi S, Kelly KR, et al. Cross-species transmission of a novel adenovirus associated with a fulminant pneumonia outbreak in a new world monkey colony. PLoS Pathog. 2011;7(7)
79. Daniel MD, Desrosiers RC, Letvin NL, et al. Simian models for AIDS. Cancer Detect Prev Suppl. 1987;1:501–7.
80. Kaup FJ, Mätz-Rensing K, Kuhn EM, Hünerbein P, Stahl-Hennig-C, Hunsmann G. Gastrointestinal pathology in rhesus monkeys with experimental SIV infection. Pathobiology. 1998;66(3–4): 159–64.
81. Song M, Mulvihill MS, Williams KD, Collins BH, Kirk AD. Fatal SV40-associated pneumonia and nephropathy following renal allotransplantation in rhesus macaque. J Med Primatol. 2018;47 (1):81–4.
82. Horvath CJ, Simon MA, Bergsagel DJ, et al. Simian virus 40-induced disease in rhesus monkeys with simian acquired immunodeficiency syndrome. Am J Pathol. 1992;140(6):1431–40.
83. Kaliyaperumal S, Dang X, Wuethrich C, et al. Frequent infection of neurons by SV40 virus in SIV-infected macaque monkeys with progressive multifocal leukoencephalopathy and meningoencephalitis. Am J Pathol. 2013;183(6):1910–7.
84. Simon MA, Ilyinskii PO, Baskin GB, Knight HY, Pauley DR, Lackner AA. Association of simian virus 40 with a central nervous system lesion distinct from progressive multifocal leukoencephalopathy in macaques with AIDS. Am J Pathol. 1999;154(2):437–46.
85. Skubic L, Hošnjak L, Staheli JP, et al. Molecular and phylogenetic characterization of novel papillomaviruses isolated from oral and anogenital neoplasms of Japanese macaques (Macaca fuscata). Viruses. 2021;13(4)
86. Chai D, Bassis CM, Bergin IL, Bell JD, Nyachieo A, Gathumbi PK. Prevalence and geographical distribution of Papio hamadryas papillomavirus 1 (PhPV1) in Kenyan baboons. J Med Primatol. 2017;46(1):13–5.

87. Hoffmann M, Schütze E, Bernhard A, et al. Disease manifestation and viral sequences in a bonobo more than 30 years after papillomavirus infection. Pathogens. 2019;8(1)
88. Silvestre RVD, de Souza AJS, Júnior ECS, et al. First new world primate papillomavirus identification in the Atlantic Forest, Brazil: Alouatta guariba papillomavirus 1. Genome Announc. 2016;4(4)
89. Hertig AT, MacKey JJ, Feeley G, Kampschmidt K. Dysplasia of the lower genital tract in the female monkey, Macaca fascicularis, the crab-eating macaque from Southeast Asia. Am J Obstet Gynecol. 1983;145(8):968–77.
90. Wood CE, Borgerink H, Register TC, Scott L, Cline JM. Cervical and vaginal epithelial neoplasms in cynomolgus monkeys. Vet Pathol. 2004;41(2):108–15.
91. Hollander CF, van Noord MJ. Focal epithelial hyperplasia: A virus-induced oral mucosal lesion in the chimpanzee. Oral Surg Oral Med Oral Pathol. 1972;33(2):220–6.
92. van Ranst M, Fuse A, Sobis H, et al. A papillomavirus related to HPV type 13 in oral focal epithelial hyperplasia in the pygmy chimpanzee. J Oral Pathol Med. 1991;20(7):325–31.
93. Glad WR, Nesland JM. Focal epithelial hyperplasia of the oral mucosa in two chimpanzees (Pan troglodytes). Am J Primatol. 1986;10(1):83–9.
94. Brown Y. The simian parvoviruses. Rev Med Virol. 1997;7(4):211–8.
95. Simon MA. Simian parvoviruses: Biology and implications for research. Comp Med. 2008;58(1):47–50.
96. O'Sullivan MG, Anderson DK, Lund JE, et al. Clinical and epidemiological features of simian parvovirus infection in cynomolgus macaques with severe anemia. Lab Anim Sci. 1996;46(3):291–7.
97. Foresman L, Narayan O, Pinson D. Progressive anemia in a pig-tail macaque with AIDS. Contemp Top Lab Anim Sci. 1999;38(4):20–2.
98. Schröder C, Pfeiffer S, Wu G, et al. Simian parvovirus infection in cynomolgus monkey heart transplant recipients causes death related to severe anemia. Transplantation. 2006;81(8):1165–70.
99. Brown KE, Liu Z, Gallinella G, Wong S, Mills IP, O'Sullivan MG. Simian parvovirus infection: A potential zoonosis. J Infect Dis. 2004;190(11):1900–7.
100. Favoretto SR, de Mattos CC, Morais NB, Alves Araújo FA, de Mattos CA. Rabies in marmosets (Callithrix jacchus), Ceará, Brazil. Emerg Infect Dis. 2001;7(6):1062–5.
101. Kotait I, de Oliveira RN, Carrieri ML, et al. Non-human primates as a reservoir for rabies virus in Brazil. Zoonoses and Public Health. 2019;66(1):47–59.
102. Moutinho FFB, de Andrade MGA, Nune VMA, et al. Rabies in callithrix sp. In the urban area of Niterói, Rio de Janeiro, Brazil. Rev Soc Bras Med Trop. 2020;53
103. Martini GA, Knauff HG, Schmidt HA, Mayer G, Baltzer G. A hitherto unknown infectious disease contracted from monkeys. "Marburg-virus" disease. Ger Med Mon. 1968;13(10):457–70.
104. Malherbe H, Strickland-Cholmley M. Human disease from monkeys (Marburg virus). Lancet. 1968;1(7557):1434.
105. Deng IM, Duku O, Gillo AL. Ebola haemorrhagic fever in Sudan, 1976. Report of a WHO/International Study Team. Bull World Health Organ. 1978;56(2):247–70.
106. Burke J, Declerq R, Ghysebrechts G. Ebola haemorrhagic fever in Zaire, 1976. Report of an international commission. Bull World Health Organ. 1978;56(2):271–93.
107. Jahrling PB, Geisbert TW, Johnson ED, Peters CJ, Dalgard DW, Hall WC. Preliminary report: isolation of Ebola virus from monkeys imported to USA. Lancet. 1990;335(8688):502–5.
108. Dalgard DW, Hardy RJ, Pearson SL, et al. Combined Simian hemorrhagic fever and Ebola virus infection in cynomolgus monkeys. Lab Anim Sci. 1992;42(2):152–7.
109. Ebola-Reston virus infection among quarantined nonhuman primates—Texas, 1996. J Travel Med; 3(3):189–189.
110. Demetria C, Smith I, Tan T, et al. Reemergence of Reston ebolavirus in Cynomolgus Monkeys, the Philippines, 2015. Emerg Infect Dis. 2018;24(7):1285–91.
111. Wyers M, Formenty P, Cherel Y, et al. Histopathological and immunohistochemical studies of lesions associated with Ebola virus in a naturally infected chimpanzee. J Infect Dis. 1999;179 (Suppl. 1)
112. Formenty P, Boesch C, Wyers M, et al. Ebola virus outbreak among wild chimpanzees living in a rain forest of Cote d'Ivoire. J Infect Dis. 1999;179(Suppl. 1)
113. Lauck M, Sibley SD, Hyeroba D, et al. Exceptional simian hemorrhagic fever virus diversity in a wild African primate community. J Virol. 2013;87(1):688–91.
114. Tauraso NM, Shelokov A, Palmer AE, Allen AM. Simian hemorrhagic fever. 3. Isolation and characterization of a viral agent. Am J Trop Med Hyg. 1968;17(3):422–31.
115. Allen AM, Palmer AE, Tauraso NM, Shelokov A. Simian hemorrhagic fever. II. Studies in pathology. Am J Trop Med Hyg. 1968;17(3):413–21.
116. Palmer AE, Allen AM, Tauraso NM, Shelokov A. Simian hemorrhagic fever. I. Clinical and epizootiologic aspects of an outbreak among quarantined monkeys. Am J Trop Med Hyg. 1968;17(3):404–12.
117. Lauck M, Alkhovsky SV, Bào Y, et al. Historical outbreaks of simian hemorrhagic fever in captive macaques were caused by distinct arteriviruses. J Virol. 2015;89(15):8082–7.
118. Mascheretti M, Tengan CH, Sato HK, et al. Yellow fever: reemerging in the state of Sao Paulo, Brazil, 2009. Revista de Saude Publica. 2013;47(5):881–9.
119. Sallis ESV, de Barros VLRS, Garmatz SL, Fighera RA, Graça DL. A case of yellow fever in a brown howler (Alouatta fusca) in Southern Brazil. J Vet Diagn Investig. 2003;15(6):574–6.
120. Kirya BG, Mukwaya LG, Sempala SDK. A yellow fever epizootic in zika forest, Uganda, during 1972: Part 1: Virus isolation and sentinel monkeys. Trans R Soc Trop Med Hyg. 1977;71(3):254–60.
121. Julander JG. Animal models of yellow fever and their application in clinical research. Curr Opin Virol. 2016;18:64–9.
122. Ølberg RA, Barker IK, Crawshaw GJ, Bertelsen MF, Drebot MA, Andonova M. West Nile virus encephalitis in a Barbary Macaque (Macaca sylvanus). Emerg Infect Dis. 2004;10(4):712–4.
123. Robertson SN, Cameron AI, Morales PR, Burnside WM. West Nile virus seroprevalence in an outdoor nonhuman primate breeding colony in South Florida. J Am Assoc Lab Anim Sci. 2021;60 (2):168–75.
124. Ratterree MS, Travassos da Rosa APA, Bohm RP, et al. West Nile virus infection in nonhuman primate breeding colony, concurrent with human epidemic, Southern Louisiana. Emerg Infect Dis. 2003;9(11):1388–94.
125. Verstrepen BE, Fagrouch Z, van Heteren M, et al. Experimental infection of rhesus macaques and common marmosets with a European strain of West Nile Virus. PLoS Negl Trop Dis. 2014;8 (4)
126. Montali RJ, Scanga CA, Pernikoff D, et al. A common-source outbreak of callitrichid hepatitis in captive tamarins and marmosets. J Infect Dis. 1993;167(4):946–50.
127. Montali RJ, Connolly BM, Armstrong DL, Scanga CA, Holmes KV. Pathology and immunohistochemistry of callitrichid hepatitis, an emerging disease of captive new world primates caused by lymphocytic choriomeningitis virus. Am J Pathol. 1995;147(5):1441–9.
128. Asper M, Hofmann P, Osmann C, et al. First outbreak of callitrichid hepatitis in Germany: genetic characterization of the causative lymphocytic choriomeningitis virus strains. Virology; 2001; 284(2):203–213.

129. Bailey C, Mansfield K. Emerging and reemerging infectious diseases of nonhuman primates in the laboratory setting. Vet Pathol. 2010;47(3):462–81.
130. Welshman MD. Measles in the cynomolgus monkey (Macaca fascicularis). Vet Rec. 1989;124(8):184–6.
131. Yamanouchi K, Shishido A, Honjo S. Natural infection of cynomolgus monkeys with measles virus. Jikken Dobutsu. 1973;22(Suppl(0)):389–93.
132. McChesney MB, Fujinami RS, Lerche NW, Marx PA, Oldstone MBA. Virus-Induced Immunosuppression: Infection of peripheral blood mononuclear cells and suppression of immunoglobulin synthesis during natural measles virus infection of rhesus monkeys. J Infect Dis. 1989;159(4):757–60.
133. Renne RA, McLaughlin R, Jenson AB. Measles virus associated endometritis, cervicitis, and abortion in a rhesus monkey. J Am Vet Med Assoc. 1973;163(6):639–41.
134. Qiu W, Zheng Y, Zhang S, et al. Canine distemper outbreak in rhesus monkeys, China. Emerg Infect Dis. 2011;17(8):1541–3.
135. Sakai K, Nagata N, Ami Y, et al. Lethal canine distemper virus outbreak in cynomolgus monkeys in Japan in 2008. J Virol. 2013;87(2):1105–14.
136. Sun Z, Li A, Ye H, Shi Y, Hu Z, Zeng L. Natural infection with canine distemper virus in hand-feeding Rhesus monkeys in China. Vet Microbiol. 2010;141(3–4, 374):–378.
137. de Vries RD, Ludlow M, Verburgh RJ, et al. Measles vaccination of nonhuman primates provides partial protection against infection with canine distemper virus. J Virol. 2014;88(8):4423–33.
138. Heidecker G, Lerche NW, Lowenstine LJ, et al. Induction of simian acquired immune deficiency syndrome (SAIDS) with a molecular clone of a type D SAIDS retrovirus. J Virol. 1987;61 (10):3066–71.
139. Lerche NW, Marx PA, Osborn KG, et al. Natural history of endemic type D retrovirus infection and acquired immune deficiency syndrome in group-housed rhesus monkeys. J Natl Cancer Inst. 1987;79(4):847–54.
140. Lerche NW, Switzer WM, Yee JL, et al. Evidence of infection with simian type D retrovirus in persons occupationally exposed to nonhuman primates. J Virol. 2001;75(4):1783–9.
141. Steve AM, Ahidjo A, Placide MK, et al. High prevalences and a wide genetic diversity of simian retroviruses in non-human primate bushmeat in rural areas of the Democratic Republic of Congo. Ecohealth. 2017;14(1):100–14.
142. Mossoun A, Calvignac-Spencer S, Anoh AE, et al. Bushmeat hunting and zoonotic transmission of simian T-lymphotropic virus 1 in Tropical West and Central Africa. J Virol. 2017;91(10)
143. Brignolo L, Spinner A, Yee JAL, Lerche NW. Subsets of T cells in healthy rhesus macaques (Macaca mulatta) infected with simian T-lymphotropic virus type 1. Comp Med. 2004;54(3):271–4.
144. Yee JL, Montiel NA, Ardeshr A, Lerche NW. Constitutive release of IFNγ and IL2 from peripheral blood mononuclear cells of Rhesus Macaques (Macaca mulatta) infected with Simian T-lymphotropic virus type 1. Comp Med. 2013;63(6):508–14.
145. Hubbard GB, Mone JP, Allan JS, et al. Spontaneously generated non-Hodgkin's lymphoma in twenty-seven simian T- cell leukemia virus type 1 antibody-positive baboons (Papio species). Lab Anim Sci. 1993;43(4):301–9.
146. D'Offay JM, Eberle R, Wolf RF, et al. Simian T-lymphotropic virus-associated lymphoma in 2 naturally infected baboons: T-cell clonal expansion and immune response during tumor development. Comp Med. 2013;63(3):288–94.
147. Pandrea I, Sodora DL, Silvestri G, Apetrei C. Into the wild: simian immunodeficiency virus (SIV) infection in natural hosts. Trends Immunol. 2008;29(9):419–28.
148. Pandrea I, Silvestri G, Apetrei C. AIDS in African nonhuman primate hosts of SIVs: a new paradigm of SIV infection. Curr HIV Res; 2009; 7(1):57–72.
149. Pandrea I, Ribeiro RM, Gautam R, et al. Simian immunodeficiency virus SIVagm dynamics in African Green Monkeys. J Virol. 2008;82(7):3713–24.
150. Mansfield KG, Lerche NW, Gardner MB, Lackner AA. Origins of simian immunodeficiency virus infection in macaques at The New England Regional Primate Research Center. J Med Primatol. 1995;24(3):116–22.
151. Sasseville VG, Smith MM, Mackay CR, et al. Chemokine expression in simian immunodeficiency virus-induced AIDS encephalitis. Am J Pathol. 1996;149(5):1459–67.
152. Ringler DJ, Wyand MS, Walsh DG, et al. Cellular localization of simian immunodeficiency virus in lymphoid tissues. I. Immunohistochemistry and electron microscopy. Am J Pathol. 1989;134(2):373–83.
153. King NW, Chalifoux LV, Ringler DJ, et al. Comparative biology of natural and experimental SIVmac infection in macaque monkeys: a review. J Med Primatol. 1990;19(2):109–18.
154. Wyand MS, Ringler DJ, Naidu YM, et al. Cellular localization of simian immunodeficiency virus in lymphoid tissues. II. In situ hybridization. Am J Pathol. 1989;134(2):385–93.
155. Chalifoux LV, Ringler DJ, King NW, et al. Lymphadenopathy in macaques experimentally infected with the simian immunodeficiency virus (SIV). Am J Pathol. 1987;128(1):104–10.
156. Chalifoux LV, Simon MA, Pauley DR, MacKey JJ, Wyand MS, Ringler DJ. Arteriopathy in macaques infected with simian immunodeficiency virus. Lab Investig. 1992;67(3):338–49.
157. Ringler DJ, Hunt RD, Desrosiers RC, Daniel MD, Chalifoux LV, King NW. Simian immunodeficiency virus—induced meningoencephalitis: natural history and retrospective study. Ann Neurol. 1988;23(1 S):S101–7.
158. Simon MA, Chalifoux LV, Ringler DJ. Pathologic features of SIV-induced disease and the association of macrophage infection with disease evolution. AIDS Res Hum Retrovir. 1992;8(3): 327–37.
159. Centers for Disease Control (CDC). Seroconversion to simian immunodeficiency virus in two laboratory workers. MMWR Morb Mortal Wkly Rep. 1992;41(36):678–81.
160. Lairmore MD, Kaplan JE, Daniel MD, et al. Guidelines for the prevention of simian immunodeficiency virus infection in laboratory workers and animal handlers. J Med Primatol. 1989;18(3–4): 167–74.
161. Heneine W, Switzer WM, Sandstrom P, et al. Identification of a human population infected with simian foamy viruses. Nat Med. 1998;4(4):403–7.
162. Gessain A. Mechanisms of viral emergence and interspecies transmission: The example of simian foamy viruses in Central Africa. Bulletin de l'Academie Nationale de Medecine. 2013;197(9): 1655–68.
163. Nandakumar S, Bae EH, Khan AS. Complete genome sequence of the African green monkey simian foamy virus serotype 3 strain FV2014 (SFVcae_FV2014). Genome Announc; 2018; 6(3).
164. Ensser A, Großkopf AK, Mätz-Rensing K, Roos C, Hahn AS. Isolation and sequence analysis of a novel rhesus macaque foamy virus isolate with a serotype-1-like env. Arch Virol; 2018; 163(9):2507–2512.
165. Schweizer M, Neumann-Haefelin D. Phylogenetic analysis of primate foamy viruses by comparison of pol sequences. Virology. 1995;207(2):577–82.
166. Sa-nguanmoo P, Thawornsuk N, Rianthavorn P, Sommanustweechai A, Ratanakorn P, Poovorawan Y. High prevalence of antibodies against hepatitis A virus among captive nonhuman primates. Primates. 2010;51(2):167–70.
167. Doroshenko NV, Shevtsova ZV, Stakhanova VM, et al. Sensitivity of Macacus fascicularis and Macacus rhesus to human hepatitis A virus. Vopr Virusol. 1990;35(4):299–303.

168. Lankas GR, Jensen RD. Evidence of hepatitis A infection in immature rhesus monkeys. Vet Pathol. 1987;24(4):340–4.
169. Shevtsova ZV, Krylova RI, Belova EG, Korzaya LI, Andzhaparidze AG. Spontaneous hepatitis A with a fatal outcome in Macaca rhesus monkeys. Vopr Virusol. 1987;32(6):686–90.
170. Shevtsova ZV, Lapin BA, Doroshenko NV, et al. Spontaneous and experimental hepatitis A in old world monkeys. J Med Primatol. 1988;17(4):177–94.
171. Reddacliff LA, Kirkland PD, Hartley WJ, Reece RL. Encephalomyocarditis virus infections in an Australian zoo. J Zoo Wildl Med. 1997;28(2):153–7.
172. Blanchard JL, Soike K, Baskin GB. Encephalomyocarditis virus infection in African green and squirrel monkeys: comparison of pathologic effects. Lab Anim Sci. 1987;37(5):635–9.
173. Hubbard GB, Soike KF, Butler TM, et al. An encephalomyocarditis virus epizootic in a baboon colony. Lab Anim Sci. 1992;42(3): 233–9.
174. Masek-Hammerman K, Miller AD, Lin KC, et al. Epizootic myocarditis associated with encephalomyocarditis virus in a group of rhesus macaques (Macaca mulatta). Vet Pathol. 2012;49(2): 386–92.
175. Krylova RI, Dzhikidze EK. Encephalomyocarditis in monkeys. Bull Exp Biol Med. 2005;139(3):355–9.
176. Hukkanen RR, Gillen M, Grant R, Liggitt HD, Kiem HP, Kelley ST. Simian varicella virus in pigtailed Macaques (Macaca nemestrina): clinical, pathologic, and virologic features. Comp Med. 2009;59(5):482–7.
177. Melendez LV, Hunt RD, Garcia FG. Herpesvirus ateles, the second lymphoma virus of monkeys. Bibl Haematol. 1973;39:410–5.
178. Reston Ebolavirus Antibodies in Bats, the Philippines. Available from: www.cdc.gov/eid
179. Cantoni D, Hamlet A, Michaelis M, Wass MN, Rossman JS. Risks posed by Reston, the forgotten ebolavirus. mSphere. 2016;1(6)

Bacterial Diseases in Nonhuman Primates

2

Peter J. Didier and Ivanela I. Kondova-Perseng

Abstract

A variety of bacterial infections are common in nonhuman primate colonies and can be a significant factor for morbidity and mortality in experimental settings and in the wild. Proper postmortem diagnosis based on macroscopic and histopathological findings along with the clinical history and results from microbiological analyses will ensure the elimination of the infection, will improve the treatment strategy and will reduce the mortality. This chapter will address the pathology and diagnostic methods of common bacterial diseases in nonhuman primates (NHPs), with addition of some uncommon pathogens that can initiate large-scale epidemics and are mostly used in experimental settings. The emphasis will be on images of macroscopic and histologic lesions of the given disease.

Keywords

Nonhuman primates · Bacterial infections · Etiologic agents · Zoonosis · Gross pathology · Histopathology · Diagnosis

2.1 Introduction

Bacterial diseases can be a major problem for the health of nonhuman primates (NHPs) living in experimental settings, zooparks, and in the wild. Some of these infections can lead to tremendous morbidity and mortality and can have a significant impact on colony management, breeding programs, and experimental research. Rapid identification of the bacterial causative pathogen is needed to limit the spread of infection. This involves recognition of etiological agent by proper diagnostic methods including microbiological, histopathological, biochemical, serological, and molecular analyses. The correct identification will guarantee the control and elimination of the infection and will minimize the risk of transmission to personnel when the disease is zoonotic.

This chapter will address the pathology and diagnostic methods of common bacterial diseases in NHPs, with addition of some uncommon pathogens that can initiate large-scale epidemics and are mostly used in experimental settings. The emphasis will be on images of macroscopic and histologic lesions of the given disease.

2.2 Gram-Positive Bacteria

2.2.1 *Mycobacteria*

Mycobacteria are a group of Gram-positive, aerobic, rod-shaped, nonmotile, acid-fast bacilli. Some atypical species as *M. avium* and *M. intracellulare* frequently cause opportunistic infections in immunocompromised animals. The species of greatest concern are *M. tuberculosis* and *M. bovis*. These two pathogens present a significant challenge in the experimental settings because of their zoonotic potential and ability to establish latent infection. The lack of a practical, sensitive, and specific antemortem diagnostic test remains a challenge in maintaining a tuberculosis-free colony. The intradermal skin test with mammalian old

P. J. Didier
Division of Comparative Pathology, Tulane National Primate Research Center (Tulane University School of Medicine), Covington, LA, USA
e-mail: pjdidier@tulane.edu

I. I. Kondova-Perseng (✉)
Division of Pathology and Microbiology, Animal Science Department, Biomedical Primate Research Centre, Rijswijk, The Netherlands
e-mail: kondova@bprc.nl

I. Kondova - Perseng et al. (eds.), *Atlas of Diagnostic Pathology in Nonhuman Primates*,
https://doi.org/10.1007/978-3-031-41280-6_2

tuberculin remains the standard tool for screening animals in the colonies. Culture and isolation is the gold standard for a positive diagnosis of *M. tuberculosis*.

Mycobacterium tuberculosis

Mycobacterium tuberculosis (*Mtb*) is a pathogenic bacterium and the causative agent of tuberculosis—an infectious disease of man which is easily transmitted to NHPs. The organism is spread by the aerosolized route often acquired from animal caretakers and may lead to an epizootic of high morbidity and mortality. Infected primates may pose a further risk to human handlers. The pathogen is a significant threat in animals imported from regions of the world with high rate of human infection [1].

Mycobacterium bovis

Mycobacterium bovis (*Mb*) is a zoonotic pathogen closely related to *Mtb* which can infect humans and a broader range of mammalian hosts including NHP. The transmission occurs by ingestion and aerosolized route. *Mtb* and *Mb* are intracellular pathogens best visualized by acid-fast stain. The difference in species susceptibility is similar to what was reported with *Mtb*.

Severe outbreaks of tuberculosis caused by *Mtb* are mostly observed in Old World primates (OWPs). New World primates (NWPs) seem to be more resistant. Clinical disease may vary dependent on the species. Tuberculosis in OWPs generally results in a rapidly progressive, debilitating disease. Clinical signs may include weight loss, anorexia, and respiratory signs including coughing, dyspnea, and hemoptysis. NHPs are important model for the disease because of their genetic, immunological, and physiological similarities to humans and ability to generate clinical correlates of infection [2, 3]. Viral-induced immunodeficiency increases the susceptibility in NHPs as is suspected in humans [4].

Because of the high susceptibility of NHPs to these pathogenic mycobacteria, intradermal skin testing with mammalian old tuberculin and more recently interferon-based immunologic tests have been done for many years in captive colonies (Fig. 2.1) [5]. Dual skin testing can be done to distinguish *Mycobacterium avium complex* (*MAC*) from *Mtb* infections [6]. Culture and isolation remain the gold standard for a positive diagnosis of *Mtb* but may take several months to complete. Of the OWPs, rhesus and cynomolgus macaques appear the best suited as animal models for TB with rhesus macaques somewhat more susceptible [7, 8] and less protected by vaccination with *Mb* bacillus Calmette-Guerin (BCG) [9].

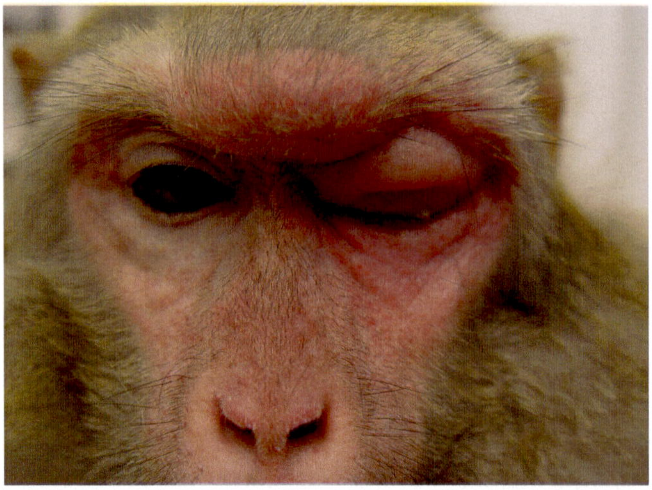

Fig. 2.1 An image of a positive reaction to Old Tuberculin skin test reagent given intradermally in the left eyelid of an infected rhesus macaque shows severe edema after 48 h

Naturally acquired infections of *Mtb* may present as solitary grey to white or hemorrhagic firm granulomas in the lung (Fig. 2.2) or as red to brown, grey multinodular lobar disease (Fig. 2.3). In natural and experimental disease, hilar and mediastinal lymph nodes are typically markedly enlarged and caseous or sometimes suppurative (Fig. 2.4a, b). Experimentally, it can be shown that low dose, low pathogenicity organisms initially produce small focal lesions that may eventually resolve into a chronic or clinically latent state, while high dose and longer duration of disease with higher pathogenic strains produce lethal lobar disease [7]. In the case of aerosol administration, miliary pulmonary lesions can be observed (Fig. 2.5). Besides the lung and hilar lymph nodes, granulomas can be found in different visceral organs (Figs. 2.6, 2.7, 2.8, and 2.9).

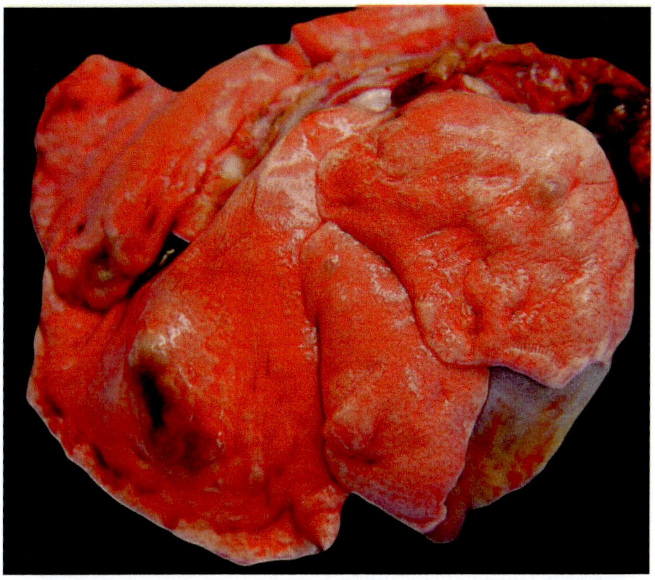

Fig. 2.2 *Mycobacterium tuberculosis.* Gross pathology image of subclinical tuberculosis in a rhesus macaque. Several greyish-white firm nodules and a firm hemorrhagic area in the right lower lobe. Skin test was positive after natural exposure to a roommate with clinical disease due to *Mtb*

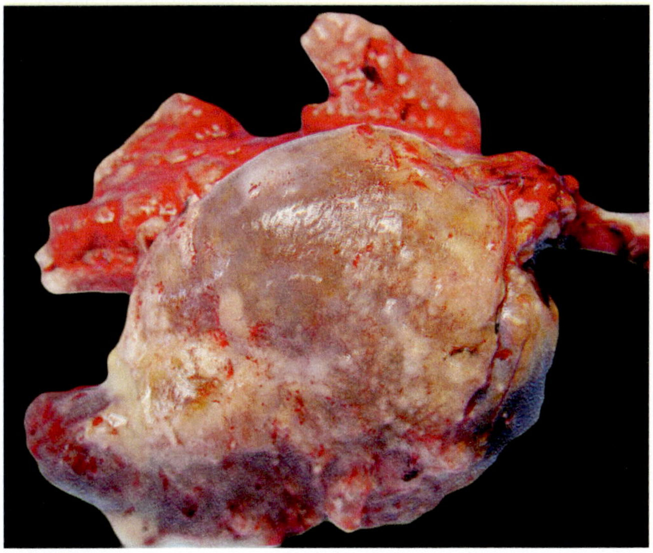

Fig. 2.3 *Mycobacterium tuberculosis.* Gross pathology image of parenchymal consolidation and tan to hemorrhagic lobar to lobular discoloration of the right pulmonary lobes (rhesus macaque with natural clinical infection of *Mtb*)

a

b

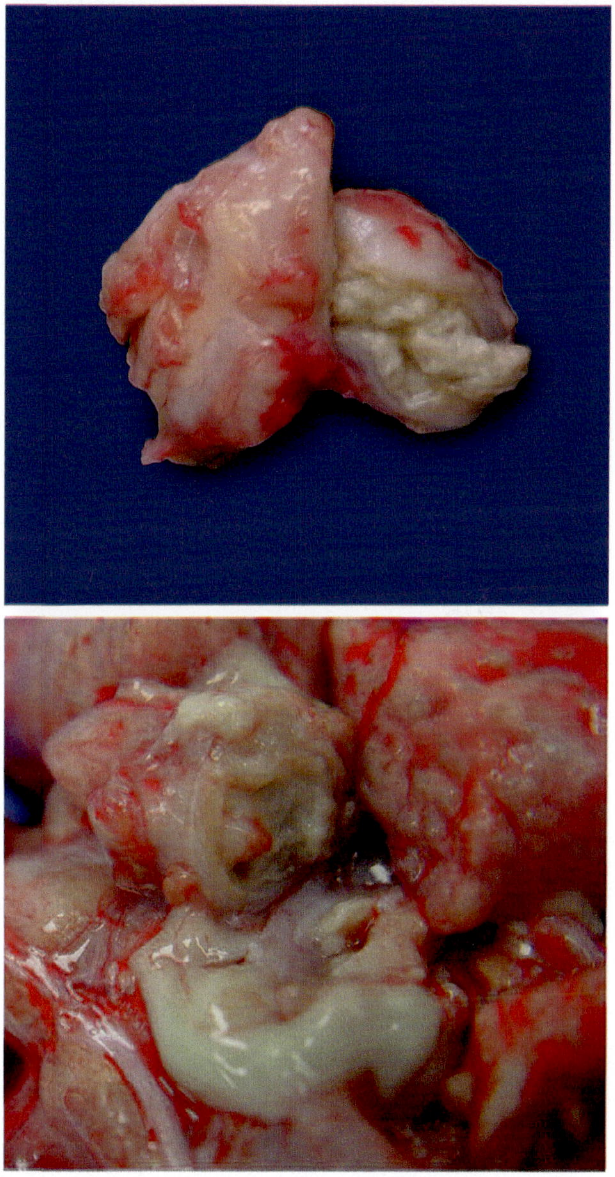

Fig. 2.4 *Mycobacterium tuberculosis*. (**a**) Macroscopic image of cross section of enlarged hilar lymph node with yellow, cheesy-appearing caseous necrosis and markedly thickened capsule (rhesus macaque experimentally infected with *Mtb*). (**b**) *Mycobacterium tuberculosis*. Macroscopic image of cross section of an enlarged bronchial lymph node with thickened, laminated capsule, and focal pockets of liquid suppuration (rhesus macaque experimentally infected with *Mtb*)

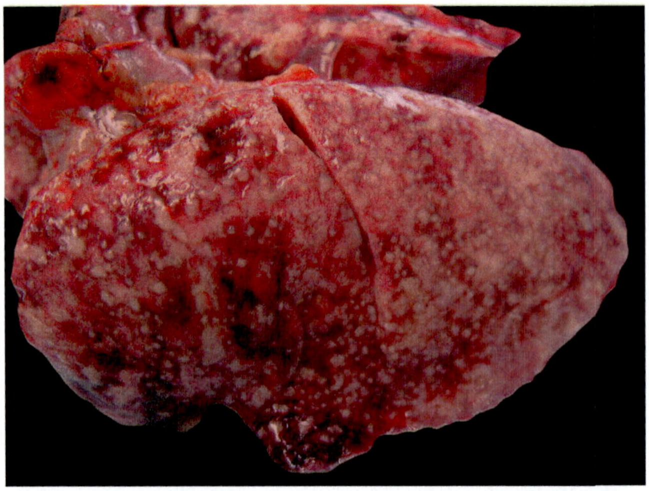

Fig. 2.5 *Mycobacterium tuberculosis*. Gross pathology image of lung with multifocal granulomatous disease. All pulmonary lobes exhibit miliary 2–3mm white granulomas and patchy areas of congestion and hemorrhage (rhesus macaque experimentally infected via aerosol with *Mtb* CDC155, courtesy of D. Kaushal)

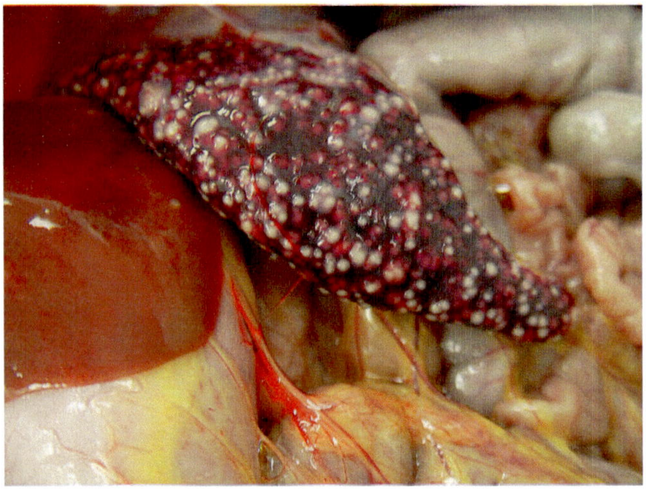

Fig. 2.6 *Mycobacterium tuberculosis*. Gross pathology image presents miliary white to red 2–4-mm granulomas in the spleen and widely scattered granulomas in the liver demonstrating septic distribution of *Mtb* from the lung after aerosol inoculation and IV challenge with SIVmac239. (rhesus macaque, spleen and liver, courtesy of D. Kaushal)

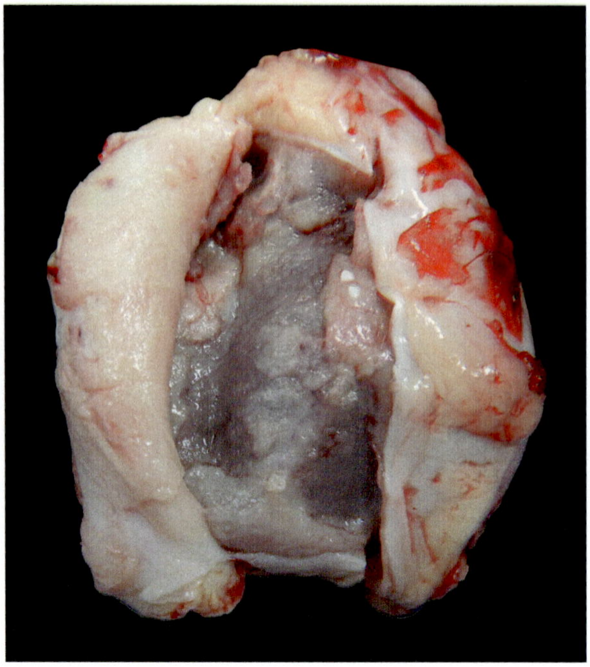

Fig. 2.7 *Mycobacterium tuberculosis*. Gross pathology image of diffusely thickened pericardial sac of rhesus macaque. The lesion is caused by granulomatous inflammation after infection via aerosol with *Mtb* CDC1551 and subsequent challenge with SIVmac239 and CD4 depletion by monoclonal antibody (rhesus macaque, pericardium, courtesy of D. Kaushal)

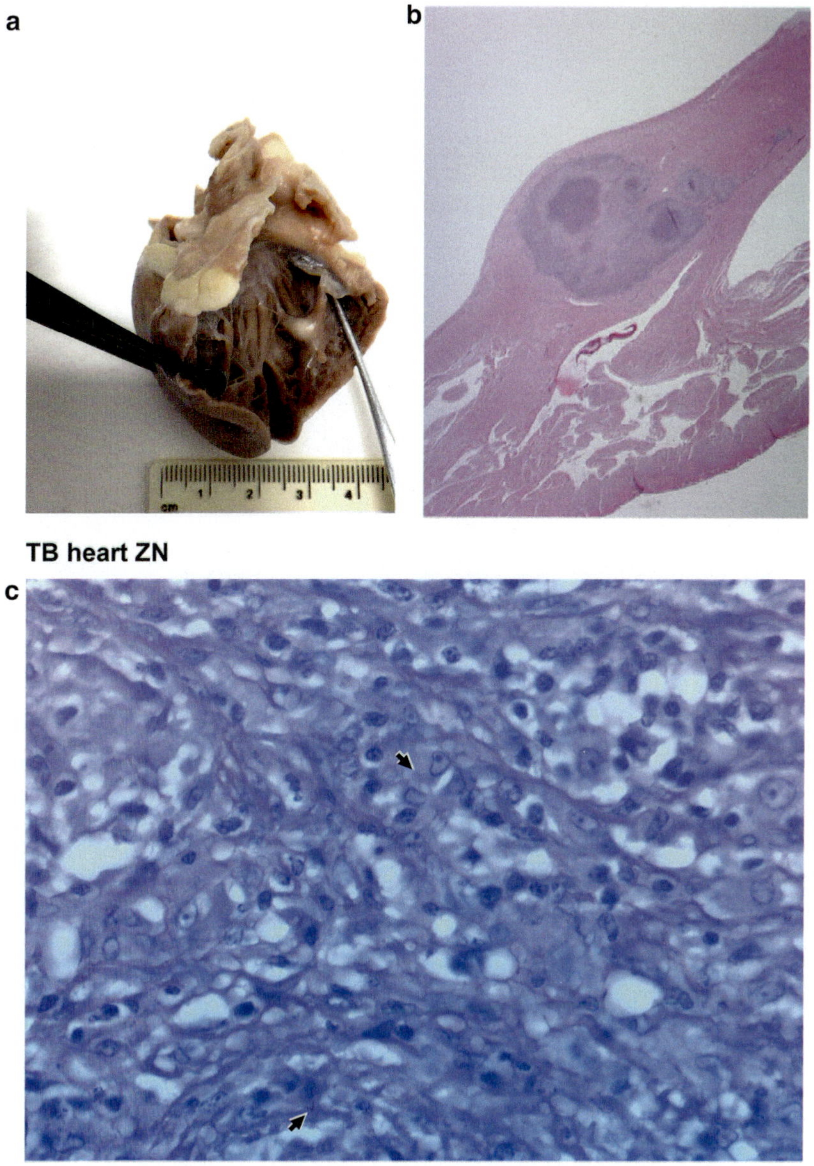

Fig. 2.8 *Mycobacterium tuberculosis*. (**a**) Macroscopic image of myocardial tubercular granuloma in the right ventricle of unvaccinated rhesus macaque, endobronchially inoculated with *Mtb* strain Erdman K01 with approximately 15 CFU (rhesus macaque, formalin fixed heart, vaccine study; courtesy of Dr. F. Verreck, BPRC, The Netherlands). (**b**) Photomicrograph of myocardial granuloma from (**a**) (H&E stain). (**c**) Visualized rare intracellular bacilli (arrows) in the same granuloma (Ziehl–Neelsen stain)

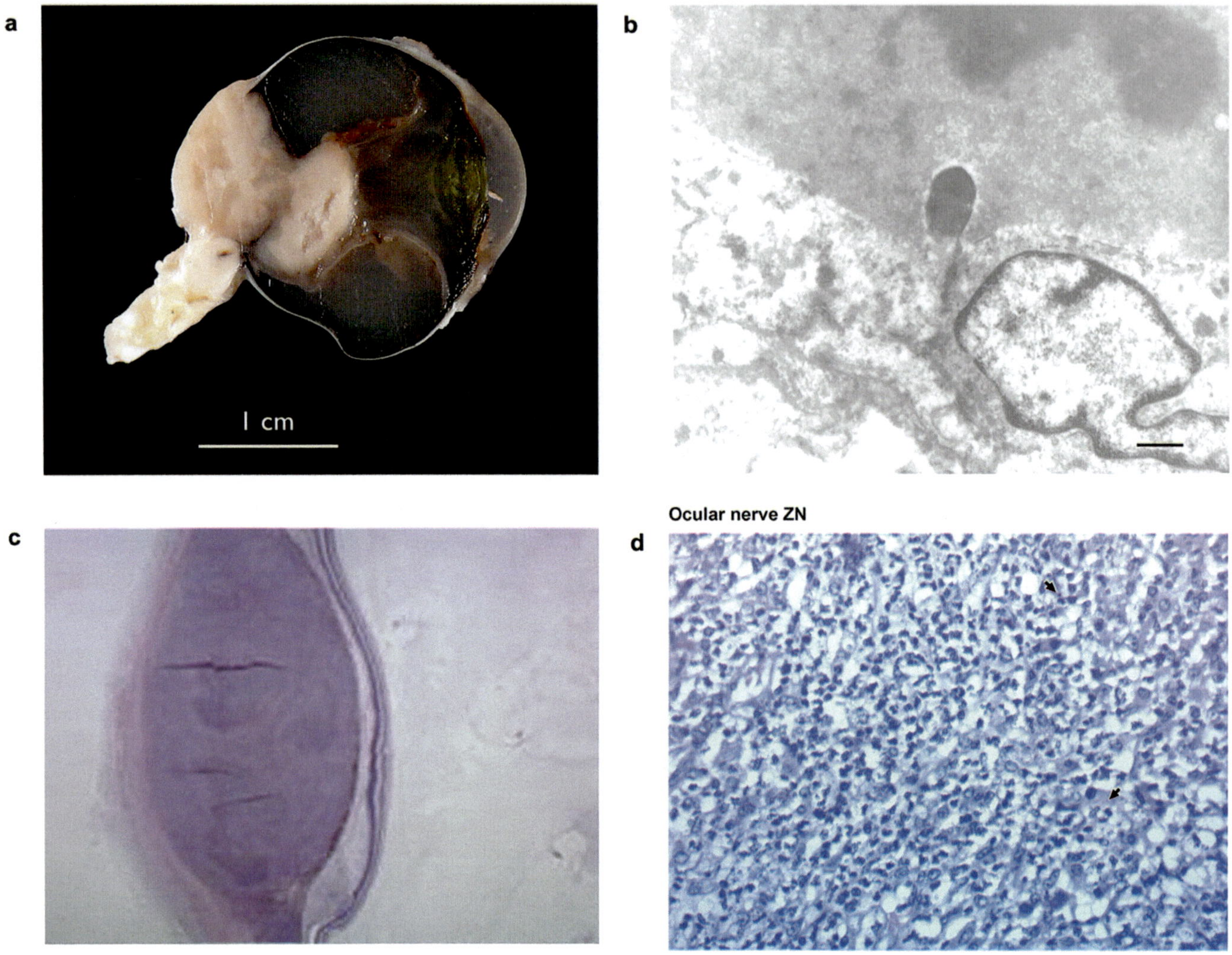

Fig. 2.9 *Mycobacterium tuberculosis*. (**a**) Macroscopic image of tuberculous, granulomatous inflammation of the eye and the optic nerve (panuveitis and optic neuritis) from unvaccinated, *Mtb* infected animal (rhesus macaque, eye, vaccine study; courtesy of Dr. F. Verreck, BPRC, The Netherlands). (**b**) Electron micrograph of (**a**) showing *Mtb* (center) surrounded by necrotic cellular debris of TB ocular granuloma (electron microscopy, magnification bar = 0.5 microns; courtesy of Prof. Eugenio Bertelli, University of Siena, Italy). (**c**) Photomicrograph of the optic nerve granuloma from (**a**) (optic nerve, H&E stain). (**d**) Visualized few acid-fast bacteria (arrows) in the necrotic center of the optic nerve granuloma. (ZN stain)

The hallmark of TB is the formation of granuloma with specific structural organization called tubercule. There are different patterns of histological presentation of the granulomas in different organs (Figs. 2.10, 2.11, 2.12, and 2.13). Histologically, TB granulomas in macaques are composed of discrete aggregates of epithelioid macrophages with a few multinucleated giant cells surrounding a variably sized central zone of necrosis with neutrophils and a peripheral layer of lymphocytes that may form lymphoid aggregates (Fig. 2.14). Many of these lymphoid cells are antigen-specific T-cell lineage with chemokine CXCR3 expression and interferon production that correlates with mycobacterial memory [8, 10, 11]. Skin test lesions in contrast are characterized by edema of palpebral or abdominal subcutaneous tissue with modest mixed inflammatory infiltrates including neutrophils, lymphocytes, some macrophages, and rare eosinophils (Fig. 2.15).

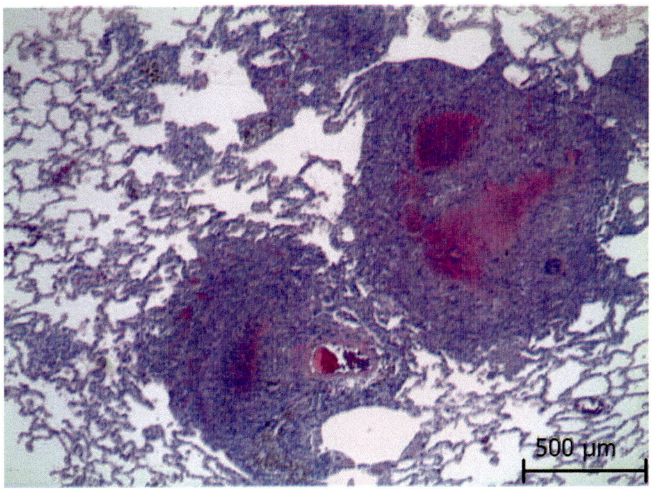

Fig. 2.10 *Mycobacterium tuberculosis*. Photomicrograph of pulmonary TB granulomas with variably sized central areas of necrosis (rhesus macaque with natural TB infection, lung, H&E stain)

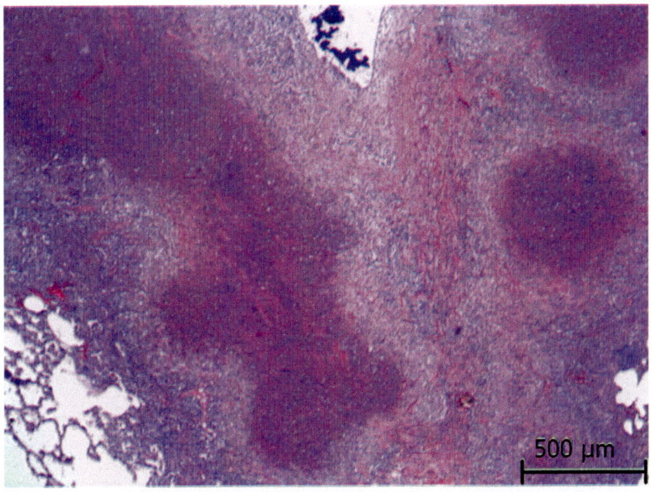

Fig. 2.11 *Mycobacterium tuberculosis*. Photomicrograph of pulmonary, multifocal to coalescing granulomas creating a serpentine pattern of necrosis (rhesus macaque, lung, H&E stain)

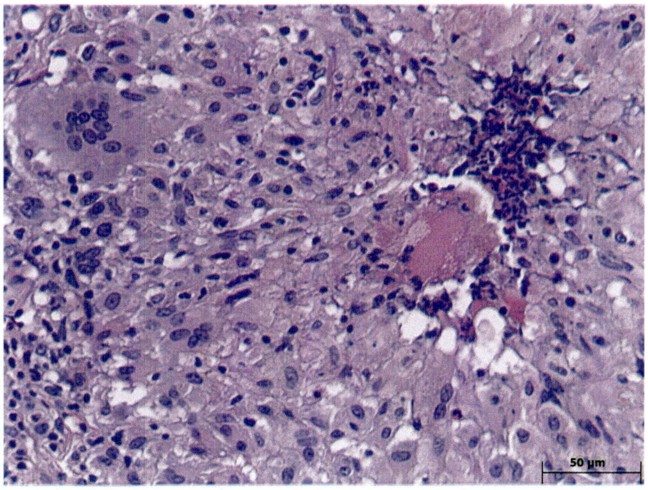

Fig. 2.12 *Mycobacterium tuberculosis*. Photomicrograph of a center of pulmonary granuloma composed of necrotic macrophages and polymorphonuclear (PMN) leukocytes adjacent to degenerating syncytial giant cell surrounded by sheets of epithelioid macrophages and a peripheral syncytial giant cell. (SIV-infected rhesus macaque, lung, H&E stain)

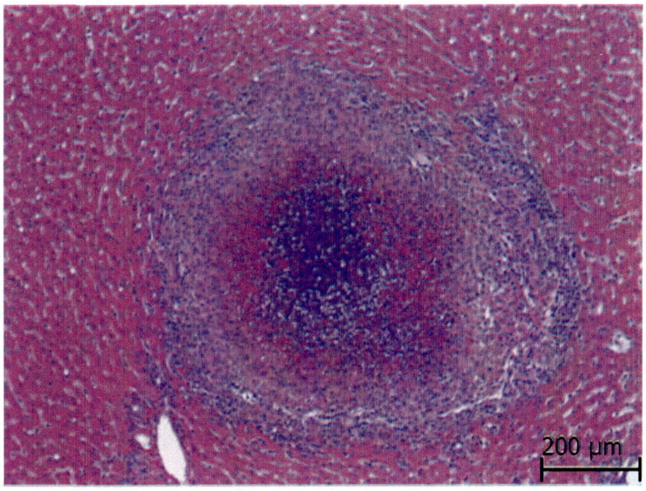

Fig. 2.13 *Mycobacterium tuberculosis*. Photomicrograph of typical TB granuloma with central area of necrosis and dystrophic mineralization, surrounded by sheets of epithelioid macrophages mixed with a few polymorphonuclear cells and a distinct peripheral collar of lymphocytes. (rhesus macaque, liver, natural infection, H&E stain)

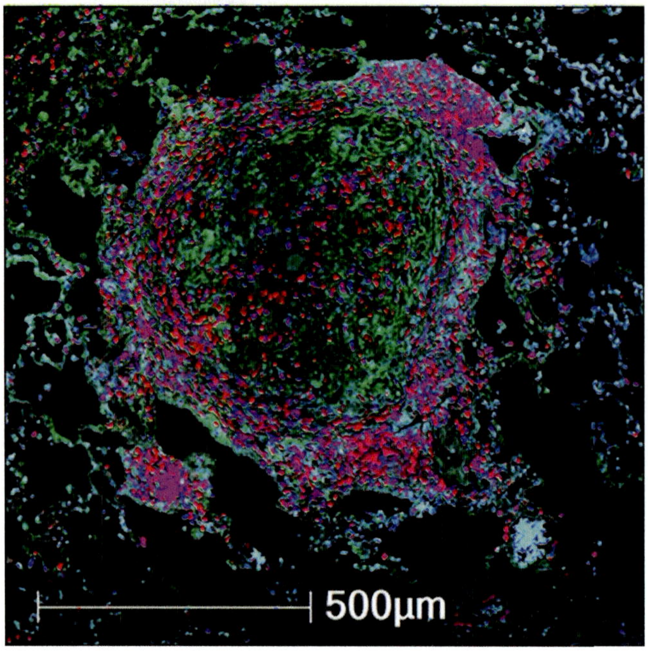

Fig. 2.14 *Mycobacterium tuberculosis*. Immunofluorescence image of a single experimental TB granuloma stained for macrophages (CD68, green), activated T lymphocytes expressing the chemokine marker CXCR3 (red), and CD4-positive lymphocytes (blue). Macrophages surround a single CD68 positive giant cell mixed with a few lymphocytes and surrounded by a collar of lymphocytes, many of which co-express CD4 and CXCR3 (pink). Peripheral aggregates of co-expressing lymphocytes are near or in the outer edge of the granuloma [10]. (rhesus macaque experimentally infected, lung, IFA, courtesy of Dr. Uma Shanmugasundaram)

Mycobacterium avium Complex (*MAC*)

Mycobacterium avium complex (*MAC*) is a group of respiratory pathogens composed of common environmental bacteria *M. avium* and *M. intracellulare* which may cause systemic infection in immunocompromised animals [12, 13] and may also occur in immunologically normal NHPs. The pathogens infect the digestive tracts and less frequently the lung in a wide range of immunodeficient mammals, birds, and reptiles.

MAC infections occur subclinically in immunocompetent NHPs with rare thoracic disease, but viable organisms can be cultured from bronchial lymph nodes [13]. Exposure is through food, water, soil, and dust. Granulomatous enteritis (Fig. 2.16) occurs in immunocompromised NHPs accompanied by enlargement of mesenteric lymph nodes (Fig. 2.17). Acid-fast bacilli in lesions caused by *Mtb* may be found within the caseous core or within the surrounding macrophages, but often multiple sections and granulomas must be examined to identify the pathogen. In contrast to TB granulomas where acid-fast Ziehl–Neelsen stain detects rare organisms, *MAC* bacteria are detectable in vast numbers within epithelioid macrophages of the gut (Fig. 2.18) and liver (Fig. 2.20). The lamina propria of the small intestine (less frequently the colon) and sinuses of the mesenteric lymph node (Fig. 2.19) are packed with epithelioid macrophages. Syncytial giant cells are rare. Lymphoid infiltration is difficult to recognize in tissues with a constitutive lymphoid population. Usually less granulomatous inflammation is found in the liver (Fig. 2.20) and spleen.

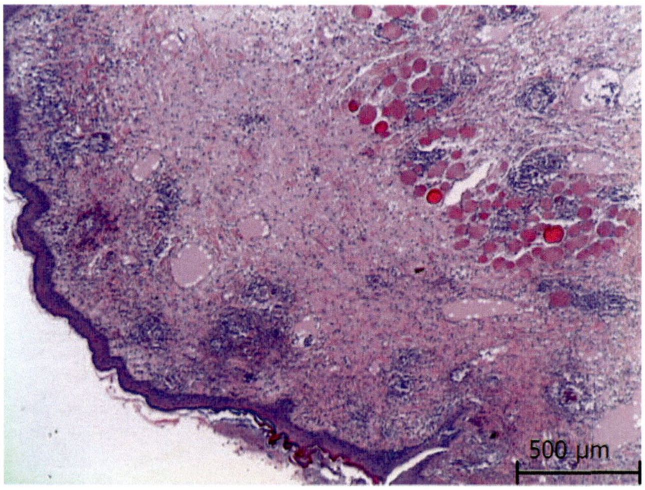

Fig. 2.15 *Mycobacterium tuberculosis*. Photomicrograph shows positive TB test site in eyelid skin which contains diffuse subcutaneous edema, widely scattered polymorphonuclear cells, macrophages, and aggregates of lymphocytes. A few adjacent skeletal muscle bundles are swollen and hyalinized (rhesus macaque, eyelid skin, natural infection, H&E stain)

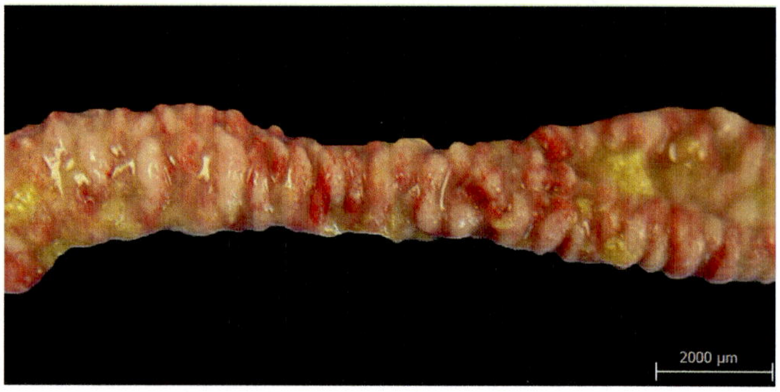

Fig. 2.16 *Mycobacterium avium*. Macroscopic image of irregularly thickened section of jejunum from rhesus macaque infected with SIVmac251 and spontaneously became infected with *M. avium*

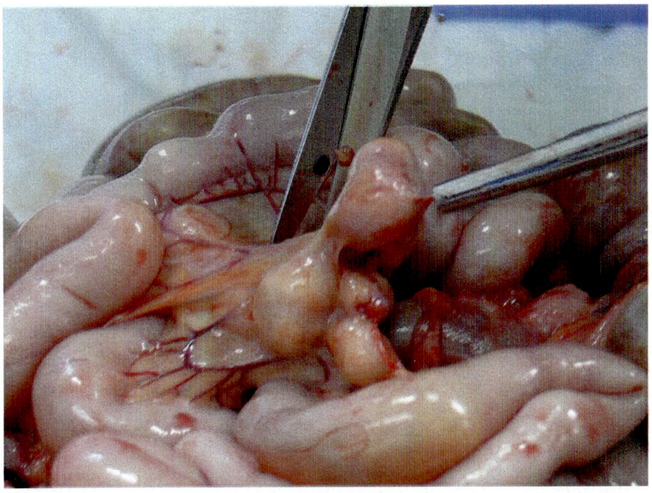

Fig. 2.17 *Mycobacterium avium*. Macroscopic image of markedly enlarged mesenteric lymph nodes from naturally infected cynomolgus macaque with *M. avium*. The imported animal had positive skin test during the quarantine period

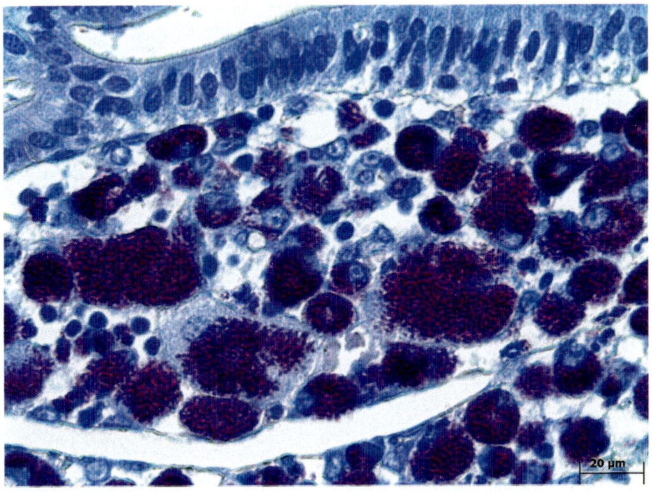

Fig. 2.18 *Mycobacterium avium*. Photomicrograph of small intestinal mucosa from macaque with an opportunistic *M. avium* infection. Epithelioid macrophages in the lamina propria of jejunum are filled with acid-fast bacilli in red (rhesus macaque, jejunum, Ziehl-Neelsen stain)

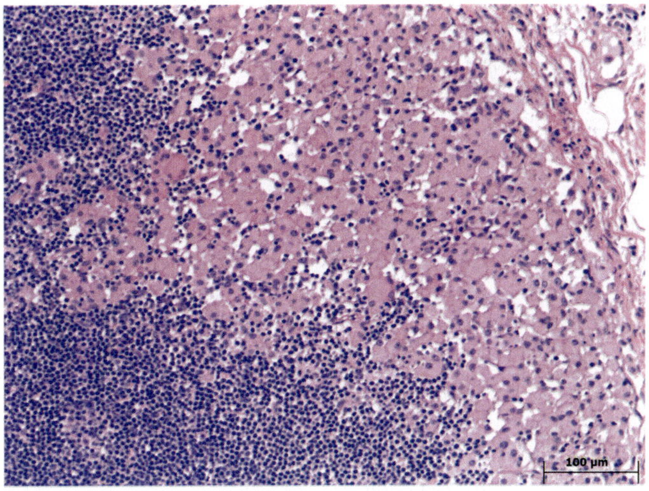

Fig. 2.19 *Mycobacterium avium*. Photomicrograph of mesenteric lymph node from rhesus macaque infected with *M. avium*. Epithelioid macrophages fill the subcapsular and cortical sinuses (H&E stain)

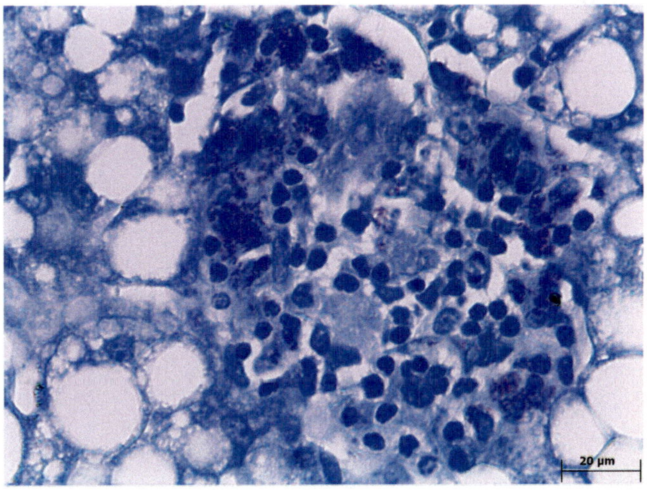

Fig. 2.20 *Mycobacterium avium*. Photomicrograph of granulomatous micronodule that contains acid-fast bacteria (red) and is composed of epithelioid macrophages and small numbers of lymphocytes (SIV-inoculated rhesus macaque, *M. avium* infected, liver, Ziehl-Neelsen stain)

Mycobacterium avium Paratuberculosis (Johne's Disease)

M. avium paratuberculosis is a common pathogen of ruminants and causative agent of paratuberculosis. The infection can be acquired from contaminated soil, water, and paratenic insects. Clinically affected animals develop diarrhea and weight loss. A positive TB test can be the only indicator of infection. An outbreak has been reported in OWPs. The lesions resemble Johne's disease in cattle. Typical macroscopic lesions are granulomatous enterocolitis and enlarged mesenteric lymph nodes. Histologically, the intestinal mucosa is infiltrated by epithelioid histiocytes in which numerous positive bacilli can be detected by acid-fast stain. Isolation of the organism is by culture, or PCR and subsequent DNA sequencing.

Mycobacterium leprae

Mycobacterium leprae is an obligate intracellular pathogen and causative agent of leprosy. Natural infections are rare, but a case was reported in a cynomolgus monkey [14]. Transmission is through respiratory route or through infected skin lesions. The disease is chronic and granulomatous and affects the dermis and track along the nerves. The involvement of the nerves is pathognomonic. Nodules are observed on the hands, face, feet, and tail and are associated with paralysis and deformities of the extremities. Acid-fast bacilli can be visualized within the nerves.

Mycobacterium kansasii

Mycobacterium kansasii is an atypical, environmental mycobacteria that may induce disease in NHPs. Infection is usually not associated with clinical signs but may cause false positive intradermal TB tests. Infection occurs by inhalation and can induce granulomatous inflammation of the tracheal-bronchial lymph nodes. The disease is frequently reported in squirrel monkeys and have been diagnosed in rhesus macaques [15].

2.2.2 *Listeria monocytogenes*

Listeria monocytogenes is ubiquitous, facultative anaerobic, intracellular Gram-positive coccobacillus. The organism is widespread in the environment living as a saprophyte in decaying plant material, soil, and water. *L. monocytogenes* is the causative agent of listeriosis, a bacterial infection that has a worldwide distribution and affects a wide range of mammals and birds, including human beings [16]. The infection can become endemic in group-housed animals. Cases of spontaneous listeriosis among nonhuman primates in captivity are sparse [17, 18], and more experimental infections have been reported recently (Fig. 2.21a–c) [19–21]. Cases in NHPs have been associated with reproductive failure manifested by miscarriage and neonatal death or cerebral inflammation. Lesions are usually present in the placenta and the aborted fetus and consist of diffuse fibrinopurulent placentitis and multifocal splenic and hepatic necrosis on the fetus. The CNS lesions due to *L. monocytogenes* are primarily superficial, involving the meninges, choroid plexus, and ependyma. In humans and macaques, the infection of the brain occurs via hematogenous route, after entry through the gastrointestinal tract and proliferation of the pathogen in the liver [17]. Following the histopathological findings, additional diagnostic methods for detection of *L .monocytogenes* include Gram-positive special stain, culture from sterile site, and PCR test.

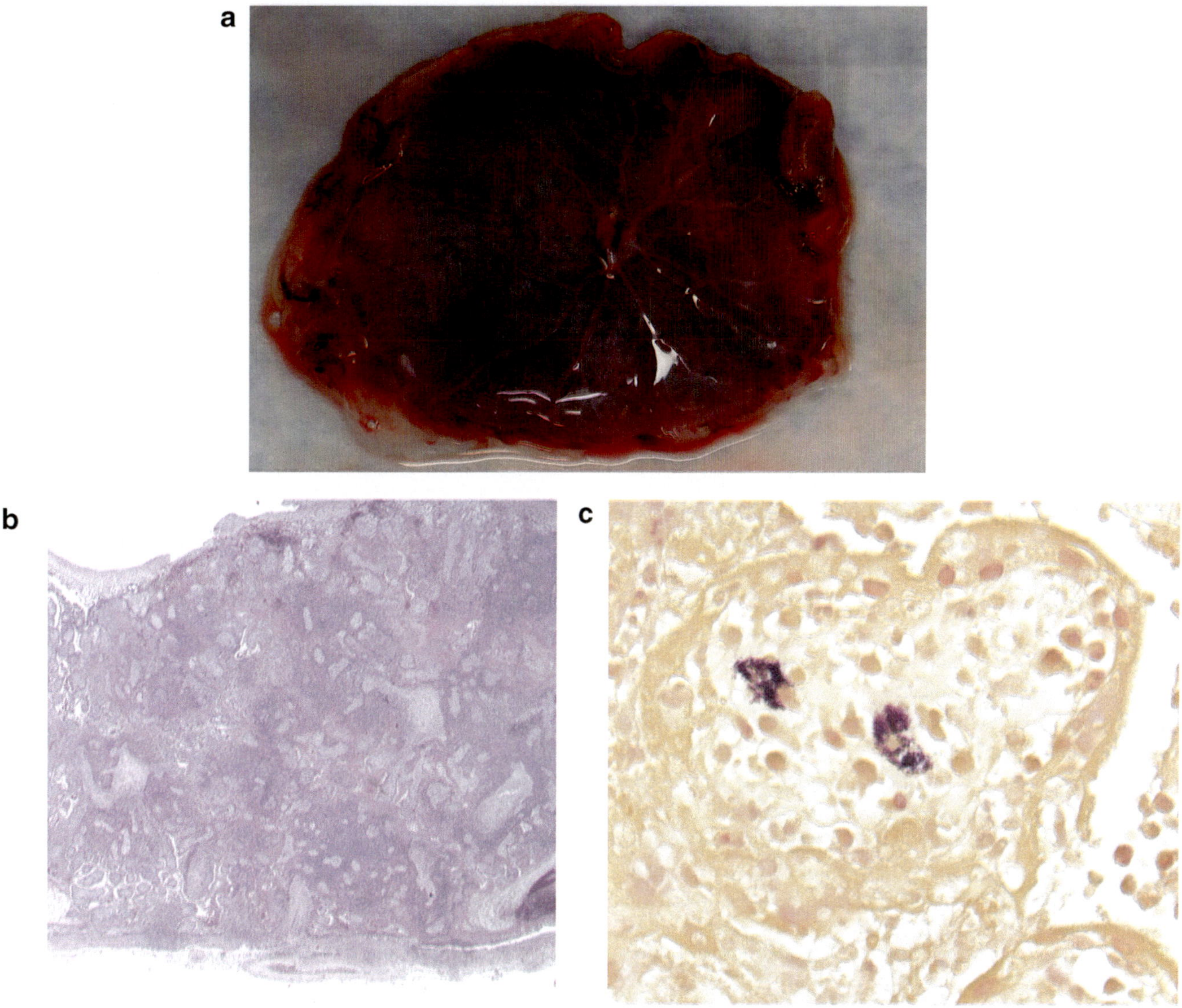

Fig. 2.21 Placental listeriosis in rhesus macaques. (**a**) A gestational day 49 primary placental disc with multifocal pallor visible through the chorionic plate. (**b**) Photomicrograph of (**a**) with marked multifocal necrotizing placentitis with loss of villous architecture (H&E). (**c**) Photomicrograph of a GD 58 rhesus macaque placenta with Gram-positive *L. monocytogenes* within the villous parenchyma. Bacteria are not noted within the syncytiotrophoblasts (Gram stain). Case material and images provided by Thaddeus Golos, Bryce Wolfe, and Heather A. Simmons

2.2.3 *Rhodococcus equi*

Rhodococcus equi is a Gram-positive, nonmotile, obligate aerobe that is an important pathogen of immunocompromised animals and humans. The organism is widespread in the soil of farms and grazing fields, and it is carried in the gut of many herbivores [22]. Major rout of acquisition for animal and human infection is likely the exposure to soil contaminated with herbivore manure. The infection can be acquired by inhalation or ingestion of contaminated material as well as traumatic inoculation of wounds or mucous membranes [23].

R. equi can cause chronic pyogranulomatous pneumonia in foals and immunocompromised humans. In foals and immunosuppressed macaques, dissemination from the lung to the intestinal tract is common, and, in the context of immunosuppressive SRV infection, *R. equi* infection can have a similar microscopic appearance to *MAC* [24]. Clinical signs may be nonspecific and include anorexia, weight loss, and diarrhea [25]. Microscopic lesions are observed primarily in the lung, large intestine, and draining lymph nodes and are characterized by pyogranulomatous inflammation. Specimens typically contain a dense, histiocytic infiltrate with an eosinophilic, granular cytoplasm, and intra-histiocytic coccobacilli. Gastrointestinal tract lesions may be associated with mesenteric lymphadenopathy. The organism can be identified in culture and tend to have a pale salmon pink color on solid media [26].

2.2.4 *Clostridium*

Bacteria of genus *Clostridium* are Gram-positive, spore-forming, and anaerobic bacilli. Species of *Clostridium* inhabit the soil and intestinal tract of animals and humans.

Clostridium tetani (Tetanus)

Clostridium tetani is a Gram-positive, spore-forming, obligate anaerobe widely found in the soil and feces of animals and humans. It is the cause of neurological disease tetanus. External abrasions, punctures, or trauma wounds can be easily contaminated with *C. tetani* spores from the environment. The optimal anaerobic conditions, tissue necrosis, and presence of foreign material may promote germination and development of the disease. The location and type of wounds can influence the risk of infection [27, 28]. The bacterium is difficult to isolate, and no pathologically characteristic lesion is present during infection. The disease has been reported in OWPs and NWPs and can be problematic in free-ranging and outdoor-housed animals [29, 30]. Tetanus diagnosis is based on tetanus-specific clinical symptoms. In the early stage of disease, animals exhibit stiffness and reluctance to move, difficulty eating and swallowing, and extensor rigidity. As disease progresses trismus, opisthotonos, and status epilepticus may occur (Figs. 2.22 and 2.23). The disease progression may take 1–10 days when respiratory paralyses and exhaustion can result in death. Tetanus is caused by a powerful neurotoxin tetanospasmin produced by *C. tetani* that inhibits the function of Renshaw cells that control the duration and intensity of motor neurons leading to continuous stimulation of skeletal muscles. Previous episodes of tetanus are not protective. As prevention for institutions housing NHPs outdoors, immunization program is suggested [31, 32].

a

b

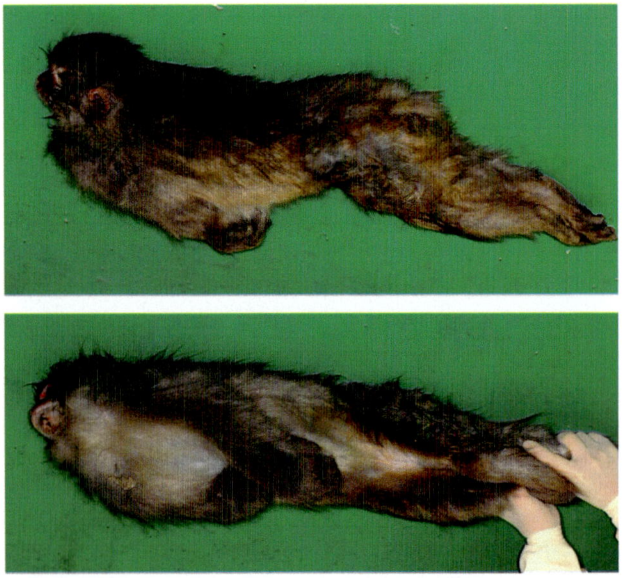

Fig. 2.22 *Clostridium tetani* Tetanus. (**a**, **b**) Opisthotonos in a Japanese macaque (*Macaca fuscata*). Case material courtesy of Professor Dr. Yumi Une, Okayama University of Science (OUS), Japan

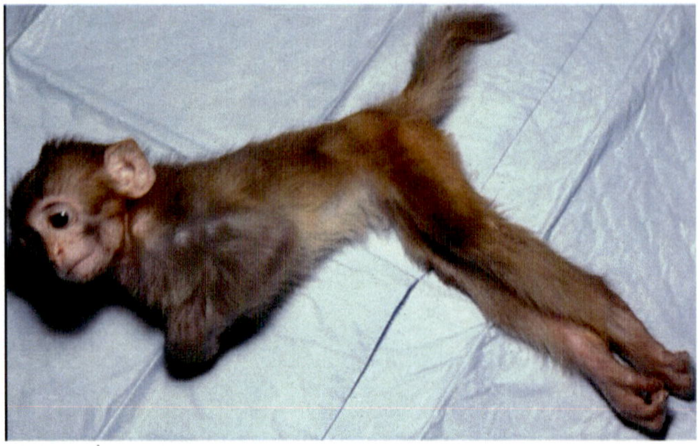

Fig. 2.23 *Clostridium tetani* Tetanus. A one-year-old female rhesus macaque exhibiting adduction of the pectoral limbs and rigid extension of the hind limbs

Clostridium difficile

Clostridium difficile is an anaerobe, Gram-positive, spore-forming bacillus that is naturally found in the gastrointestinal tract of humans, animals including NHPs, and in the environment. The organism is the causative agent of pseudomembranous colitis. Clinical disease occurs secondary to antibiotic exposure when alteration of the normal flora leads to proliferation of the organism and toxin production [33]. In OWPs the history of prolonged treatment with beta-lactam antibiotics precedes the onset of diarrhea. In NWPs disease is sometimes associated with stress or chronic maladies such as diabetes [34, 35]. Clinical signs of disease can range from mild to life-threatening pseudomembranous colitis. Infected animals with mild disease develop profuse watery diarrhea that may resolve after antibiotic therapy is withdrawn. In severe cases gross findings are characteristic and diagnostic and consist of multifocal pseudomembranes that appear as pale plaques adhered to ulcerated areas of the colon. The adjacent mucosa is hyperemic and inflamed. Histologically, areas of micro-ulcerations are accompanied by sloughed mucosa, fibrinous exudate admixed with neutrophils, erythrocytes, and cellular debris covering the mucosa with flat pseudomembranes. Microscopically the lesion appears as a volcano. Diagnostic methods include toxigenic culture (TC), gross and histological findings, and PCR test.

Clostridium piliforme (Tyzzer's Disease)

Clostridium piliforme is an obligate intracellular spore-forming pathogen and the etiologic agent of Tyzzer's disease [36]. The organism has been isolated from domestic, wild, and experimental animals with intestinal and hepatic disease [37]. Cases of Tyzzer's disease in humans and nonhuman primates are rare, with one report in Old World nonhuman primates [38]. Clinical disease has been reported in cotton-top tamarins [39, 40]. Affected animals developed disease at

2–6 months of age and died acutely or developed rapidly progressive diarrhea. Microscopic examination revealed necrotic and marked inflammatory lesions in the gastrointestinal tract (predominantly cecum, colon), liver, and myocardium accompanied by abundant *C. piliforme* bacteria in the periphery of the lesions visualized by silver stains. Histologically lesions are similar to those described in other species. Cases of animals with compromised immune system were recognized.

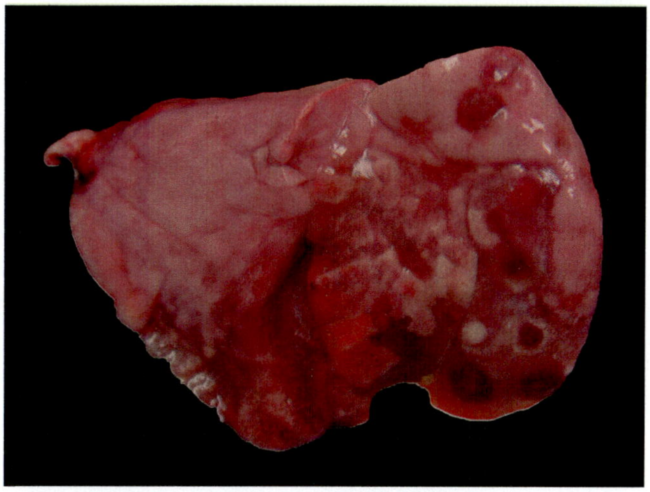

Fig. 2.24 *Bacillus anthracis*. Gross pathology image of lung with patchy hemorrhage and congestion with septal edema in monkey that received aerosolized *B. anthracis* 6 days earlier (pig-tailed macaque, lung)

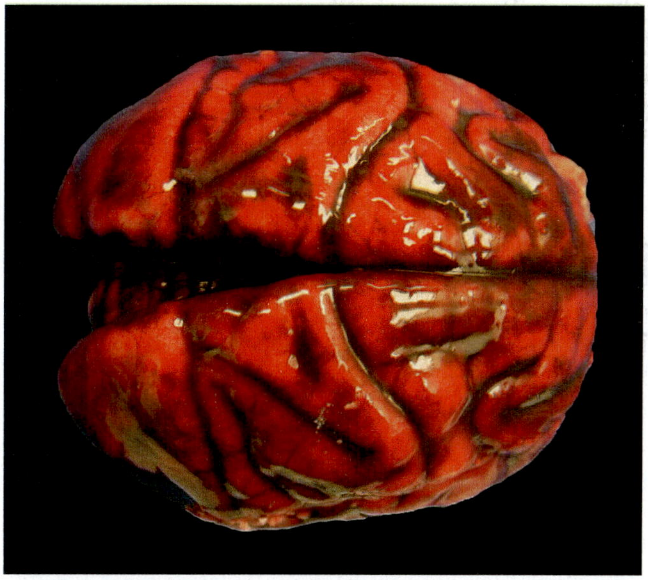

Fig. 2.25 *Bacillus anthracis*. Gross pathology image of brain from macaque 7 days after aerosol exposure. Diffuse hemorrhage in the meninges replicates the classic "Cardinal's Cap" sign of anthrax in humans. More often, meningeal hemorrhage is mild, patchy, or not apparent grossly (pig-tailed macaque, brain)

2.2.5 *Bacillus anthracis* (Anthrax)

Bacillus anthracis is a soil-dwelling spore-forming Gram-positive bacillus typically causing a septic condition called anthrax in grazing livestock, livestock keepers and processors, and free-ranging ungulates in Eurasia, Africa, and North America [41]. The pathogenesis of the disease depends on the route of infection. There are three routes reported: inoculation, ingestion, and inhalation. Cutaneous exposure is by inoculation of anthrax spores through a break in the skin with low mortality that accounts for most human cases. Gastrointestinal exposure with higher mortality rates occurs when wildlife and livestock are exposed through ingestion of spores, and local epidemics lead to zoonotic infections in humans. In tropical locations, synanthropic flies (the house fly, family *Muscidae*, and blowflies, family *Calliphoridae*) have been shown to be mechanical vectors for a number of pathogens including *B. anthracis* in wild nonhuman primate social groups of mangabeys and chimps [42]. Anthrax has been linked by flies to deaths of multiple species of monkeys and other mammals in the rain forest [43]. Respiratory exposure, known as wool-sorter's disease in man, is rare [44] but important now for addressing threats from bioterrorism [45, 46].

In an experimental setting, the clinical responses of the rhesus macaques to respiratory exposures of *B. anthracis* are described as undramatic and inconstant. Some animals develop fevers, others develop respiratory distress prior to death, and some show depression and convulsions before death. Signs are similar in rhesus macaques [47], African green monkeys [48], and cynomolgus macaques [49] with lethality at higher dose levels in less than 10 days. Thoracic lymph node and splenic enlargements two- to three fold are typical.

Animals in experiments with pulmonary exposure to 10^4 to 10^7 spores of *B. anthracis* exhibit edema, congestion, hemorrhage, fibrin, and hyaline membrane deposition in the lung (Fig. 2.24) and pleural and pericardial serosanguinous effusions, subcutaneous edema, mediastinal edema, and hemorrhage, frequent ecchymotic, and severe hemorrhages in the meninges (Fig. 2.25), gastric mucosa, adrenal glands (Fig. 2.26), liver, and near ovaries and testes. Infected animals died with terminal bacteremia.

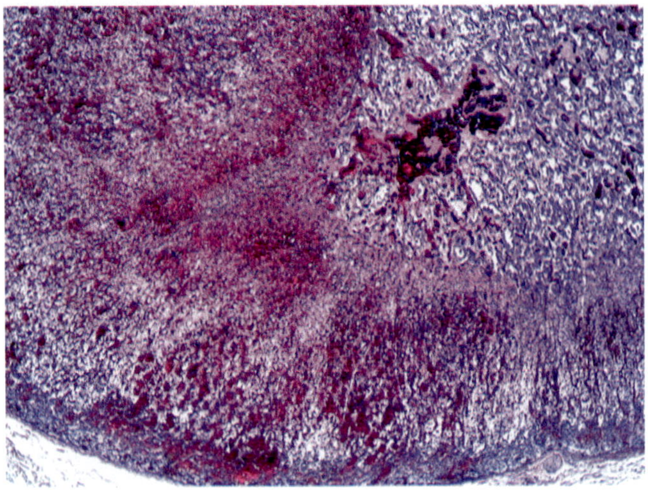

Fig. 2.26 *Bacillus anthracis*. Photomicrograph of adrenal gland from pigtail macaque with aerosol exposure to *B. anthracis*. The adrenal cortex exhibits marked multifocal hemorrhages (pig-tailed macaque, H&E stain)

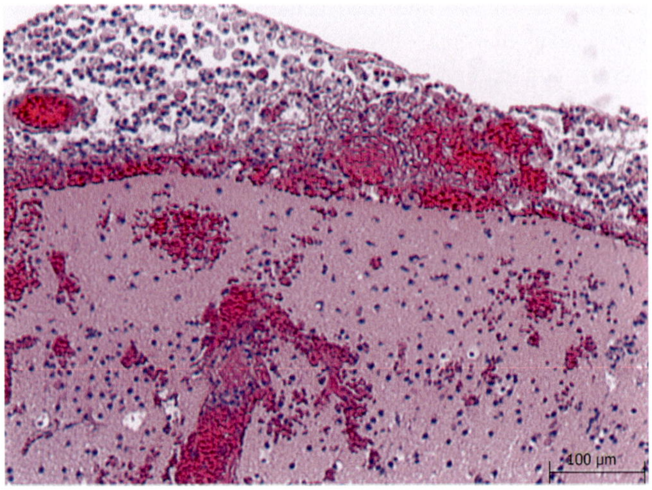

Fig. 2.27 *Bacillus anthracis*. Photomicrograph of acute hemorrhage accompanied by mixed inflammatory cells (neutrophils and macrophages) in the subarachnoid space and around small vessels in the outer cortex of the occipital lobe (pig-tailed macaque, brain, H&E stain)

Microscopically, hemorrhage is the most common finding in many organs including the brain (Figs. 2.27 and 2.28), bronchial and mediastinal lymph nodes, gastrointestinal tract, heart, and adrenal often accompanied by edema and fibrin deposition in the lung (Fig. 2.29). Sepsis can be identified in multiple tissues, particularly vessels in the meninges (Fig. 2.30), sinusoids of the lymph nodes and spleen, the glomeruli (Fig. 2.31), and the heart (Fig. 2.32). Suppurative responses are minimal, scattered, and found in groups of alveoli, localized portions of bronchial nodes, and meninges. Lymphocytolysis is present sporadically in bronchial and mediastinal nodes and splenic corpuscles. Vasculitis is noted in the lung and occasionally in the brain where it may account for thrombosis and focal spongiosis, gliosis, and neuronal necrosis [48].

A review of older literature [47] suggests that macrophages transport spores via lymphatics to regional lymph nodes and sites of unrelated previous injury (natural lung mite lesions) where bacilli multiply and disseminate via vascular necrosis into the circulation. Sepsis due to anthrax induces disseminated intravascular coagulation (DIC) when bacterial peptidoglycan activates tissue factor which activates plasma coagulation Factor VII producing TF-VIIa tenase that inhibits coagulation [50]. The result can produce hemorrhage and edema in the meninges, lung, mediastinum, and spleen whether the challenge originated in the lung, skin, or gut. Anthrax is diagnosed by cultures from blood, respiratory secretion, or skin sores and by PCR test.

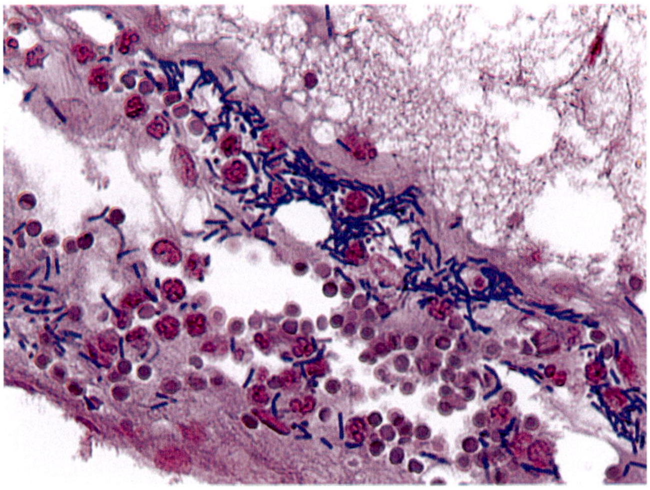

Fig. 2.28 *Bacillus anthracis*. Photomicrograph of subarachnoid space contains extravasated erythrocytes, numerous mononuclear, and polymorphonuclear leukocytes and extravascular *B. anthracis* (pig-tailed macaque, Gram stain, University of Nottingham method)

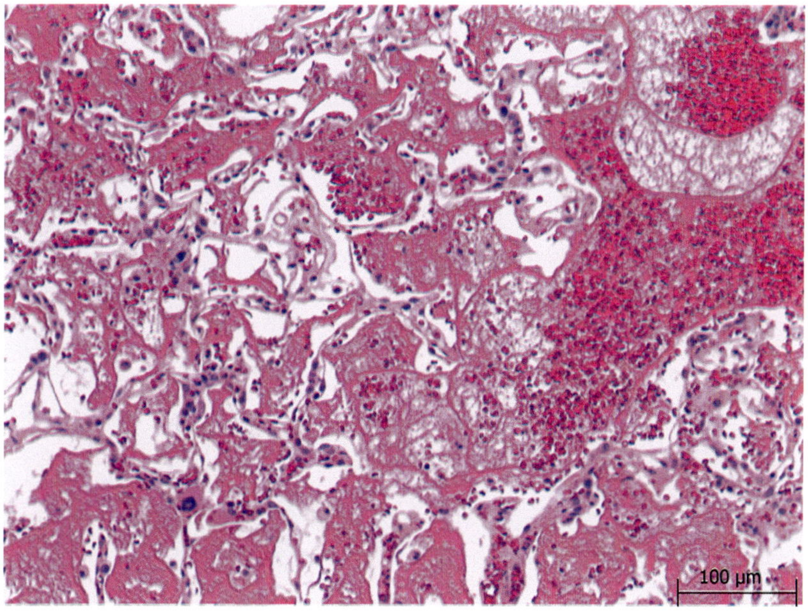

Fig. 2.29 *Bacillus anthracis*. Photomicrograph of pulmonary lesions in macaque 6 days after aerosol exposure. Alveolar spaces are filled with fibrin, edema, hemorrhage, and few inflammatory cells (pig-tailed macaque, lung, H&E stain)

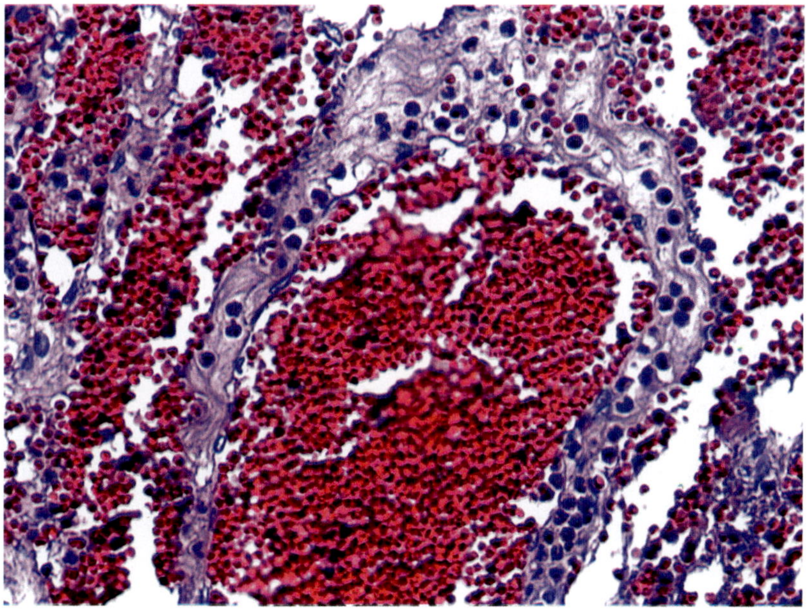

Fig. 2.30 *Bacillus anthracis*. Photomicrograph of brain with vasculitis demonstrated in a congested meningeal vessel with an edematous wall infiltrated with polymorphonuclear cells and a few macrophages in a hemorrhagic subarachnoid space (pig-tailed macaque, brain, H&E stain)

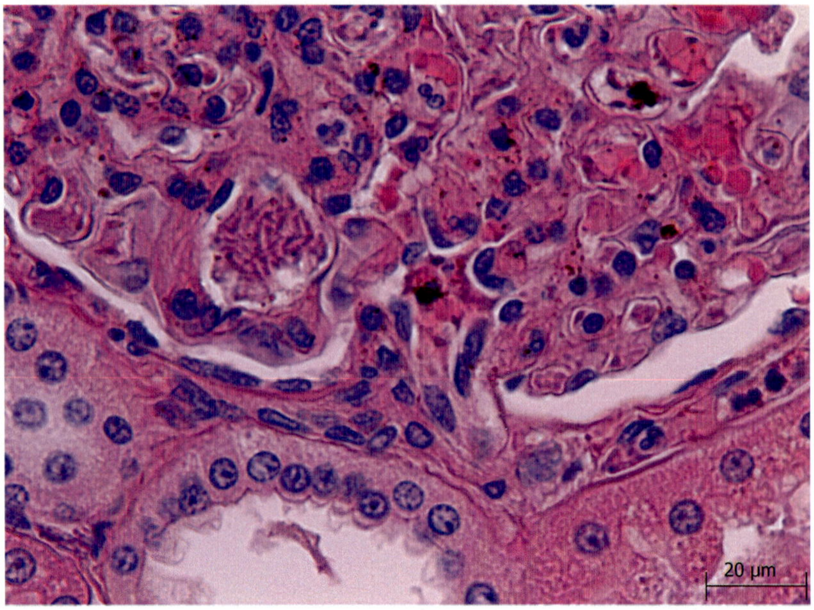

Fig. 2.31 *Bacillus anthracis*. Photomicrograph of renal lesion showing aggregated intravascular bacilli in the glomerulus without other inflammatory changes (pig-tailed macaque, kidney, H&E stain)

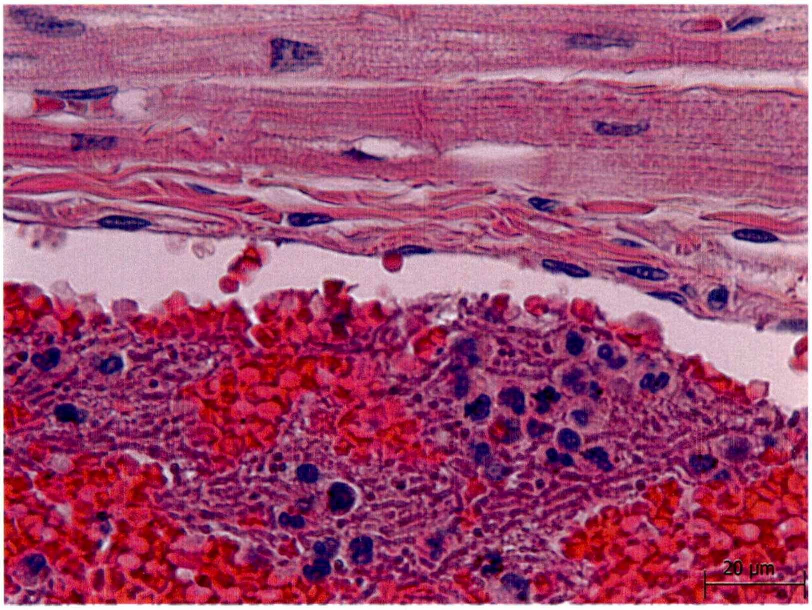

Fig. 2.32 *Bacillus anthracis*. Photomicrograph of heart from macaque with sepsis inoculated via aerosol with *B. anthracis*. Aggregates of bacilli are found within the left ventricle (pig-tailed macaque, heart, H&E stain)

2.2.6 *Streptococcus* Genus

Streptococcus spp. are Gram-positive coccoid, chain-forming bacteria and some of the most invasive species in man and animals. The group includes alpha- and beta-hemolytic organisms. *Streptococci* are part of the normal flora of the skin, nasal and oral cavities, upper respiratory tract, and intestine. As pathogens they cause meningitis, sepsis, otitis, pneumonia, serositis and arthritis [51–53].

Streptococcus pneumoniae

*Streptococcus pneumonia*e is a Gram-positive encapsulated bacterium that produces alpha-hemolysin on blood agar and is the most frequent cause of meningitis in NHPs. When meningitis develops in some cases, the course of disease is rapidly progressive, and death could occur without prodromal clinical signs [54]. Animals are often febrile and show neutrophilia with a left shift. Usually, the neurological signs include ataxia, cervical rigidity, nystagmus, blindness, and paresis. *S. pneumoniae* could be anthropozoonotic and is responsible for 95% of the pneumonia in NHPs. The organism resides asymptomatically in healthy carriers typically colonizing the nasopharynx. Predisposing conditions for the host may include stress, transportation, and viral infections such as RSV, PIV-3, and influenza [55]. *S. pneumoniae* may cause epizootics of bronchopneumonia. Infected animals may exhibit respiratory signs of cough and dyspnea which may be followed by the development of arthritis or meningitis. Transmission of the organism occurs by contaminated fomites or by the aerosolized route. Diagnosis in culture is confirmed by optochin disk sensitivity (Fig. 2.33). Pulmonary lesions are consistent with fibrinous and suppurative pleuritis and bronchopneumonia (Figs. 2.34 and 2.35). Histology confirms fibrinopurulent exudate with hemorrhage and often accompanied by necrosis (Figs. 2.36, 2.37, 2.38, and 2.39) and thrombosis (Fig. 2.39a) [51, 55, 56].

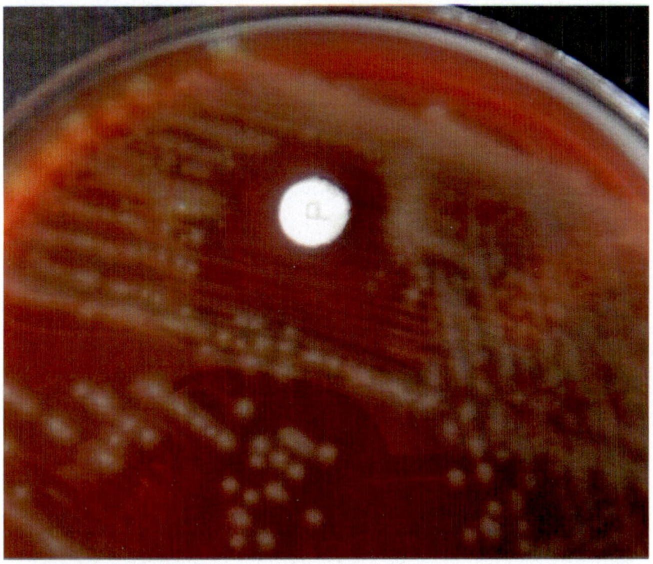

Fig. 2.33 *Streptococcus pneumoniae*. confirmed by culture on blood agar plate demonstrates alpha hemolysis and zone of inhibition created by diffusion from an optochin (ethylhydrocuprein hydrochloride) disk

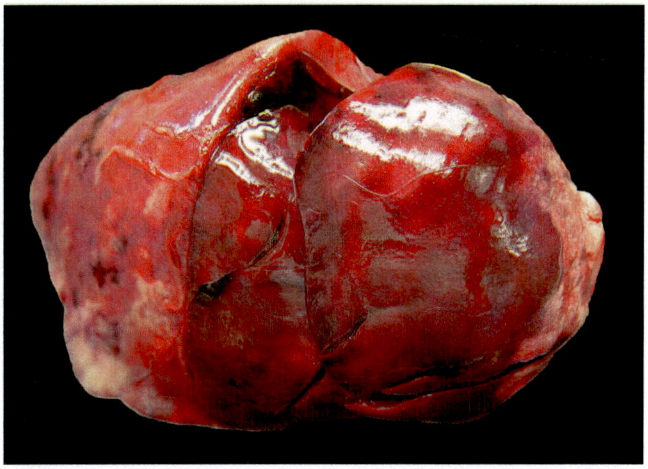

Fig. 2.34 *Streptococcus pneumoniae*. Gross pathology image of hemorrhagic lobar pneumonia 4 days postinfection with *S. pneumoniae* (rhesus macaque, lung)

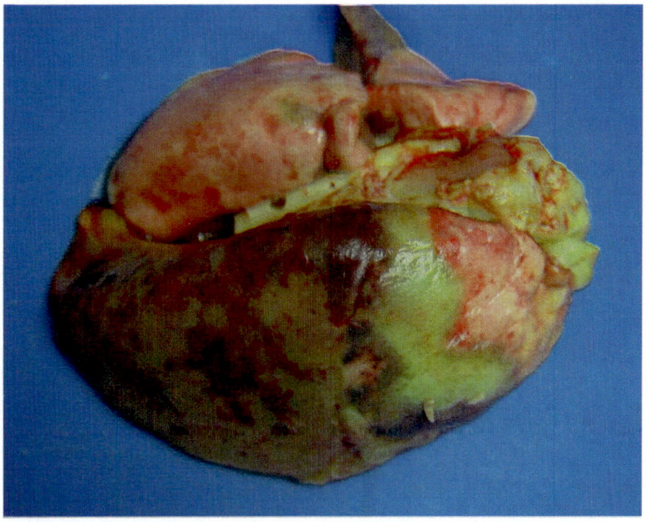

Fig. 2.35 *Streptococcus pneumoniae*. Gross pathology image of lobar pneumonia and fibrinous pleuritis 8 days postinfection with *S. pneumoniae* (rhesus macaque, lung)

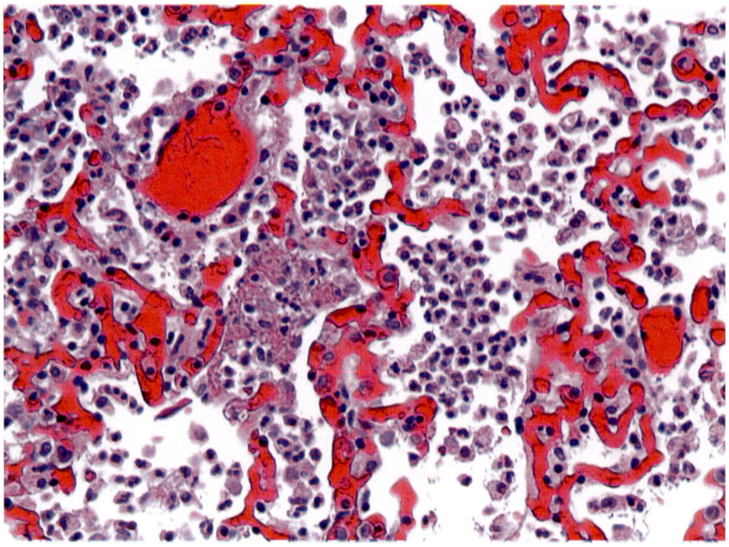

Fig. 2.36 *Streptococcus pneumoniae*. Photomicrograph of lung from infant rhesus macaque with fibrinopurulent bronchopneumonia due to naturally acquired *S. pneumoniae* infection (H&E stain)

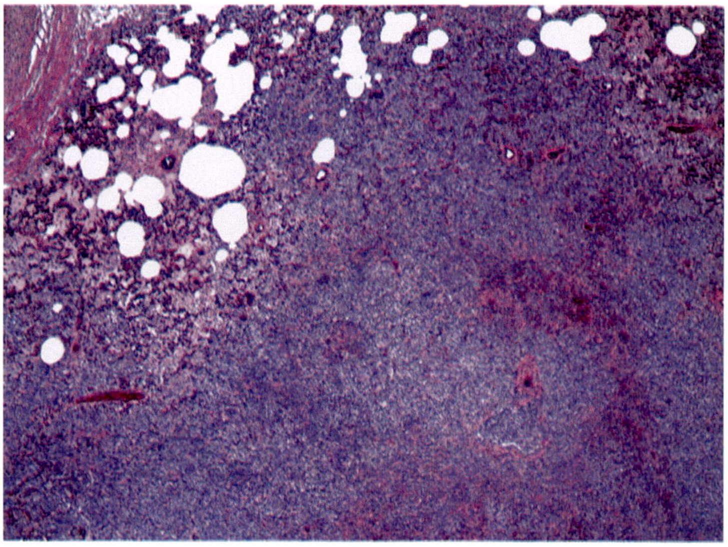

Fig. 2.37 *Streptococcus pneumoniae*. Photomicrograph of the lung from immunodeficient infant rhesus macaque infected with SIV that developed lobar pneumonia with alveolar edema, dense infiltration of neutrophils and macrophages, and areas of hemorrhage due to opportunistic infection with *S. pneumoniae* (H&E stain)

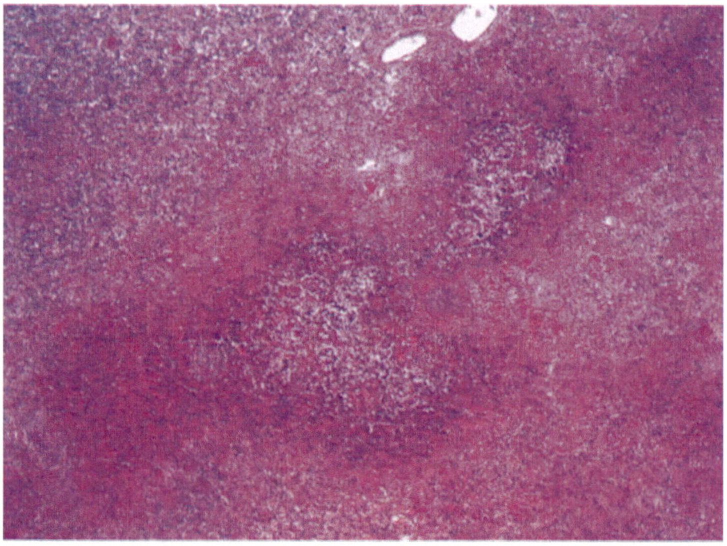

Fig. 2.38 *Streptococcus pneumoniae*. Photomicrograph of lung from rhesus macaque. Multiple confluent zones of necrosis within consolidated lobes after naturally acquired *S. pneumoniae* infection (H&E stain)

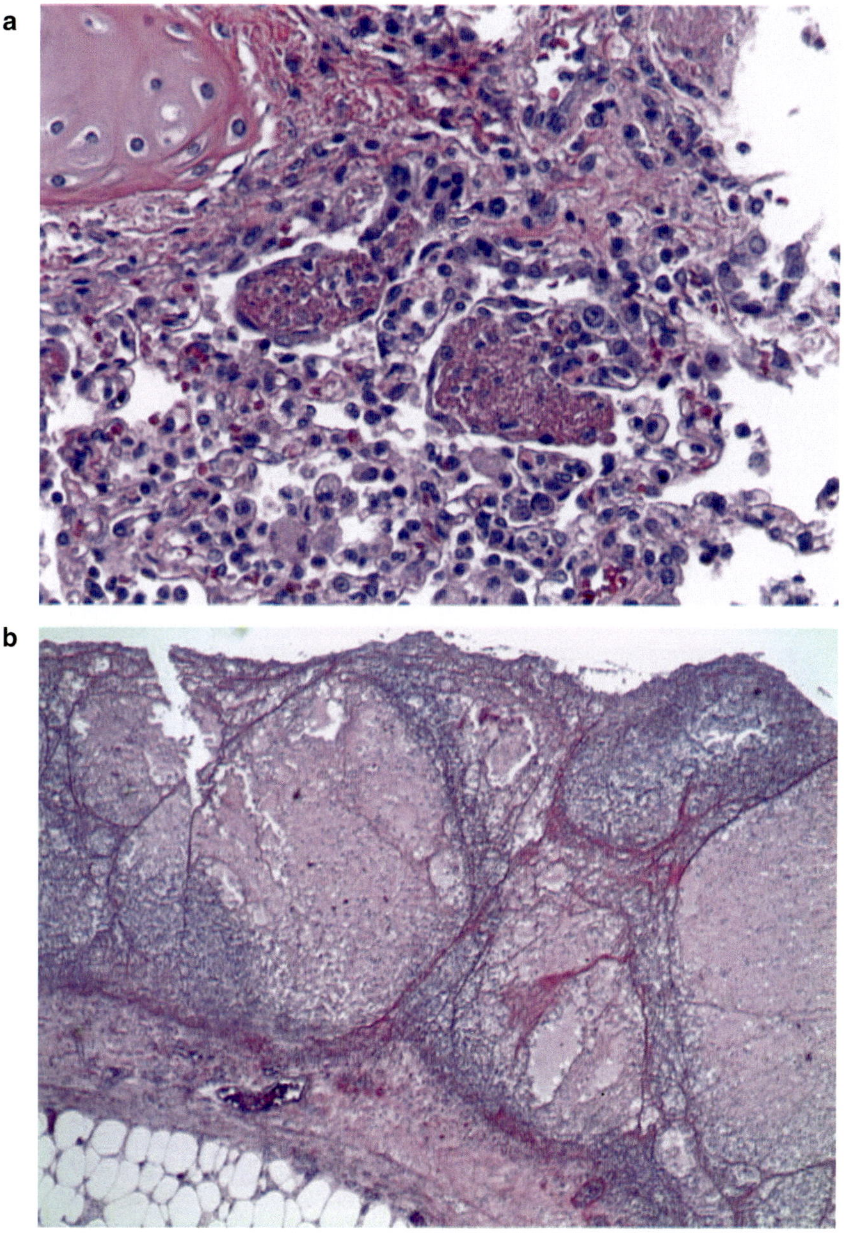

Fig. 2.39 *Streptococcus pneumoniae*. (**a**) Photomicrograph of lung from rhesus macaque with fibrin thrombi in alveolar microvasculature after naturally acquired *S. pneumoniae* infection (H&E stain). (**b**) Photomicrograph of the lung from the same rhesus macaque. The pleura is markedly expanded by thick fibrinous exudate forming layers and cavities filled with fluid and neutrophils (H&E stain)

In cases with meningitis, meninges of the brain are cloudy to hemorrhagic (Figs. 2.40, 2.41, 2.42, and 2.43) with areas of malacia and abscess formation. Histology confirms fibrinopurulent exudate with hemorrhage and abscess formation often accompanied by necrosis.

In addition, peritonitis, epicarditis, corneal ulcers, and otitis media may be observed in multiple nonhuman species. In marmosets, disease is rapidly progressive involving meninges, serosal surfaces, and joints [53]. In human patients that survive long term, adverse cardiac events are now described and associated with formation of perivascular microcolonies without inflammatory response. The lack of response in the myocardium is unexplained compared to the vigorous response in the pericardium, lung, middle ear, and CNS of the same patient [57, 58]. We have not observed bacterial colonies without inflammation in rhesus monkeys although the incidence of myocardial abscess (Fig. 2.44) and myocarditis (Fig. 2.45a, b) due to pneumococcus is low (unpublished, Didier). *Streptococci* are readily identified by culture.

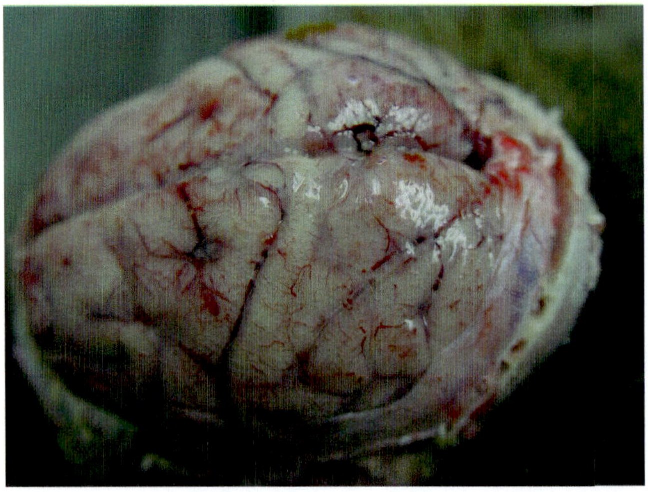

Fig. 2.40 *Streptococcus pneumoniae*. Macroscopic image of brain from a 1-year-old rhesus macaque with suppurative meningitis due to naturally acquired *S. pneumoniae* infection

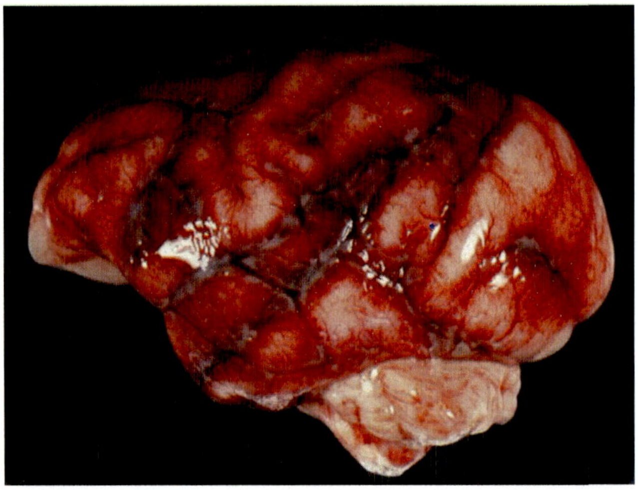

Fig. 2.41 *Streptococcus pneumoniae*. Macroscopic image of brain from a 26-day-old rhesus macaque with fibrinopurulent hemorrhagic meningitis due to naturally acquired *S. pneumoniae* infection

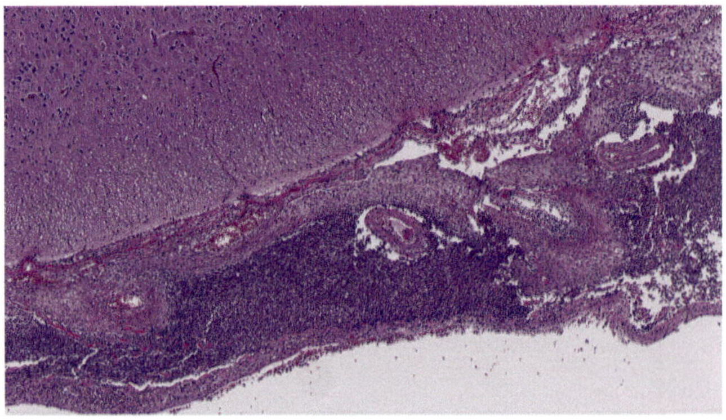

Fig. 2.42 *Streptococcus pneumoniae*. Photomicrograph of brain stem and meninges from infant rhesus macaque. The meninges are densely infiltrated with polymorphonuclear cells due to naturally acquired *S. pneumoniae* infection (H&E stain)

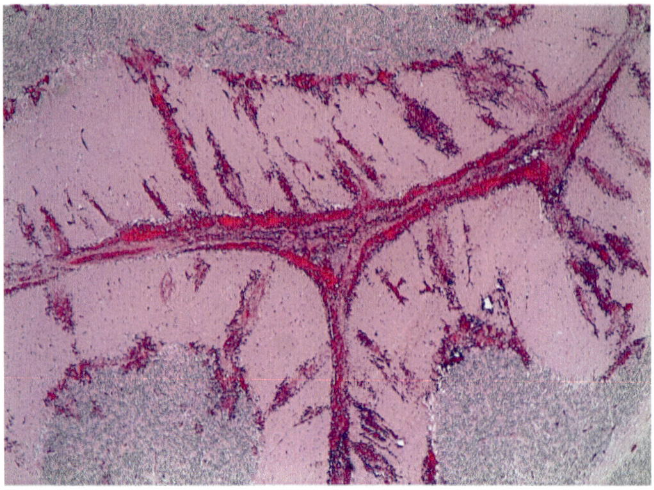

Fig. 2.43 *Streptococcus pneumoniae*. Photomicrograph of the cerebellum from an infant rhesus macaque with hemorrhagic meningoencephalitis due to naturally acquired *S. pneumoniae* infection (H&E stain)

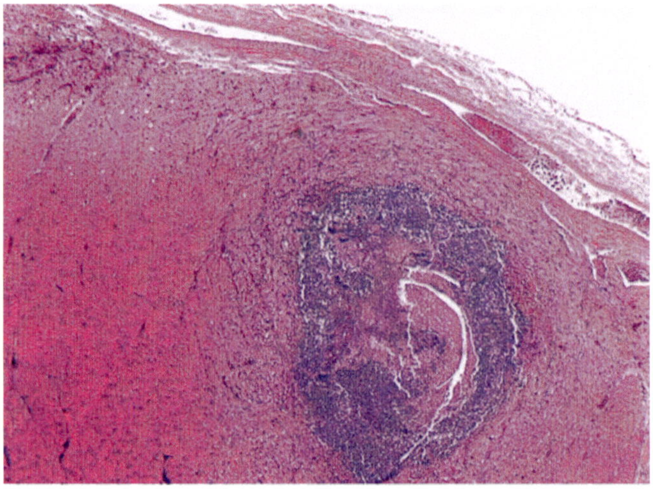

Fig. 2.44 *Streptococcus pneumoniae*. Photomicrograph of the heart from rhesus macaque with a subendocardial abscess filled with neutrophils and colonies of *S. pneumoniae* (H&E stain)

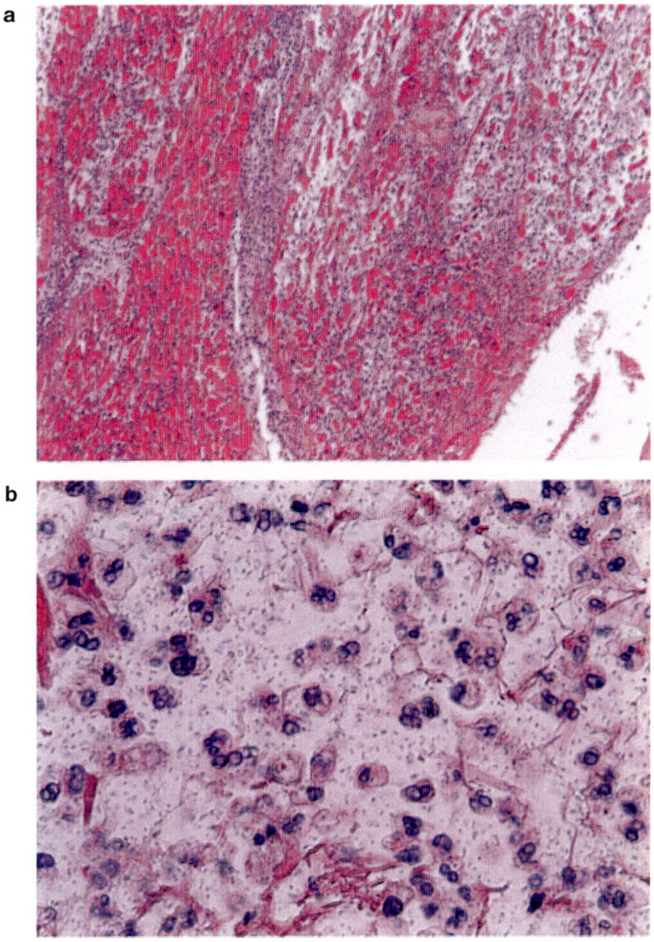

Fig. 2.45 *Streptococcus pneumoniae*. (**a**) Photomicrograph of heart form rhesus macaque showing edema, myocardial degeneration, and necrosis with infiltration of polymorphonuclear cells and macrophages after naturally acquired infection with *S. pneumoniae* (H&E stain). (**b**) Higher magnification shows polymorphonuclear cells and numerous diplococci in edematous spaces between myocardiocytes (H&E stain)

Streptococcus equi Subspecies *Zooepidemicus*

Streptococcus equi subspecies *zooepidemicus* is a beta-hemolytic streptococcus which can be found in the nasopharynx and respiratory tract of healthy horses and cattle but can cause infections in foals and young horses. The organism can be transmitted through direct contact or droplets. Outbreaks can affect NWPs and OWPs. Transmission in zoo-housed callitrichids may occur via ingestion of infected horse meat being fed to other species in exhibits [59] or from animal caretakers in contact with horses carrying the organism. Infected animal die within a few days after developing signs of infection [60].

2.2.7 *Staphylococcus* Genus

Staphylococcus spp. are aerobic, nonsporing, Gram-positive bacteria with spherical shape which form clusters, pairs, or short chains in culture. They are common commensals identified in skin and nasopharyngeal cultures of NHPs.

Staphylococcus aureus

Staphylococcus aureus is common secondary pathogen and can be characterized by the production of coagulase. The organism is colonizing the nares, axillae, vagina, pharynx, and damaged skin in macaques [61]. Predisposing conditions like compromised immunity, stress, diabetes mellitus, ectoparasites, endocrinopathies, renal and heart disease, and indwelling catheters allow infection of the skin and deep tissues [62, 63]. Mortality in humans can be 20–50% despite treatment. The organism often contaminates surgical sites and complicates bite wounds. It is one of the most common bacteria isolated from purulent inflammation. *S. aureus* evolve along different animal and human lineages [64] by point mutation of the chromosome and selection and horizontal transfer of prophages, plasmids, transposons and larger pathogenicity islands [65].

Diagnosis of *S. aureus* is made by obtaining a culture from the area of suspected infection. Cytologically the presence of degenerate neutrophils with intracellular bacteria is required.

The pathogenicity of *S. aureus* is related to variety of factors and has evolved a complex regulatory network. One of the main functions of this interconnected network is to sense various environmental cues and respond by altering the production of virulence factors. Potential virulence factors include (1) surface proteins that promote colonization of host tissues, (2) factors that probably inhibit phagocytosis (capsule, immunoglobulin-binding protein A and (3) toxins that damage host tissues and cause disease symptoms [66]. These complex systems engage when *S. aureus* transitions from carrier state to infection of tissue or blood. In blood stream infections, hepatic Kupffer cells phagocytize bacteria, but survivors redistribute and infect peritoneal macrophages and circulating neutrophils that amplify the inflammatory response. Toxins and superantigens activate and kill leukocytes, activate complement and T cells which promote platelet aggregation, endothelial damage, and exposure of underlying collagen, and reinforce bacterial adherence and thrombosis [63].

The ability to bind to endothelium and infect leukocytes is key to spread *S. aureus* from skin and mucus membranes [67]. Pustular dermatitis and furunculosis [62] are suppurative conditions described for skin and hair follicles (Fig. 2.46a, b). Isolation of *S. aureus* from NHP mammary gland is far more frequent than disease (Didier unpublished), but abscess formation with suppuration and fibrous capsule, lobular mixed inflammation, and adjacent cellulitis are comparable to published reports (Fig. 2.47).

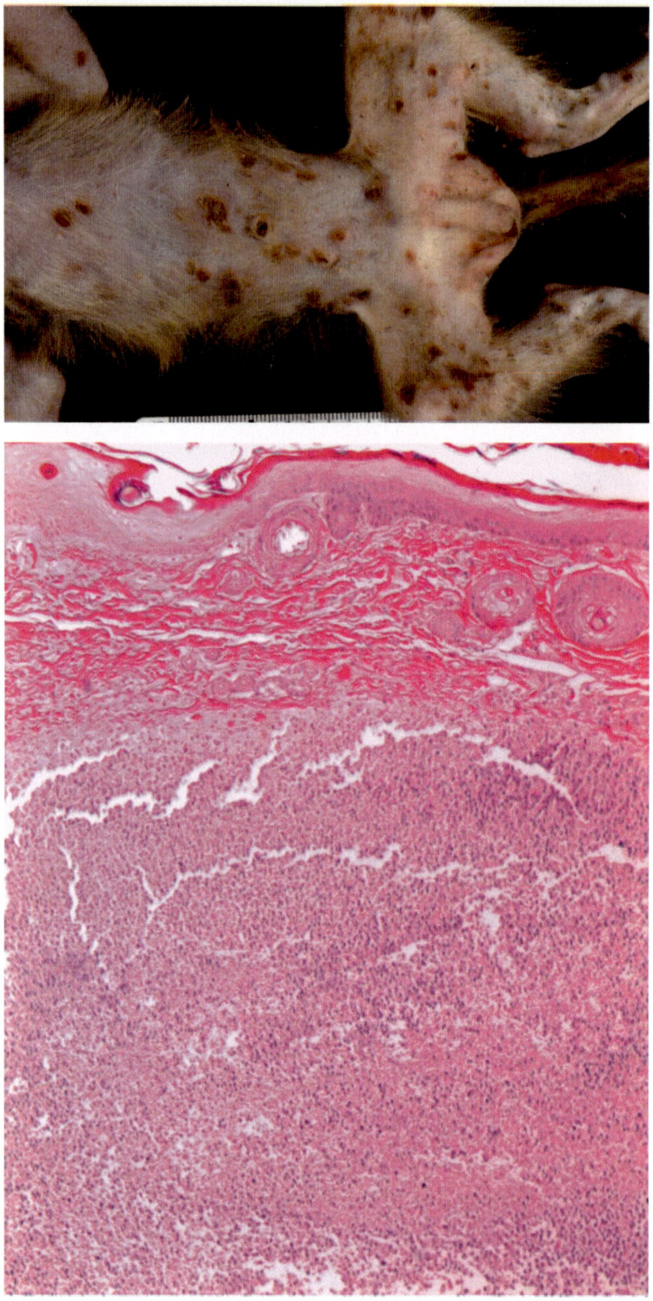

Fig. 2.46 *Staphylococcus aureus*. (**a**) Gross pathology image of infant rhesus macaque with ulcerative, pustular dermatitis. *S. aureus* is cultured from the exudate. (**b**) Photomicrograph of (**a**) with subcutaneous abscess (rhesus macaque, H&E stain)

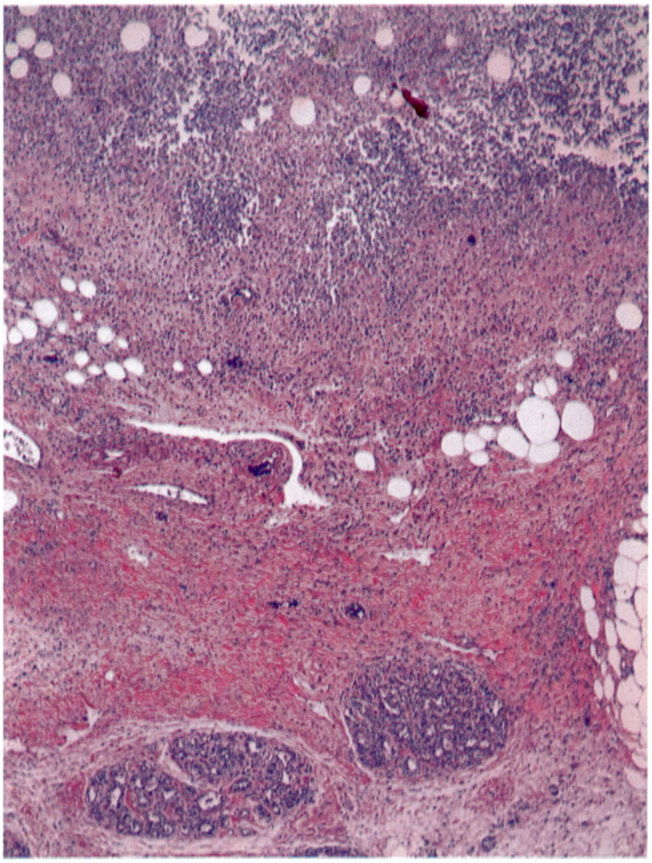

Fig. 2.47 *Staphylococcus aureus*. Photomicrograph of skin abscess obliterating scattered mammary gland lobules in a 17-year-old female rhesus macaque infected with SIV (*S. aureus* cultured, H&E stain)

Acute necrotizing bronchopneumonia, furunculosis, cutaneous abscess, and severe necrotic skin infections have been linked with production of Panton-Valentine leukocidin and could be enabled by exposure of basement membrane collagens by prior viral infection (Figs. 2.48, 2.49, and 2.50) [68].

Airsacculitis is described in macaques [69]. In an experimental setting with a single virulence factor, aerosol exposure of rhesus macaques to Staphylococcal enterotoxin B (SEB) produces extremely edematous lungs with multifocal hemorrhage, severe diffuse interstitial edema with fluid and fibrin in alveolar spaces accompanied by foamy macrophages, and small numbers of polymorphonuclear cells (Fig. 2.51a, b) [70]. Endocarditis is described in NHP particularly with the use of indwelling catheter (Fig. 2.52) [71]. Distribution to the meninges is reported in a tamarin and a cynomolgus monkey [67]. Identification of *S. aureus* in multiple surgical sites suggests the possibility that toxic shock [72] could have complicated trauma cases before suppurative lesions become apparent. Similarly, postpartum intrauterine infections and chorioamnionitis could be the source of or sequala of staphylococcal sepsis.

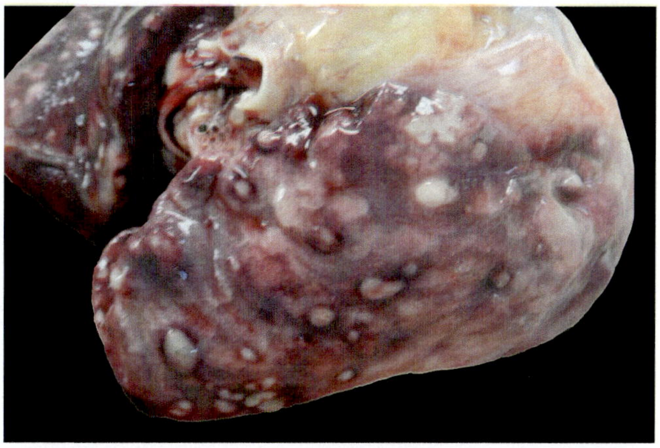

Fig. 2.48 *Staphylococcus aureus*. Macroscopic image of the lung from rhesus macaque with multifocal pulmonary abscesses secondary to trauma due to fighting. Abscesses are ringed by hemorrhage. *S. aureus* was cultured from the lung

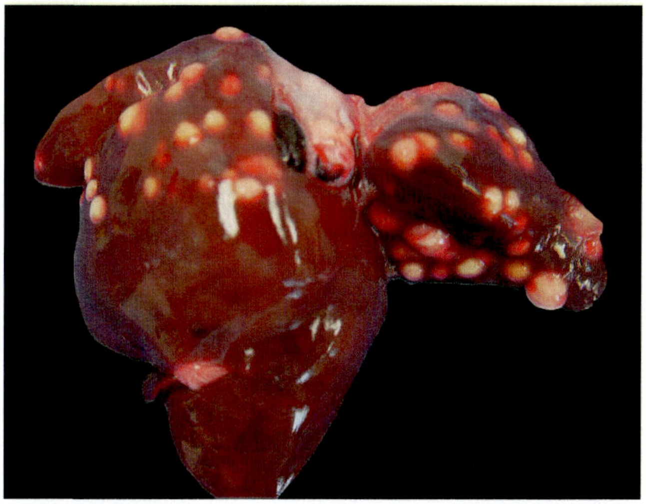

Fig. 2.49 *Staphylococcus aureus*. Macroscopic image of the liver abscesses in infant rhesus macaque with infected joint and skin wound. All sites of infection contained hemolytic coagulase positive *S. aureus*

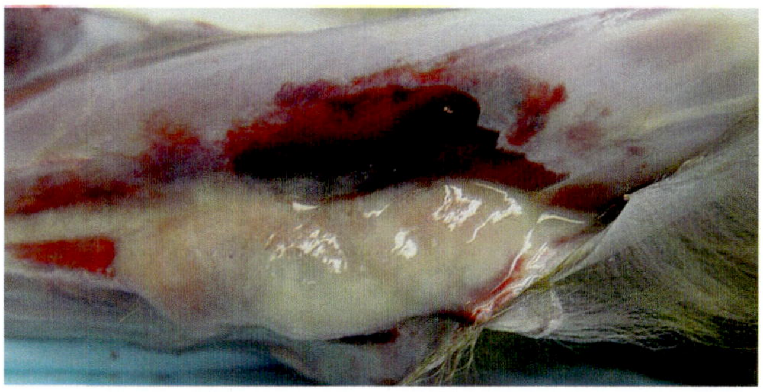

Fig. 2.50 *Staphylococcus aureus*. Macroscopic image of the lateral aspect of the left leg (below the knee) from rhesus macaque infected with SIV. In the center large subcutaneous abscess involves muscle fascia and fills the subcutis. Hemolytic coagulase positive *S. aureus* was cultured from the exudate

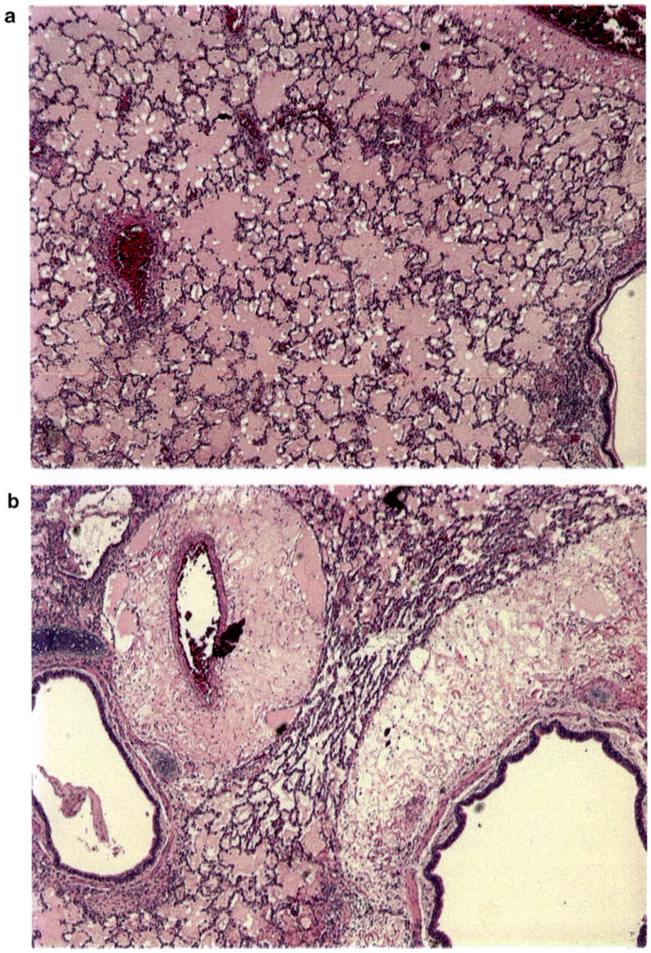

Fig. 2.51 *Staphylococcus aureus*. Photomicrographs of pulmonary lesions in rhesus macaque 48 h after receiving aerosolized SEB toxin. (**a**) Diffuse pulmonary edema and fibrin with mild interstitial inflammation. (**b**) Marked peribronchial and perivascular edema accompanied by fibrinous strands and small numbers of mixed inflammatory cells (H&E stain)

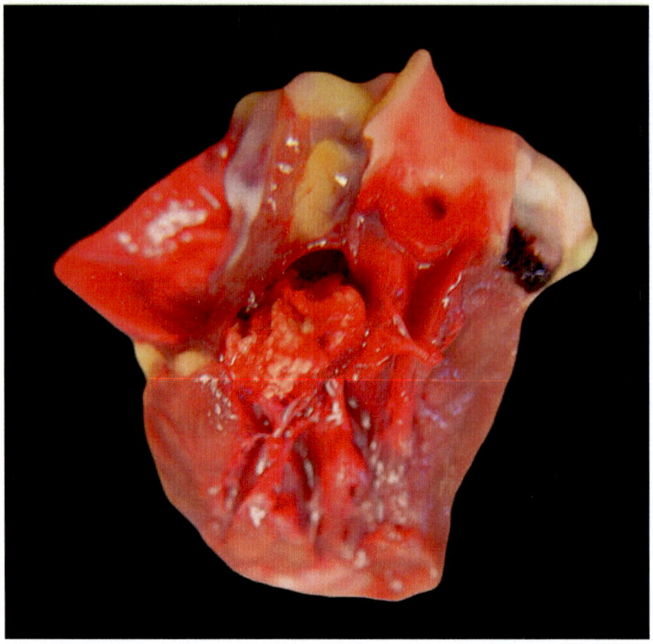

Fig. 2.52 *Staphylococcus aureus*. Gross pathology image of a heart from rhesus macaque inoculated with SIV. Nodular fibrinous to fibrous growth attached to the right atrioventricular valve (tricuspid) and extending into the pulmonary artery. The animal has a history of multiple long intravenous procedures, multiple isolations of hemolytic coagulase positive *S. aureus* from the nasal cavity, and recovery of non-hemolytic coagulase negative *Staphylococci* spp. from the lung and brain at necropsy

2.3 Gram-Negative Bacteria

2.3.1 *Escherichia coli*

Escherichia coli is a diverse species of Gram-negative bacteria in the family *Enterobacteriaceae*, living in the gut of vertebrates often as a commensal and as a free-living organism in water and sediment. The genus is genetically diverse due to gene gains and losses, and through horizontal transfer of bacteriophages, plasmids, genomic islands, transposons, and insertion elements [73]. The types of *E.coli* that are capable of causing disease in healthy individuals are pathogenic *E.coli* [74]. Different pathogenic *E. coli* strains cause diverse intestinal and extraintestinal diseases by means of virulence factors that affect a wide range of cellular processes. Members of this group are responsible for gastroenteritis with watery or bloody diarrhea, pneumonia (Figs. 2.53 and 2.54) and myocardial infarction, urinary and renal infections (Fig. 2.55), septic shock, meningitis and wound infection, prostatitis, osteomyelitis, and peritonitis [75–77]. Pathogenicity is determined by the presence of a broad range of factors including toxins, adhesins, lipopolysaccharides, polysaccharide capsule proteases, and invasins that may increase competitiveness and ability to

a
b
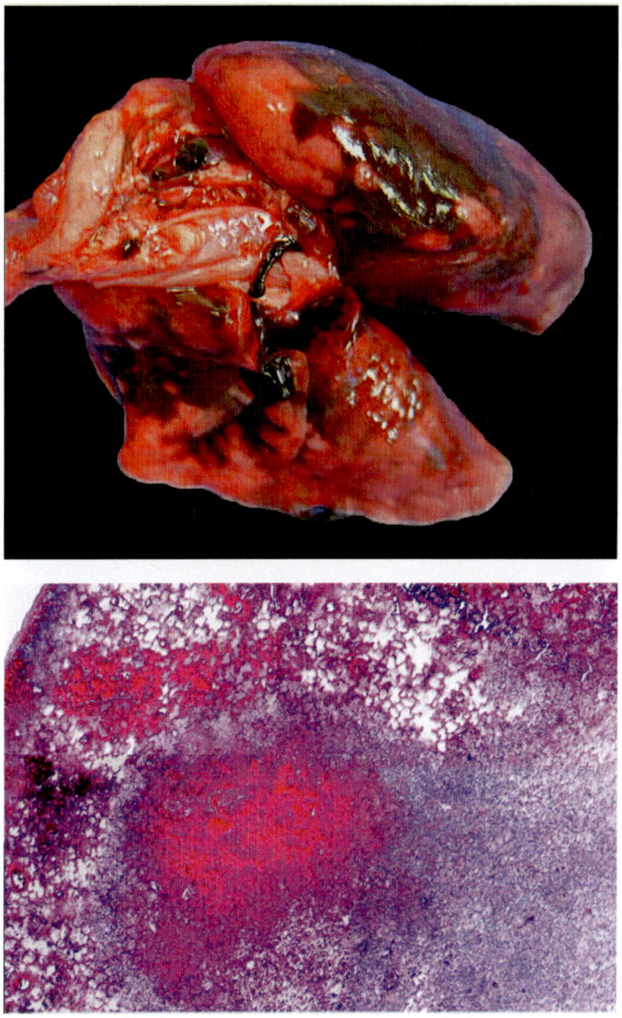

Fig. 2.53 *E. coli*. (**a**) Macroscopic image of hemorrhagic bronchopneumonia in a 3-month-old female found dead without previous clinical signs (rhesus macaque, lung). (**b**) Photomicrograph of the lung shows multifocal hemorrhage and congestion surrounded by dense infiltration of polymorphonuclear cells and macrophages adjacent to a necrotic center. *E. coli* was isolated in pure culture (rhesus macaque, lung, H&E stain)

a
b
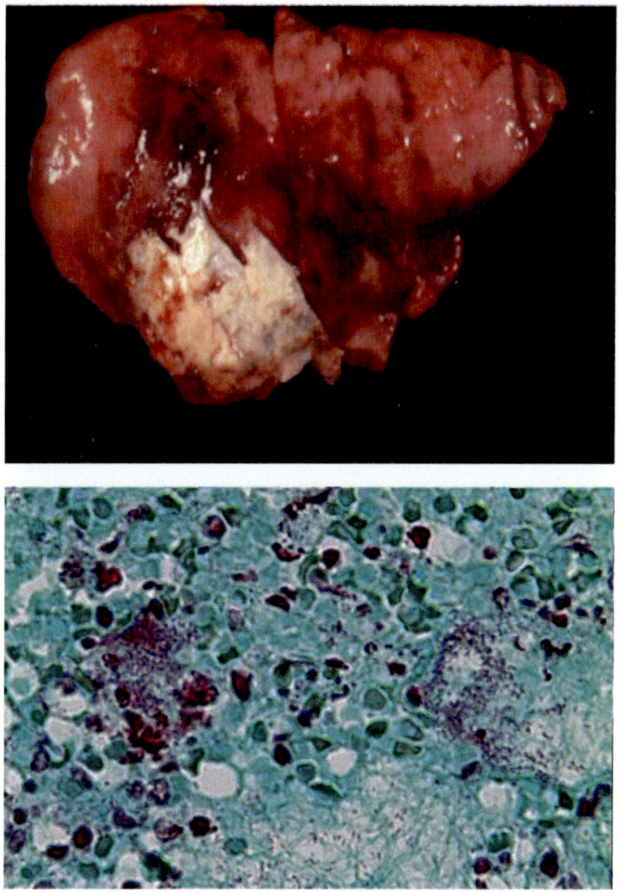

Fig. 2.54 *E. coli*. (**a**) Macroscopic image of fibrinous pleuritis and bronchopneumonia in macaque complicated by coinfection with SRV and herpes virus in the pancreas (rhesus macaque, lung). (**b**) Colonies of *E. coli* within cellular exudate of the lung (rhesus macaque, lung, Brown and Brenn Gram stain)

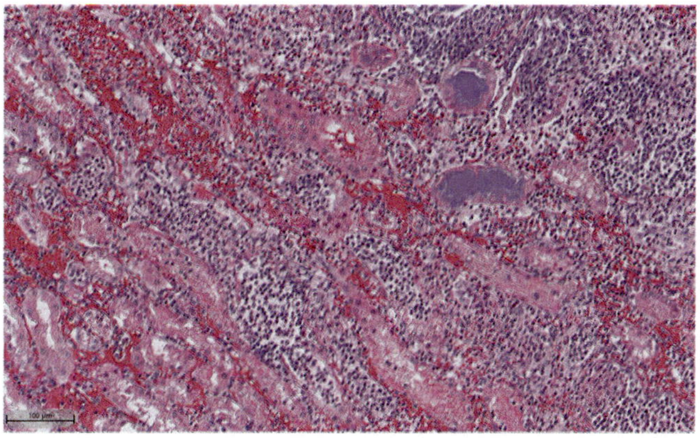

Fig. 2.55 *E.coli*-UPEC. Photomicrograph of severe suppurative pyelonephritis. Renal tubules are lined by swollen and degenerated epithelium, and some are filled with polymorphonuclear infiltrates, sloughed epithelial cells, and isolated colonies of bacteria. The interstitium contains lymphocytic and neutrophilic infiltrates accompanied by hemorrhage. (cynomolgus macaque infected with Uropathogenic *E. coli*—UPEC, kidney, H&E stain)

colonize, as well as to kill cells and rearrange the cytoskeleton [77, 78].

Among the diarrheagenic *E.coli*, there are six categories: enteropathogenic *E. coli* (EPEC), enterohaemorrhagic *E. coli* (EHEC), enterotoxigenic *E. coli* (ETEC), enteroaggregative *E. coli* (EAEC), enteroinvasive *E. coli* (EIEC), and diffusely adherent *E. coli* (DAEC). These groups are defined based on the underlying mechanism of disease pathogenesis, presence of specific genes encoding virulence factors, and in vivo and in vitro growth [74].

Enteropathogenic E. coli

Enteropathogenic *E. coli* (EPEC) is a major cause of infant diarrhea often causing mortality in children worldwide [79]. In macaques infection may be associated with a persistent nonhemorrhagic diarrhea seen primarily in infant or neonatal animals. Infected NWPs frequently exhibit clinical signs of acute hemorrhagic diarrhea accompanied by severe blood loss and hypovolemia. In OWPs and NWPs, a chronic and persistent form is also observed [80–82]. The disease is zoonotic and the transmission of the pathogen is by the fecal oral route.

The EPEC infection is characterized by unique intestinal histopathology known as "attaching and effacing" (A/E) lesions. The bacteria adhere to the intestinal epithelial cells causing cytoskeletal changes. The arrangement of colonic surface epithelium is altered by the intimate association with the bacterium. In young immunodeficient rhesus macaques and in NWPs with experimental EPEC infections, histological lesions include adherence of bacilli to surface epithelium (Figs. 2.56 and 2.57) [80–82, 83,84] accompanied by individual cell necrosis, crypt hyperplasia, "cobblestone" appearance of surface epithelium, flattened squamous morphology, and effacement of microvilli [82].

Adherent bacteria have pilus antigens that form a localized attachment to the surface epithelium which is enhanced by intimin, an outer membrane colonization factor, encoded by the locus of enterocyte effacement (LEE) chromosomal island, and injected into the host enterocyte by type III secretion system [74]. This process produces attachment and effacement of enterocyte microvilli, increased permeability of tight junctions, and fluid loss.

The diagnosis in immunodeficient humans and animals may be obscured by the presence of multiple agents, and EPEC is likely to be missed. Appropriate identification of EPEC requires a systematic approach including the use of

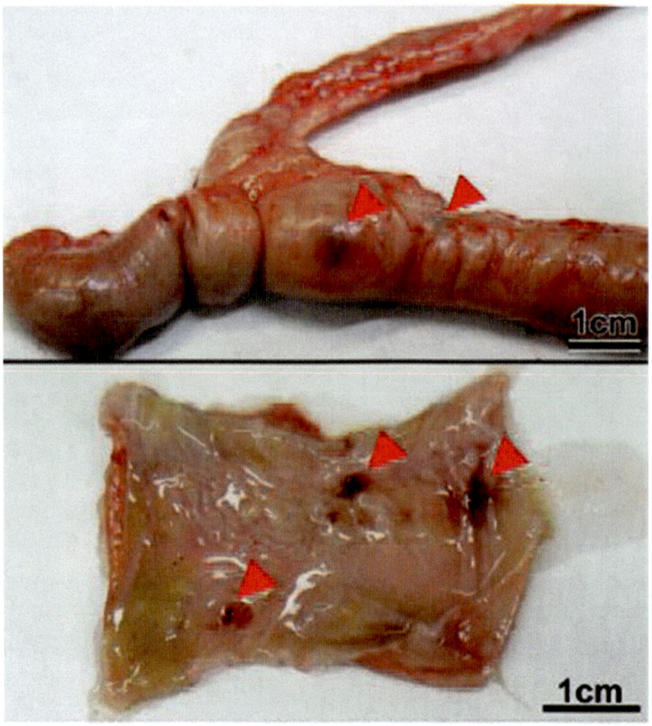

Fig. 2.56 *E. coli*-EPEC. Petechia of colon mucosa in common marmosets experimentally infected with Enteropathogenic *E. coli* (ID no. 2 in the high-concentration group) at 3 days post-inoculation (top, petechia from the serosal side; bottom, hemorrhage from the mucosal side). (*From* Hayashimoto et al. [84]; Creative Commons Attribution 4.0 International [CC BY 4.0]; https://creativecommons.org/licenses/by/4.0)

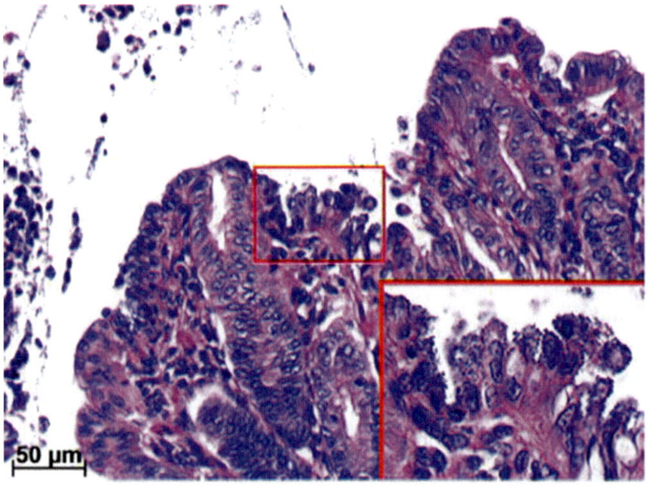

Fig. 2.57 *E. coli*-EPEC. Photomicrograph of a section of cecum from common marmoset no. 1 in the high-concentration group at 3 days post-inoculation. *Bacilli* were attached to epithelial apical membranes (inset) and desquamated epithelial cells, as visualized by H-E staining. (*From* Hayashimoto et al. [84]; Creative Commons Attribution 4.0 International [CC BY 4.0]; https://creativecommons.org/licenses/by/4.0)

histological analyses of biopsies, adhesion assays, and molecular identification of virulence genes.

2.3.2 *Salmonella*

Salmonellae are Gram-negative facultative anaerobic motile bacteria that occur worldwide, inhabiting variety of vertebrate species. The commonly encountered isolates that produce disease in nonhuman primates are *S. typhimurium* and *S. enteritidis*. The pathogen is acquired by NHPs in captivity from exposure to infected humans or contaminated water. Transmission is by the fecal-oral route in both humans and NHPs. Salmonellosis is rarely reported in established colonies [83].

Clinical signs include watery diarrhea secondary to enterocolitis, fever, and extraintestinal infections as neonatal septicemia, abortion, osteomyelitis, and pyelonephritis. In SIV-infected animals and macaques with AIDS, necrotizing thymitis and mediastinal abscessation have been observed. Colonization of the small and large intestine by the bacteria alter the normal electrolyte and fluid balance of the intestinal mucosa. The combination of the inflammation caused by bacterial-mediated endocytosis and the disruption of tight junctions is thought to contribute significantly to the induction of diarrhea [85].

Microscopic lesions in the intestines are nonspecific and nondiagnostic and include villi shortening within the small intestine accompanied by epithelial hyperplasia and edema in the lamina propria and villous tips. In the colon, hyperplasia of the crypt epithelial cells with microabscess formation are observed, and the lamina propria often contain large numbers of neutrophils and edema. Diagnosis is through stool culture and biochemical typing of isolates.

2.3.3 *Shigella*

Shigellae are Gram-negative, nonmotile facultatively anaerobic bacteria without flagella or adherence factors that are closely related to *Escherichia coli*. Based on 0-specific polysaccharide typing of their lipopolysaccharides (LPS), there are four subgroups: *S. dysenteriae*, *S. boydii*, *S. flexneri*, and *S. sonnei* [86]. *S. flexneri* is the most prevalent isolated strain associated with disease in NHPs and the common cause of hemorrhagic colitis (bacillary dysentery). Disease has been described in different species of OWPs and only rarely in NWPs [87]. Humans and nonhuman primates are considered natural hosts [88]. *Shigella* is a highly infectious microorganism since as few as 10–100 bacteria are sufficient to produce diarrhea in humans [89]. The infection is transmitted via the fecal-oral route, through direct contact or indirectly through contaminated food, water, or fomites, and by fly transmission [86, 90].

Infection may result in mild clinical signs with production of loose stools containing blood and mucus and histologic evidence of multifocal superficial colonic erosions with edema and acute inflammation that are more severe distally [91]. In severe cases infection results in bacillary dysentery in NHPs characterized clinically by weakness, prostration, and liquid stool with mucus and blood (Fig. 2.58). Histologically purulent, hemorrhagic exudate with pseudomembrane and shallow circumscribed ulcers (Figs. 2.59, 2.60, and 2.61) can develop in 24 hours and can become diffuse by 48 hours [92]. Thickening around ulcers is produced by edema, muscular hypertrophy, fibrosis, and lymphoid hyperplasia (Didier, unpublished) (Figs. 2.62 and 2.63). Ulcers may extend to the serosa and rarely perforate [86].

Complications in primates include ulcerative gastritis with gingival inflammation and progression to necrotizing gingivitis with alveolar bone loss [93]. Gingival forms may persist for prolonged periods and such animals can be carriers. Some animals that recover from the enteric form of disease may develop a neutrophilic mono- or polyarthritis.

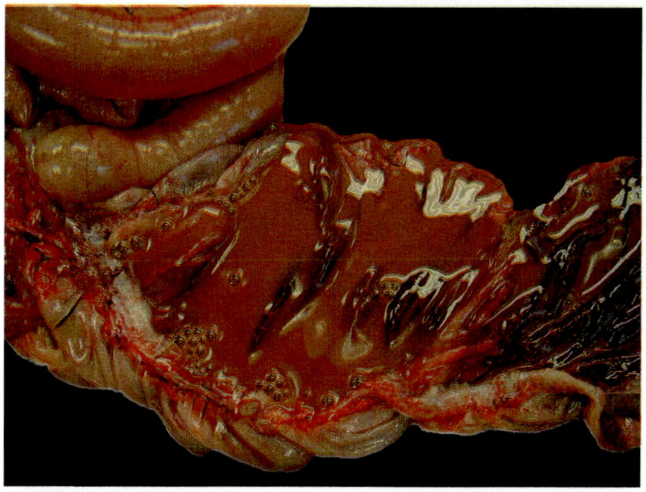

Fig. 2.58 *Shigella flexneri*. Gross pathology image of colon from rhesus macaque presented with weight loss and expanded abdomen due to gaseous distention of the colon. The mucosa is covered by thick necro-hemorrhagic purulent exudate. *S. flexneri* was cultured

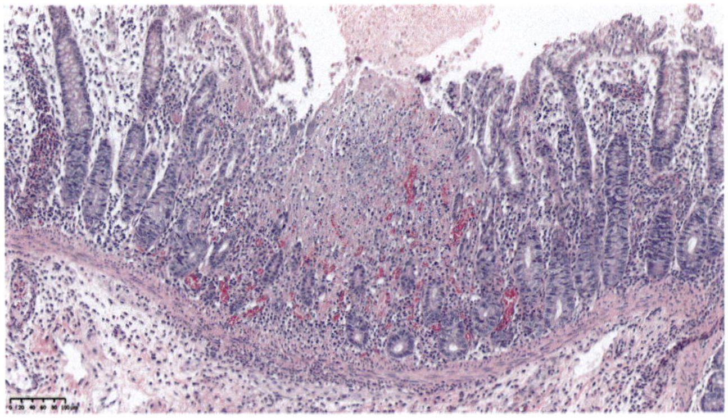

Fig. 2.59 *Shigella flexneri*. Photomicrograph of thickened colonic mucosa with focal necrosis infiltrated by polymorphonuclear cells and extending into the muscularis mucosa and submucosa and accompanied by multifocal hemorrhage and edema (rhesus macaque, colon, H&E stain)

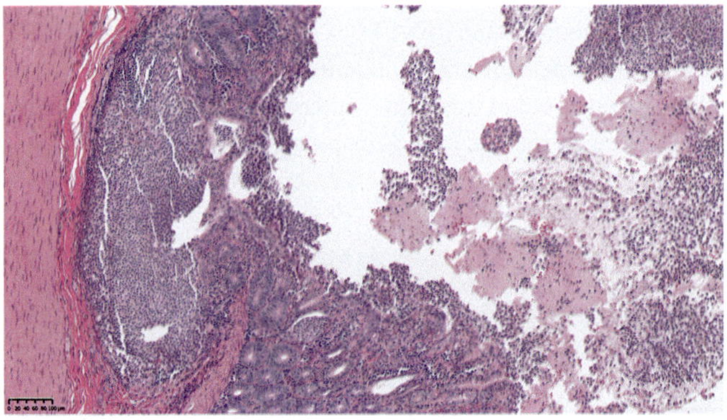

Fig. 2.60 *Shigella flexneri*. Photomicrograph of submucosal abscess subjacent to ulcerated colonic mucosa with fragments of a fibrinopurulent pseudomembrane in the lumen (rhesus macaque, colon, H&E stain)

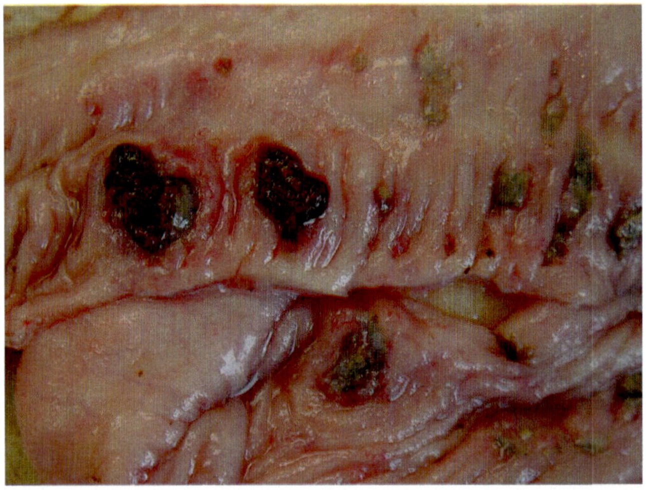

Fig. 2.61 *Shigella flexneri*. Macroscopic image shows multifocal, deep, chronic ulcerations of the middle and descending colon in rhesus macaque

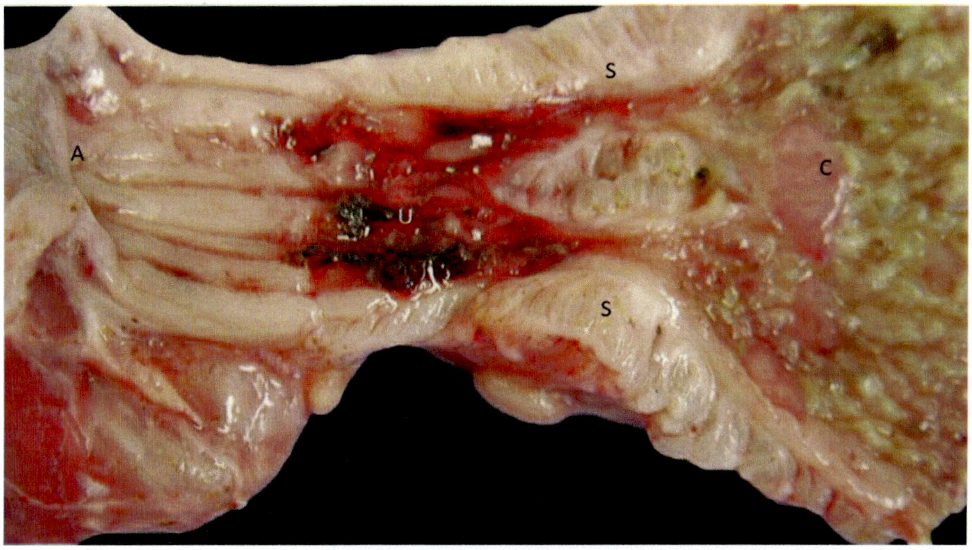

Fig. 2.62 *Shigella flexneri*. Macroscopic image of a stricture in the descending colon (S) of rhesus macaque with compensatory proximal dilatation of the middle and ascending colon (C). Hemorrhagic ulcerated mucosa (U) is noted between the stricture and the anus (A). (rhesus macaque, *S. flexneri* was cultured)

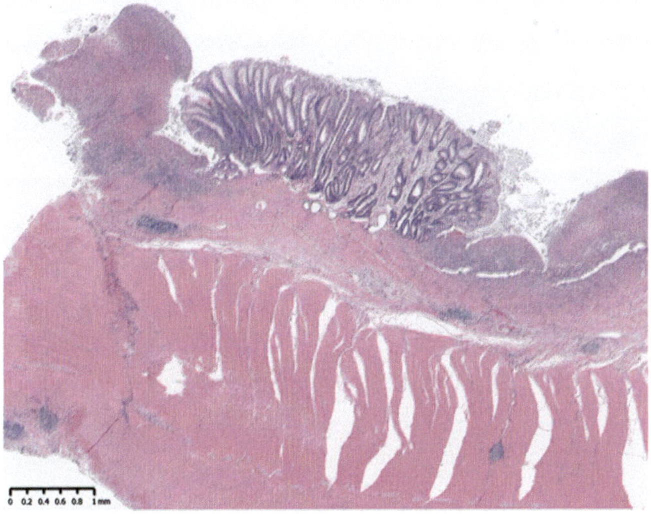

Fig. 2.63 *Shigella flexneri*. Photomicrograph of the stricture from Fig. 2.62. Ulcerated areas are covered by pseudomembrane surrounding residual hyperplastic mucosa. Submucosa is infiltrated with mixed inflammatory cells and thickened by accumulation of fibroblasts and collagen (fibrosis). Smooth muscle layers are hyperplastic and contain prominent lymphoid follicles. (rhesus macaque, colon, H&E stain)

The association in humans of shigellosis and spondyloarthropathies in individuals with HLA-B27 genetic markers has not been found in NHPs [94].

Diagnostic of shigellosis can be established by culture and PCR test.

2.3.4 *Leptospira*

Leptospiras are Gram-negative, tightly coiled, spiral-shaped bacteria. They are obligate aerobes which are found throughout the world and can occupy diverse environments and habitats with warm, humid conditions, high stagnant water, and during rainy seasons in warm-climate regions. The genus *Leptospira* contains pathogenic species, (e.g., *L. interrogans*), nonpathogenic saprophytes (e.g., *L. biflexa*), and species of indeterminate pathogenicity (e.g., *L. inadai*). Transmission of the spirochetes is through ingestion and direct contact with infected animals, skin wounds, and mucous membranes or placental transfer. Possible mechanism of infection is also the indirect transmission through contact with contaminated fomites, soil and water, and inhalation of microscopic droplets [95, 96].

Pathogenic *Leptospira* spp. are the cause of worldwide distributed zoonotic disease leptospirosis which now has been identified as emerging infectious disease. Rodents and other animals are chronic carriers that excrete pathogenic *Leptospiras* in their urine and contaminate the environment [97]. The organism can infect most mammalian species, including wildlife, domestic animals, humans, and nonhuman primates. Natural-acquired disease in NHPs is uncommon, but cases have been reported in squirrel monkeys, black-tufted marmoset, and capuchin monkeys [98–100]. Leptospiral seroprevalence was found in captive neotropical nonhuman primates suggestive that they are asymptomatic carriers [97, 101, 102]. The report of an outbreak in squirrel monkeys described acute illness with jaundice, hemorrhagic syndrome, and miscarriages in pregnant monkeys [98]. Natural cases of leptospirosis in humans and nonhuman primates often include flu-like symptoms and jaundice. Left untreated, leptospirosis can lead to multiorgan failure and ultimately, death. In a report of *Leptospira interrogans* infection in a free-ranging, black-tufted marmoset in Brazil gross and histopathological findings of icterus, pulmonary hemorrhage, interstitial nephritis, and hepatocellular dissociation are main pathological changes observed (Figs. 2.64 and 2.65). Diagnostic conformation is based on specific immunohistochemical and PCR assays for *Leptospira species* [99].

a

b

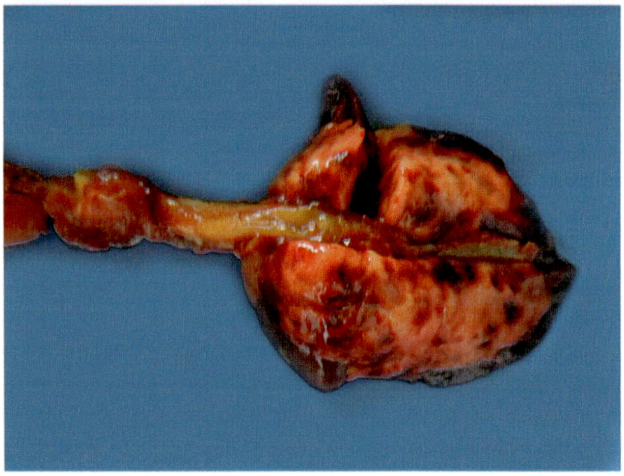

Fig. 2.64 *Leptospira interrogans*. Gross pathology images of fatal leptospirosis in a free-ranging marmoset. (**a**) Diffuse subcutaneous yellow discoloration-jaundice (asterisk). (**b**) Multifocal hemorrhages of the lung. (Courtesy of Prof. Marcio Botelho De Castro, College of Agronomy and Veterinary Medicine, University of Brasília, Brazil)

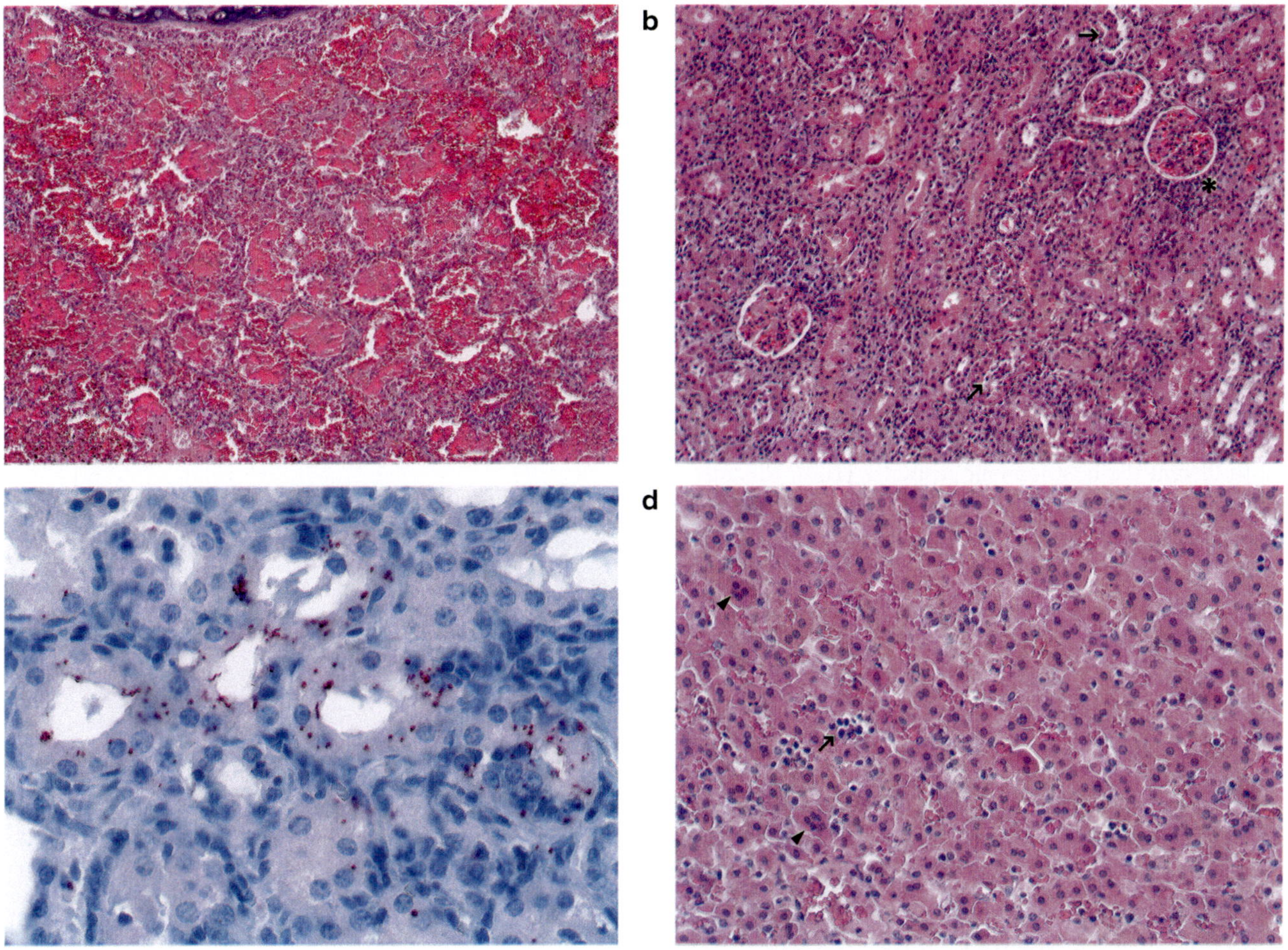

Fig. 2.65 *Leptospira interrogans*. Photomicrographs of pulmonary, hepatic, and renal lesions in the same free-ranging marmoset. (**a**) Severe intra-alveolar hemorrhage, edema, and fibrin in the lung (H&E). (**b**) Lymphoplasmacytic tubulointerstitial nephritis (asterisk), with tubular degeneration, necrosis, and luminal neutrophils and necrotic debris (arrowheads) (H&E). (**c**) Immunostaining of spirochetes and fragments within renal tubular epithelium and interstitium (IHC). (**d**) Mild hepatocellular dissociation with mild sinusoidal leucocytosis (arrow), and multinucleated hepatocytes (arrowhead) (H&E). (Courtesy of Prof. Márcio Botelho De Castro, College of Agronomy and Veterinary Medicine, University of Brasília, Brazil)

Mechanisms that *Leptospira* spp. use to cause disease are not clearly understood. Potential virulence factors include immune mechanisms, toxin production, adhesins, and other surface proteins. Diagnostic can be done by culture of the organism from affected animal, PCR test, and demonstration of rising antibody titers. Histologic diagnosis by silver impregnation staining and immunohistochemical staining offers high sensitivity and specificity. Most leptospirosis cases are diagnosed by serology. The reference standard assay is the microscopic agglutination test (MAT) [95].

2.3.5 *Yersinia*

Yersinia pseudotuberculosis

Yersinia enterocolitica

Yersinia spp. are Gram-negative pleomorphic facultative anaerobes in the family *Yersiniaceae*. *Y. pseudotuberculosis* and *Y. enterocolitica* are two species of *Yersinia* associated with diseases in nonhuman primates. Infections are described in variety of NWPs and OWPs. Transmission is by the fecal-oral route through contaminated food and water in a wide range of species including fish, reptiles, birds, nonhuman primates, and other mammals [103, 104]. Bird and rodent feces may be an important reservoir in NHP colonies housed outdoors [105, 106]. Clinical signs may be nonspecific and include diarrhea, dehydration, and depression. Acute morbidity and mortality are prominent in NWPs, but chronic debilitating cases are often observed. The infection can be associated with abortion and stillbirth. The disease can become enzootic in zoological settings with seasonal occurrence. Outbreaks of yersiniosis in NHPs are reported frequently and are usually characterized by ulcerative enterocolitis, mesenteric lymphadenopathy, and hepatic and/or splenic necrosis and abscess formation [104]. When bacteria cross the intestinal epithelium via M cells and replicate in the Peyer's patches and draining lymph nodes of the gut, this results in transmural intestinal ulceration and through the hepatic portal system and lymphatics leads to septicemic dissemination [107, 108]. Unique virulence factors of *Yersinia* spp. enable them to invade the enterocytes and to disarm the host immune response [109]. Macroscopically the lesions in the spleen and liver are recognized as spleno- or hepatomegaly with multifocal small (1–4 mm)

a

b

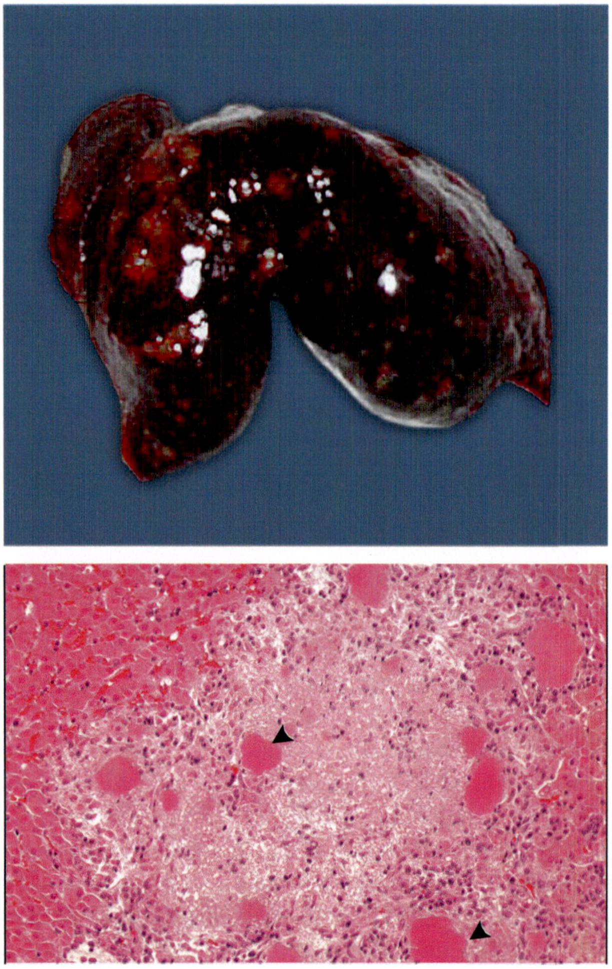

Fig. 2.66 *Yersinia enterocolitica*. (**a**) Macroscopic image of liver with multifocal hepatic necroses evident as white-yellow foci (common marmoset, liver). (**b**) Microscopic findings of necrotizing hepatitis. Focal area of necrosis comprised of granular eosinophilic and karyorrhectic debris accompanied by mixed inflammatory cells in the periphery, and large lobulated colonies (arrowheads) of coccobacilli (common marmoset, liver, H&E stain)

a

b

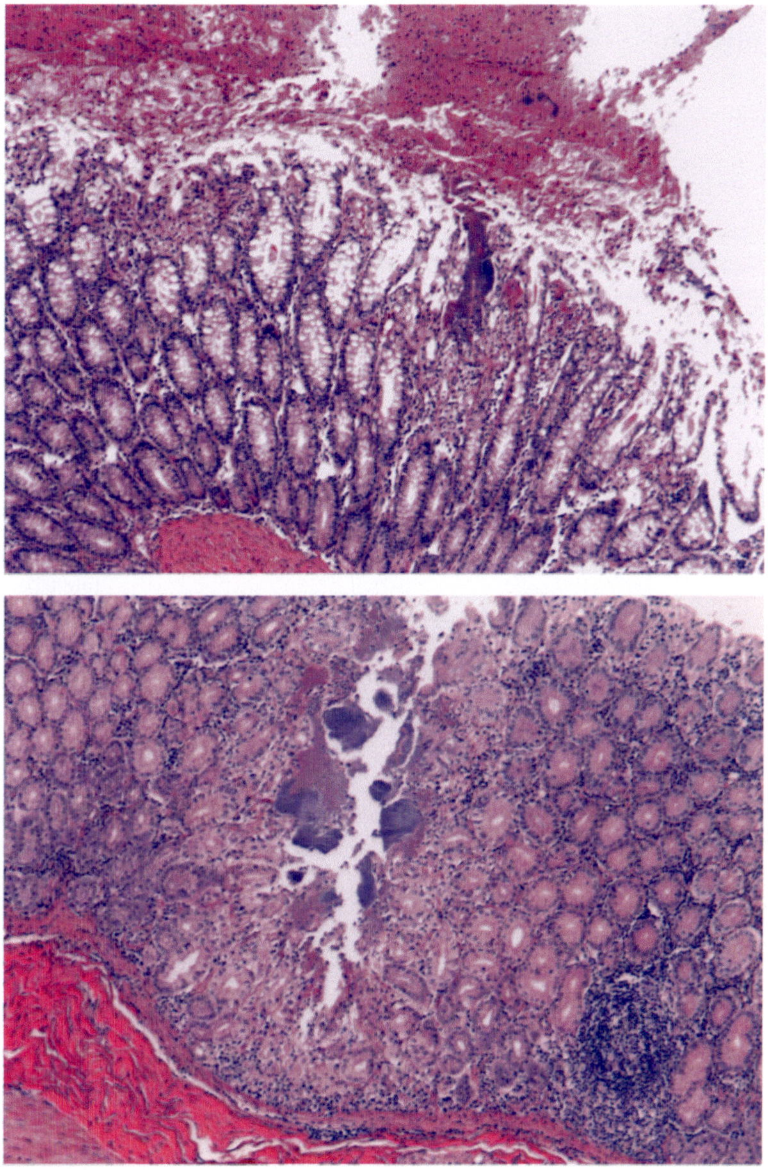

Fig. 2.67 *Yersinia pseudotuberculosis*. (**a**) Photomicrograph of necrosuppurative enteritis with superficial mucosal fibrinous pseudomembrane and large colonies of bacteria (pig-tailed macaque, jejunum, H&E stain). (**b**) Photomicrograph of necrosuppurative gastritis with penetrating ulcer and numerous large bacterial colonies (the same animal, stomach, H&E stain)

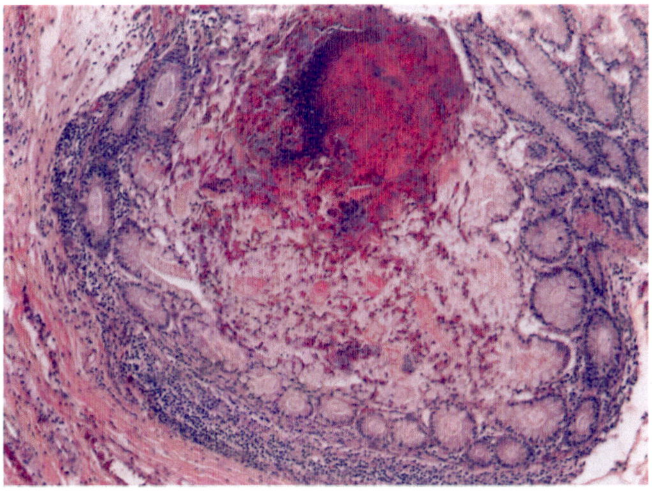

Fig. 2.68 *Yersinia pseudotuberculosis*. Microscopic findings of necrotizing colitis with prolapsed necrotic mucosa extended into the submucosal lymphoid tissue, with large colonies of coccobacilli (baboon, colon, H&E stain)

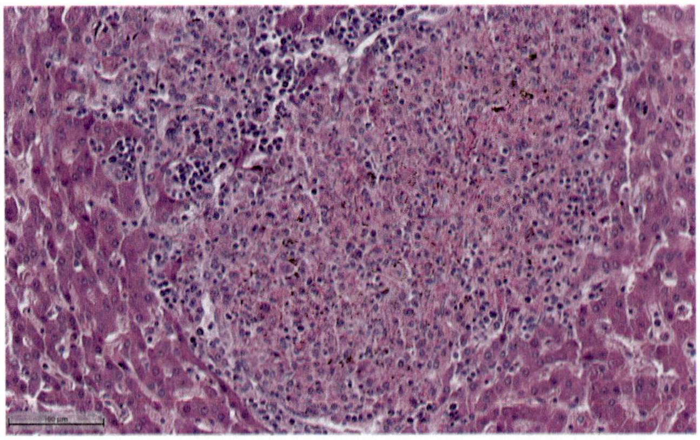

Fig. 2.69 *Yersinia pseudotuberculosis*. Photomicrograph of necrotic foci in the liver comprised of numerous epithelioid macrophages (granulomatous) admixed with cellular and necrotic debris, and inflammatory cells in the periphery (rhesus macaque, liver, H&E stain)

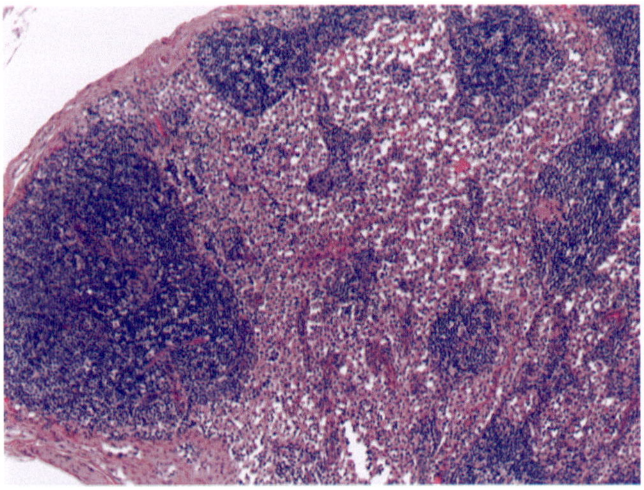

Fig. 2.70 *Yersinia pseudotuberculosis*. Photomicrograph of lymphadenopathy due to yersiniosis. Medullary sinuses contain increased numbers of histiocytes. Culture of the colon identified *Y. pseudotuberculosis*. Histiocytosis is not a specific response and may be observed with *Y. enterocolitica* and other pathogens (rhesus macaque, mesenteric lymph node, H&E stain)

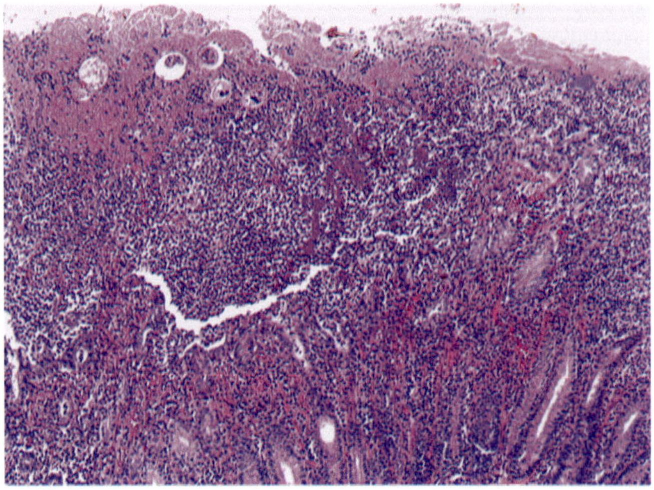

Fig. 2.71 *Yersinia enterocolitica*. Photomicrograph of ulcerative fibrinohemorrhagic colitis with intralesional large bacterial colonies (pig-tailed macaque, colon, H&E stain)

white-yellowish foci and histologically are characterized by necrosuppurative hepatitis (Fig. 2.66a, b), splenitis, and lymphadenitis [110]. Gross lesions of ulcerative enterocolitis involve reddened mucosa, confluent erosions, and ulcerations in the gut. Histologically, mucosal necrosis often overlays lymphoid follicles and contains necrotic debris, admixed with large colonies of coccobacilli (Figs. 2.67, 2.68, 2.69, 2.70, and 2.71) [111, 112]. Inflammatory response to *Y. pseudotuberculosis* is granulomatous rather than a histiocytic response to *Y. enterocolitica* [105]. Occasionally, the organism may be associated with pneumonitis and meningitis. The presence of large intralesional colonies of Gram-negative organisms within regions of necrosis is indicative of the diagnosis. Isolation (culture) and identification are required for confirmation and differentiation of *Y. enterocolitica* from *Y. pseudotuberculosis*.

2.3.6 *Klebsiella pneumoniae*

Klebsiella pneumoniae is a Gram-negative, nonmotile, lactose-fermenting bacillus with prominent capsule.

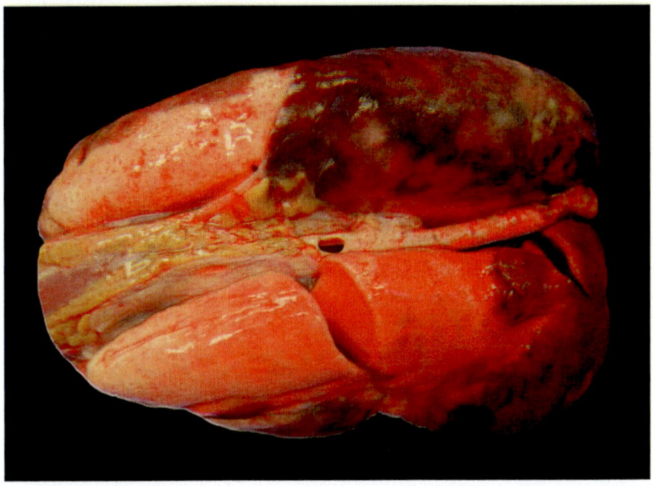

Fig. 2.72 *Klebsiella pneumoniae*. Gross pathology image of lung (dorsal aspect) from rhesus macaque inoculated with SIVmac251. The right lower pulmonary lobe is hemorrhagic and consolidated with similar but smaller areas of consolidation in the left lung

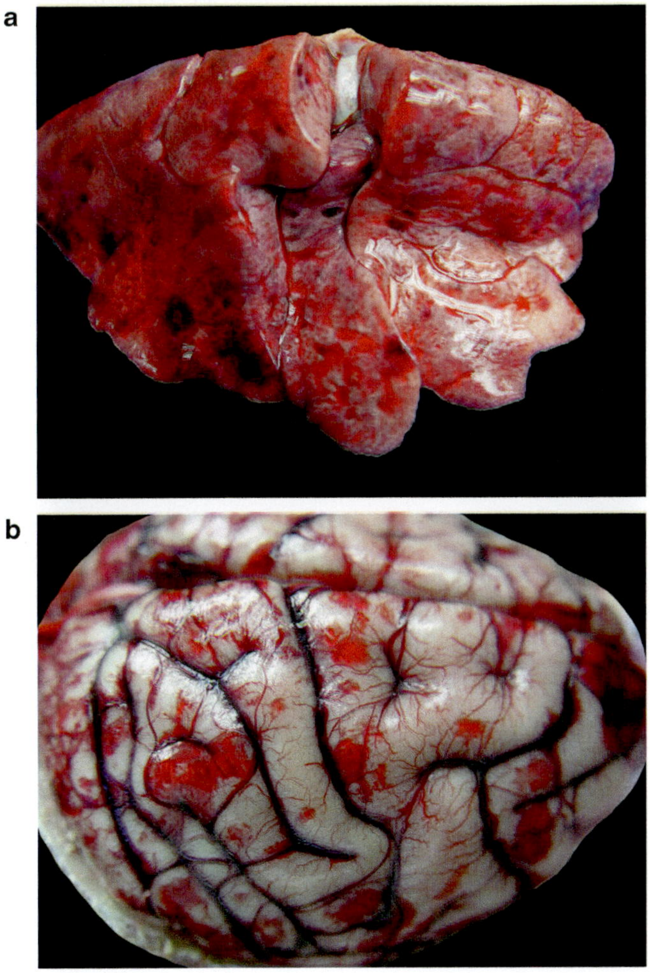

Fig. 2.73 *Klebsiella pneumoniae*. (**a**) Gross pathology image of lung from a mangabey with petechial and ecchymotic hemorrhages in all lobes. (**b**) Brain from the same animal with ecchymotic hemorrhages in the meninges

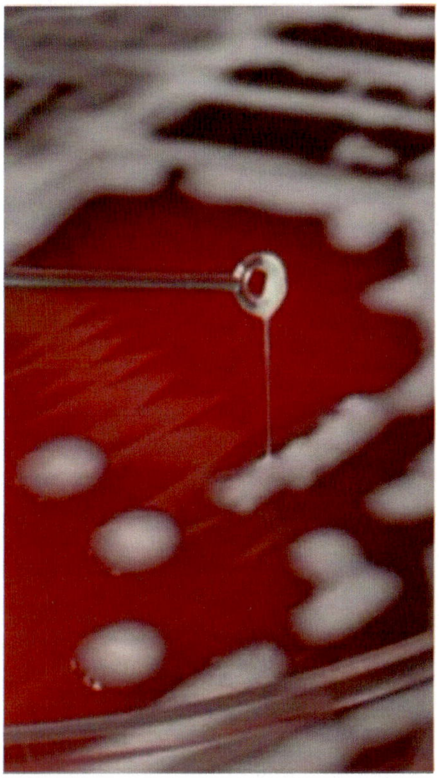

Fig. 2.74 *Klebsiella pneumoniae*. Hypermucoviscosity phenotype can be detected by classic "String Test" formation of mucoviscous string from a new colony in culture that is greater than 5 mm in length

Typically, it is part of the normal fecal and oral flora, found worldwide on surface mucosa of animals and in the environment but can act as an opportunistic agent causing septicemia, arthritis, peritonitis, pneumonia (Figs. 2.72 and 2.73a), and meningitis (Fig. 2.73b) in both OWPs and NWPs [113–115]. The organism usually is contracted from environmental sources such water and fomites. Pathogenic strains of invasive organisms like hypervirulent *K. pneumoniae* (hvKp) have been reported in humans and in some NHPs and have been associated with severe complications [113, 114, 116, 117]. These strains expressing virulence factors like rmpA (regulator of mucoid phenotype), mag4 (mucoviscosity gene), and $K_2$4 (capsule-associated gene) [118, 119] may acquire drug-resistance and can increase the risk of recurrent morbidity and difficult elimination. The hypermucoviscosity phenotype can be diagnosed by a positive string test (Fig. 2.74). *Klebsiella* infections are diagnosed by bacterial culture test and PCR-ELISA methods.

Klebsiella pneumoniae infections are reported in different NHPs including OWPs and NWPs [120–122]. In OWPs, *K. pneumoniae* usually affects individual animals and can cause secondary bacterial infection in animals with compromised immune system due to stress or concurrent disease. The organism uses virulence factors like pili, capsular polysaccharide to form biofilms that protect bacterial colonies from host compliment and antimicrobial peptides. There are known resistance genes for aminoglycosides, quinolones, B lactam, polymyxin, and tigecycline [113].

In OWP a clinical syndrome defined by pneumonia and meningitis produces nonspecific clinical signs like dyspnea, anorexia, and lethargy. Lungs contain multiple abscesses,

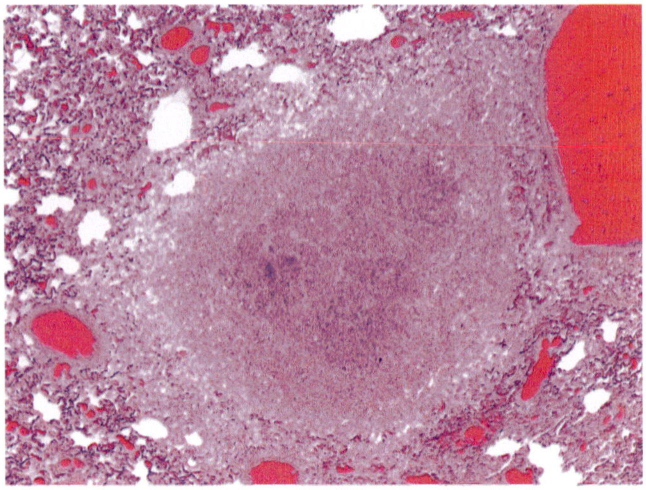

Fig. 2.75 *Klebsiella pneumoniae*. Photomicrograph of necrotic abscess formation with central coagulative necrosis in the lung of rhesus macaque (H&E stain)

a

b

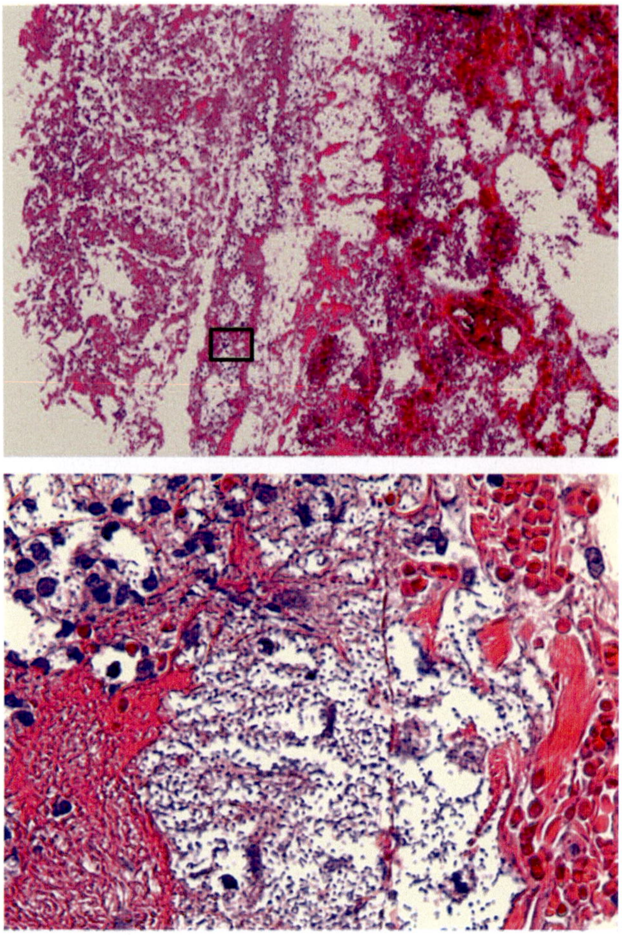

Fig. 2.76 *Klebsiella pneumoniae*. (**a**) Photomicrograph of the lung from a SIV-infected rhesus macaque with hemorrhagic pneumonia and fibrinous pleuritis (H&E stain). (**b**) Enlargement of box in (**a**) demonstrating large numbers of Individualized bacteria within subpleural alveolar spaces with characteristic individual spacing due to the capsule (H&E stain)

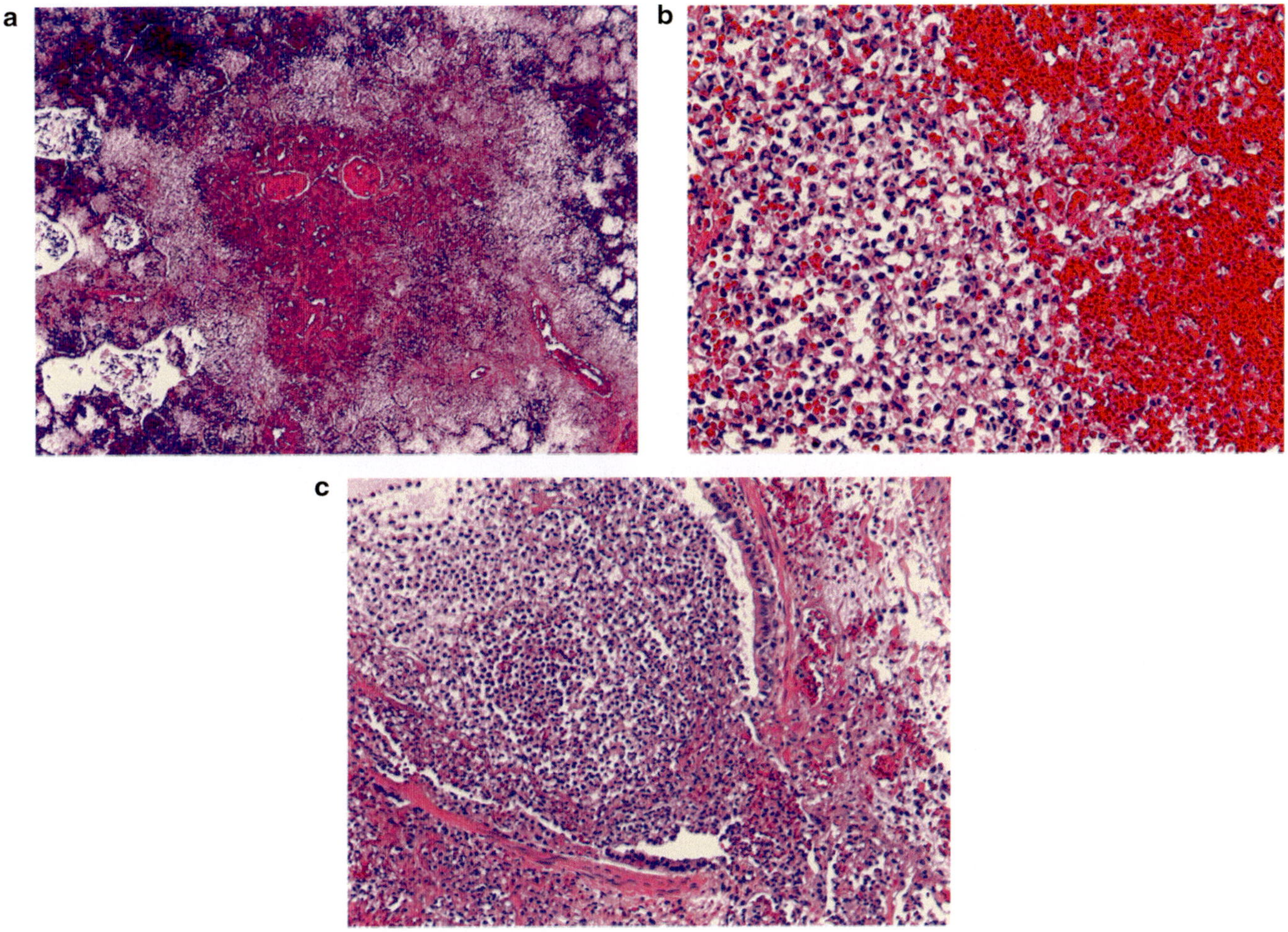

Fig. 2.77 *Klebsiella pneumoniae*. (**a**) Photomicrograph of abscess formation in the lung of 5-year-old rhesus macaque shows aggregation of polymorphonuclear infiltrates in an area of severe hemorrhage (H&E stain). (**b**) Inset. Enlargement of (**a**) shows collection of degenerating polymorphonuclear cells and macrophages on the edge of the hemorrhagic zone (H&E stain). (**c**) Photomicrograph of distended large airway with suppurative exudate which demonstrates airway disease as well as sepsis (H&E stain)

lobar areas of dark consolidation, and sheets of fibrin on pleural surfaces. Suppurative peritonitis accompanies pneumonia with thick yellow to tan fluid. Abscess in the lung and liver have extensive central necrosis filled with fluid, necrotic debris, and bacteria. Surrounding parenchyma is infiltrated with polymorphonuclear cells and macrophages (Figs. 2.75, 2.76, and 2.77) [122]. In the brain meninges are hemorrhagic accompanied by encephalitis, meningitis, optic neuritis, and all associated with suppuration, intralesional bacteria, perivascular cuffing, and fibrinoid vasculitis. In these cases, the organism may be more difficult to observe but can be visualized in impression smears obtained from exudates. In NWPs common clinical syndromes include pneumonia, enterocolitis, and neonatal septicemia. Environmental contamination with virulent strains in facilities housing NWPs may result in epizootics of high morbidity and mortality. A gastrointestinal route of entry is most frequently suspected.

2.3.7 *Francisella tularensis* (Tularemia)

Francisella tularensis is a pleomorphic Gram-negative coccobacillus, and it is the causative agent of tularemia, known as rabbit fever or deer-fly fever. Some nonpathogenic members of the genera are found in the environment. Natural infections of NHPs has not been reported often but cases has been described in some NWPs [123, 124] and OWPs [125, 126]. The organism is transmitted by arthropods and biting ticks and fleas, ingestion of contaminated food and water, and aerosols from infected tissues fluids [123, 127]. Because of its low infectious dose, ability to be aerosolized, and potentially fatal outcome, *F. tularensis* is considered a Class A select agent [127, 128].

Multiple clinical forms are described based on disease characteristics linked to the route of exposure: ulceroglandular, glandular, oculoglandular, oropharyngeal, pneumonic, typhoidal, and septic. The typhoidal form describes a systemic illness without anatomic location [123, 128]. In squirrel monkeys white necrotic foci are present in the liver, spleen, and mesenteric fat with hemorrhage in the lung and kidney that can include fibrinous to suppurative foci [123]. In cynomolgus monkeys lingual ulcers and abscess with submandibular lymphadenopathy accompanied multifocal necrotizing hepatitis, splenitis, and granulomatous bronchopneumonia [129]. In young rhesus macaque, white necrotic variably sized nodules in the liver, spleen, and lymph nodes were accompanied by lymphatic and intravascular thrombosis, and multifocal septal necrosis in the lung with neutrophilic and histiocytic infiltration [126].

Several NHP species have been used to characterize an aerosol model of infection. Necrotizing to pyogranulomatous lesions were produced in the lung, lymph nodes, spleen, and bone marrow [130]. In the marmoset model, severe hemorrhage and suppurative bronchopneumonia, multifocal pyogranulomatous hepatitis, splenitis, and lymphadenitis were observed. Primates are the only experimental model that develop ulcerations of the skin and lymphadenitis [131]. Diagnosis of tularemia requires a high degree of clinical suspicion and can be confirmed by serology and culture.

2.3.8 *Campylobacter*

Campylobacter spp. are Gram-negative, microaerophilic, motile bacteria with curved or spiral shape and one of the most common causes of human gastroenteritis worldwide. The disease caused by the organism is zoonotic, and transmission occurs by the fecal-oral route through ingestion. The sources of infections include wild and domestic birds, mammals, and contaminated water [132, 133]. In primate colonies, *Campylobacter* spp. are associated with recurring enterocolitis and are a main cause of morbidity [134]. The infection closely mimics the disease and immune response seen in humans. The bacteria are commonly isolated from OWPs especially macaques. The incidence of *C. jejuni* in sick monkeys is higher (70% v 51%) in multiple species and 92% in sick rhesus macaques [135]. However, it must be kept in mind that *Campylobacter* spp. are commonly isolated from healthy macaques not associated with disease or pathology. The most important isolates from NHPs are *C. jejuni* and *C. coli*. The clinical spectrum of *Campylobacter* varies from mild watery diarrhea to severe dysentery according to the present strain and toxins released.

Culture results from natural infections in the colony of Tulane Primate Center tend to support partial immunity observations. Frequently, rhesus macaques with persistent or periodic episodes of diarrhea often produce initial isolates of *C. jejuni* followed by subsequent isolation of *C. coli* and rarely *C. fetus* as individual or combined isolates (Didier, unpublished). A lack of normal flora may be noted. The typical lesions in these animals are mild to moderate lymphoplasmacytic colitis with suppurative cryptitis and lymphoid hyperplasia in submucosa and mesenteric lymph nodes (Fig. 2.78). The small intestine is involved occasionally and histologically may exhibit villous blunting and fusion (Fig. 2.79). It is also observed that some animals presenting with these signs and lesions lack identifiable pathogens in their gut. These animals fail to respond to

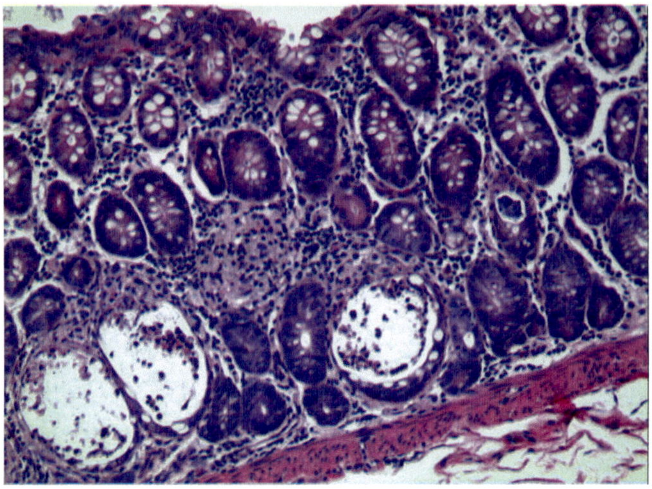

Fig. 2.78 *Campylobacter coli.* Photomicrograph of colonic mucosa with lymphoplasmacytic and histiocytic colitis, suppurative cryptitis, and reduction in goblet cells in 1.6-year-old animal with clinical history of profuse, watery diarrhea (rhesus macaque, colon, H&E stain)

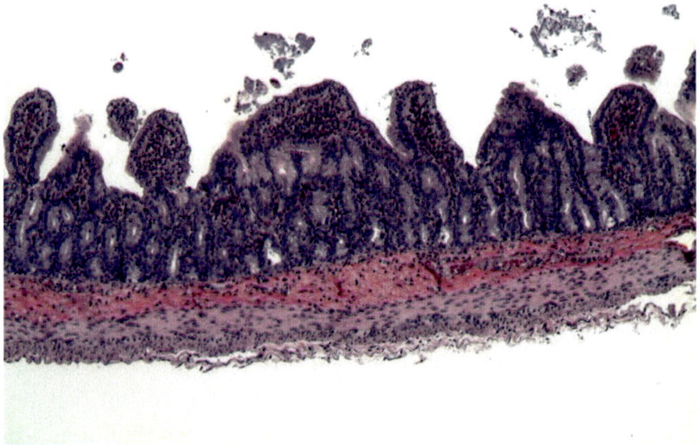

Fig. 2.79 *Campylobacter jejuni.* Photomicrograph of lymphoplasmacytic enteritis in 3.5-year-old animal. Small intestinal mucosa exhibits severe villous blunting and fusion (rhesus macaque, jejunum, H&E stain)

treatment and the term idiopathic chronic diarrhea (ICD) has been used. Diseases such a campylobacteriosis in NHP may be due as much to distortions in commensal flora that cross feed and support invasion by opportunists prolonging inflammation.

In NWPs *Campylobacter* is often associated with disease [136]. Clinically affected animals have watery diarrhea that may contain blood or mucous. The animals most likely become infected from human handlers or through direct contact with OWPs or contaminated fomites. Histological lesions reveal neutrophilic colitis with occasional crypt abscesses. In captivity strict separation of NWPs and OWPs should be maintained.

Diagnosis of *Campylobacter* infections can be challenging because the organism is difficult to isolate, grow, and identify. Several methods for detection of *Campylobacter* include culture, stool immunoassays for *Campylobacter*-specific antigen, and molecular tests [132].

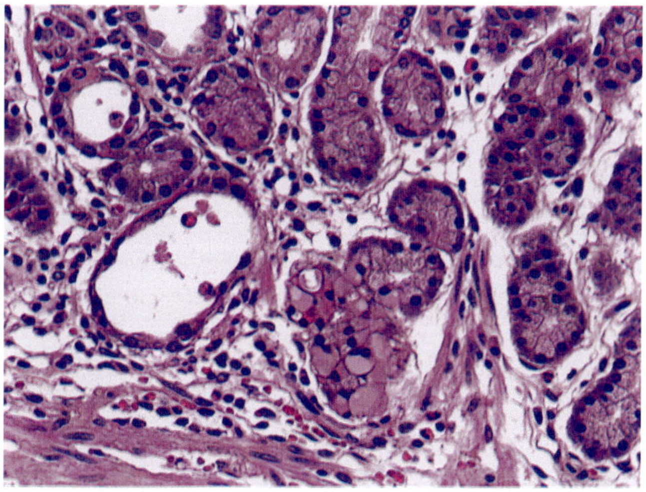

Fig. 2.81 *Helicobacter* sp. Photomicrograph of gastric mucosa from a 12-year-old rhesus macaque with ectatic glands lined by metaplastic epithelium. Adjacent glandular epithelium has severe hydropic swelling (rhesus macaque, stomach, H&E stain)

2.3.9 *Helicobacter*

Helicobacter is a genus of Gram-negative, microaerophilic, curved or spiral bacteria that has been isolated from the gastrointestinal mucosa of many mammals including humans and NHPs. The number of species in the genus is rapidly expanding, and the frequent changes in the nomenclature are causing perplexity [137]. NHPs can be infected with diverse helicobacter organisms, some of which may represent commensals that may cause disease in stressed or immunocompromised animals (Figs. 2.80, 2.81, 2.82, 2.83, and 2.84).

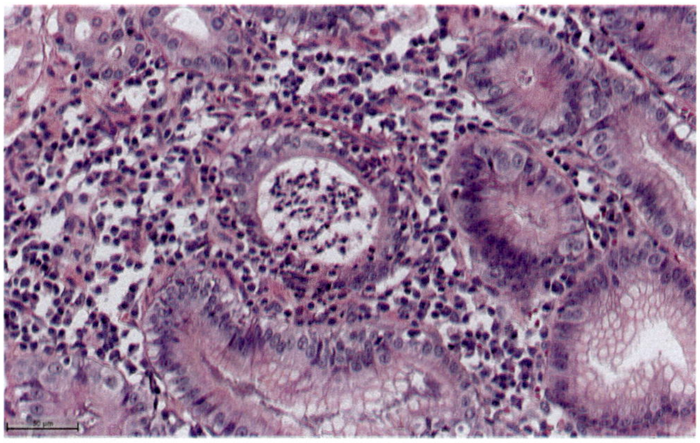

Fig. 2.80 *Helicobacter* sp. Photomicrograph of gastric mucosa from SIV-infected rhesus macaque. The mucosa exhibits polymorphonuclear (PMN) infiltrates in the lumen of a gastric gland, intraepithelial PMN cells, and squamous metaplasia of glandular epithelium (rhesus macaque, stomach, H&E stain)

a

b

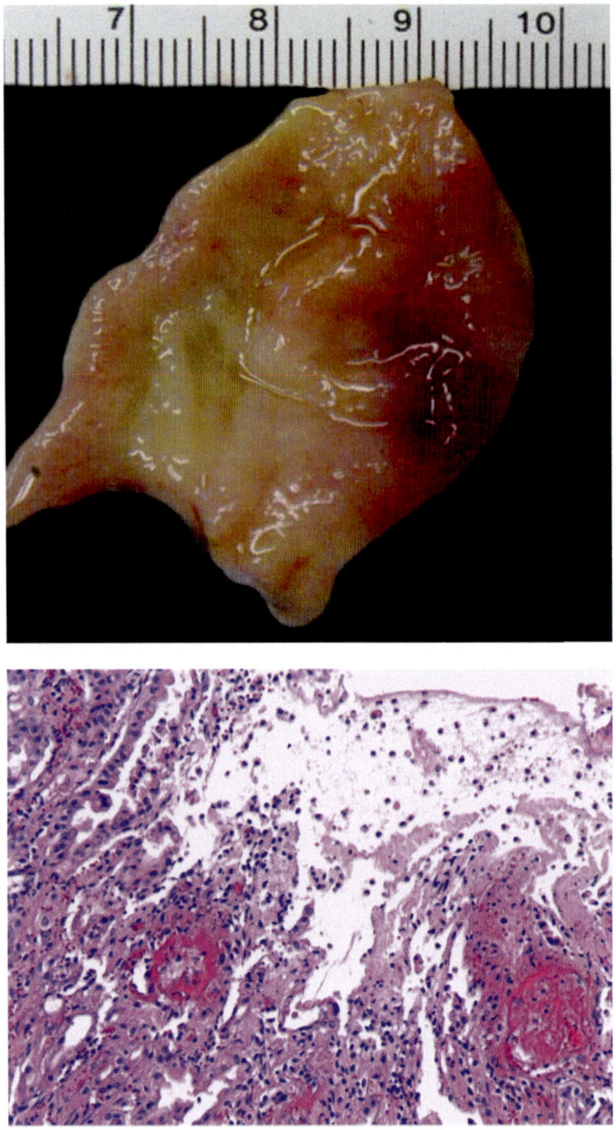

Fig. 2.82 *Helicobacter* sp. (**a**) Macroscopic image of stomach from SIV-infected rhesus macaque presented with a focal hemorrhagic ulcer. (**b**) Histological examination reveals necrosis of superficial and glandular epithelium, accompanied by hemorrhage and fibrin deposition, and infiltration of polimorphonuclear cells and macrophages (rhesus macaque, stomach, H&E stain)

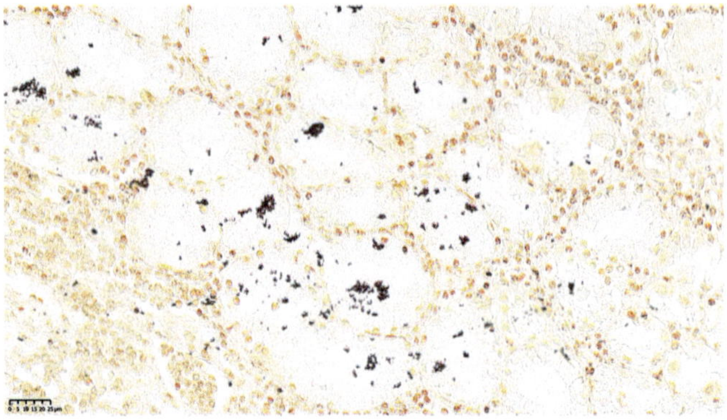

Fig. 2.83 *Helicobacter* sp. Photomicrograph of gastric mucosa from baboon with individual and clumps of *Helicobacter* in the lumen and on the surface of glandular epithelium (baboon, stomach, Warthin-Starry stain)

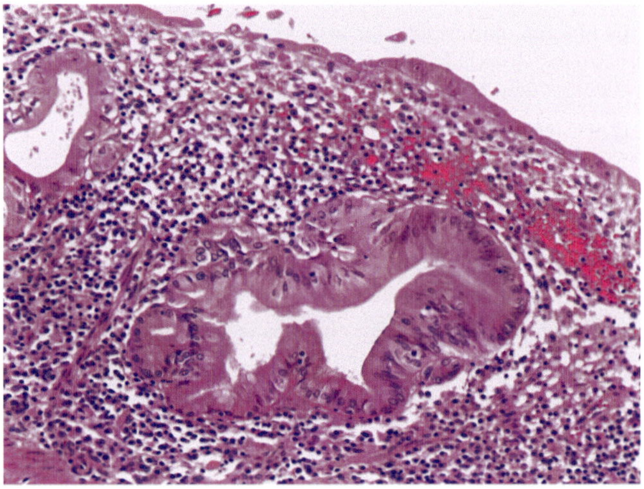

Fig. 2.84 *Helicobacter* sp. Photomicrograph of gastric mucosa from SIV-infected rhesus macaque with a partially healed ulcer covered with metaplastic squamous epithelium. Glandular epithelium is reduced and residual glands are hyperplastic. Lamina propria contains large numbers of lymphocytes and plasma cells (rhesus macaque, stomach, H&E stain)

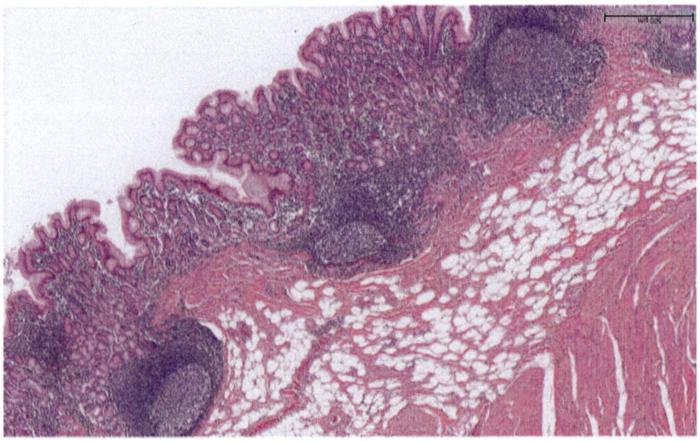

Fig. 2.85 *Helicobacter* sp. Photomicrograph of gastric mucosa from rhesus macaque with moderate lymphoplasmacytic gastritis and prominent nodular lymphoid hyperplasia, a sequel of *Helicobacter* infection (rhesus macaque, stomach, H&E stain)

H. pylori, a spiral-shaped bacteria 3–4 μm × 0.5–1 μm, oxidase and urease positive, was first found in humans and implicated as a causative organism of mild to moderate gastritis, duodenal ulcer, and a risk factor for adenocarcinoma and mucosa-associated lymphoma [137, 138]. Infections in the stomach of OWPs due to *H. pylori* are frequently asymptomatic. Natural acquisition of the infection occurs early during the first year of life with 40% of the newborn infected by 12 weeks and more than 90% of animals infected by 1 year of age. The infection is through close contact with infected mothers, and it is most consistent with an oral-oral means of transmission [139]. In NHP lesions include minimal to moderate lymphoplasmacytic gastritis with intraglandular and luminal spiral bacteria, infrequent PMN infiltration of the lamina propria, and lumina of the gastric glands [140]. Sometimes, changes like gastric gland hyperplasia, depleted mucin production, squamous metaplasia, cytoplasmic vacuolation, glandular atrophy, and lymphonodular hyperplasia are observed (Figs. 2.81, 2.83, and 2.85). The loss of mucin can be transient [141].

Other helicobacter species, like *H. heilmannii* (formerly *Gastrospirillum hominis*), are larger tightly coiled gastric helicobacter, 7–9 μm × 0.1–1 μm in size located in the mucus of the surface epithelium, in the lumina of gastric pits and glands of the stomach fundus and body. *H. heilmannii* are reported in OWPs and produce minimal gastritis or parietal cell damage and hyperchlorhydria [142, 143]. Typically, there is less intense gastritis, absence of epithelial damage, infrequent lymphoid hyperplasia, and epithelial metaplasia. There are more than 32 species with this highly coiled morphology that can be grouped by 16 SrRNA analysis into two types: Type 1 is exemplified by *H. suis* reported in rhesus macaque [140] and Type 2, a larger group that includes *H. felis*, *H. bizzozeronii*, and *H. salomonis* [137].

In addition to the gastric helicobacter species, there are reports of helicobacter isolates from intestinal and hepatobiliary tracts that are called enterohepatic *Helicobacter species* (EHS) [144]. A novel EHS isolated from cotton-top tamarins was associated with chronic colitis and colon cancer [145]. Multiple EHS including *H. macacae* [146] of rhesus and cynomolgus monkeys were linked to idiopathic colitis and years later to adenocarcinoma suggesting that persistent EHS infection in aged NHP may have a role in intestinal neoplasia [144].

Pathogenicity features of helicobacter are encoded in the Cag pathogenicity island that utilize 27 genes and a Type IV secretion system [147]. Factors include outer membrane proteins (OMP), BabA, a binding protein for fucosylated blood group antigens on gastric epithelium, LPS, phospholipase, mucinase that disrupts gastric mucus coat, and urease.

Different *Helicobacter* spp. can be histologically distinguished on the base of their shape (*H. heilmannii* appear more tightly coiled), size, and location in the gastric lesions. Additional diagnostic tests include silver stain, IHC, and PCR.

2.3.10 *Burkholderia*

The genus *Burkholderia* contains two species of medical significance, *B. pseudomallei* and *B. mallei*. Both are classified as Tier 1 select agents and share 99% genetic homology [148, 149]. The *Burkholderia* genome codes for a long list of virulence factors including capsule, flagella, pili, LPS, quorum sensing, and secretory systems. *B. pseudomallei* and *B. mallei* are diagnosed by culture from blood, sputum, urine, and throat swab and by serology.

Burkholderia pseudomallei

B. pseudomallei is a Gram-negative aerobic, motile bacterium, and environmental saprophyte found in soils and water in tropical and subtropical locations most frequently in Asia and Australia. The organism is the causative agent of an important zoonotic infectious disease melioidosis [150, 151]. Incubation period can be up to 21 days with a mortality rate from 16 to 50%. Melioidosis may also remain latent for years before symptoms develop. Transmission can

a

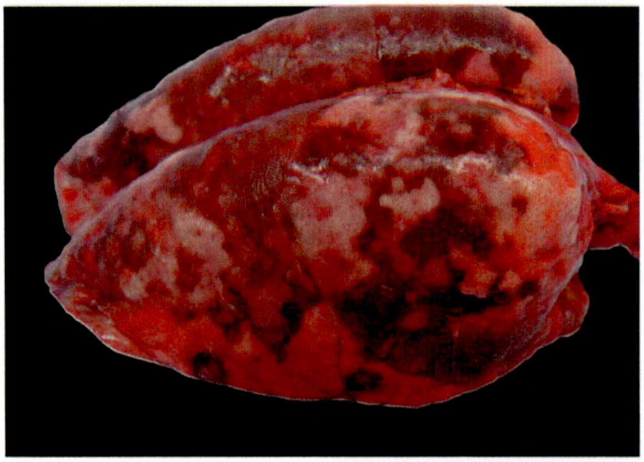

b

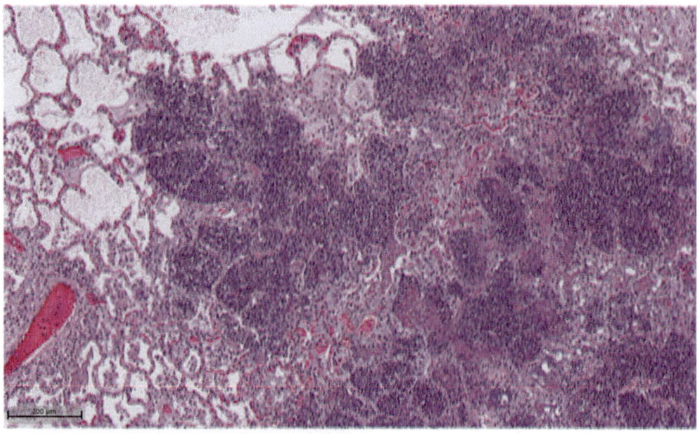

c

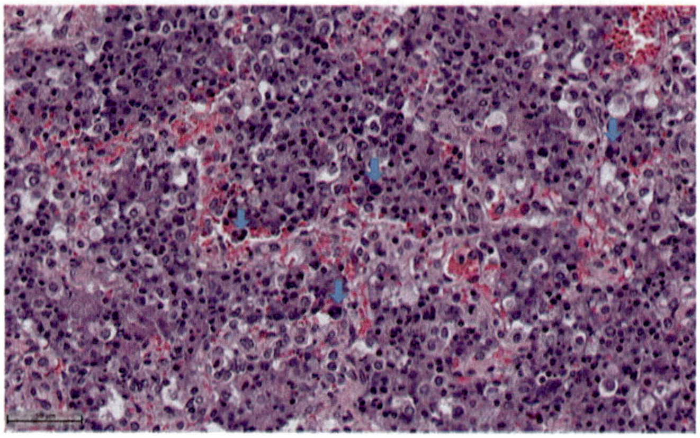

Fig. 2.86 *Burkholderia*. (**a**) Gross image of fibrinopurulent, hemorrhagic bronchopneumonia in rhesus macaque experimentally challenged by aerosol with *B. pseudomallei* 16 days earlier. (**b**) Photomicrograph shows marked pulmonary polimorphonuclear infiltrates and macrophages accompanied by fibrin thrombi, edema, hemorrhage, and congestion (rhesus macaque, lung, H&E stain). (**c**) Alveoli are filled with degenerated and necrotic PMN infiltrates, macrophages, and degenerated multinucleated giant cells (arrows) (rhesus macaque, lung, H&E stain)

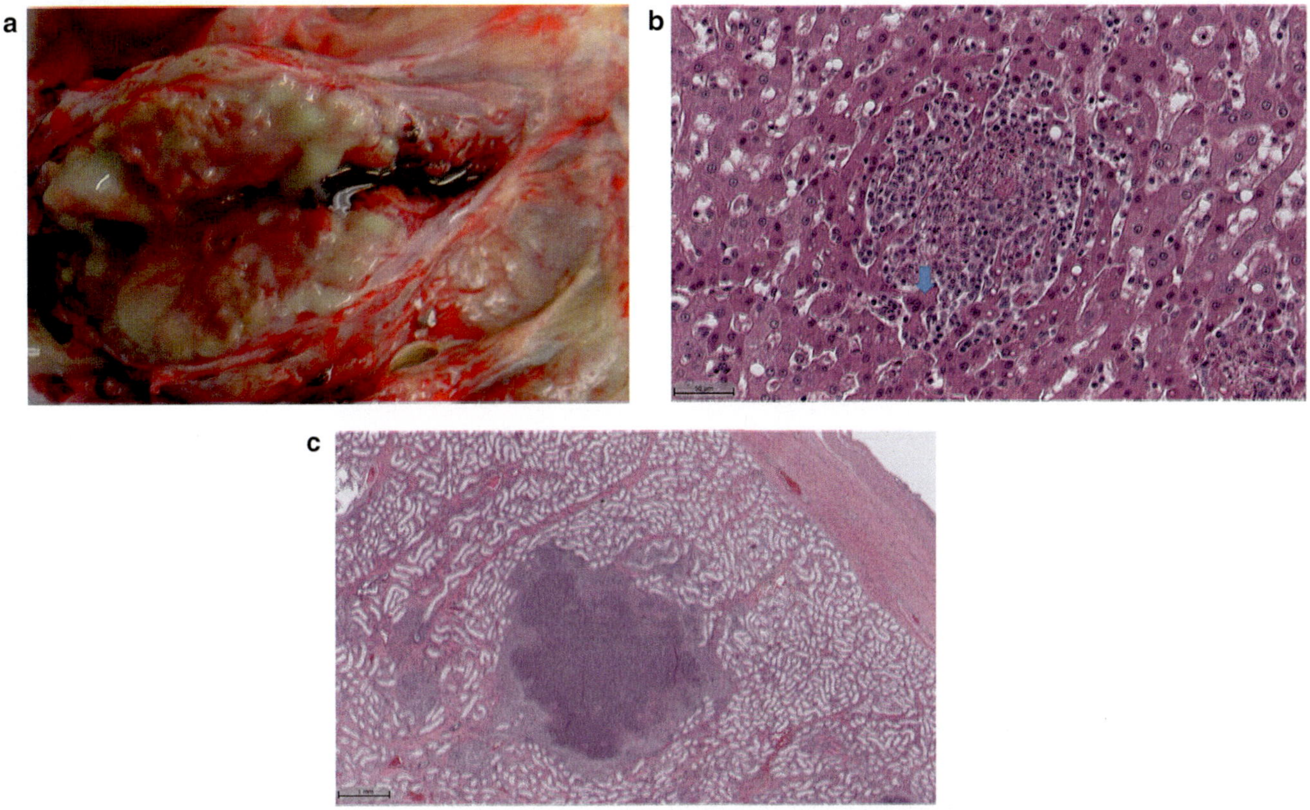

Fig. 2.87 *Burkholderia*. (**a**) Gross image of seminal vesicle from rhesus macaque that developed a spontaneous *B. pseudomallei* infection with septicemia and draining abscess in the inguinal skin. The gland is swollen and contains numerous suppurative foci. (**b**) Photomicrograph of necrotizing hepatic granuloma with multinucleated cells (arrow) (rhesus macaque, liver, H&E stain). (**c**) Photomicrograph of necrotizing granulomatous orchitis with effacement of seminiferous tubules and stroma (rhesus macaque, testis, H&E stain)

be through cutaneous wounds, inhalation, and ingestion [148, 151]. The disease is characterized by necrotizing abscesses in multiple tissues and is common in zoo animals in endemic areas with cases reported in multiple species of NHPs including stump-tailed macaques, rhesus macaques, and apes [151].

Clinical signs vary and depend on the primary system involved. In a zoological setting, NHPs with melioidosis exhibit lethargy, fever, lymphadenopathy, fibrinopurulent pneumonia, and coalescing abscesses in the spleen, liver, and less often the heart and testicle (Fig. 2.86a–c) and (Fig. 2.87a–c). Orangutans seem highly susceptible with death occurring within a week of signs [151]. Histological lesions include suppurative inflammation and multifocal abscesses. Exudate in the lung is purulent to mucopurulent with the presence of multinucleated giant cells.

Burkholderia mallei

B. mallei is a Gram-negative, aerobic, nonmotile bacillus that infects horses, mules, and donkeys. It is a host-adapted bacterium in its equine reservoir. The organism is the causative agent of zoonotic disease glanders. Transmission is through skin wounds and the respiratory route. Humans, goats, dogs, cats, rabbits, and carnivorous predators can become infected.

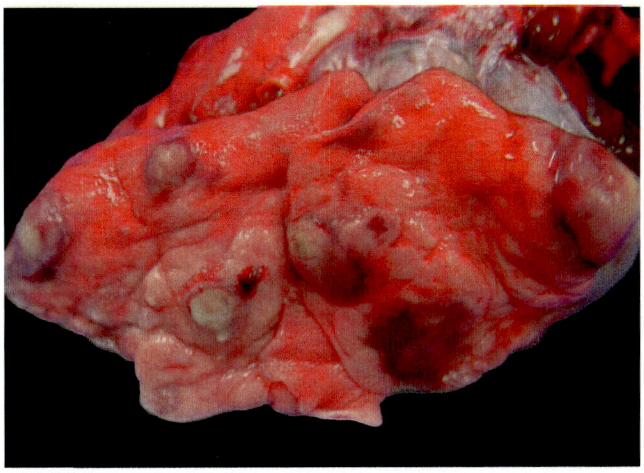

Fig. 2.88 *Burkholderia*. Gross image of the lung from rhesus macaque experimentally challenged with *B. mallei* 12 days earlier. Multiple large pyogranulomas often associated with hemorrhage are present in all lobes

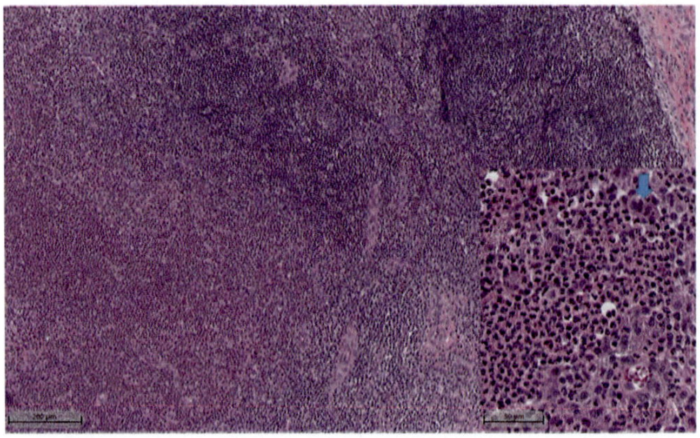

Fig. 2.89 *Burkholderia*. Photomicrograph of bronchial lymph node from rhesus macaque experimentally challenged by aerosol with *B. mallei* 20 days prior to collection (H&E stain). Medullary sinuses contain large numbers of polymorphonuclear cells mixed with macrophages and occasional multinucleated giant cells (arrow, inset, H&E stain)

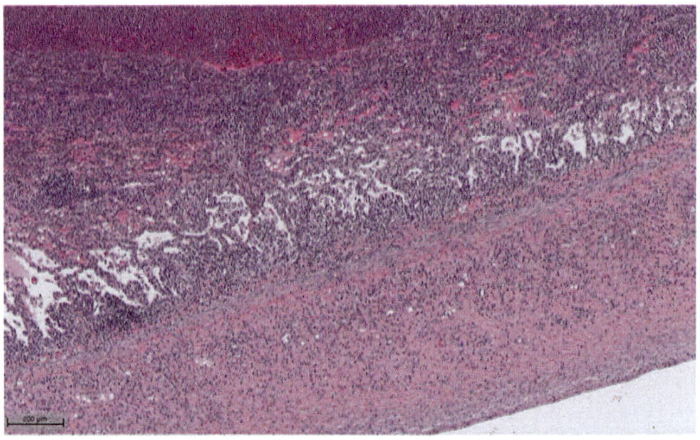

Fig. 2.90 *Burkholderia*. Photomicrograph of lung from rhesus macaque experimentally challenged by aerosol with *B. mallei* 20 days earlier. Pleura is thickened by mononuclear inflammatory cells and fibrosis. Adjacent alveolar epithelium is hyperplastic, and the underlying granuloma contains a central core of PMN infiltrates (rhesus macaque, lung, H&E stain)

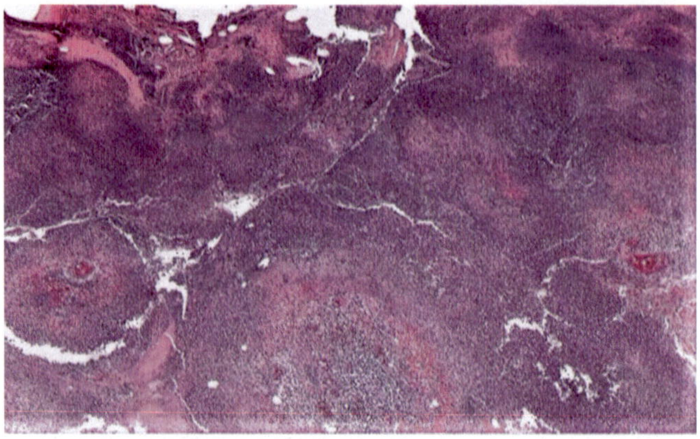

Fig. 2.91 *Burkholderia*. Photomicrograph of necrotizing pyogranulomatous dermatitis. The animal received aerosolized *B. mallei* 22 days prior to collection and developed a focal ulcer on the back. Necrotic epithelium overlies multiple pyogranulomas in the upper dermis (rhesus macaque, skin, H&E stain)

Glanders is difficult to diagnose and treat. Reports of *B. mallei* infection in NHPs are limited to experimental work. Four days after respiratory exposure, marmosets develop signs of fever, myalgia, fatigue, and lymphadenopathy and progress to pneumonia with necrotizing abscesses and dissemination to the liver, spleen, nasopharynx, and lymph nodes [152]. Lesions are mucopurulent to pyogranulomatous with multinucleated giant cells. In rhesus macaques clinical signs include fever, myalgia, head ache, and fatigue followed by diarrhea and weight loss before pneumonia and pulmonary abscess become apparent (Figs. 2.88, 2.89, 2.90, and 2.91) [153]. Histologically on day 1 polymorphonuclear infiltrates are present in alveoli followed by necrosuppurative bronchopneumonia and pleuritis. Longer surviving rhesus macaques have lesions with lymphoplasmacytic inflammation, fibrosis, and type II pneumocytes hyperplasia [154].

a

b

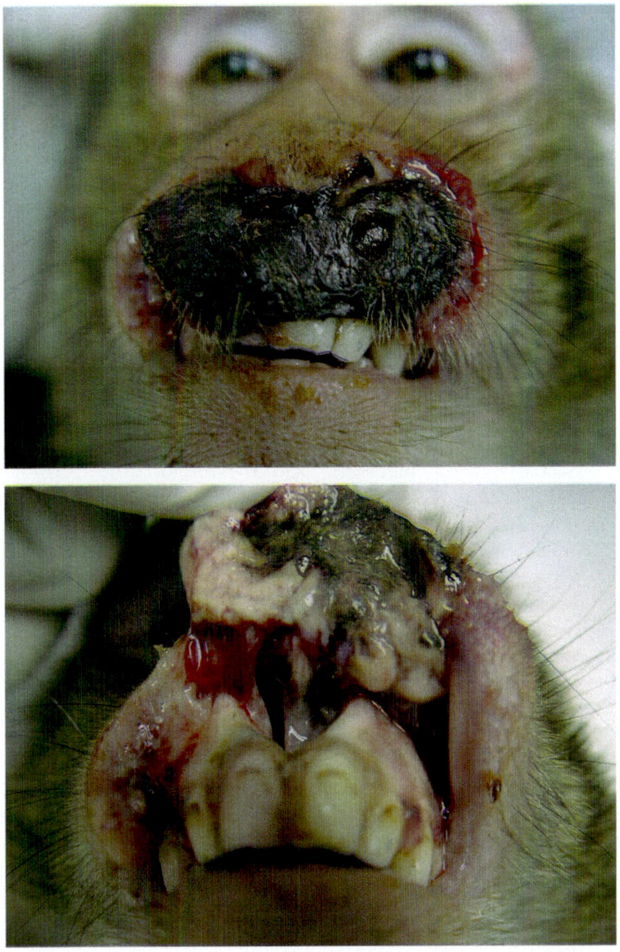

Fig. 2.92 (**a**) Macroscopic image of NOMA. Avascular necrosis of the face and lip in a 6-year-old rhesus macaque. Hemolytic coagulase positive *S. aureus* along with *Providencia rettgeri*, *Stenotrophomonas maltophilia*, and *Moraxella* spp. were cultured from the nasal and lip wound. (**b**) Retraction of the lip exhibits retraction of the gums and dissolution of the nasal septum

2.4 Polymicrobial (NOMA)

Acute necrotic stomatitis, called NOMA, is described in rhesus macaques, tamarins, and cynomolgus monkeys and is attributed to combination infections by staphylococcus, streptococcus, micrococcus, *Fusobacterium necrophorum* with viral initiators like SRV. NOMA presents as expansive necrotizing ulcerative gingivitis with destruction of facial soft tissue and oronasofacial bone. Poor nutrition or poor dental hygiene are identified as underlying diseases (Fig. 2.92a, b) [155]. Histologically, ulcerative lesions progress to coagulative necrosis with infiltration of polymorphonuclear cells, macrophages, and giant cells in underlying collagen, muscle, and bone [156, 157]. Methicillin-resistant *S. aureus* (MRSA) have been associated with NOMA and reported in rhesus monkeys [158]. Mandibular osteomyelitis is described in a squirrel monkey and derived from an infected tooth [159].

Acknowledgment Peter Didier would like to acknowledge his colleagues Carol Coyne and Chris Mabee for the histology support.
Ivanela Kondova-Perseng would like to acknowledge Boudewijn Ouwerling for providing histology and microbiology support; Tom Haaksma, Wim Collignon, and Monica Almagro for their assistance during necropsies; Saxwin G. Brouwer and Billy B. van Diemen for their excellent IT services; Francisca van Hassel for the technical support; and Prof. Jan Langermans and Prof. Ronald Bontrop for the support and discussions.

References

1. Bennett BT, Abee CR, Henrickson R. Nonhuman primates in biomedical research: diseases. New York: Elsevier; 1998.
2. Kaushal D, Mehra S, Didier PJ, Lackner AA. The non-human primate model of tuberculosis. J Med Primatol. 2012;41(3):191–201. https://doi.org/10.1111/j.1600-0684.2012.00536.x.
3. Flynn JL, Gideon HP, Mattila JT, Lin PL. Immunology studies in non-human primate models of tuberculosis. Immunol Rev. 2015;264(1):60–73.
4. Rodgers MA, Ameel C, Ellis-Connell AL, Balgeman AJ, Maiello P, Barry GL, Friedrich TC, Klein E, O'Connor SL, Scanga CA. Preexisting Simian immunodeficiency virus infection increases susceptibility to tuberculosis in Mauritian Cynomolgus macaques. Infect Immun. 2018;86(12):e00565.
5. Vervenne RAW, Jones SL, van Soolingen D, van der Laan T, Andersen P, Heidt PJ, Thomas AW, Langermans JAM. TB diagnosis in non-human primates: comparison of two interferon-gamma assays and the skin test for identification of Mycobacterium tuberculosis infection. Vet Immunol Immunopath. 2004;100:61–71.
6. Von Reyn CF, Williams DE, Horsburgh CR Jr, Jaeger AS, Marsh BJ, Haslov K, Magnusson M. Dual skin testing with Mycobacterium avium sensitin and purified protein derivative to discriminate pulmonary disease due to M. avium complex from pulmonary disease due to Mycobacterium tuberculosis. J Infect Dis. 1998;177:730–6.
7. Pena JC, Ho WZ. Non-human primate models of tuberculosis. Microbiol Spectrum. 2016;4(4):TBTB2-0007.
8. Maiello P, DiFazio RM, Cadena AM, Rodgers MA, Lin PL, Scanga CA, Flynn JL. Rhesus macaques are more susceptible to progressive tuberculosis than cynomolgus macaques: a quantitative comparison. Infect Immun. 2018;86(2):505–15.
9. Langermans JAM, Andersen P, van Soolingen D, Vervenne RAW, Frost PA, van der Laan T, van Pinxteren LAH, van den Hombergh J, Kroon S, Peekel I, Florquin S, Thomas AW. Divergent effect of bacillus Calmette-Guerin (BCG) vaccination on Mycobacterium tuberculosis infection in highly related macaque species: implications for primate models in tuberculosis vaccine research. PNAS. 2001;98(20):11497–502.
10. Shanmugasundaram U, Bucsan AN, Ganatra S, Ibegbu C, Quezada M, Blair RV, Alvarez VV, Kaushal D, Rengarajan J. Pulmonary Mycobacterium tuberculosis control associates with CXCR3- and CCR6-expressing antigen-specific Th1 and Th17 cell recruitment. JCI Insight. 2020;2020:137858. https://doi.org/10.1172/jci.insight.137858.
11. Kaushal D, Foreman TW, Gautam US, Alvarez X, Adekambi T, Rangel-Moreno J, Golden NA, Johnson AF, Phillips BL, Ahsan M, Russell-Lodrigue KE, Doyle LA, Roy CJ, Didier PJ, Blanchard J,

Rengarajan J, Lackner AA, Khader SA, Mehra S. Mucosal vaccination with attenuated Mycobacterium tuberculosis induces strong central memory responses and protects against tuberculosis. Nat Commun. 2015;6:8533. https://doi.org/10.1038/ncomms9533.
12. Mansfield KG, Pauley D, Young HL, Lackner AA. Mycobacterium avium complex in macaques with AIDS is associated with a specific strain of simian immunodeficiency virus and prolonged survival after primary infection. J Infect Dis. 1995;172:1149–52.
13. Maslow JN, Brar I, Smith G, Newman GW, Mehta R, Thornton C, Didier P. Latent infection as a source of disseminated disease with organisms of the Mycobacterium avium complex in SIV-infected rhesus macaques. J Infect Dis. 2003;149(Pt 11):3193–202.
14. Valverde CR, et al. Spontaneous leprosy in a wild-caught cynomolgus macaque. Int J Lepr Other Mycobact Dis. 1998;66 (2):140–8.
15. Brammer DW, et al. Mycobacterium kansasii infection in squirrel monkeys (Saimiri sciureus). J Med Primatol. 1995;24:231–5.
16. Lamond NM, Freitag NE. Vertical transmission of Listeria monocytogenes: probing the balance between protection from pathogens and fetal tolerance. Pathogens. 2018;7(2):52.
17. Lemoy MJ, Lopes DA, Reader JR, Westworth DR, Tarara RP. Meningoencephalitis due to Listeria monocytogenes in a pregnant rhesus macaque (Macaca mulatta). Comp Med. 2012;62(5): 443–7.
18. Struthers JD, Kucerova Z, Finley A, Goe A, Huffman J, Phair K. Septicaemic Listeriosis in a white-faced Saki (Pithecia pithecia). J Comp Pathol. 2022;194:7–13. https://doi.org/10.1016/j.jcpa.2022.03.003.
19. Li M, Brokaw A, Furuta AM, Coler B, Obregon-Perko V, Chahroudi A, Wang HY, Permar SR, Hotchkiss CE, Golos TG, Rajagopal L, Adams Waldorf KM. Non-human primate models to investigate mechanisms of infection-associated fetal and pediatric injury, teratogenesis and stillbirth. Front Genet. 2021;12:680342. https://doi.org/10.3389/fgene.2021.680342.
20. Wolfe B, Wiepz GJ, Schotzko M, Bondarenko GI, Durning M, Simmons HA, Mejia A, Faith NG, Sampene E, Suresh M, Kathariou S, Czuprynski CJ, Golos TG. Acute fetal demise with first trimester maternal infection resulting from listeria monocytogenes in a nonhuman primate model. mBio. 2017 Feb 21;8(1):e01938-16. https://doi.org/10.1128/mBio.01938-16.
21. Smith MA, Takeuchi K, Anderson G, Ware GO, McClure HM, Raybourne RB, Mytle N, Doyle MP. Dose-response model for Listeria monocytogenes-induced stillbirths in nonhuman primates. Infect Immun. 2008 Feb;76(2):726–31. https://doi.org/10.1128/IAI.01366-06.
22. Ayoade F, Alam MU. Rhodococcus Equi. [Updated 2022 Jul 18]. In: StatPearls [Internet]. Treasure Island, FL: StatPearls Publishing. 2022 Jan. https://www.ncbi.nlm.nih.gov/books/NBK441978/
23. Weinstock DM, Brown AE. Rhodococcus equi: an emerging pathogen. Clin Infect Dis. 15 May 2002;34(10):1379–85. https://doi.org/10.1086/340259.
24. Hondalus MK. Pathogenesis and virulence of Rhodococcus equi. Vet Microbiol. 1997;56(3–4):257–68.
25. Stein FJ, Stott G. Corynebacterium equi in the cottontop marmoset (Saguinus oedipus): a case report. Lab Anim Sci. 1979 Aug;29(4): 519–20.
26. Lin WV, et al. Diagnosis and management of pulmonary infection due to Rhodococcus equi. Clin Microbiol Infect. 2019;25(3): 310–5.
27. Stammen RL, Cohen JK, Meeker TL, Crane MM, Amara RR, Hicks SL, et al. Effect of chronic social stress on prenatal transfer of antitetanus immunity in captive breeding rhesus macaques (Macaca mulatta). J Am Assoc Lab Anim Sci. 2018;57(4):357–67.
28. Springer DA, Phillippi-Falkenstein K, Smith G. Retrospective analysis of wound characteristics and tetanus development in captive macaques. J Zoo Wildl Med. 2009 Mar;40(1):95–102. https://doi.org/10.1638/2008-0055.1.
29. Nakano T, Nakamura S, Yamamoto A, Takahashi M, Une Y. Tetanus as cause of mass die-off of captive Japanese macaques, Japan, 2008. Emerg Infect Dis. 2012 Oct;18(10):1633–5. https://doi.org/10.3201/eid1810.120503.
30. Rawlins RG, Kessler MJ. A five-year study of tetanus in the Cayo Santiago rhesus monkey colony: behavioral description and epizootiology. Am J Primatol. 1982;3(1–4):23–39.
31. Kessler MJ, Berard JD, Rawlins RG, Bercovitch FB, Gerald MS, Laudenslager ML, Gonzalez-Martinez J. Tetanus antibody titers and duration of immunity to clinical tetanus infections in free-ranging rhesus monkeys (Macaca mulatta). Am J Primatol. 2006;68(7):725–31.
32. Kessler MJ, Hernández Pacheco R, Rawlins RG, Ruiz-Lambrides-A, Delgado DL, Sabat AM. Long-term effects of tetanus toxoid inoculation on the demography and life expectancy of the Cayo Santiago rhesus macaques. Am J Primatol. 2015 Feb;77(2): 211–21. https://doi.org/10.1002/ajp.22323.
33. Arnon SS, Mills DC, Day PA, Henrickson RV, Sullivan NM, Wilkins TD. Rapid death of infant rhesus monkeys injected with Clostridium difficile toxins A and B: physiologic and pathologic basis. J Pediatr. 1984;104(1):34–40.
34. Rolland RM, Chalifoux LV, Snook SS, Ausman LM, Johnson LD. Five spontaneous deaths associated with Clostridium difficile in a colony of cotton-top tamarins (Saguinus oedipus). Comp Med. 1997;47(5):472–6.
35. Armstrong AR, Wünschmann A, Rigatti LH, Klein EC. Clostridium difficile enterocolitis in a captive Geoffroy's spider monkey (Ateles geoffroyi) and common marmosets (Callithrix jacchus). Vet Pathol. 2019 Nov;56(6):959–63. https://doi.org/10.1177/0300985819864307.
36. Fujiwara K. Tyzzer's disease. Jpn J Exp Med. 1978;48:467–80.
37. Harkness JE, Murray KA, Wagner JE. Biology and diseases of guinea pigs. Lab Anim Med. 2007;2002:203–46. https://doi.org/10.1016/B978-012263951-7/50009-0.
38. Niven JS. Tyzzer's disease in laboratory animals. Z Versuchstierkd. 1968;10(1):168–74.
39. Sasseville VG, Simon MA, Chalifoux LV, Lin KC, Mansfield KG. Naturally occurring Tyzzer's disease in cotton-top tamarins (Saguinus oedipus). Comp Med. 2007;57(1):125–7.
40. Snook S, Reimann K, King N. Neonatal mortality in cotton top tamarins: failure of passive immunoglobin transfer. Vet Pathol. 1992;29:444.
41. Carlson CJ, Kracalik IT, Ross N, Alexander KA, Hugh-Jones ME, Fegan M, Elkin BT, Epp T, Shury TK, Zhang W, Bagirova M, Getz WM, Blackburn JK. The global distribution of Bacillus anthracis and associated anthrax risk to humans, livestock and wildlife. Nat Microbiol. 2019 Aug;4(8):1337–43. https://doi.org/10.1038/s41564-019-0435-4.
42. Gogarten JF, et al. Tropical rainforest flies carrying pathogens form stable associations with social nonhuman primates. Mol Ecol. 2019;28(18):4242–58.
43. Hoffmann C, Zimmermann F, Biek R, Kuehl H, Nowak K, Mundry R, Agbor A, Angedakin S, Arandjelovic M, Blankenburg A, Brazolla G, Corogenes K, Couacy-Hymann E, Deschner T, Dieguez P, Dierks K, Düx A, Dupke S, Eshuis H, Formenty P, Yuh YG, Goedmakers A, Gogarten JF, Granjon AC, McGraw S, Grunow R, Hart J, Jones S, Junker J, Kiang J, Langergraber K, Lapuente J, Lee K, Leendertz SA, Léguillon F, Leinert V, Löhrich T, Marrocoli S, Mätz-Rensing K, Meier A, Merkel K, Metzger S, Murai M, Niedorf S, De Nys H, Sachse A, van Schijndel J, Thiesen U, Ton E, Wu D, Wieler LH, Boesch C, Klee SR, Wittig RM, Calvignac-Spencer S, Leendertz FH. Persistent anthrax as a major driver of wildlife mortality in a

tropical rainforest. Nature. 2017 Aug 2;548(7665):82–6. https://doi.org/10.1038/nature23309.
44. Gochenour WS Jr, Sawyer WD, Henderson JE, Gleiser CA, Kuehne RW, Tigertt WD. On the recognition of Woolsorter's disease. J Hyg (Lond). 1963 Sep;61(3):317–22. https://doi.org/10.1017/s0022172400039590.
45. Borio L, Frank D, Mani V, Chiriboga C, Pollanen M, Ripple M, Ali S, DiAngelo C, Lee J, Arden J, Titus J, Fowler D, O'Toole T, Masur H, Bartlett J, Inglesby T. Death due to bioterrorism-related inhalational anthrax: report of 2 patients. JAMA. 2001 Nov 28;286(20):2554–9. https://doi.org/10.1001/jama.286.20.2554.
46. Bush LM, Abrams BH, Beall A, Johnson CC. Index case of fatal inhalational anthrax due to bioterrorism in the United States. N Engl J Med. 2001 Nov 29;345(22):1607–10. https://doi.org/10.1056/NEJMoa012948.
47. Gleiser CA, Berdjis CC, Hartman HA, Gochenour WS. Pathology of experimental respiratory anthrax in Macaca mulatta. Br J Exp Pathol. 1963 Aug;44(4):416–26.
48. Twenhafel NA, Leffel E, Pitt ML. Pathology of inhalational anthrax infection in the african green monkey. Vet Pathol. 2007 Sep;44(5):716–21. https://doi.org/10.1354/vp.44-5-716.
49. Vasconcelos D, Barnewall R, Babin M, Hunt R, Estep J, Nielsen C, Carnes R, Carney J. Pathology of inhalation anthrax in cynomolgus monkeys (Macaca fascicularis). Lab Investig. 2003 Aug;83(8):1201–9. https://doi.org/10.1097/01.lab.0000080599.43791.01.
50. Popescu NI, Girton A, Burgett T, Lovelady K, Coggeshall KM. Monocyte procoagulant responses to anthrax peptidoglycan are reinforced by proinflammatory cytokine signaling. Blood Adv. 2019 Aug 27;3(16):2436–47. https://doi.org/10.1182/bloodadvances.2019000513.
51. Davis KL, Gonzalez O, Kumar S, Dick EJ Jr. Pathology associated with Streptococcus spp. infection in Baboons (Papio spp.). Vet Pathol. 2020 Sep;57(5):714–22. https://doi.org/10.1177/0300985820941496.
52. Krzyściak W, Pluskwa KK, Jurczak A, Kościelniak D. The pathogenicity of the Streptococcus genus. Eur J Clin Microbiol Infect Dis. 2013 Nov;32(11):1361–76. https://doi.org/10.1007/s10096-013-1914-9.
53. Mansfield KJ, Fox JG. Bacterial diseases. In: Marini R, Wachtman L, Tardif S, Mansfield K, Fox J, editors. Common marmoset in captivity and biomedical research. New York: Elsevier; 2019. p. 265–87. https://doi.org/10.1016/B978-0-12-811829-0.00016-9.
54. Graczyk TK, et al. Fulminant Streptococcus pneumoniae meningitis in a lion-tailed macaque (Macaca silenus) without detected signs. J Wildl Dis. 1995;31(1):75–8.
55. Lowenstein LJ, Osborn KG. Respiratory system diseases of nonhuman primates, Chapter 9. In: Abee CR, Mansfield K, Tardif S, Morris T, editors. Nonhuman primates in biomedical research. New York: Academic Press; 2012. p. 413–69.
56. Reyes LF, Restrepo MI, Hinojosa CA, Soni NJ, Shenoy AT, Gilley RP, Gonzalez-Juarbe N, Noda JR, Winter VT, de la Garza MA, Shade RE, Coalson JJ, Giavedoni LD, Anzueto A, Orihuela CJ. A non-human primate model of severe pneumococcal pneumonia. PLoS One. 2016 Nov 17;11(11):e0166092. https://doi.org/10.1371/journal.pone.0166092.
57. Brown AO, Millett ER, Quint JK, Orihuela CJ. Cardiotoxicity during invasive pneumococcal disease. Am J Respir Crit Care Med. 2015 Apr 1;191(7):739–45. https://doi.org/10.1164/rccm.201411-1951PP.
58. Reyes LF, Restrepo MI, Hinojosa CA, Soni NJ, Anzueto A, Babu BL, Gonzalez-Juarbe N, Rodriguez AH, Jimenez A, Chalmers JD, Aliberti S, Sibila O, Winter VT, Coalson JJ, Giavedoni LD, Dela Cruz CS, Waterer GW, Witzenrath M, Suttorp N, Dube PH, Orihuela CJ. Severe pneumococcal pneumonia causes acute cardiac toxicity and subsequent cardiac remodeling. Am J Respir Crit Care Med. 2017 Sep 1;196(5):609–20. https://doi.org/10.1164/rccm.201701-0104OC.
59. Schiller CA, Wolff MJ, Munson L, Montali RJ. Streptococcus zooepidemicus infections of possible horsemeat source in red-bellied tamarins and Goeldi's monkeys. J Zoo Wildl Med. 1989;1989:322–7.
60. Mätz-Rensing K, Winkelmann J, Becker T, Burckhardt I, Van Der Linden M, Köndgen S, et al. Outbreak of Streptococcus equi subsp. zooepidemicus infection in a group of rhesus monkeys (Macaca mulatta). J Med Primatol. 2009;38(5):328–34.
61. Carrier CA, Elliott TB, Ledney GD. Resident bacteria in a mixed population of rhesus macaque (Macaca mulatta) monkeys: a prevalence study. J Med Primatol. 2009 Dec;38(6):397–403. https://doi.org/10.1111/j.1600-0684.2009.00366.x.
62. Coughlin P, Bradford C, Montali RJ, Bronson E. Pustular dermatitis caused by impetigo in red-tailed monkeys (cercopithecus ascanius). J Zoo Wildl Med. 2018 Mar;49(1):206–9. https://doi.org/10.1638/2012-0293R1.1.
63. Kwiecinski JM, Horswill AR. Staphylococcus aureus bloodstream infections: pathogenesis and regulatory mechanisms. Curr Opin Microbiol. 2020 Feb;53:51–60. https://doi.org/10.1016/j.mib.2020.02.005.
64. Schaumburg F, Alabi AS, Köck R, Mellmann A, Kremsner PG, Boesch C, Becker K, Leendertz FH, Peters G. Highly divergent Staphylococcus aureus isolates from African non-human primates. Environ Microbiol Rep. 2012 Feb;4(1):141–6. https://doi.org/10.1111/j.1758-2229.2011.00316.x.
65. Lindsay JA. Genomic variation and evolution of Staphylococcus aureus. Int J Med Microbiol. 2010 Feb;300(2–3):98–103. https://doi.org/10.1016/j.ijmm.2009.08.013.
66. Jenul C, Horswill AR. Regulation of Staphylococcus aureus Virulence. Microbiol Spectr. 2019 Apr 5;7(2) https://doi.org/10.1128/microbiolspec.GPP3-0031-2018.
67. Lin Z, Zhang L, Zhang D, Huo G, Zhou X, Yang YW, Huo Y, Li B, Geng XC. A case report of spontaneous staphylococcal meningitis in a cynomolgus monkey. J Med Primatol. 2018 Apr;47(2):132–5. https://doi.org/10.1111/jmp.12330.
68. Chertow DS, Kindrachuk J, Sheng ZM, Pujanauski LM, Cooper K, Nogee D, Claire MS, Solomon J, Perry D, Sayre P, Janosko KB, Lackemeyer MG, Bohannon JK, Kash JC, Jahrling PB, Taubenberger JK. Influenza A and methicillin-resistant Staphylococcus aureus co-infection in rhesus macaques - a model of severe pneumonia. Antivir Res. 2016 May;129:120–9. https://doi.org/10.1016/j.antiviral.2016.02.013.
69. Cullin CO, Colgin LM, Lewis AD. Air sacculitis in three rhesus macaques (Macaca mulatta) and one Japanese macaque (M. fuscata). J Med Primatol. 2017 Apr;46(2):48–50. https://doi.org/10.1111/jmp.12257.
70. Mattix ME, Hunt RE, Wilhelmsen CL, Johnson AJ, Baze WB. Aerosolized staphylococcal enterotoxin B-induced pulmonary lesions in rhesus monkeys (Macaca mulatta). Toxicol Pathol. 1995 May–Jun;23(3):262–8. https://doi.org/10.1177/019262339502300304.
71. Kelly J, Barnass S. Staphylococcus aureus endocarditis presenting as meningitis and mimicking meningococcal sepsis. Int J Clin Pract. 1999 Jun;53(4):306–7.
72. Hinshaw LB, Emerson TE Jr, Taylor FB Jr, Chang AC, Duerr M, Peer GT, Flournoy DJ, White GL, Kosanke SD, Murray CK, et al. Lethal Staphylococcus aureus-induced shock in primates: prevention of death with anti-TNF antibody. J Trauma. 1992 Oct;33(4):568–73.
73. Clermont O, Olier M, Hoede C, Diancourt L, Brisse S, Keroudean M, Glodt J, Picard B, Oswald E, Denamur E. Animal and human pathogenic Escherichia coli strains share common genetic backgrounds. Infect Genet Evol. 2011 Apr;11(3):654–62. https://doi.org/10.1016/j.meegid.2011.02.005.

74. Kaper JB, Nataro JP, Mobley HL. Pathogenic Escherichia coli. Nat Rev Microbiol. 2004;2(2):123–40.
75. Mueller M, Tainter CR. *Escherichia coli*. 2023 Feb 5. In: StatPearls [Internet]. Treasure Island (FL): StatPearls Publishing; 2023 Jan.
76. Riley LW. Extraintestinal foodborne pathogens. Annu Rev Food Sci Technol. 2020;11:275–94. https://doi.org/10.1146/annurev-food-032519-051618.
77. Tapader R, Basu S, Pal A. Secreted proteases: a new insight in the pathogenesis of extraintestinal pathogenic Escherichia coli. Int J Med Microbiol. 2019 May–Jun;309(3–4):159–68. https://doi.org/10.1016/j.ijmm.2019.03.002.
78. Pitout JD. Extraintestinal pathogenic Escherichia coli: a combination of virulence with antibiotic resistance. Front Microbiol. 2012 Jan;19(3):9. https://doi.org/10.3389/fmicb.2012.00009.
79. DeVinney R, Gauthier A, Abe A, Finlay BB. Enteropathogenic Escherichia coli: a pathogen that inserts its own receptor into host cells. Cell Mol Life Sci. 1999 Jun;55(6–7):961–76. https://doi.org/10.1007/pl00013202.
80. Mansfield KG, Lin KC, Newman J, Schauer D, MacKey J, Lackner AA, Carville A. Identification of enteropathogenic Escherichia coli in simian immunodeficiency virus-infected infant and adult rhesus macaques. J Clin Microbiol. 2001 Mar;39(3):971–6. https://doi.org/10.1128/JCM.39.3.971-976.2001.
81. Mansfield KG, et al. Enteropathogenic Escherichia coli and ulcerative colitis in cotton-top tamarins (Saguinus oedipus). J Infect Dis. 2001;184:803–7.
82. Kolappaswamy K, Nazareno J, Porter WP, Klein HJ. Outbreak of pathogenic Escherichia coli in an outdoor-housed non-human primate colony. J Med Primatol. 2014 Apr;43(2):122–4. https://doi.org/10.1111/jmp.12099.
83. National Research Council. Identifying infectious hazards associated with the use of nonhuman primates in research. Occupational health and safety in the care and use of nonhuman primates. New York: National Academies Press; 2003.
84. Hayashimoto N, Inoue T, Morita H, Yasuda M, Ueno M, Kawai K, Itoh T. Survey and experimental infection of Enteropathogenic Escherichia coli in common marmosets (Callithrix jacchus). PLoS One. 2016 Aug 8;11(8):e0160116. https://doi.org/10.1371/journal.pone.0160116.
85. Haraga A, Ohlson MB, Miller SI. Salmonellae interplay with host cells. Nat Rev Microbiol. 2008;6(1):53–66.
86. Kotloff KL, Riddle MS, Platts-Mills JA, Pavlinac P, Zaidi AKM. Shigellosis. Lancet. 2018 Feb 24;391(10122):801–12. https://doi.org/10.1016/S0140-6736(17)33296-8.
87. Cooper JE, Needham JR. An outbreak of shigellosis in laboratory marmosets and tamarins (family: Callithricidae). J Hyg (Lond). 1976 Jun;76(3):415–24. https://doi.org/10.1017/s0022172400055340.
88. Islam D, Ruamsap N, Khantapura P, Aksomboon A, Srijan A, Wongstitwilairoong B, Bodhidatta L, Gettayacamin M, Venkatesan MM, Mason CJ. Evaluation of an intragastric challenge model for Shigella dysenteriae 1 in rhesus monkeys (Macaca mulatta) for the pre-clinical assessment of Shigella vaccine formulations. APMIS. 2014 Jun;122(6):463–75. https://doi.org/10.1111/apm.12168.
89. Sansonetti PJ. Rupture, invasion and inflammatory destruction of the intestinal barrier by Shigella, making sense of prokaryote-eukaryote cross-talks. FEMS Microbiol Rev. 2001 Jan;25(1):3–14. https://doi.org/10.1111/j.1574-6976.2001.tb00569.x.
90. Brady AG, Morton DG. Digestive system. In: Nonhuman primates in biomedical research: diseases, Chapter 10. Academic Press; 1998. p. 377–411.
91. Katakura S, Reinholt FP, Kärnell A, Huan PT, Trach DD, Lindberg AA. The pathology of Shigella flexneri infection in rhesus monkeys: an endoscopic and histopathological study of colonic lesions. APMIS. 1990 Apr;98(4):313–9. https://doi.org/10.1111/j.1699-0463.1990.tb01038.x.
92. Takeuchi A, Jervis HR, Formal SB. Animal model of human disease. Bacillary dysentery, shigellosis, Shigella dysentery. Animal model: monkey shigellosis or dysentery. Am J Pathol. 1975 Oct;81(1):251–4.
93. Armitage GC, Banks TA, Newbrun E, Greenspan JS, Hoover CI, Anderson JH. Immunologic observations in macaques with Shigella-associated periodontal disease. J Periodontal Res. 1983 Mar;18(2):139–48. https://doi.org/10.1111/j.1600-0765.1983.tb00346.x.
94. Urvater JA, McAdam SN, Loehrke JH, Allen TM, Moran JL, Rowell TJ, Rojo S, López de Castro JA, Taurog JD, Watkins DI. A high incidence of Shigella-induced arthritis in a primate species: major histocompatibility complex class I molecules associated with resistance and susceptibility, and their relationship to HLA-B27. Immunogenetics. 2000 Apr;51(4–5):314–25. https://doi.org/10.1007/s002510050625.
95. Levett PN, Haake DA. Leptospira species (leptospirosis). In: Principles and practice of infectious diseases. Philadelphia: Churchill Livingstone Elsevier; 2010. p. 3059–65.
96. Sun A-H, Liu X-X, Yan J. Leptospirosis is an invasive infectious and systemic inflammatory disease. Biomed J. 2020;43(1):24–31.
97. Woolf D, Sanchez C, Gonzalez-Astudillo V, Navarro M, Tapia CC, Franco M, et al. Leptospira species status of captive nonhuman primates and free-ranging rodents at the Barranquilla zoo, Colombia, 2013. J Zoo Wildl Med. 2021;51(4):780–8.
98. Perolat P, Poingt JP, Vie JC, Jouaneau C, Baranton G, Gysin J. Occurrence of severe leptospirosis in a breeding colony of squirrel monkeys. Am J Trop Med Hyg. 1992 May;46(5):538–45. https://doi.org/10.4269/ajtmh.1992.46.538.
99. Wilson TM, Ritter JM, Martines RB, Gonçalves AAB, Fair P, Galloway R, Weiner Z, Romano APM, Costa GRT, Melo CB, Zaki SR, Castro MB. Pathology and one health implications of fatal Leptospira interrogans infection in an urbanized, free-ranging, black-tufted marmoset (Callithrix penicillata) in Brazil. Transbound Emerg Dis. 2021 Nov;68(6):3207–16. https://doi.org/10.1111/tbed.14287.
100. Zonya B, Agudelo-Flórez P, Ramírez M, Moreno N, Ko AI. An outbreak of severe leptospirosis in capuchin (Cebus) monkeys. Vet J. 2011 May;188(2):237–9. https://doi.org/10.1016/j.tvjl.2010.05.002.
101. Aliaga-Samanez GG, Lescano J, Quevedo Urday MJ, Salvatierra Rodríguez GS, Erkenswick Watsa M, Calderon Escalante JE, Erkenswick GA. First detection of antibodies against Leptospira among free-ranging neotropical non-human primates in the Peruvian Amazon lowland rainforest. Transbound Emerg Dis. 2022 May;69(3):1458–65. https://doi.org/10.1111/tbed.14112.
102. Pinna MH, Martins G, Pinheiro AC, Almeida DS, Oriá AP, Lilenbaum W. Detection of anti-Leptospira antibodies in captive nonhuman primates from Salvador. Braz Am J Primatol. 2012 Jan 1;74(1):8–11. https://doi.org/10.1002/ajp.21005.
103. Zhao N, Li M, Amer S, et al. Mortality in captive rhesus monkeys (Macaca mulatta) in China due to infection with Yersinia pseudotuberculosis serotype O:1a. Ecohealth. 2016;13(3):597–601. https://doi.org/10.1007/s10393-016-1148-2.
104. Walker D, Gibbons J, Harris JD, et al. Systemic Yersinia pseudotuberculosis as a cause of osteomyelitis in a captive ring-tailed lemur (Lemur catta). J Comp Pathol. 2018;164:27–31. https://doi.org/10.1016/j.jcpa.2018.08.004.
105. Krylova RI, Dzhikidze EK. Yersinia infection in monkeys. Bull Exp Biol Med. 2000;129:179–83. https://doi.org/10.1007/BF02434805.
106. Kageyama T, Ogasawara A, Fukushima R, Narita Y, Miwa N, et al. Yersinia pseudotuberculosis infection in breeding monkeys:

detection and analysis of strain diversity by PCR. J Med Primatol. 2002;31:129–35.
107. Clark MA, Hirst BH, Jepson MA. M-cell surface β1 integrin expression and invasin-mediated targeting of Yersinia pseudotuberculosis to mouse Peyer's patch M cells. Infect Immun. 1998;66 (3):1237–43.
108. Barnes PD, et al. Yersinia pseudotuberculosis disseminates directly from a replicating bacterial pool in the intestine. J Exp Med. 2006;203(6):1591–601.
109. Cornelis GR, Boland A, Boyd AP, Geuijen C, Iriarte M, Neyt C, et al. The virulence plasmid of Yersinia, an antihost genome. Microbiol Mol Biol Rev. 1998;62(4):1315–52.
110. Nakamura S, Hayashidani H, Iwata T, Namai S, Une Y. Pathological changes in captive monkeys with spontaneous yersiniosis due to infection by Yersinia enterocolitica serovar O8. J Comp Pathol. 2010;143(2–3):150–6. https://doi.org/10.1016/j.jcpa.2010.01.017.
111. Bronson RT, May BD, Ruebner BH. An outbreak of infection by Yersinia pseudotuberculosis in nonhuman primates. Am J Pathol. 1972;69(2):289–308.
112. MacArthur JA, Wood M. Yersiniosis in a breeding unit of Macaca fascicularis (cynomolgus monkeys). Lab Anim. 1983;17(2):151–5. https://doi.org/10.1258/002367783780959367.
113. Wang G, Zhao G, Chao X, Xie L, Wang H. The characteristic of virulence, biofilm and antibiotic resistance of Klebsiella pneumoniae. Int J Environ Res Public Health. 2020 Aug 28;17 (17):6278. https://doi.org/10.3390/ijerph17176278.
114. Burke RL, Whitehouse CA, Taylor JK, Selby EB. Epidemiology of invasive Klebsiella pneumoniae with hypermucoviscosity phenotype in a research colony of nonhuman primates. Comp Med. 2009 Dec;59(6):589–97.
115. Gozalo AS, Elkins WR, Lambert LE, Stock F, Thomas ML 3rd, Woodward RA. Genetic diversity of Klebsiella pneumoniae isolates during an outbreak in a non-human primate research colony. J Med Primatol. 2016 Dec;45(6):312–7. https://doi.org/10.1111/jmp.12229.
116. Burke RL, West MW, Erwin-Cohen R, Selby EB, Fisher DE, Twenhafel NA. Alterations in cytokines and effects of dexamethasone immunosuppression during subclinical infections of invasive Klebsiella pneumoniae with hypermucoviscosity phenotype in rhesus (Macaca mulatta) and cynomolgus (Macaca fascicularis) macaques. Comp Med. 2010 Feb;60(1):62–70.
117. Russo TA, Marr CM. Hypervirulent Klebsiella pneumoniae. Clin Microbiol Rev. 2019 May 15;32(3):e00001–119. https://doi.org/10.1128/CMR.00001-19.
118. Twenhafel NA, Whitehouse CA, Stevens EL, Hottel HE, Foster CD, Gamble S, Abbott S, Janda JM, Kreiselmeier N, Steele KE. Multisystemic abscesses in African green monkeys (Chlorocebus aethiops) with invasive Klebsiella pneumoniae--identification of the hypermucoviscosity phenotype. Vet Pathol. 2008 Mar;45(2):226–31. https://doi.org/10.1354/vp.45-2-226.
119. Kasuya K, Takayama K, Bito M, Shimokubo N, Kawashima R, Shibahara T. Septicemic invasive Klebsiella pneumoniae infection in a cynomolgus monkey (Macaca fascicularis) with severe diffused suppurative meningoencephalitis. J Vet Med Sci. 2017 Jul 7;79(7):1167–71. https://doi.org/10.1292/jvms.17-0126.
120. Sasaki E, Tokiwa T, Tsugo K, Higashi Y, Hori H, Une Y. Peracute bacterial meningitis due to infection with Klebsiella pneumoniae in captive-bred ruffed lemurs (Varecia variegate). J Comp Pathol. 2017 Feb–Apr;156(2–3):281–5. https://doi.org/10.1016/j.jcpa.2016.12.003.
121. Guerra JM, Fernandes NCCA, Morales Dos Santos AL, Barrel JSP, Petri BSS, Milanelo L, Tiba-Casas MR, Liserre AM, Gonçalves CR, Sacchi CT, Catão-Dias JL, Camargo CH. Hypervirulent Klebsiella pneumoniae as unexpected cause of fatal outbreak in captive marmosets, Brazil. Emerg Infect Dis. 2020 Dec;26(12):3039–43. https://doi.org/10.3201/eid2612.191562.
122. Keesler RI, Colagross-Schouten A, Reader JR. Clinical and pathologic features of spontaneous Klebsiella pneumoniae infection in 9 rhesus macaques (Macaca mulatta). Comp Med. 2020 Apr 1;70 (2):183–9. https://doi.org/10.30802/AALAS-CM-19-000067.
123. Guthrie AL, Gailbreath KL, Cienava EA, Bradway DS, Munoz Gutierrez JF. Septic tularemia in 2 cottontop tamarins (Saguinus oedipus). Comp Med. 2012 Jun;62(3):225–8.
124. Splettstoesser WD, Matz-Rensing K, Seibold E, Tomaso H, Al Dahouk S, Grunow R, Essbauer S, Buckendahl A, Finke EJ, Neubauer H. Re-emergence of Francisella tularensis in Germany: fatal tularaemia in a colony of semi-free-living marmosets (Callithrix jacchus). Epidemiol Infect. 2007;135:1256–65.
125. Sammak RL, Rejmanek DD, Roth TM, Christe KL, Chomel BB, Foley JE. Investigation of tularemia outbreak after natural infection of outdoor-housed rhesus macaques (Macaca mulatta) with Francisella tularensis. Comp Med. 2013;63(2):183–90.
126. Ferrecchia CE, Colgin LM, Andrews KR, Lewis AD. An outbreak of tularemia in a colony of outdoor-housed rhesus macaques (Macaca mulatta). Comp Med. 2012;62(4):316–21.
127. Foley JE, Nieto NC. Tularemia. Vet Microbiol. 2010 Jan 27;140 (3–4):332–8. https://doi.org/10.1016/j.vetmic.2009.07.017.
128. Dennis DT, Inglesby TV, Henderson DA, et al. Tularemia as a biological weapon: medical and public health management. JAMA. 2001;285(21):2763–73. https://doi.org/10.1001/jama.285.21.2763.
129. Mätz-Rensing K, Floto A, Schrod A, et al. Epizootic of tularemia in an outdoor housed group of cynomolgus monkeys (Macaca fascicularis). Vet Pathol. 2007;44(3):327–34. https://doi.org/10.1354/vp.44-3-327.
130. Glynn AR, Alves DA, Frick O, et al. Comparison of experimental respiratory tularemia in three nonhuman primate species. Comp Immunol Microbiol Infect Dis. 2015;39:13–24. https://doi.org/10.1016/j.cimid.2015.01.003.
131. Nelson M, Lever MS, Dean RE, et al. Characterization of lethal inhalational infection with Francisella tularensis in the common marmoset (Callithrix jacchus). J Med Microbiol. 2010;59(Pt 9): 1107–13. https://doi.org/10.1099/jmm.0.020669-0.
132. Fitzgerald C. Campylobacter. Clin Lab Med. 2015;35:289–98
133. Clemmons EA, Jean SM, Machiah DK, Breding E, Sharma P. Extraintestinal campylobacteriosis in rhesus macaques (Macaca mulatta). Comp Med. 2014 Dec;64(6):496–500.
134. Dassanayake RP, Zhou Y, Hinkley S, Stryker CJ, Plauche G, Borda JT, Sestak K, Duhamel GE. Characterization of cytolethal distending toxin of campylobacter species isolated from captive macaque monkeys. J Clin Microbiol. 2005 Feb;43(2):641–9.
135. Kalashnikova VA, Dzhikidze EK, Stasilevich ZK, Chikobava MG. Detection of Campylobacter jejuni in healthy monkeys and monkeys with enteric infections by PCR. Bull Exp Biol Med Sep. 2002;134(3):299–300.
136. Johnson LD, Ausman LM, Rolland RM, Chalifoux LV, Russell RG. Campylobacter-induced enteritis and diarrhea in captive cotton-top tamarins (Saguinus oedipus) during the first year of life. Comp Med. 2001;51(3):257–61.
137. Haesebrouck F, Pasmans F, Flahou B, et al. Gastric helicobacters in domestic animals and nonhuman primates and their significance for human health. Clin Microbiol Rev. 2009;22(2):202–23. https://doi.org/10.1128/CMR.00041-08.
138. Dubois A, Fiala N, Heman-Ackah LM, et al. Natural gastric infection with helicobacter pylori in monkeys: a model for spiral bacteria infection in humans. Gastroenterology. 1994;106(6):1405–17. https://doi.org/10.1016/0016-5085(94)90392-1.
139. Solnick JV, et al. Acquisition of Helicobacter pylori infection in rhesus macaques is most consistent with oral-oral transmission. J Clin Microbiol. 2006;44(10):3799–803.

140. Marini RP, Patterson MM, Muthupalani S, et al. Helicobacter suis and Helicobacter pylori infection in a colony of research macaques: characterization and clinical correlates. J Med Microbiol. 2021;70 (3) https://doi.org/10.1099/jmm.0.001315.
141. Cooke CL, An HJ, Kim J, et al. Modification of gastric mucin oligosaccharide expression in rhesus macaques after infection with Helicobacter pylori. Gastroenterology. 2009;137(3):1061. https://doi.org/10.1053/j.gastro.2009.04.014.
142. Dubois A, Tarnawski A, Newell DG, et al. Gastric injury and invasion of parietal cells by spiral bacteria in rhesus monkeys. Are gastritis and hyperchlorhydria infectious diseases? Gastroenterology. 1991;100(4):884–91. https://doi.org/10.1016/0016-5085(91)90260-r.
143. Stolte M, Wellens E, Bethke B, Ritter M, Eidt H. Helicobacter heilmannii (formerly Gastrospirillum hominis) gastritis: an infection transmitted by animals. Scand J Gastroenterol. 1994;29(12): 1061–4. https://doi.org/10.3109/00365529409094888.
144. Lertpiriyapong K, Handt L, Feng Y, et al. Pathogenic properties of enterohepatic helicobacter spp. isolated from rhesus macaques with intestinal adenocarcinoma. J Med Microbiol. 2014;63(Pt 7): 1004–16. https://doi.org/10.1099/jmm.0.072462-0.
145. Saunders KE, Shen Z, Dewhirst FE, Paster BJ, Dangler CA, Fox JG. Novel intestinal Helicobacter species isolated from cotton-top tamarins (Saguinus oedipus) with chronic colitis. J Clin Microbiol. 1999;37(1):146–51. https://doi.org/10.1128/JCM.37.1.146-151.1999.
146. Fox JG, Boutin SR, Handt LK, et al. Isolation and characterization of a novel helicobacter species, "Helicobacter macacae," from rhesus monkeys with and without chronic idiopathic colitis. J Clin Microbiol. 2007;45(12):4061–3. https://doi.org/10.1128/JCM.01100-07.
147. Skoog EC, Deck SL, Entwistle HD, Hansen LM, Solnick JV. Characterization of the cag pathogenicity Island in helicobacter pylori from naturally infected rhesus macaques. FEMS Microbiol Lett. 2016;363(24):fnw275. https://doi.org/10.1093/femsle/fnw275.
148. Hatcher CL, Muruato LA, Torres AG. Recent advances in Burkholderia mallei and B. pseudomallei research. Curr Trop Med Rep. 2015;2(2):62–9. https://doi.org/10.1007/s40475-015-0042-2.
149. Trevino SR, Dankmeyer JL, Fetterer DP, et al. Comparative virulence of three different strains of Burkholderia pseudomallei in an aerosol non-human primate model. PLoS Negl Trop Dis. 2021;15 (2):e0009125. Published 2021 Feb 11
150. Galyov EE, Brett PJ, DeShazer D. Molecular insights into Burkholderia pseudomallei and Burkholderia mallei pathogenesis. Annu Rev Microbiol. 2010;64:495–517. https://doi.org/10.1146/annurev.micro.112408.134030.
151. Kasantikul T, Sommanustweechai A, Polsrila K, et al. Retrospective study on fatal Melioidosis in captive zoo animals in Thailand. Transbound Emerg Dis. 2016;63(5):e389–94. https://doi.org/10.1111/tbed.12315.
152. Jelesijevic T, Zimmerman SM, Harvey SB, et al. Use of the common marmoset to study Burkholderia mallei infection. PLoS One. 2015;10(4):e0124181. Published 2015 Apr 10. https://doi.org/10.1371/journal.pone.0124181.
153. Yingst SL, Facemire P, Chuvala L, Norwood D, Wolcott M, Huzella L. Pathological findings and diagnostic implications of a rhesus macaque (Macacca mulatta) model of aerosol exposure to Burkholderia mallei (glanders). J Med Microbiol. 2015;64(6): 646–53. https://doi.org/10.1099/jmm.0.000065.
154. Waag DM, Chance TB, Trevino SR, et al. Comparison of three non-human primate aerosol models for glanders, caused by Burkholderia mallei. Microb Pathog. 2021;155:104919. https://doi.org/10.1016/j.micpath.2021.104919.
155. Buchanan W, Sehgal P, Bronson RT, Rodger RF, Horton JE. Noma in a nonhuman primate. Oral Surg Oral Med Oral Pathol. 1981 Jul;52(1):19–22. https://doi.org/10.1016/0030-4220(81)90166-3.
156. Kimura T. An oro-facial disease 'noma (cancrum oris)' in a Japanese monkey (Macaca fuscata): clinical signs, clinicopathological features, and response to treatment. J Med Primatol. 2008;37:217–22. https://doi.org/10.1111/j.1600-0684.2008.00312.x.
157. Pittet B, Jaquinet A, Montandon D. Clinical experience in the treatment of Noma sequelae. J Craniofac Surg. 2001 May;12(3): 273–83. https://doi.org/10.1097/00001665-200105000-00014.
158. Kolappaswamy K, Shipley ST, Tatarov II, DeTolla LJ. Methicillin-resistant staphylococcus non-aureus infection in an irradiated rhesus macaque (Macaca mulatta). J Am Assoc Lab Anim Sci. 2008 May;47(3):64–7.
159. Silverman J, Loftus MJ, Sharpless M. Mandibular osteomyelitis in a squirrel monkey. Contemp Top Lab Anim Sci. 1997;36(1):94–6.

Parasitic and Fungal Infections in Nonhuman Primates

3

Amanda L. Johnson, Andrew N. Cartoceti, and Keith G. Mansfield

Abstract

Historically, a number of parasitic and fungal infections impacted nonhuman primate animal and colony health. While many of these infections have been controlled through improved husbandry and new chemotherapeutics, such infections may still be present in free-ranging animals that have greater exposure to the environment and wildlife. Such infections can occasionally still be imported from country of origin for non-domestically sourced animals. In addition, enzootic infections with low-virulence organisms may still cause disease in immunosuppressed or immunodeficient animals. Routine histological evaluation of tissues accompanied by special stains or molecular pathology techniques play a critical role in the diagnosis of infections and in elucidating their relationship to disease.

Keywords

Parasitic infections · Fungal infections · Opportunistic infections · Protozoa · Nematodes

Historically, a number of parasitic and fungal infections impacted nonhuman primate animal and colony health. While many of these infections have been controlled through improved husbandry and new chemotherapeutics, such infections may still be present in free-ranging animals that have greater exposure to the environment and wildlife. Such infections can occasionally still be imported from country of origin for non-domestically sourced animals. In addition, enzootic infections with low-virulence organisms may still cause disease in immunosuppressed or immunodeficient animals. Routine histological evaluation of tissues accompanied by special stains or molecular pathology techniques play a critical role in the diagnosis of infections and in elucidating their relationship to disease.

A. L. Johnson (✉)
Inotiv - Boulder, Boulder, CO, USA
e-mail: ajohnson@inotivco.com

A. N. Cartoceti
Zoetis Reference Laboratories, Parsippany-Troy Hills, NJ, USA

K. G. Mansfield
Investigative Pathology, Novartis Biomedical Research, Cambridge, MA, USA
e-mail: keith.mansfield@novartis.com

3.1 Protozoa

3.1.1 Enteric Flagellates

Giardia

*Giardia lamblia (*synonym *G. lamblia, duodenalis)* is a pathogenic enteric flagellate found within the small intestine of humans and nonhuman primates, including macaques, baboons, marmosets, and chimpanzees [1–4]. Transmission occurs via ingestion of food or water contaminated with *Giardia* sp. cysts. Infected animals can develop a subclinical carrier state or exhibit clinical signs including vomiting and diarrhea [5]. *Giardia* sp. trophozoites are microscopically identified within the small intestine, but do not typically result in macroscopic lesions. Trophozoites are characterized as bilaterally symmetrical pear-shaped organisms approximately 5–15μm wide and 9–21μm long, with an anterior ventral sucker disk, two axostyles, eight flagella, and two anterior nuclei located above a pair of curved median bodies resulting in the appearance of a face with eyes and mouth [1, 6]. *Giardia* cysts are approximately 7–10μm wide and 8–12μm long [6].

Hexamita

Hexamita pitheci is an enteric flagellate identified in the large intestines of rhesus macaques [7, 8]. Transmission presumably occurs through ingestion of contaminated water. Gross and microscopic lesions in nonhuman primates are not described. Trophozoites are approximately 1.5–4μm wide

I. Kondova - Perseng et al. (eds.), *Atlas of Diagnostic Pathology in Nonhuman Primates*,
https://doi.org/10.1007/978-3-031-41280-6_3

and 2.5–6μm long [6–8]. A related flagellate belonging to the genus *Spironucleus* can cause disseminated disease in immunodeficient macaques. Colitis, multifocal abdominal abscessation, and lymphadenitis suggest an enteric origin of the infection and have also been associated with extensive pyogranulomatous pneumonia (Figs. 3.1, 3.2 and 3.3). The characteristic morphology can be most readily identified in impression smears (Fig. 3.4).

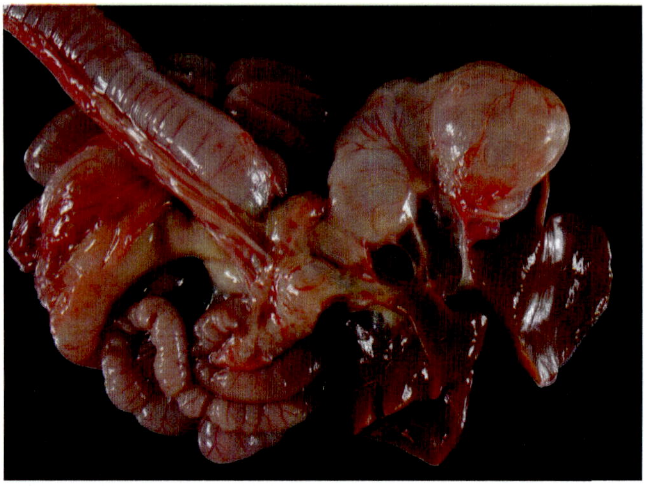

Fig. 3.1 *Spironucleus* sp. Enlarged mesenteric lymph nodes within the root of the mesentery (*Macaca mulatta*, mesenteric lymph nodes)

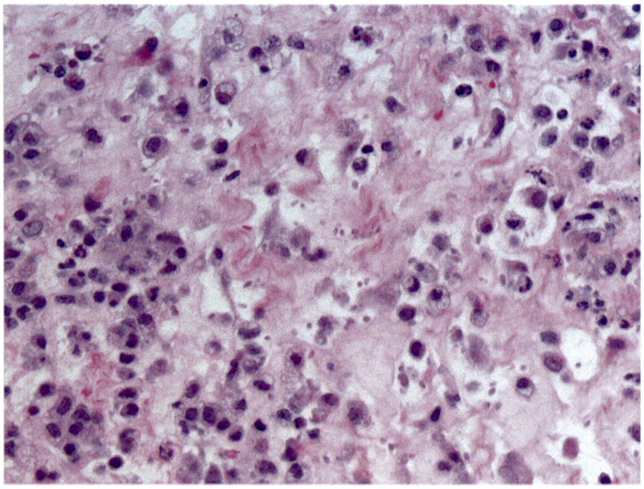

Fig. 3.2 *Spironucleus* sp. Pyogranulomatous lymphadenitis with intralesional basophilic protozoa (*Macaca mulatta*, lymph node, H&E)

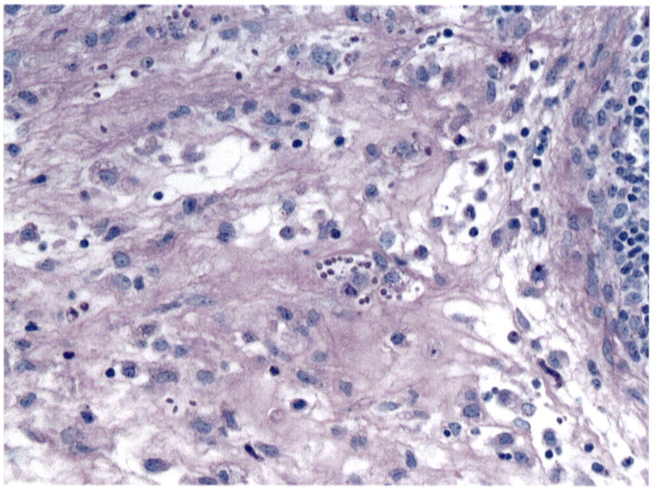

Fig. 3.3 *Spironucleus* sp. Pyogranulomatous lymphadenitis with intralesional Periodic acid-Schiff (PAS) positive protozoa (*Macaca mulatta*, lymph node, PAS)

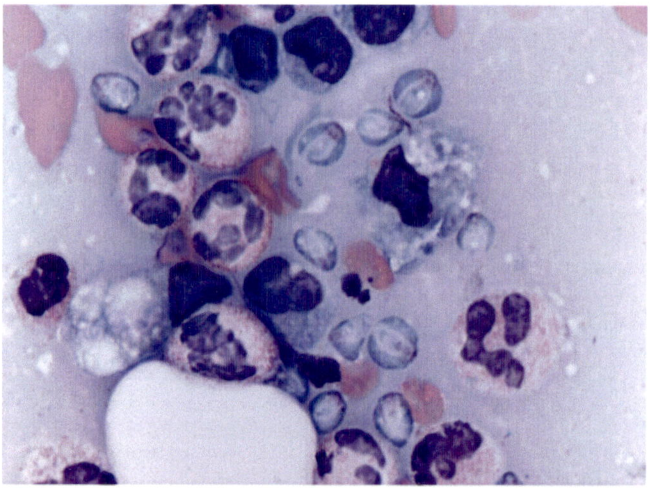

Fig. 3.4 *Spironucleus* sp. Protozoa demonstrating apically oriented nuclei and bilateral cytopharynxes which appear as negatively staining lines (*Macaca mulatta*, lymph node impression, Wright–Giemsa stain)

Trichomonas

Tritrichomonas mobilensis is a pathogenic enteric flagellate identified within the large intestine of squirrel monkeys [9, 10]. Transmission occurs through oral ingestion. 100% of squirrel monkeys appear to be infected by 8 weeks of age and infection is thought to be lifelong [11]. Trichomonads frequently fill crypt lumina and invade the mucosal epithelium, where they are occasionally associated with mucosal ulceration, cryptitis, and epithelial necrosis [9, 10]. However, they are frequently identified in areas void of inflammation [10]. *T. mobilensis* is thought to be responsible for two manifestations of gastritis in SIV-infected rhesus macaques. The first is a lymphoplasmacytic gastritis with intraglandular trophozoites, while the second is a necrosuppurative gastritis with intralesional protozoa [12]. Trophozoites are approximately 1.3–3μm by 7–10μm, with three anterior flagella and a trailing posterior flagellum [9].

Trichomonas tenax (synonyms *Tetratrichomonas buccalis, Trichomonas buccalis*) is a nonpathogenic enteric flagellate identified in between the gums and teeth of macaques, baboons, and chimpanzees [9]. Trophozoites are ovoid, ellipsoid, or pyriform and approximately 2–15μm by 4–16μm, with four anterior flagella [6, 9].

Pentatrichomonas hominis

Pentatrichomonas hominis is a nonpathogenic enteric flagellate that has been identified in the cecum and colon of numerous nonhuman primates, including macaques, marmosets, chimpanzees, and orangutans [2, 13–15]. Transmission occurs through oral ingestion of trophozoites or pseudocysts excreted in feces and reproduces via longitudinal binary fission [9, 13]. While this organism is commonly identified in the feces of diarrheic animals, it is not recognized as a cause of diarrhea [6]. Trophozoites are approximately 3–14μm wide and 8–20μm long, with four anterior flagella, one posterior flagellum, and one free-trailing flagellum [6, 9].

Chilomastix spp.

Chilomastix mesnili (synonyms *Chilomastix suis, Chilomastix hominis, Macrostoma mesnili*) is a nonpathogenic enteric flagellate that has been identified in the cecum and colon of chimpanzees, orangutans, marmosets, macaques, sooty mangabeys, colobus [9, 16, 17]. Transmission occurs through fecal–oral route or ingestion of contaminated food or water. Pear-shaped trophozoites are approximately 3–10μm wide and 6–24μm long, with three anterior flagella and a spiral groove across the middle half of the body [9]. Cysts are lemon-shaped and contain a single nucleus and cytostome and are approximately 6–10μm in diameter [9].

Enteromonas hominis

Enteromonas hominis (synonyms *Octomitus hominis, Tricercomonas intestinalis, Enteromonas bengalensis*) is a nonpathogenic enteric flagellate that has been identified in the cecum of chimpanzees and macaques [9]. Transmission occurs through ingestion of the organism passed in feces [18]. Spherical or piriform trophozoites are approximately 3–6μm wide and 4–10μm long and have three short anterior flagella and one posterior flagellum, as well as multiple food vacuoles [6, 9]. Mature cysts are ovoid and have four nuclei [6].

Retortamonas intestinalis

Retortamonas intestinalis (synonyms *Embadomonas intestinalis, Waskia intestinalis*) is a nonpathogenic enteric flagellate that has been identified in the cecum of chimpanzees and rhesus macaques [6, 9]. Transmission occurs through ingestion of contaminated stagnant water [9]. Pyriform or fusiform trophozoites possess an anterior nucleus, two anterior flagella, a posterior flagellum that arises from the cytostomal groove, and a large anterior cytostome [6, 9]. Trophozoites are approximately 3–4μm wide and 4–9μm long [6]. Pyriform or ovoid cysts are approximately 3–5μm wide and 4–7μm long and have one to two nuclei [6].

Dientamoeba fragilis

Dientamoeba fragilis is generally a nonpathogenic flagellate that was originally classified as an amoeba and is now classified as a trichomonad [9]. It is identified in the cecum and colon of rhesus macaques, cynomolgus macaques, and baboons [18–20]. Transmission occurs through ingestion of organisms in feces. Trophozoites are approximately 3–22μm in diameter, with 1–2 vesicular nuclei [9].

Enterocytozoon bieneusi

Enterocytozoon bieneusi is a microsporidia identified in the biliary tract, intestines, or feces of rhesus macaques, cynomolgus macaques, pigtailed macaques, Formosan rock macaque, and orangutans [6, 21, 22]. Enterocytozoon is generally considered a subclinical infection in immunocompetent nonhuman primates, however can be identified as an opportunistic infection in the gallbladder and biliary tract in simian immunodeficiency virus-infected macaques, contributing to the development of diarrhea [9, 22–24]. Transmission is fecal–oral, with ingestion of contaminated food or water. Gross and microscopic lesions include cholecystitis and cholangitis, with organisms primarily identified in the cytoplasm of epithelial cells of the gallbladder, common bile duct, and intrahepatic bile ducts where the organism is found adjacent to the host cell nucleus [23]. Other lesions include biliary epithelial hyperplasia with periportal fibrosis (Figs. 3.5, 3.6, 3.7 and 3.8) [6]. Sporozoites are approximately 1μm by 1.5μm and can be identified in the cytoplasm of infected cells as small ring-like basophilic structures on hematoxylin and eosin-stained sections (Figs. 3.9 and 3.10) [9].

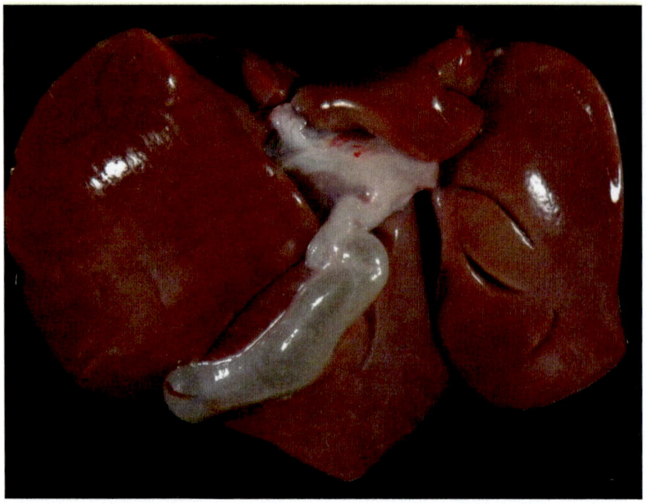

Fig. 3.5 *Enterocytozoon bieneusi*. Diffuse thickening of the common bile duct (*Macaca mulatta*, liver and gallbladder)

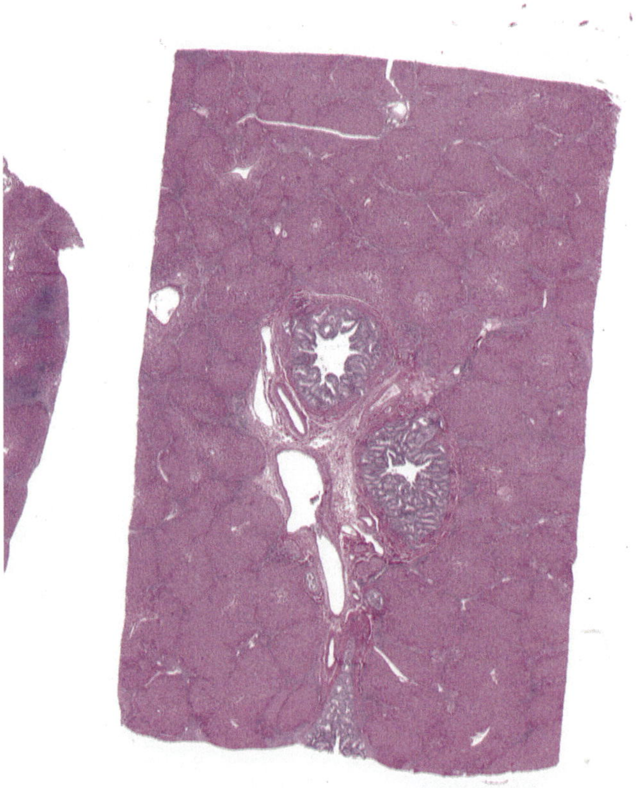

Fig. 3.6 *Enterocytozoon bieneusi*. Liver demonstrating bridging biliary hyperplasia and epithelial thickening of septal ducts (*Macaca mulatta*, liver, subgross H&E stain)

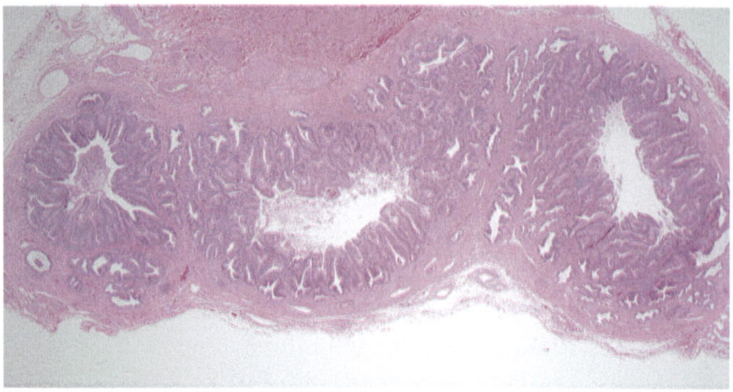

Fig. 3.7 *Enterocytozoon bieneusi*. Proliferative choledochitis, with lymphocytic, plasmacytic, and eosinophilic inflammatory infiltrates (*Macaca mulatta*, common bile duct, H&E)

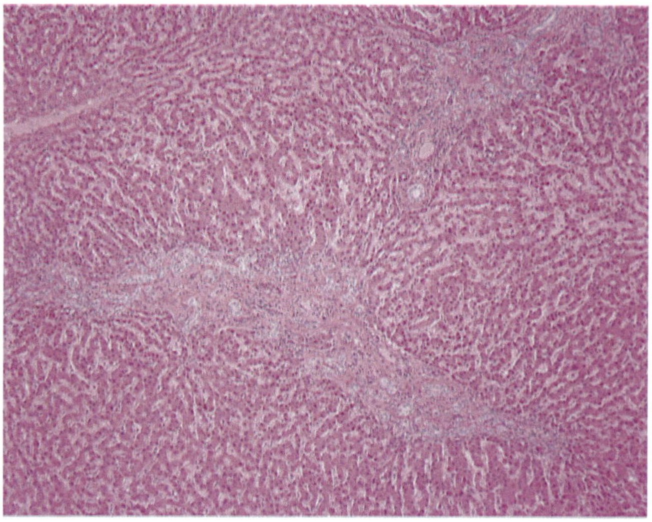

Fig. 3.8 *Enterocytozoon bieneusi*. Bridging biliary hyperplasia accompanied by scant lymphoplasmacytic infiltrate (*Macaca mulatta*, liver, H&E stain)

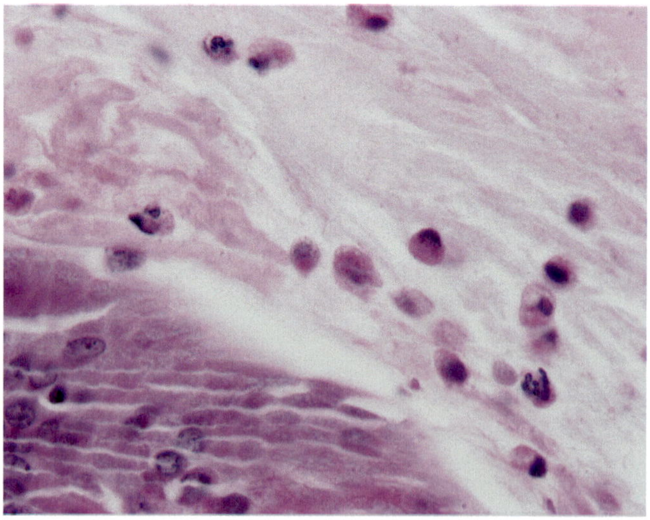

Fig. 3.9 *Enterocytozoon bieneusi*. Exfoliated biliary epithelial cells present within mucous layer and demonstrating intracellular ring-like structures (sporozoites) (*Macaca mulatta*, gall bladder, H&E stain)

3.1.2 Hemoflagellates

Trypanosoma cruzi

Trypanosoma cruizi (synonyms *Schizotrypanum cruzi, Trypanosoma lesourdi, Trypanosoma rhesii, Trypanosoma prowazekii, Trypanosoma vickersae*) is a hemoflagellate and the causative agent of Chagas' disease [9]. *T. cruzi* has been identified in several species of wild and captive-born nonhuman primates including rhesus macaques, pigtailed macaques, ring-tailed lemurs, squirrel monkey, yellow baboon, and chimpanzees [6, 25–32]. Transmission occurs via the insect vector triatomine bugs or "kissing bugs" [9]. Clinical signs include anemia, generalized edema, hepatosplenomegaly, and lymphadenitis [9]. Microscopic lesions in the heart are characterized by infiltration of lymphocytes and plasma cells, with disruption of myocardial fibers by pseudocysts (Fig. 3.10) [9, 33]. In acute cases, the trypomastigote for4m can be identified in blood smears stained with Giemsa [9]. Trypanosomes are approximately 16–20μm long with a trailing flagellum and a large subterminal kinetoplast [9]. Amastigotes are approximately 1.5–4μm in diameter and can be found in skeletal and cardiac muscle, as well as the reticuloendothelial system [9].

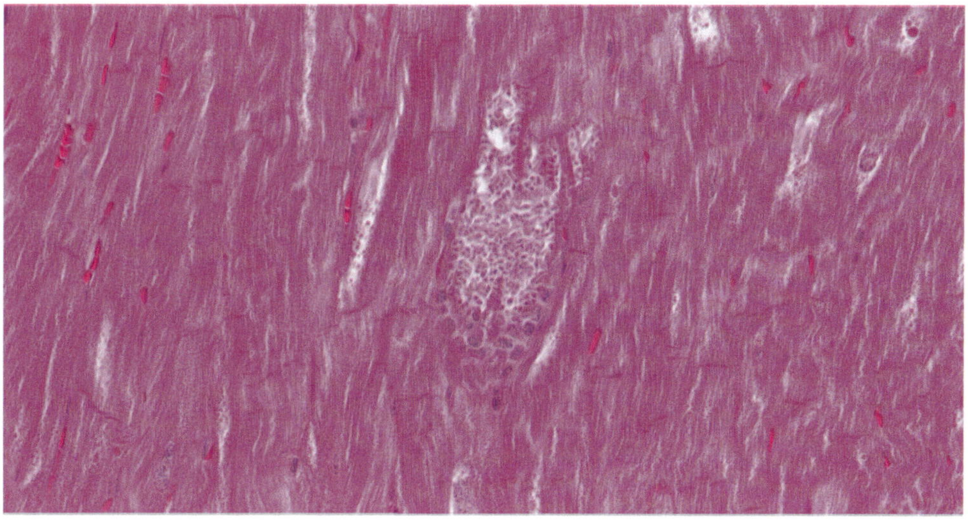

Fig. 3.10 *Trypanosoma cruzi*. Numerous protozoal amastigotes within the myocardium (*Macaca fascicularis*, heart, H&E)

3.1.3 Ciliates

Balantidium coli

Balantidium coli (synonyms *Balantidium aragaoi, B. cunhamunizi, B. philippiensis, B. rhesum, B. simile, B. suis, B. wenrichi)* is a ciliated protozoan commonly identified in the cecum and colon in numerous species of nonhuman primates, including rhesus macaques, cynomolgus macaques, bonnet macaques, baboons, orangutans, chimpanzees, and gorillas [2, 6, 19, 34–38]. Transmission occurs through the ingestion of cysts [6]. *B. coli* is a secondary invader in lesions resulting from pathogenic bacteria [6, 39]. Pleomorphic motile trophozoites are approximately 25–120μm diameter and 30–150μm long and are covered with longitudinal rows of cilia [6]. Cysts are approximately 40–60μm diameter and contain a large kidney-shaped macronucleus [6]. Generally, *B. coli* is not considered pathogenic in macaques and can often be found in the lumen of the colon in considerable numbers and may invade the mucosa and submucosa postmortem (Figs. 3.11 and 3.12). This should not be mistaken for a pathogenic infection.

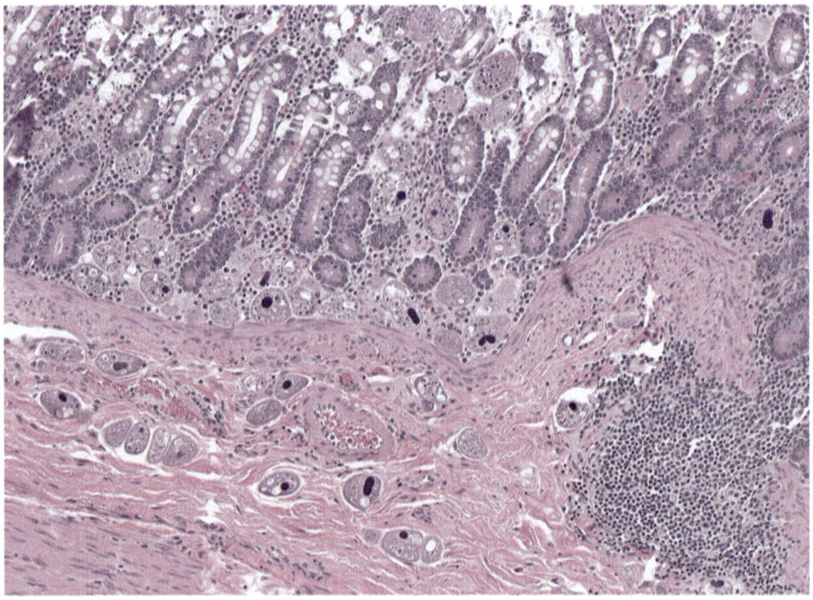

Fig. 3.11 *Balantidium coli.* Large number of *Balantidium coli* present within the colonic mucosa and submucosa likely occurring as a postmortem artifact due to the lack of necrosis and inflammation (*Macaca fascicularis*, colon, H&E stain)

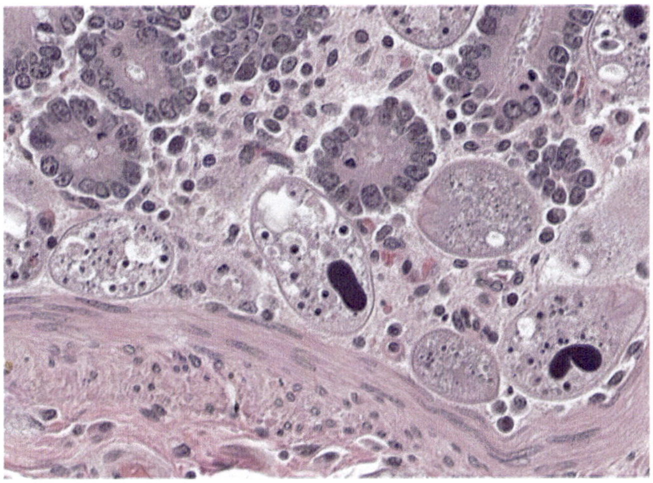

Fig. 3.12 *Balantidium coli*. *B. coli* trophozoites demonstrating ciliated morphology, large basophilic nucleosome, and cytostome (*Macaca fascicularis*, colon, H&E stain)

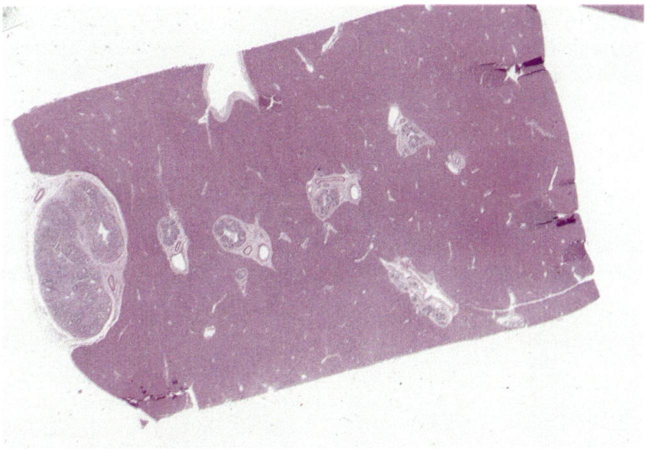

Fig. 3.13 *Cryptosporidium* sp. Liver demonstrating concentric fibrosis and inflammatory infiltrates surrounding septal ducts. Contrast with *E. bieneusi* infection in Fig. 3.3 (*Macaca mulatta*, liver, subgross H&E stain)

3.1.4 Coccidians

Cryptosporidium spp.

Cryptosporidium spp. is a coccidian parasite found in the small intestine of several nonhuman primates including macaques, baboons, marmosets, and lemurs [9, 40, 41]. Infections are common in juvenile animals and those infected with simian immunodeficiency virus, where organisms have been identified in the stomach, intestines, gallbladder, bile and pancreatic ducts, and respiratory tract (Figs. 3.13 and 3.14) [9, 42–44]. Hepatic involvement is common in immunodeficient macaques and most be differentiated from *E. bieneusi* (see Sect. 3.1.1.9). In contrast to the bridging biliary hyperplasia observed with *E. bieneusi*, hepatic *C. parvum* is more necrotizing and often accompanied by concentric (onion skin) fibrosis and neutrophilic infiltrates. The organism may be identified on the luminal aspect biliary epithelial cells as small basophilic spherical bodies (Figs. 3.15 and 3.16). Transmission is animal to animal via a fecal–oral route [9]. Clinical signs include dehydration, depression, weight loss, and diarrhea [9, 42]. Gross lesions include intestines distended with gas and liquid feces, as well as mesenteric lymphadenomegaly [42]. Microscopic lesions include small intestinal villous blunting and fusion, with enterocyte necrosis [9]. The apical aspect of enterocytes is frequently lined by variably round approximately 1–4μm trophozoites [9, 42]. Ovoid or spheroidal oocysts are approximately 3.5 by 5.0μm and can be identified with Ziehl–Neelsen stained fecal smears or hematoxylin and eosin-stained histology sections [9].

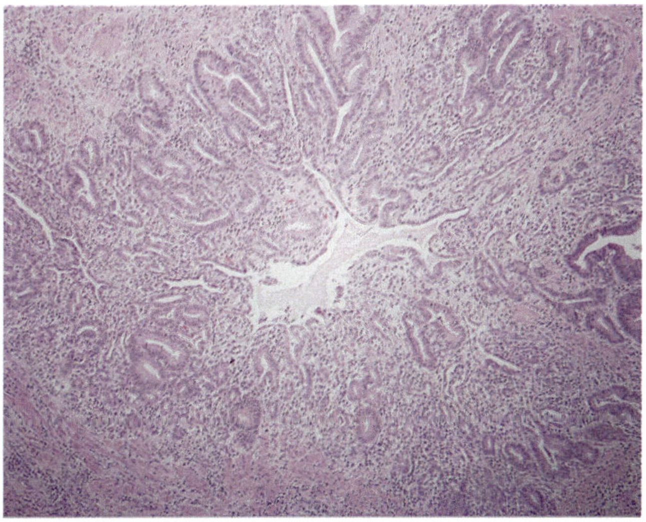

Fig. 3.14 *Cryptosporidium* sp. Septal duct demonstrating mixed inflammatory cell infiltrate with attenuation and erosion of luminal biliary epithelium (*Macaca mulatta*, liver, H&E stain)

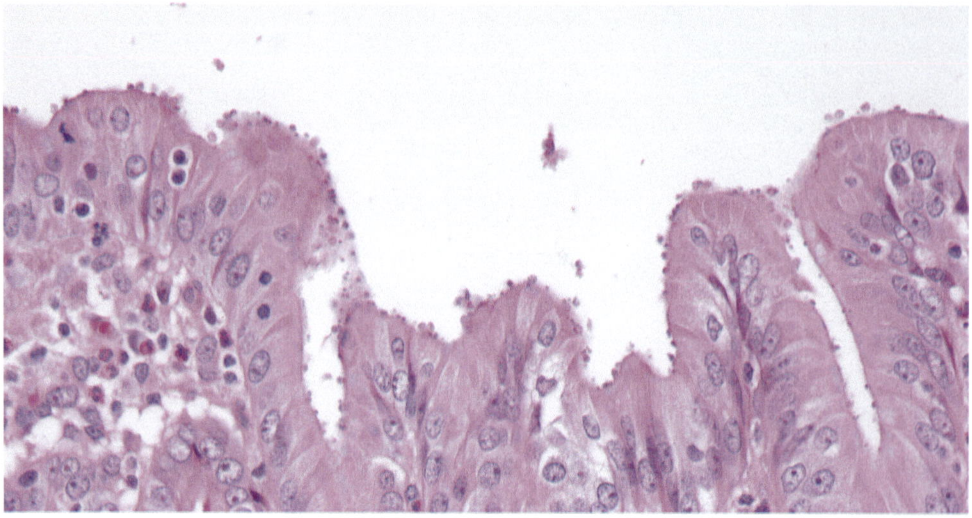

Fig. 3.15 *Cryptosporidium* sp. Basophilic protozoa line the apical epithelial cells of the common bile duct (*Macaca mulatta*, common bile duct, H&E)

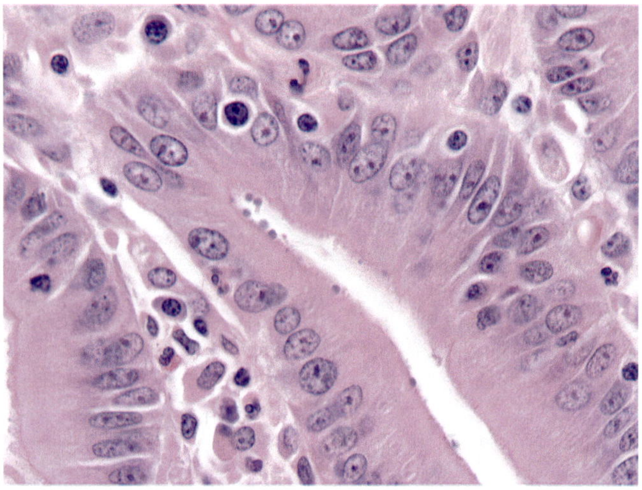

Fig. 3.16 *Cryptosporidium* sp. Small basophilic organisms located on the luminal surface of biliary epithelial cells (*Macaca mulatta*, gall bladder, H&E stain)

Toxoplasma gondii

Toxoplasma gondii is a protozoan parasite identified in the brain, lungs, liver, heart, lymph nodes, kidney, and blood of prosimians, Old World monkeys and New World monkeys; the latter are more sensitive to infection [33, 45–47]. Transmission occurs via transplacental transmission, ingestions of tissue cysts or oocysts [33]. Clinical signs are nonspecific and include weakness, respiratory distress, anorexia, malaise, hypothermia, and sudden death [45, 46]. Gross lesions include pulmonary congestion, pulmonary edema, splenomegaly, and lymphadenomegaly [45, 46]. Microscopic lesions include necrosis within the liver (Figs. 3.17 and 3.18) and spleen, myocarditis, lymphadenitis, and interstitial pneumonia with necrosis [45, 46]. Frequently, organisms were identified within necrotic foci [45, 46]. Round to oval oocysts are approximately 11–15μm by 8–12μm, while sporulated oocysts contain two ellipsoidal sporocysts approximately 6 × 8.5μm that each contains four sporozoites [9]. Tachyzoites develop within vacuoles and are approximately 6–8μm throughout multiple cell types, including hepatocytes, myocardial cells, reticular cells, and fibroblasts [9]. Tissue cysts are found within the liver, lung, muscle, and brain and are approximately 100μm in diameter and contain several lancet-shaped bradyzoites [9].

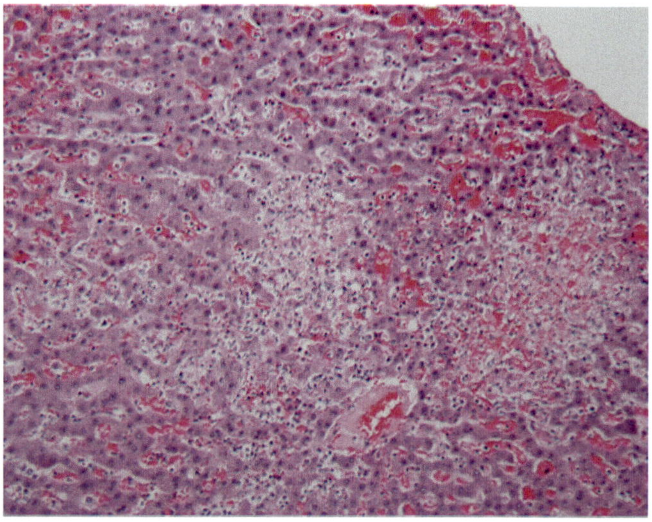

Fig. 3.17 *Toxoplasma gondii*. Multifocal hepatic necrosis (*Saimiri sciureus*, liver, H&E stain)

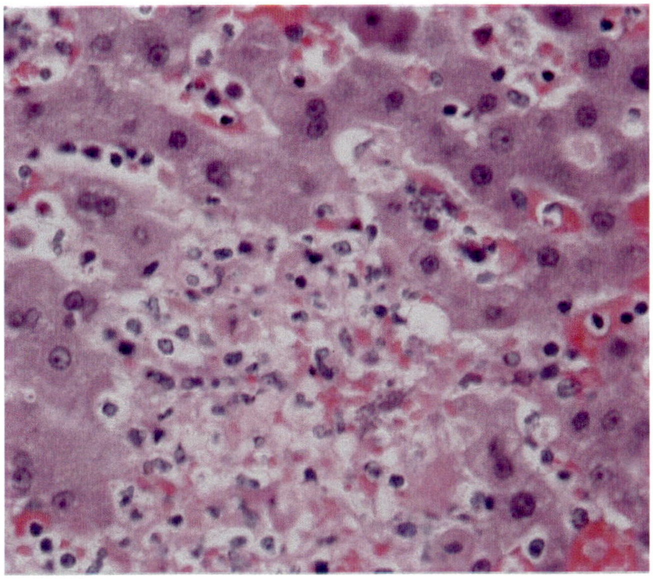

Fig. 3.18 *Toxoplasma gondii*. Focal hepatic necrosis with intralesional tachyzoites (*Saimiri sciureus*, liver, H&E stain)

Eimeria spp.

Eimeria spp. are not thought to be a significant source of disease in nonhuman primates [9].

- *Eimeria ferruginea*: prosimians
- *Eimeria galago:* prosimians
- *Eimeria lemuris:* prosimians
- *Eimeria modesta:* prosimians
- *Eimeria otolicni:* prosimians
- *Eimeria pachylepyron:* prosimians
- *Eimeria tupaiae:* prosimians

Isospora spp.

Cystoisospora arctopitheci (synonym *Isospora arctopitheci*) is a coccidian of unknown pathogenicity found in the small intestine of New World monkeys [6, 48]. Transmission occurs through ingestion of sporulated oocysts, which are approximately 25–30μm by 23–25μm and ellipsoidal [6, 49]. Sporocysts are also ellipsoidal and are approximately 10 by 10μm, with four elongate sporozoites [49]. Developmental stages are found within the distal two-thirds of the jejunal villi [50].

Other species of *Cystoisopora* include [6, 50–55]:

- *Cystoisospora callimico:* New World monkeys, small intestines
- *Cystoisospora papionis:* Old World monkeys, small intestines
- *Cystoisospora endocallimici:* Goeldi's marmosets, feces
- *Cystoisospora scorzai*: Uakari monkey, feces
- *Cystoisospora cebi*: *Cebus albifrons*, feces
- *Cystoisospora* sp. in apes

Plasmodium spp.

Plasmodium spp.. are haemoprotozoal parasites that cause malaria in nonhuman primates in tropical and sei-tropical regions [9]. These organisms are classified based on the species of host infected, parasite morphology, and the type of cyclical fever induced [9]. Fevers produced include quotidian which lasts for 24 h, tertian which lasts for 48 h, and quartan which lasts for 72 h [9, 33, 56]. Transmission occurs via a mosquito insect vector, who injects a sporozoite into the vertebrate host [9, 56]. Once in the liver the sporozoite produces pre-erythrocytic schizonts which contain numerous merozoites that are released into the bloodstream and where they invade erythrocytes and take on a small ring-shaped appearance [56]. The shape and size of this ring form vary between species [56]. The ring stage is followed by the trophozoite stage, characterized by a large uninucleate appearance that fills, enlarges, and distorts the red blood cells [56]. Trophozoites undergo nuclear division to form an erythrocytic schizont that contains numerous merozoites that are eventually released into the bloodstream to infect additional red blood cells [56]. Gross lesions include hepatosplenomegaly and lymphoid hyperplasia [9, 33]. Microscopic lesions include myeloid hyperplasia in the bone marrow, as well as hemozoin pigment deposition in Kupffer cells of the liver and macrophages in the splenic red pulp and bone marrow (Fig. 3.19) [9, 33]. This hemozoin pigment may impart a brown coloration to tissues such as the liver, brain, lung, and spleen (Figs. 3.20, 3.21, 3.22, and 3.23, respectively) Intraerythrocytic organisms are visible with Giemsa or Wright–Giemsa stained blood smears (Fig. 3.24 and 3.25) [9, 33].

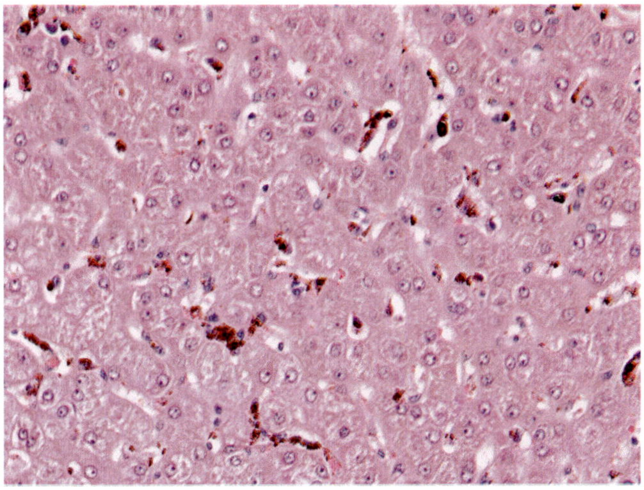

Fig. 3.19 *Plasmodium* sp. Hemozoin pigment in Kupffer cells of liver (*Macaca fascicularis*, liver, H&E stain)

Fig. 3.20 *Plasmodium cynomolgi* (Berok strain). Diffuse dark brown malarial (hemazoin) pigment deposition (*Macaca mulatta*, liver)

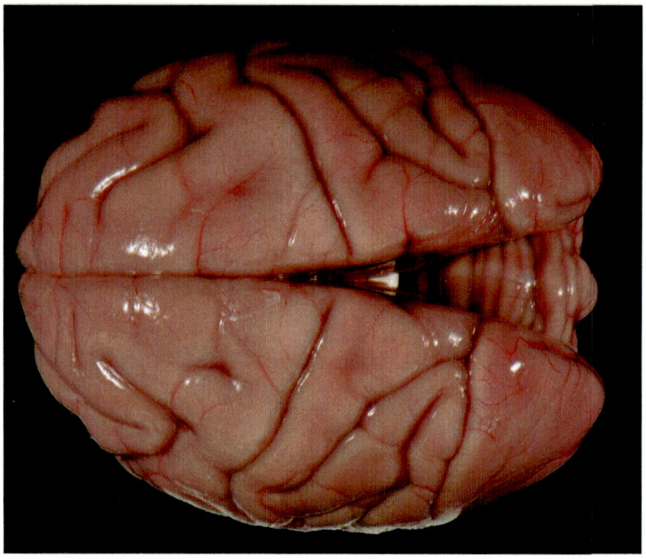

Fig. 3.21 *Plasmodium cynomolgi* (Berok strain). Diffuse faint brown malarial (hemazoin) pigment deposition (*Macaca mulatta*, brain)

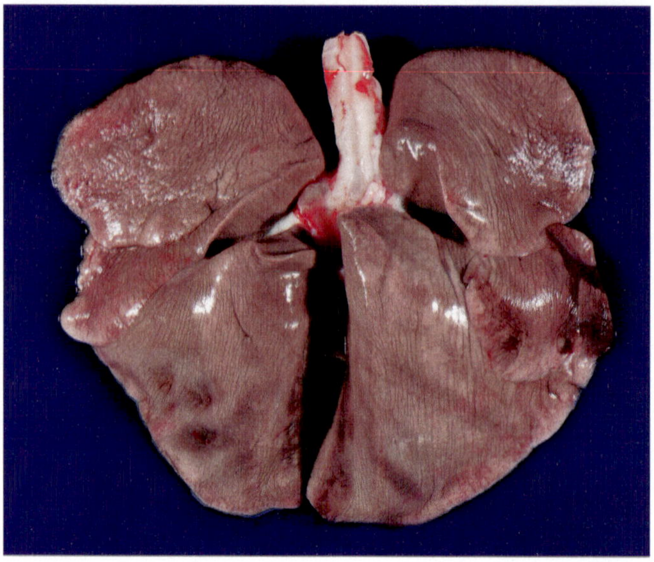

Fig. 3.22 *Plasmodium cynomolgi* (Berok strain). Diffuse dark brown malarial (hemazoin) pigment deposition (*Macaca mulatta*, lung)

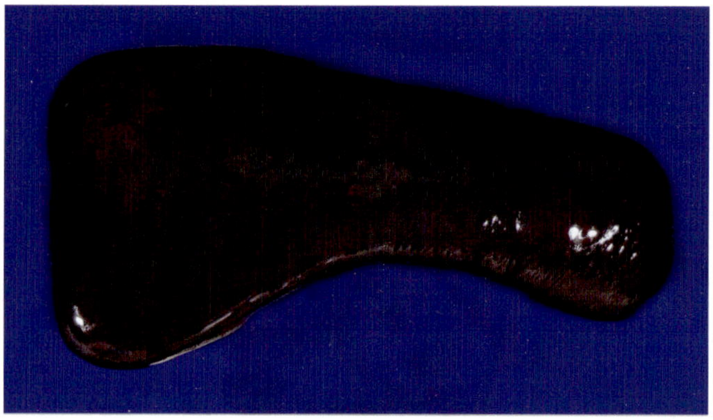

Fig. 3.23 *Plasmodium cynomolgi* (Berok strain). Diffuse dark brown malarial (hemazoin) pigment deposition (*Macaca mulatta*, spleen)

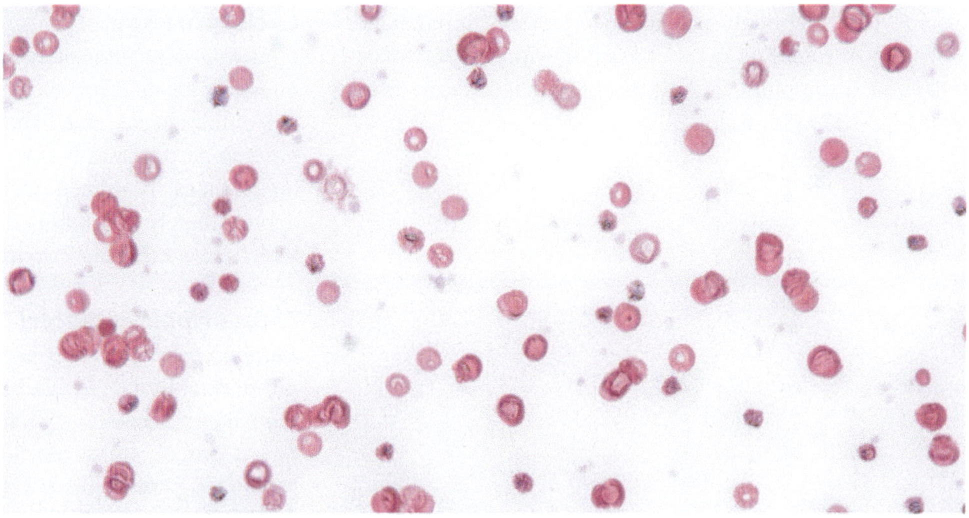

Fig. 3.24 *Plasmodium cynomolgi* (Berok strain). Numerous intraerythrocytic trophozoites, with variable amounts of malarial (hemazoin) pigment, and a single schizont containing high numbers of merozoites (*Macaca mulatta*, blood smear, Wright–Giemsa)

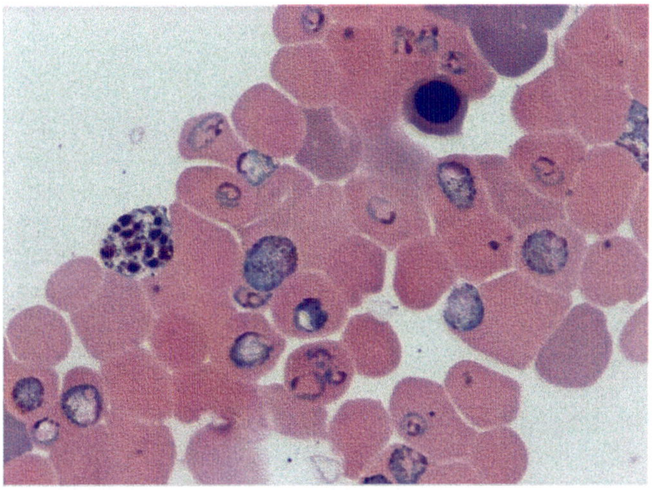

Fig. 3.25 *Plasmodium* sp. Intraerythrocytic ring-form trophozoites observed in blood smear of immunosuppressed cynomolgus macaque (*Macaca fascicularis*, blood smear, Wright–Giemsa stain)

- **Malaria of prosimians**
  - *Plasmodium hylobati:* gibbons, quartan malaria
  - *Plasmodium jefferyi:* gibbons, tertian malaria
  - *Plasmodium eylesi:* gibbons, tertian malaria
  - *Plasmodium youngi:* gibbons, severe tertian malaria
  - *Plasmodium girardi:* lemurs, quartan malaria
  - *Plasmodium lemuris:* lemurs, tertian malaria
- **Malaria of Old World monkeys**
  - *Plasmodium knowlesi:* Old World monkeys, quotidian malaria
  - *Plasmodium cynomolgi:* Old World monkeys, tertian malaria
  - *Plasmodium fieldi:* pigtailed macaques, tertian malaria
  - *Plasmodium gonderi:* mangabeys and mandrills, tertian malaria
  - *Plasmodium fragile:* macaques, tertian malaria
  - *Plasmodium simiovale:* Old World monkeys, tertian malaria
  - *Plasmodium coatneyi:* macaques, tertian malaria
  - *Plasmodium inui:* macaques, quartan malaria
  - *Plasmodium shorti:* toque monkey and bonnet macaques, quartan malaria
- **Malaria of New World monkeys**
  - *Plasmodium simium:* New World monkeys (Howler monkeys, spider monkeys, capuchin, woolly monkeys, squirrel monkeys), tertian malaria
  - *Plasmodium brasilianum:* New World monkeys (Howler monkeys, spider monkeys, capuchin, woolly monkeys, squirrel monkeys), quartan malaria
- **Malaria of apes**
  - *Plasmodium rodhaini* (synonym *Plasmodium malariae*): Chimpanzees and gorillas, quartan malaria
  - *Plasmodium pitheci:* orangutans, tertian malaria
  - *Plasmodium silvaticum:* orangutans, unknown form of malaria
  - *Plasmodium reichenowi:* apes, tertian malaria
  - *Plasmodium schwetzi:* chimpanzees and gorillas, tertian malaria

Sarcocystis spp.

Sarcocystis sp. are generally considered nonpathogenic coccidian parasites found in the striated, cardiac, and smooth muscle fibers of the rhesus macaque (*Sarcocystis kortei, Sarcocystis nesbitti*), baboons, and New World monkeys [6, 18, 56–60]. Transmission occurs through ingestion of the organisms in muscle or oocysts in feces [6]. Mature sarcocysts are cylindrical, spindle, or ellipsoidal shaped and are approximately 1.5 mm long, thin-walled, with protrusions approximately 13μm long (Fig. 3.26) [9]. The size of the banana-shaped mature trophozoites varies based on the species [6]. Although not always present, inflammatory infiltrates surrounding cysts consist of lymphocytes, plasma cells, and eosinophils, as well as possible scar tissue formation (Fig. 3.27) [33, 61].

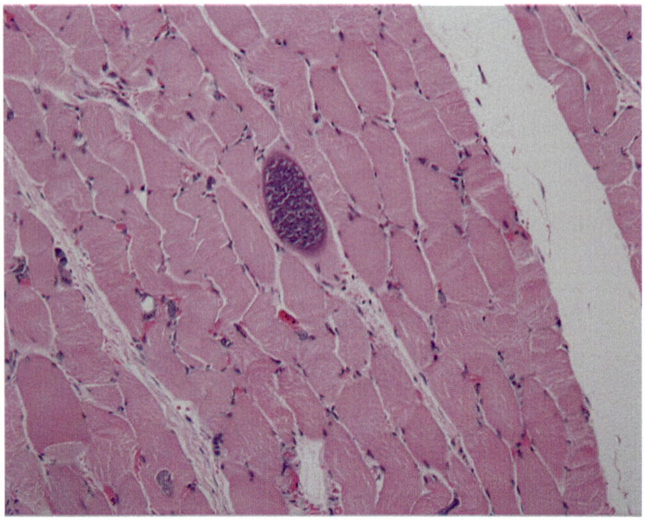

Fig. 3.26 *Sarcocystis* sp. Tissue cyst present in skeletal muscle myocyte with minimal host reaction (*Macaca mulatta*, skeletal muscle, H&E stain)

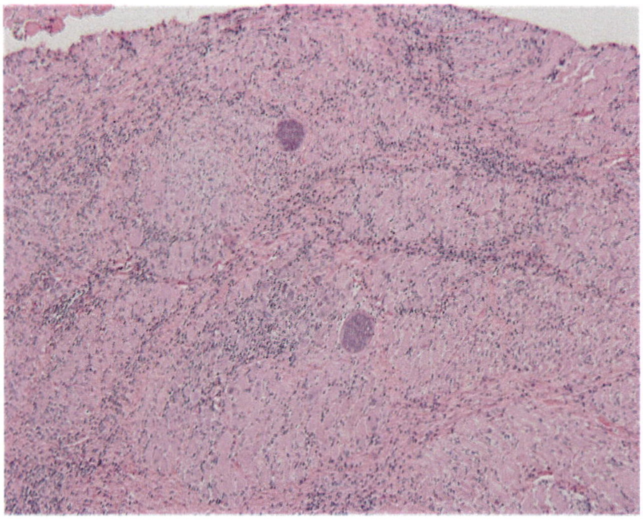

Fig. 3.27 *Sarcocystis* sp. Tissue cysts present in myocardium accompanied by myocardial degeneration and lymphoplasmacytic infiltrates (*Macaca fascicularis*, heart, H&E stain)

Cyclospora spp.

Cyclospora cayetanensis is a coccidia found in the small intestines of multiple species of nonhuman primates, including chimpanzees and baboons [9]. Clinical significance in nonhuman primates is not clear, but transmission is through ingestion of sporulated infective oocytes [9, 33]. Oocysts are approximately 8–10μm in diameter and contain 2 sporocysts, which each contains 2 sporozoites [9].

Entopolypoides macaci

Entopolypoides macaci is a mildly pathogenic *Babesia*-like piroplasm found in the erythrocytes of baboons, patas monkeys, macaques, guenons, and chimpanzees [6, 62–66]. Clinical signs are limited to mild anemia, despite prolonged parasitemia [6]. Early forms of *E. macaci* include fine rings with a large vacuole, whereas later stages are consistent with appliqué forms and Maltese cross formation [6].

Babesia spp.

Babesia pitheci is a piroplasmid parasite found in the erythrocytes of macaques, baboons, mangabeys, and marmosets [6, 33, 67]. Transmission occurs through an arthropod vector, presumably a tick [6]. Clinical signs include anemia and fever [56, 68, 69]. Gross and microscopic lesions included pericardial effusion and leukocytic infiltration of the liver and kidney, as well as hemosiderin deposition [56, 68, 69]. Pyriform piroplasms are approximately 2–6μm long [9].

Hepatocystis kochi

Hepatocystis kochi (synonym *Hepatocystis simiae*) is an intraerythrocytic parasite found in Old World monkeys, gibbons, and orangutans [6, 9]. Transmission occurs via the insect vector, the *Culicoides spp.* midge [6, 9]. Gross lesions include multiple gray-white foci (mature merocysts) on the surface of the liver [9]. Microscopically, intact cysts are surrounded by neutrophils, while ruptured cysts are associated with granulomatous inflammation, as well as lymphocytic infiltrates (Figs. 3.28 and 3.29) [9]. Intraerythrocytic parasites are apparent with a Giemsa stain revealing a large, pink, oval nucleoplasm that contains abundant red chromatin granules [9].

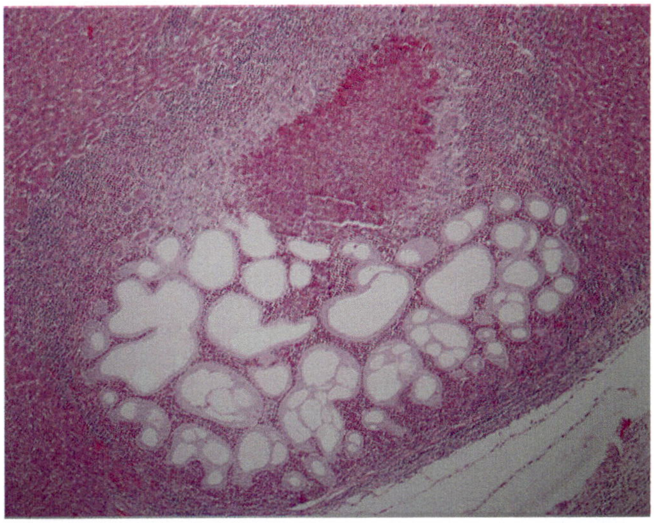

Fig. 3.28 *Hepatocystis kochi*. Myriad of schizonts present within mature merocyst (*Papio* sp., liver, H&E stain)

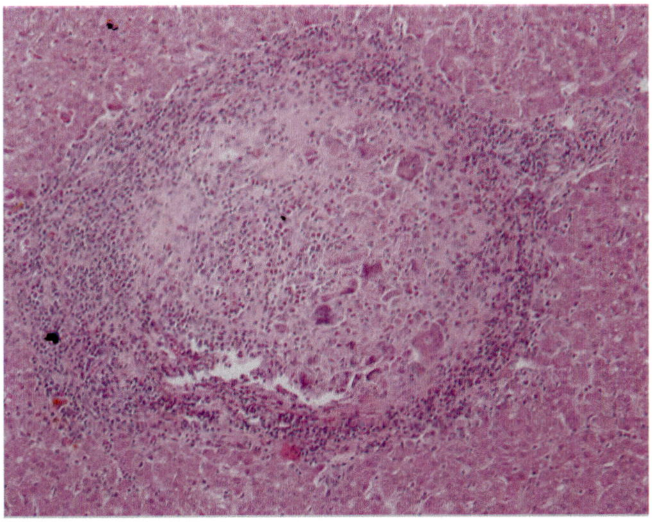

Fig. 3.29 *Hepatocystis kochi*. Remnant *H. kochi* infection demonstrating pyogranulomatous inflammation with multinucleated giant cells and eosinophils and absence of visible organisms (*Papio* sp., liver, H&E stain)

Klossiella

Klossiella sp. is classified in the suborder Adeleorina in the family Klossiellidae; it is generally considered nonpathogenic and has been identified in a galago [9, 70]. Oocysts are found within renal tubules and contain numerous sporocysts approximately 9–10μm in diameter, each containing multiple sporozoites approximately 1.3–1.9μm wide and 8–10μm long [70]. Transmission occurs through ingestion of sporocysts passed in the urine [9].

3.1.5 Amoeba

Entamoeba (*E. histolytica, E. dispar, E. gingivalis*)

Entamoeba histolytica (synonyms *Entamoeba dysenteriae, Endamoeba histolytica*) is a pathogenic protozoa found in the large intestine, liver, lungs, spleen, stomach, and/or rarely the brain of rhesus macaques, cynomolgus macaques, baboons, chimpanzees, spider monkeys, proboscis monkeys, colobus monkeys, silver leaf monkeys, and langurs [9, 71]. Transmission is through the ingestion of food and/or water contaminated with cysts shed in the feces. Infections range from asymptomatic to animals that experience diarrhea, dehydration, weight loss, anorexia, and emesis [9]. Microscopic lesions include small colonies of amoebae in the mucosa, submucosa, and muscularis of the large intestine exacerbated by co-infection with pathogenic bacteria where they can be associated with neutrophilic infiltrates (Fig. 3.30) [9].

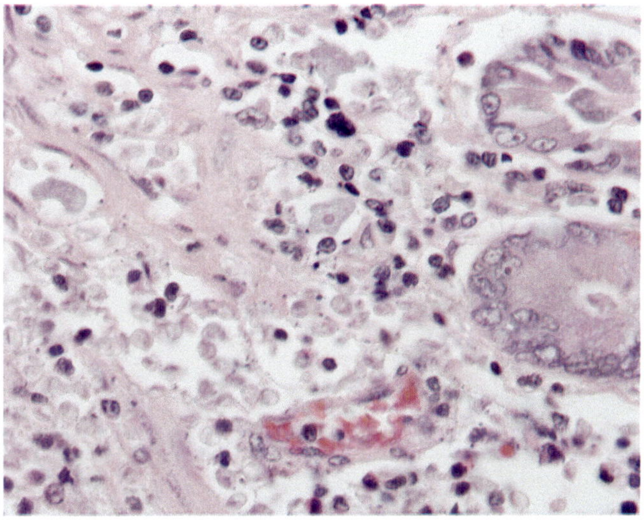

Fig. 3.30 *Entamoeba histolytica*. Invasive amoebiasis with irregularly shaped trophozoites present with the submucosa (*Macaca mulatta*, colon, H&E stain)

Amoeba can enter the lymphatics and migrate to the liver and other organs, where they form abscesses [9]. *E. histolytica* trophozoites range from 12–15μm to 20–30μm in diameter with a nucleus with a small central endosome and few scattered chromatin granules, as well as mature cysts approximately 10–12μm with four nuclei [9]. *E. dispar* is morphologically identical to *E. histolytica but* is considered nonpathogenic [6].

E. gingivalis is a nonpathogenic protozoa found in the oral cavity of rhesus macaques, cynomolgus macaques, baboons, and chimpanzees with underlying gingivitis [6, 18]. Transmission occurs through oral contact [6]. Trophozoites range from approximately 5–35μm long to 10–20μm long, have a nucleus with a small central endosome and a ring of peripheral granules, and lack a cyst form [9].

Balamuthia mandrillaris

Balamuthia mandrillaris is a pathogenic amoeba recognized in the brains of a mandrill, white-cheeked gibbon, western lowland gorillas, and a colobus monkey [6, 72]. Transmission occurs through ingestion of food or water contaminated with amoeba. Clinical signs included progressive limb paresis or paralysis, depression, and weakness [72]. Gross lesions within the central nervous system were either absent or included cerebral malacia and hemorrhage [72]. Microscopic lesions range from acute to subacute necrotizing meningoencephalitis to granulomatous amoebic meningoencephalitis with extraneural fibrogranulomatous inflammation (Figs. 3.31 and 3.32) [72]. Disseminated infections were evident as multifocal masses throughout the abdomen or within multiple organs [72]. Round trophozoites are approximately 15–30µm in diameter with a 3–5µm eccentric nucleus [72]. Cysts are approximately 10–15µm in diameter [72].

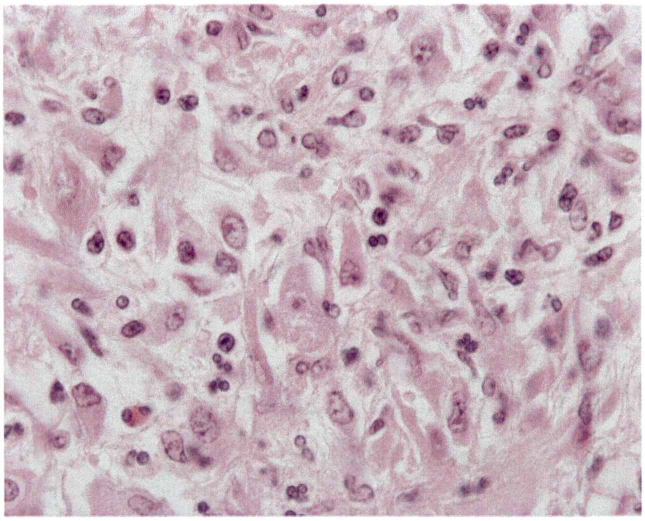

Fig. 3.31 *Balamuthia mandrillaris*. Rare trophozoite identified accompanied by extensive gliosis and mixed inflammatory cell infiltrate (*Gorilla gorilla*, brain, H&E stain)

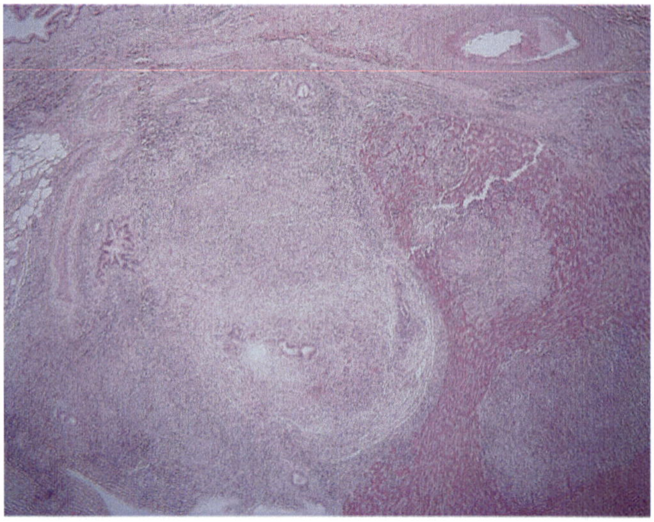

Fig. 3.32 *Balamuthia mandrillaris*. Multifocal chronic pyogranulomatous hepatitis (*Gorilla gorilla*, liver, H&E stain)

Endolimax nana

Endolimax nana (synonyms *Amoeba limax, Entamoeba nana, Endolimax intestinalis, Endolimax suis, Endolimax ratti*) is a nonpathogenic amoeba found in the cecum and colon of several nonhuman primates including rhesus macaques, cynomolgus macaques, baboons, chimpanzees, and gorillas [6]. Transmission occurs through ingestion of food or water contaminated with cysts [73]. Trophozoites are approximately 6–15µm diameter and mature cysts are oval and approximately 8–10µm long with four nuclei [9].

3.2 Metazoa

3.2.1 Nematodes

Ancylostomids (Hookworms)

Ancylostoma, Necator

Ancylostoma duodenale and *Necator americanus* are hookworms that are primarily found in humans but have been uncommonly reported in wild and laboratory primates (baboons, chimpanzees, gibbons, gorillas, mandrills, patas monkeys, and spider monkeys), which appear to be only incidental hosts [6, 33, 74–77]. Infection occurs through ingestion or larval penetration of the skin [6]. Gross lesions include worms attached to the intestinal mucosa, thickening and hemorrhage of the intestinal wall, and organ pallor due to anemia [6, 74, 75, 77]. *Ancylostoma duodenale* worms are 8–13 mm long; eggs are 60 by 40 μm. *Necator* worms are 7–11 mm long; eggs are 64–76 μm by 36–40 μm [6, 74]. Ancylostomid worms (platymyarian, smooth and thick cuticle, vacuolated lateral chords, and an intestine that is large and lined by few multinucleate enterocytes with a brush border) and eggs (elliptical, thin-shelled and contain a morula) are very similar to strongylids and are easily misidentified [78, 79].

Ascarids (Roundworms)

Ascarids are a common finding in the digestive tract of nonhuman primates and worms are indistinguishable from *Ascaris lumbricoides* in humans [33]. They have been reported in chimpanzees, gibbons, gorillas, orangutans and rhesus macaques [6, 33, 74, 75, 77]. Infections occur through ingestion and are typically subclinical, consisting simply of worms within the intestinal lumen [6]. Rare fatalities thought to be due to high parasite load, intestinal obstruction and migration of worms into the liver and bile duct exist in chimpanzees and macaques [80, 81]. Visceral migration of ascarid larvae, especially *Baylisascaris procyonis*, in paratenic primate hosts can also cause severe disease, especially when larvae migrate through the central nervous system. Published and unpublished reports exist from a wide range of primate species, including lemurs, bushbabies, marmosets, tamarin, titi monkeys, spider monkeys, macaques, gibbons, and orangutans [82–90]. Histologic lesions consist of granulomatous inflammation and larvae in non-central nervous system viscera. In the brain and spinal cord, migration tracts consist of malacia, microcavitation, foamy macrophages (Gitter cell), hemorrhage, and leukocyte infiltration (need to characterize). Mononuclear and eosinophilic perivascular cuffing, gliosis, astrocytosis, astrocytic scarring, and granulomatous or lymphoeosinophilic encephalitis can also occur [6, 33, 74, 75]. Larvae can be difficult to capture in histologic section in the brain and may be distant to migration tracts. Adult ascarids are large worms (in the range of 15–50 cm long by 2–6 mm wide) with coelomyarian musculature and an intestine that is large and lined by many uninucleate cuboidal to columnar cells with a low brush border. Many ascarid larvae (including *Baylisascaris procyonis*) have prominent lateral alae and an excretory column between the intestinal tract and the lateral chords. *Baylisascaris procyonis* larvae specifically are 60–80 μm in midbody diameter. Ascarid eggs are typically approximately 90 μm by 40–50 μm with a thick shell and uninucleate zygote [6, 78, 79].

Filariids

Filariids are long, thin worms that live outside of the digestive tract. Females produce microfilariae that live in the peripheral blood, lymph, skin, or intermuscular tissue of the definitive host. Microfilariae can be sheathed or unsheathed and may have a diurnal or nocturnal periodicity to their peripheral blood circulation in the host [6, 33, 74]. Infection occurs when the intermediate host, a biting or sucking insect, transmits microfilaria while feeding on the definitive host [6]. Most often, filariae live in the subcutaneous tissues and induce little to no inflammation. Adults can also live in the peritoneal or pleural cavity where heavy infections can result in fibrinous adhesions and fibrinopurulent pleuritis/peritonitis [6, 74, 91]. Filariids have coelomyarian musculature and most have a very small intestine. Free microfilariae may be visible in the uteri of females, although some "primitive" filariids have thick-shelled eggs containing the larvae [78, 79].

Dirofilaria

Dirofilaria corynodes is the most commonly reported species of African Old World monkeys (colobus monkeys, guenons, mangabeys, patas monkeys, and vervet monkeys) and has also been found in langurs [6, 33, 74, 77, 91–93]. Worms typically reside in the subcutis of the trunk and lower extremities with minimal tissue reaction. Male worms are approximately 8 cm long while females are 20–30 cm long; microfilaria are 250–290 μm long [6]. Other reported species and hosts include *D. magnilarvatum* and *D. macacae* in the subcutis and peritoneal cavity of cynomolgus and rhesus macaques; *D. immitis* (syn. *D. pongoi*) in the heart of an orangutan and saki monkey and the peritoneal cavity of an orangutan [33, 92].

Dipetalonema, Mansonella, Sandnema, Tetrapetalonema

This family of worms has been challenging for taxonomists to classify and species have been moved between related genera over the years [94]. *Dipetalonema caudispina*, *D. gracile*, *D. marmosetae*, and *D. tamarinae* are most commonly reported with fewer reports of *D. graciliformis*, *D. freitasi*, *D. obtusa*, *D. parvum*, *D. robini*, and *D. tenue*. Many species of Mansonella have been reported from primates, including *M. atelense* (syn. *D. atelense*), *M. colombiensis*, *M. mystaxi*, *M. ozzardi*, *M. panamensis*, *M. perstans* (syn. *Dipetalonema perstans*), *M. rodhaini* (syn. *Dipetalonema rodhain*), *M. streptocerca* (syn. *Dipetalonema streptocerca*), and *M. vanhoofi* (syn. *Dipetalonema vanhoofi*). These genera of filariids primarily infect New World monkeys (callitrichids, capuchins, lorises, owl monkeys, spider monkeys, squirrel monkeys, and woolly monkeys), although *M. streptocerca*, *M. rodhaini*, and *M. vanhoofi* are found in chimpanzees and *Sandnema digitatum* is reported from gibbons [6, 74, 77, 94]. Mixed infections can occur. *M. streptocerca* and *M. rodhaini* are unique in that microfilariae are present in the subcutis and not the peripheral blood. Adults range from approximately 2–20 cm long with females being much larger than males; microfilaria are 130–430 μm long [6].

Other Filariids

Edesonfilaria malayensis is described in cynomolgus and rhesus macaques [6, 33, 74]. Adults typically reside free in the peritoneal cavity or within the subserosal connective tissue. In few reports, worms are associated with retroperitoneal or serosal nodules consisting of fibrosis/fibroplasia, granulation tissue, hemorrhage and lymphocytic and eosinophilic inflammation [33]. *Macacanema formosana* is reported in the peritracheal connective tissue and diaphragm of macaques [6, 33, 74, 95]. *Loa papionis* and *L. loa* are reported in the subcutis and peritoneal cavity of baboons, drills, guenons, gorillas, and mangabeys, occasionally microfilaria can elicit granulomatous nodules on the superficial red pulp of the spleen. *Brugia malayi* and *B. pahangi* are reported in the lymph nodes and lymphatics of macaques, leaf monkeys, and slow lorises with no associated lesions. *Meningonema peruzzii* are reported in the subarachnoid space of the dorsal brainstem of vervet monkeys and talapoin monkeys with no associated lesions [96]. *Onchocerca volvulus* has been reported in a subcutaneous nodule from a gorilla [6, 33, 74].

Oxyurids (Pinworms)

Enterobius, Trypanoxyuris

Trypanoxyuris, Oxyuronema, and *Enterobius* spp. are most commonly reported and have been found in a range of New and Old world monkeys, great apes and prosimians, including baboons, callitrichids, capuchins, chimpanzees, gibbons, gorillas, howler monkeys, lemurs, lorises, night monkeys, orangutans, rhesus monkeys, spider monkeys, squirrel monkeys, and woolly monkeys [6, 33, 74–77, 97, 98]. Infection occurs through ingestion of infective eggs [6]. Adults reside in the cecum and colon. Oxyurid infections are typically subclinical and gross lesions are often restricted to self-mutilation of the perineum due to perineal/anal pruritus. Subclinical infections lack histologic lesions aside from adult worms in the intestinal lumen. Fatal Enterobius infections have been reported in chimpanzees due to ulcerative enterocolitis and granulomatous peritonitis, lymphadenitis, and hepatitis [97, 98]. Adult worms are approximately 2–13 mm long by 100–500 μm wide with platymyarian musculature, many uninucleate cells lining the intestine and most have a rhabditoid esophagus with a corpus, isthmus and bulb and lateral alae. Eggs are 50–60 μm by 20–30 μm, slightly flattened on one side, thick-shelled and contain an embryo [6, 78, 79].

Rhabditoids

Strongyloides

Strongyloides cebus, *S. fuelleborni*, and *S. stercoralis* are the most commonly reported species with fewer reports of *S. simiae* and *S. papillosus*. Reported host animals for *strongyloides* sp. include baboons, capuchins, cebus monkeys, chimpanzees, cynomolgus and rhesus macaque, gibbons, gorillas, guenons, Japanese monkeys, lemurs, lorises, marmosets, orangutans, patas monkeys, spider monkeys, squirrel monkeys, and woolly monkeys [6, 33, 74–77, 99]. Infection occurs when larvae penetrate the host's skin or digestive tract mucosa after ingestion [6]. Only adult females and larvae are present in the gastrointestinal tract and migrating filariform larvae are found in lungs and other parenchymal organs. Gross lesions include catarrhal, hemorrhagic, or necrotizing enterocolitis occasionally with secondary peritonitis and pulmonary hemorrhage. Histologic lesions include erosive and ulcerative enteritis with adults, larvae, and eggs that are mainly in intraepithelial tunnels in the mucosa or in intestinal gland lumina. Neutrophilic inflammation is common with fewer eosinophils and neutrophils and lymphoid hyperplasia. Villous blunting, loss, and fusion may occur in severe infections. Invasion of larvae into the deeper lymphatics and mesenteric lymph nodes may result in granulomatous lymphangitis, lymphangiectasia, edema, and fibrosis. Larvae that migrate to parenchymal organs may be unaccompanied by a tissue response or can produce hemorrhage and serosal granulomas. Larvae in the lungs are associated with interstitial pneumonia and mild to severe hemorrhage [6, 33, 75, 99]. Adult female worms are 2–5 cm long by 30–80 μm wide with a short pointed tail. First-stage larvae are 150–800 μm in length and 12–20 μm in width. Eggs are 40–70 μm by 20–30 μm, thin-shelled and initially uninucleate but embryonate in the tissue in most hosts [6]. Worms have platymyarian musculature, an intestine with uninucleate cells, a paired genital tract, and a rhabditiform muscular esophagus with a corpus, isthmus, and bulb; however, the small size of these worms may make it difficult to appreciate these features or capture them in histologic sections [78, 79].

Spirurids

Trichospirura

Trichospirura leptostoma have been reported from New World monkeys, including marmosets, tamarins, owl monkeys, and tamarins [6, 33, 74, 77, 100]. Route of infection is presumed to be ingestion [6]. Infections are often subclinical, but heavy worm burdens are associated with pancreatitis in marmosets and owl monkeys and wasting syndrome in common marmosets [100]. Gross lesions are typically absent, and worms are found in the inter- and intralobular pancreatic ducts incidentally on histologic examination (Fig. 3.33).

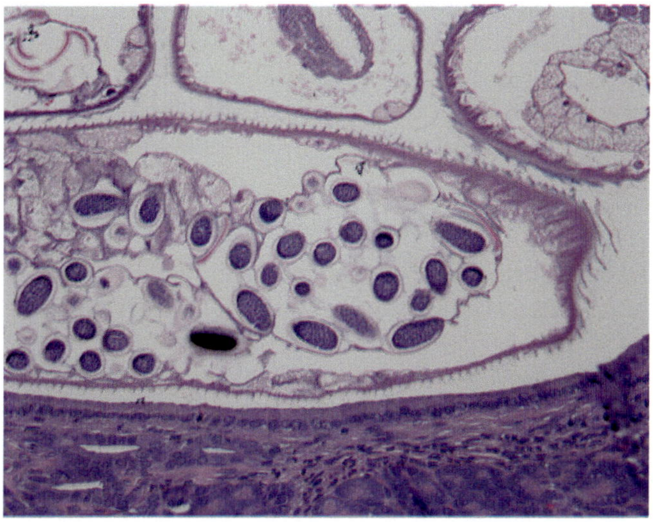

Fig. 3.33 *Trichospirura leptostoma.* Spirurid nematode present in pancreatic duct of common marmoset with minimal host reaction (*Callithrix jacchus*, pancreas, H&E stain)

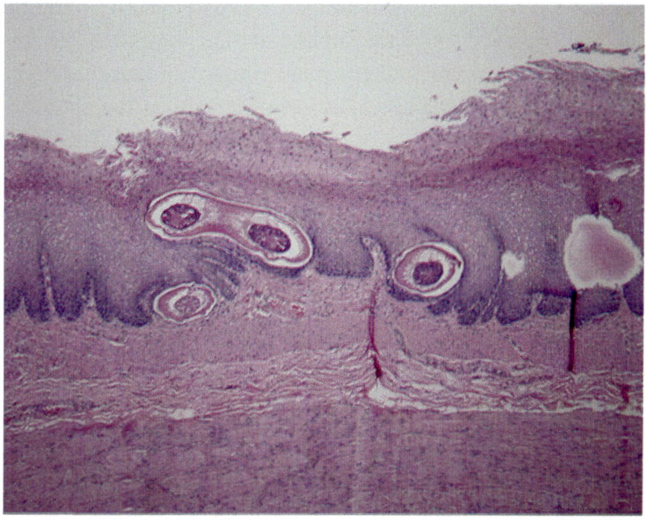

Fig. 3.34 *Gongylonema pulchrum.* Intraepithelial nematode present with oral mucosa (*Macaca fascicularis*, H&E stain)

There is mild, if any, tissue reaction (duct ectasia, epithelial attenuation, periductal fibrosis, lymphoeosinophilic peridochitis, and parenchymal necrosis and atrophy) [6, 33, 74]. Adult worms are up to 20 mm long [33, 74]. They have typical spirurid features, including coelomyarian musculature, eosinophilic fluid in the pseudocoelom, lateral chords, an esophagus with an anterior muscular portion and posterior glandular portion, an intestine lined by many uninucleate cuboidal to columnar cells, and extensive uteri containing embryonated eggs [78, 79]. Eggs are 50–55 μm by 23–30 μm, thick-shelled and contain an embryo/larva with a large hook and several rows of spines on the anterior end [74].

Gongylonema

Gongylonema macrogubernaculum and/or *G. pulchrum* have been reported New and Old World monkeys and prosimians, including baboons, callitrichids, capuchins, gibbons, guenons, lemurs, macaques, patas monkeys, spider monkeys, squirrel monkeys, and varis [6, 33, 74, 76, 101, 102]. Route of infection is ingestion, often of an insect intermediate host [6]. Infections are typically subclinical; callitrichids can exhibit facial pruritus and ptyalism. Gross lesions are rare and worms are often only found incidentally on histology. Esophageal squamous cell carcinoma was attributed to *Gongylonema* infection in a vari [101]. Adults are present in tunnels in the stratified squamous epithelium of the lips, oral cavity, and esophagus with little to no tissue reaction (Fig. 3.34).

Empty tunnels may contain eggs and few granulocytes. Swallowed or aspirated eggs may be found in the stomach and bronchi. Cytological diagnosis can be made via tongue scraping [102]. Adults are 5–55 mm long and the anterior end has several rows of conspicuous, round to oval, cuticular bosses [74]. They have typical spirurid features, including coelomyarian musculature, eosinophilic fluid in the pseudocoelom, lateral chords, an esophagus with an anterior muscular portion and posterior glandular portion, an intestine lined by many uninucleate cuboidal to columnar cells, and extensive uteri containing embryonated eggs [78, 79]. Eggs are 50–70 μm by 25–35 μm, ovoid, thick-shelled and contain an embryo/larva with a large hook and several rows of spines on the anterior end [74].

Pterygodermatites

Pterygodermatites (syn. *Riticularia*) *nycticebi* and/or *P. alphi* have been reported in capuchins, guenons, gibbons, marmosets, tamarins, and slow lorises [6, 33, 75, 76, 103]. Route of infection is ingestion, often of an insect intermediate host [33]. In golden lion tamarins, adult worms can be found grossly in the intestinal tract lumen. On histology, adult worms are anchored in the mucosa by their anterior end and larvae are in the deeper submucosa. Rarely, worms may be present in the tunica muscularis and pancreatic ducts. Small intestinal villi may be blunted and overlain by a necrotic pseudomembrane containing spirurid eggs [33, 103]. Adults are approximately 9–30 mm long and 0.5–1 mm wide; they have typical spirurid features, including coelomyarian musculature, eosinophilic fluid in the pseudocoelom, lateral chords, an esophagus with an anterior muscular portion and posterior glandular portion, an intestine lined by many uninucleate cuboidal to columnar cells, and extensive uteri containing embryonated eggs [78, 79]. *P. nycticebi* have sublateral alae. Eggs are 39–45 μm by 26–36 μm, thick-shelled and contain an embryo/larva with a large hook and several rows of spines on the anterior end [103].

Streptopharagus

Streptopharagus armatus and/or *S. pigmentatus* have been reported in baboons, gibbons, guenons, macaques, and patas monkeys [6, 33, 75, 104]. Adults are found in the lumen of the stomach. Route of infection is ingestion, often of an insect intermediate host [104]. Pathogenic effects are not well-documented. Adults are 30–56 mm long by 0.75–1.3 mm wide; they have typical spirurid features, including coelomyarian musculature, eosinophilic fluid in the pseudocoelom, lateral chords, an esophagus with an anterior muscular portion and posterior glandular portion, an intestine lined by many uninucleate cuboidal to columnar cells, and extensive uteri containing embryonated eggs [6, 78, 79]. Eggs are 28–38 μm by 17–22 μm, thick-shelled, asymmetrical and contain an embryo/larva with a large hook and several rows of spines on the anterior end [6].

Physaloptera

Nine species within the genus Physaloptera have been reported in nonhuman primates, including *Physaloptera dilatata* in the stomach of capuchins, lemurs, marmosets, saki monkeys, titi monkeys, and woolly monkeys; *P. tumefaciens* in the stomach of Asian macaques; *Physaloptera* (syn. *Abbreviata*) caucasica in esophagus, stomach, and small intestine of baboons, orangutans, and rhesus macaques; and *Physaloptera* (syn. *Abbreviata*) *poicilometra* in the stomach of guenons and mangabeys [6, 33, 74, 76]. Route of infection is ingestion, often of an insect intermediate host [6]. Worms found at necropsy in the lumen of the infected organs. Attachment to the wall can produce local inflammation, erosion, and ulceration. Hyperplasia, lumen obstruction, perforation, and death are rarely reported [6, 33, 74, 76]. Worms are thick and muscular, resembling ascarids, and approximately 15–60 mm long [6]. They have typical spirurid features, including coelomyarian musculature, eosinophilic fluid in the pseudocoelom, lateral chords, an esophagus with an anterior muscular portion and posterior glandular portion, an intestine lined by many uninucleate cuboidal to columnar cells, and extensive uteri containing embryonated eggs [78, 79]. Eggs are 39–50 μm by 23–34 μm, thick-shelled, asymmetrical and contain an embryo/larva with a large hook and several rows of spines on the anterior end [6].

Other Spirurids

Other spirurids have been reported from nonhuman primates, including *Spirura guianensis* in the esophagus of tamarins and squirrel monkeys, *Chitwoodspirura serrata* in the stomach and small intestine of chimpanzees and gorillas, *Protospirura muricola* in the esophagus and stomach of Cebus monkeys, *Metathelazia ascaroides* in the lungs of guenons, and *Thelazia callipaeda* in the eyes of monkeys [6, 33].

Strongylids

Oesophagostomum (Nodular worm)

Oesophagostomum aculeatum, O. apiostomum, O. bifurcum, and *O. stephanostomum* are the most common species (although there are many others) and have been reported from baboons, chimpanzees, gibbons, gorillas, guenons, macaques, mangabeys, and orangutans [6, 33, 74, 75, 105]. These worms are rare in New World monkeys. Infection occurs through ingestion of infective larvae [6, 33]. Infection is typically mild and subclinical. Within the host, worms reside within a 2–4 mm diameter, smooth, firm, nodule within the intestinal wall, most often within the serosa or mesentery of the cecum or colon. Less frequently they may be found in the diaphragm, kidney, liver, lung, omentum, peritoneum, or small intestine. Nodules may be red, brown, or black when they contain hemorrhage or white or yellow if they are older and contain caseous material or mineralization. In the acute stages, intact and viable worms are visible within nodules. Occasionally, a mucosal ulcer and a fistulous tract can be found connecting the site of mucosal penetration to the nodule. Secondary bacterial infection can result in ulceration, perforation, peritonitis with fibrous adhesions and septicemia. When infection results in hypoproteinemia, hydropericardium and hydrothorax may occur. On histology, worms are surrounded by necrotic cellular debris, hemorrhage, and/or mixed inflammatory cells (neutrophils, macrophages, multinucleated giant cells, eosinophils, lymphocytes, and plasma cells), with a variably thick and mature fibrous capsule depending on the chronicity (Fig. 3.35). In rhesus macaques, granulomatous lesions lacking caseation have been reported in the colon, kidney, prostate, pancreas, and heart [6, 33, 74, 75, 105]. Worms are 8–11 mm long by 200–300 μm wide; eggs are approximately 60–85 μm by 25–55 μm, elliptical, thin-shelled and contain a morula [6]. Oesophagostomum has a thick and smooth cuticle, platymyarian musculature, vacuolated lateral chords, and an intestine that is large and lined by few multinucleated cells with a prominent brush border [78, 79].

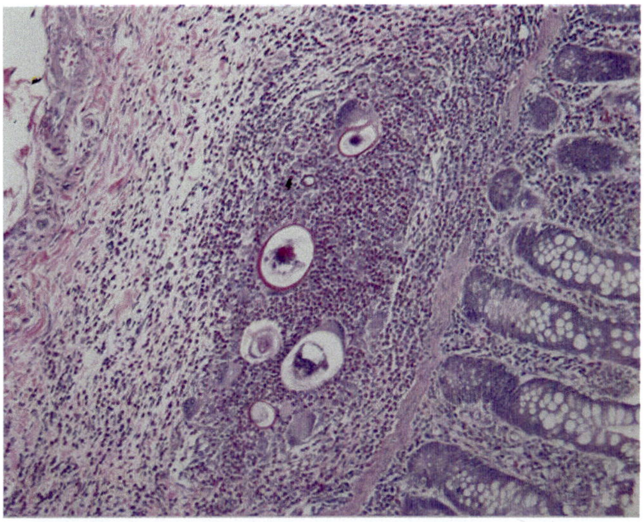

Fig. 3.35 *Oesophagostomum* sp. Pyogranulomatous inflammation with eosinophils and intralesional nematode present in submucosa of colon (*Macaca fascicularis*, colon, H&E stain)

Ternidens

Ternidens deminutus has been reported in baboons, chimpanzees, gorillas, guenons, and macaques [6, 33, 74]. Infection occurs through ingestion of infective larvae [6]. Infection is typically subclinical and lesions are often limited to worms in the small and large intestinal lumen that occasionally cause mucosal damage, cystic nodules, luminal hemorrhage, and anemia [6, 33, 74]. The morphology of *Ternidens* is similar to *Oesophagostomum* (eggs 57–65 μm by 36–45 μm) [6].

Trichostrongylids

Molineus

Molineus elegans, *M. torulosus*, *M. vexillarius*, and *M. vogelianus* are all reported, although only *M. torulosus* appears to be pathogenic. Infection has been documented in capuchins, night monkeys, pottos, squirrel monkeys, and tamarins [6, 33, 77, 106]. Route of transmission is unknown. Worms are small, slender (3–5 mm long), pale red and reside on (but unattached to) the mucosa of the stomach and duodenum, and less often in the pancreas and intestinal mesentery. *M. torulosus* produces necrotic ulcers and hemorrhage in the intestinal wall [6]. Serosal nodules that contain worms and communicate with the intestinal lumen through an ulcer have been documented in capuchins [106]. Histologically, nodules consist of a granuloma with a central core of worms and eggs surrounded by eosinophilic debris and mixed leukocytes. In pancreatic infections, chronically inflamed pancreatic ducts contain worms and eggs [33]. Worms have platymyarian musculature, evenly spaced external longitudinal cuticular ridges, and an intestine that is large and lined by few multinucleated cells with a prominent brush border [78, 79]. Eggs are elliptical, 40–52 μm by 20–30 μm, thin-shelled and contain a morula [6].

Nochtia

Nochtia nochti has been reported in rhesus, cynomolgus, and stump-tailed macaques [6, 33, 74, 77]. The worms reside within the mucosa at the junction of the fundic and pyloric stomach. Route of infection is presumed to be ingestion and infections are typically subclinical [6]. The gross lesion is a reddened, bosselated, polypoid mass. Histologically, the polyp consists of benign papillary projections of hyperplastic gastric mucosa supported by a fibrovascular stalk. In the gastric glands, there is marked hyperplasia of mucous neck cells that replace parietal cells (metaplasia). There may be proprial or submucosal infiltrates of lymphocytes, plasma cells and eosinophils, edema and fibrosis [6, 33, 74]. Worms and eggs are in the deep mucosa or submucosa. Worms are bright red, small, and slender (5–10 mm long by 100–170 μm wide) with typical trichostrongylids features (platymyarian musculature, evenly spaced external longitudinal cuticular ridges, and an intestine that is large and lined by few multinucleated cells with a prominent brush border [6, 78, 79]. Eggs are thin-shelled, elliptical, 60–80 μm by 35–42 μm and contain a morula [6].

Other Trichostrongylids

Additional reports of trichostrongylids include *Trichostrongylus colubriformis* in baboons, chimpanzees, and rhesus macaques, *Pithecostrongylus alatus* in guenons and orangutans, *Graphidiodes berlai* in spider monkeys and woolly monkeys, *Nematodirus weinbergi* in chimpanzees, and *Longistriata dubia* in howler monkeys, squirrel monkeys, and tamarins [6]. In all of these cases, worms inhabit the intestinal lumen of the host but produce minimal disease.

Metastrongylids

Filaroides, Filariopsis

Filaroides barretoi, *F. gordius*, and *Filariopsis arator* are lungworms that are most commonly reported from New World monkeys, including capuchins, cebus monkeys, howler monkeys, marmosets, squirrel monkeys, and tamarins. Route of infection is unknown. Most infections are subclinical. Gross lesions are subtle and include small, slightly elevated subpleural nodules in the lung that may be darkly pigmented. First-stage larvae can be found in the feces. Histologically, worms are within bronchioles and alveoli and there may be variable levels of atelectasis as well as little or no chronic inflammation in the alveolar lumina and interstitium. Worms are slender and fragile [6, 33, 74]. Morphological descriptions of the species affecting primates are lacking, but the more well-studied *Filaroides osleri* worms are 5–14 mm long with eggs measuring 80 by 50 μm [6]. Like all other metastrongylids, these worms have coelomyarian musculature, an intestine that is large and lined by few multinucleated cells with a prominent brush border, and typically have accessory hypodermal chords [78, 79].

Angiostrongylus

Angiostrongylus (syn. *Parastrongylus*) *costaricensis* and *A. cantonensis* (*P. cantonensis*) have been reported to produce disease in callitrichids, gibbons, night monkeys, orangutans, owl monkeys, and siamangs similar to that in rats and humans [6, 33, 75, 107–109]. Grossly, thickening, discrete granulomas and/or hemorrhage are present in the small intestinal wall and mesenteric artery (*A. costaricensis*) or pulmonary vessels (*A. cantonensis*). Intact worms can sometimes be recovered from the mesenteric vessels, pulmonary vessels, or heart. Histologically, microgranulomas/granulomatous inflammation in all tunics of the intestine was accompanied by larvae and eggs that are within blood vessels or in the interstitium. Larvae can also be found within portal triads [33, 109]. *A. cantonensis* can also produce eosinophilic meningoencephalitis and/or myelitis [109]. Morphological descriptions of *Angiostrongylus costaricensis* are lacking, but metastrongylids in general have coelomyarian musculature, an intestine that is large and lined by few multinucleated cells with a prominent brush border, and typically have accessory hypodermal chords [78, 79].

Trichurids

Trichuris (Whipworms)

Trichuris sp. have been reported from a wide range of New and Old World monkeys, prosimians, and apes, including African green monkeys, baboons, chimpanzees, gibbons, howler monkeys, lemurs, lorises, macaques, squirrel monkeys, woolly monkeys [6, 33, 74]. The route of infection is ingestion and adult worms reside in the mucosa of the cecum and colon. Infections are usually subclinical unless worm burden is heavy, in which typhlocolitis can occur and sometimes result in death [6]. Adults are approximately 30–50 mm long and the anterior end is embedded in the intestinal mucosa and is much thinner than the thicker and more blunt proximal portion (Fig. 3.36 and 3.37) [6, 74]. Like all aphasmids, *Trichuris* has bacillary bands rather than lateral chords and a row of basophilic esophageal glands that forms a stichosome surrounding the esophagus [78, 79]. Eggs are brown, ovoid with bipolar plugs and 50 × 22 μm [6, 74].

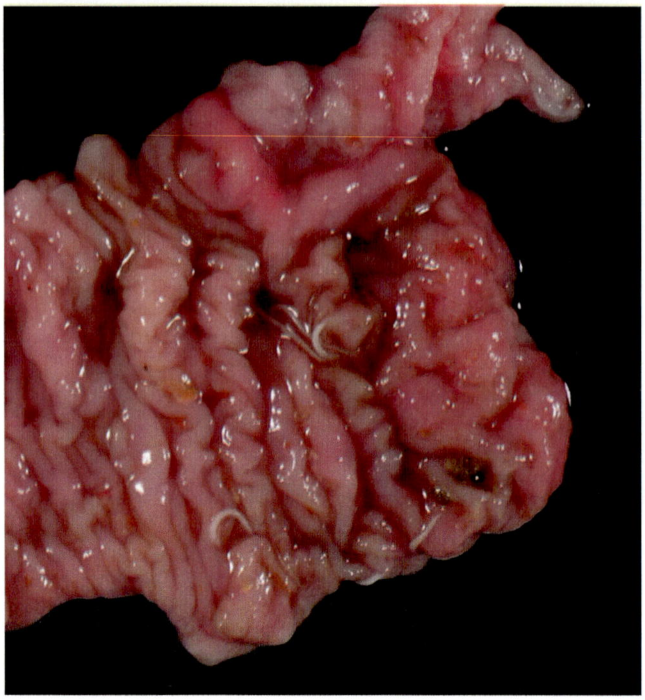

Fig. 3.36 *Trichuris trichiura*. Numerous thin white nematodes embedded within the cecal and ascending colonic mucosa. Mucosal thickening was attributed to concurrent infection with *Campylobacter coli* (Macaca mulatta, cecum and ascending colon)

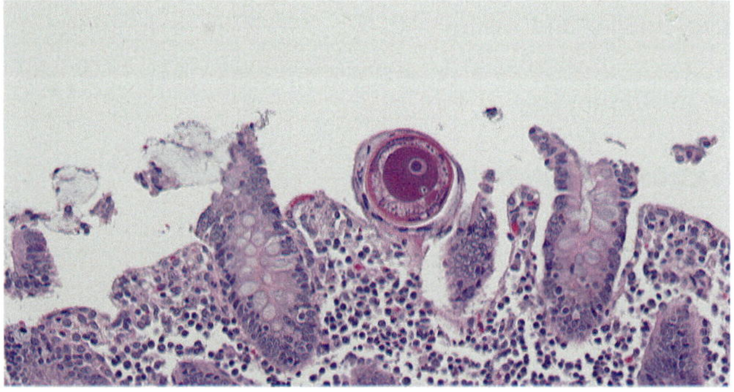

Fig. 3.37 *Trichuris trichiura*. Cross section of the anterior end of a trichurid embedded within the superficial ascending colonic mucosa. The section demonstrates a stichosome surrounding the esophagus, stichosome nucleus, bacillary bands (*Macaca mulatta*, ascending colon, H&E)

Capillaria

Capillaria hepatica (syn. *Calodium hepatica*) has been reported from a wide range of New and Old World monkeys and apes, including callitrichids, chimpanzees, baboons, cebus monkeys, gorillas, lemurs, rhesus macaques, saki monkeys, and spider monkeys [6, 33, 74, 75, 110, 111]. The route of infection is ingestion of embryonated eggs, typically from environments that are contaminated by decomposed liver or feces of predated natural rodent hosts

[6]. Adults reside in the liver parenchyma where they deposit eggs. Gross lesions include white or yellow plaques or nodules on the liver. Histologic lesions include adult worms and eggs that are surrounded by fibrous connective tissue, chronic inflammatory cell infiltrates, granulomatous inflammation, and multinucleated giant cells. Eventually, hepatic lesions are replaced by fibrosis, and cirrhosis can occur [6, 33, 74, 75, 110, 111]. Fatal hepatitis has been reported [74]. Adults are 17–100 mm long by 40–200 μm wide [6]. The anterior end of the worm is thinner than the posterior, similar to *Trichuris* but less pronounced. Like all aphasmids, *Capillaria* has bacillary bands rather than lateral chords and a row of basophilic esophageal glands that forms a stichosome surrounding the esophagus [78, 79]. Eggs are ovoid, 48–62 μm long by 29–37 μm wide, with bipolar plugs and many small perforations in the shell giving it a pitted appearance [6, 74].

Anatrichosoma

Anatrichosoma cynomolgi (syn. *A. cutaneum, A. rhina, or A. nacepobi*) had been reported from a wide range of New and Old World monkeys and apes, including baboons, gibbons, langurs, mangebeys, marmosets, orangutans, patas monkeys, siamangs, talapoin monkeys, and vervets [6, 33, 74, 75, 77]. The route of infection is unknown. Adults migrate through the stratified squamous epithelium of the nares or skin of the extremities, where they form tunnels and deposit eggs. Gross lesions include white serpentine tracks in the palmar and plantar skin that may have seropurulent discharge or vesicle, bulla or pustule formation. Regional lymph nodes may be enlarged. Histologic lesions include squamous hyperplasia, parakeratotic hyperkeratosis, and lymphoplasmacytic dermatitis/rhinitis associated with intraepithelial adults and larvated eggs [6, 33, 74]. Adults are 60–200 μm wide and like all aphasmids have bacillary bands rather than lateral chords and a row of basophilic esophageal glands that forms a stichosome surrounding the esophagus. Eggs are ovoid, 60–67 μm by 40–48 μm with bipolar plugs and contain a larva (unlike *Trichuris* and *Capillaria* that lack larvae) [6, 78, 79].

3.2.2 Cestodes (Tapeworms)

Many cestodes, or tapeworms, have been reported from nonhuman primates, often with little to no clinical disease or host response; only the most commonly reported and significant are discussed below. Adults reside in the intestinal tract where they rarely cause disease. Larval cestodiasis occurs when nonhuman primates act as the intermediate host for several tapeworm species. Larvae can occur in solid (sparganum) or bladder forms (cysticercus, coenurus, hydatid cysts, and tetrathyridium) depending on the species.

Anoplocephalids, Davaineids, Dilepidids, Hymenolopids

Adults of these families of cestodes live in the small intestine of nonhuman primates, but rarely cause clinical disease or pathological lesions. *Bertiella* spp., *Hymenolepis* spp., *Raillietina* spp., and *Dilepis* sp. are the most commonly reported [6, 33, 75–77, 112]. *Hymenolepis nana* can cause catarrhal enteritis, obstruction, and abscessation of mesenteric lymph nodes. The route of infection for most is ingestion of an intermediate host; *H. nana* can be transmitted directly through ingestion of eggs passed by the definitive host [6, 33]. Adult cestodes have body segmentation including an anterior segment, called a scolex, with suckers and/or hooks that act as a holdfast, followed by a series of developing proglottid segments. Each proglottid has an external tegument and a parenchymus body that lacks a cavity and digestive tract but has muscle fibers, excretory canals, basophilic calcareous corpuscles, and both male and female reproductive organs [78, 79].

Diphyllobothrids

Diphyllobothrium and *Spirometra* are the most important genera, whose plerocercoid larva, called spargana/sparganum, infect the intermediate nonhuman primate host. Infection (sparganosis) has been reported in New and Old World monkeys, including baboons, cynomolgus macaques, marmosets, rhesus macaques, squirrel monkeys, talapoin monkeys, and vervet monkeys [6, 33, 112–115]. The route of infection is ingestion of infected primary (crustacean) or secondary intermediate hosts (amphibian, reptile) [6]. Most infections are subclinical. Grossly, there are small nodules in the subcutis, skeletal muscle, pleural and peritoneal cavities, retroperitoneal tissues, and parenchymal organs. Often, live spargana are encased in a thin connective tissue capsule, sometimes called a larval cyst, and elicit minimal host reaction; when larva die or migrate through tissues they can cause local inflammation, hemorrhage, necrosis, and/or edema. Spargana are white, ribbon-like and range from a few millimeters to 30 cm long [6, 33, 112, 113]. Histologically, spargana have a solid, parenchymus body that is covered in tegument and contains muscles, basophilic calcareous corpuscles, and excretory canals but lacks scolices, suckers, a body cavity, and a digestive tract [78, 79]. They are typically between 250 and 500 μm in diameter. The capsule wall encasing spargana is often infiltrated by lymphocytes and plasma cells. Macrophages, multinucleated giant cells, lymphocytes, and eosinophils may surround the intact and degenerate spargana within the capsule. Free larva may result in a more florid granulomatous reaction with lymphocytes and eosinophils [113].

Mesocestoidids

The tetrathyridial larva of *Mesocestoides* spp. uncommonly parasitize Old World monkeys and apes, including baboons, gibbons, guenons, rhesus and cynomolgus macaques, and vervet monkeys, as intermediate hosts [6, 33, 77, 112, 116]. Infection is typically subclinical with little to no host response. Tetrathyridiosis is similar to sparganosis, in which larvae are loose within body cavities or are surrounded by a thin fibrous capsule (i.e., encysted) within the thoracic and abdominal organs. Tetrathyridial larvae are 2–70 mm long, flat and have a contractile body. The anterior end is knob-like and has an invaginated holdfast with four suckers and no hooks [6, 33, 112]. Larva have typical histologic characteristics of cestodes (a parenchymus body that is covered in tegument and contains muscles, basophilic calcareous corpuscles, and excretory canals but lacks a true body cavity and a digestive tract) [78, 79].

Taeniids

The larval stage of these cestodes is either a cysticercus, coenurus, or hydatid depending on the species. Larva resemble a fluid-filled cyst with an anterior scolex and neck that are invaginated into the posterior, fluid-filled, bladder end of the worm. Nonhuman primates act as intermediate hosts and become infected following ingestion of eggs shed in the feces of the definitive host. Infections are typically subclinical. Cysts may compress adjacent tissue as they enlarge and as they degenerate they may elicit more florid inflammation or become mineralized. Histologically, cysts are surrounded by a fibrous connective tissue capsule and variable infiltrates of lymphocytes, plasma cells, macrophages, multinucleated giant cells, and eosinophils [6, 33, 112, 117–119]. Larva have typical histologic characteristics of cestodes, including a parenchymus body that is covered in tegument and contains muscles, basophilic calcareous corpuscles (that are numerous in the scolex and neck but rare in the bladder), and excretory canals but lacks a true body cavity and a digestive tract [78, 79].

Taenia crassiceps, *T. crocutae*, *T. hydatigena*, and *T. solium* have a cysticercoid larva that has been reported in New and Old World monkeys, apes, and prosimians, including baboons, chimpanzees, gibbons, langurs, lemurs, mangabeys, marmosets, patas monkeys, squirrel monkeys, and vervet monkeys [6, 33, 75, 77, 112, 119]. Cysticerci have a dilated, bladder-like posterior end into which a single scolex with four suckers is invaginated [78, 79]. They are found in the pleural and peritoneal cavities, subcutis, heart, liver, and central nervous system [33, 112].

Taenia (syn. *Multiceps*) *multiceps*, *T. serialis*, and *T. brauni* have a coenurus larvae that have been reported in Old World monkeys and prosimians, including baboons, guenons, lemurs, macaques, and vervet monkeys [6, 33, 77, 112]. Coenuri are similar to cysticerci but contain multiple invaginated scolices in the bladder segment. Coenuri can produce both internal and external daughter coenuri. They are found in the subcutis, skeletal muscle, pleural and peritoneal cavities, liver, brain, and other organs [6, 33, 112].

Echinococcus granulosus, *E. multilocularis*, and *E. vogeli* have hydatid larvae that have been reported in a wide variety of New and Old World monkeys, apes, and prosimians, including baboons, chimpanzees, colobus monkeys, gibbons, galagos, gorillas, guenons, lemurs, macaques, mandrills, mangabeys, marmosets, and orangutans [6, 33, 75–77, 112, 117, 118]. Hydatid cysts can be found in the subcutis, pleural, or peritoneal cavities, liver, lung, retrobulbar tissue, and other organs. *E. granulosus* cysts are unilocular and have a thick, hyaline, laminated cyst wall. *E. multilocularis* has multiloculated cysts. The inner layer of the hydatid wall is lined by germinal epithelium that buds inward to produce numerous brood capsules/secondary cysts that are surrounded by clear fluid. Each brood capsule contain multiple protoscolices. The size of the cyst and the nature of the host response depend on the age and location of the hydatid cyst. Distension or swelling of the affected cavity/organ often occurs without clinical signs [6, 33, 112]. Hepatic infections may have granulomatous hepatitis and abundant fibrosis that cause atrophy, necrosis, or replacement of large sections of the liver [117, 118]. Occasionally, cysts will rupture releasing hydatid sand containing free hooklets, protoscolices, and calcareous corpuscles into the abdomen or thorax which may be identified on cytological preparations of effusions (Fig. 3.38).

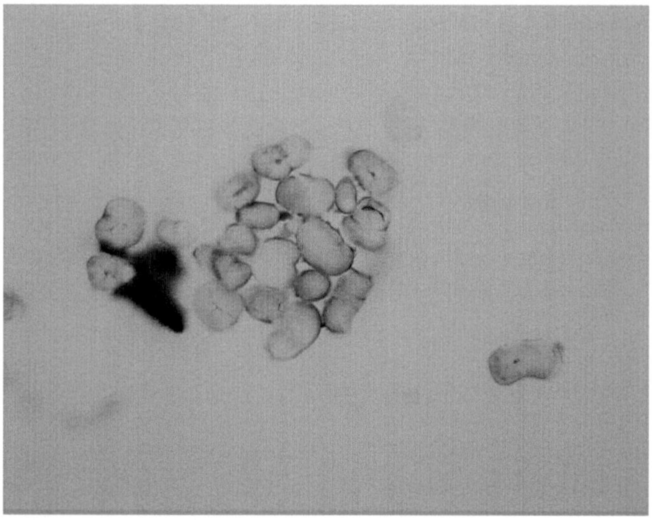

Fig. 3.38 Hydatid sand. (*Macaca fascicularis*, abdominal effusion, Wright–Giemsa stain)

3.2.3 Trematodes (Flukes)

Many trematodes, or flukes, have been reported from nonhuman primates, often with little to no clinical disease or host response; only the most commonly reported and significant are discussed below. The histological features that are characteristic of trematodes include a tegument that may or may not contain spines; an oral sucker at the anterior end (some having additional ventral suckers); parenchymatous tissue that surrounds the internal organs rather than a body cavity; paired ceca that often contain brown to black pigment from blood breakdown; and peripherally located reproductive organs consisting of an ovary, two testes, and vitellaria (yolk producing glands) [78, 79]. Most trematodes have both male and female reproductive organs (i.e., hermaphrodites), with the exception being schistosomes.

Athesmia

Athesmia foxi, sometimes referred to as *A. heterolecithodes*, is reported in New World monkeys, including cebus monkeys, tamarins, and titi monkeys [6, 33, 77]. The route of infection is unknown. Adults, which are long and thin (8.5 by 0.7 mm) live in the bile ducts. Infections are typically subclinical, but heavy infections can result in duct enlargement, biliary obstruction and stasis, biliary epithelial hyperplasia, and peribiliary fibrosis. Scant epithelial necrosis and inflammation rarely occur. Histologic features of adult trematodes are described above. Eggs are ovoid, golden brown, and 27–34 μm by 17–21 μm [6, 33].

Gastrodiscoides hominis

Gastrodiscoides hominis is reported in rhesus and cynomolgus macaques [6, 33, 77]. The route of infection is ingestion of metacercariae [6]. Infection is typically subclinical with low numbers of worms, but heavy infections can produce chronic typhlocolitis. Adults live in the cecum and colon and are orange-red, 6 by 3 mm with a conical anterior end and a cupped discoid posterior. Histologic features of adult trematodes are described above. Eggs are 150–152 μm by 60–72 μm, spindloid with rounded ends and a single operculum. Histologically, the mucosa at the site of attachment may be hyperemic, ulcerated, necrotic, and/or infiltrated by neutrophils. The submucosa may contain fibrosis and lymphoplasmacytic infiltrates [6, 33].

Paragonimus

Paragonimus westermani is reported from cynomolgus macaques [6, 33, 120]. The route of infection is ingestions, typically of infected raw crustaceans or mollusks that act as the intermediate hosts for metacercariae [6]. Adults typically live in the lungs, but occasionally are found in the brain and liver. Gross lesions include pulmonary emphysema, pleural adhesions, and small (2–3 cm diameter), soft, red-brown cysts that are scattered throughout the lungs and contain one or more worms. Adults have a plump, ovoid, brown body with many scale-like spines and measure 7.5–12.0 mm long by 4–6 mm wide by 3.5–5.0 mm thick. Histologic features of adult trematodes are described above. Eggs are ovoid, golden brown, a partly flattened operculum on one end and measure 80–118 μm by 48–60 μm. Histologically, cysts consist of a fibrous capsule that contains the adult worms and is surrounded by erythrocytes, neutrophils, and fluke eggs. There may also be lymphoplasmacytic and eosinophilic bronchopneumonia, and bronchial epithelial and submucosal gland hyperplasia [6, 33, 120].

Schistosoma

The most commonly reported species of schistosomes include *Schistosoma mansoni*, *S. haematobium*, and *S. mattheei*, with fewer reports of *S. japonicum* and *S. bovis*. They have been reported from a wide range of New and Old World monkeys and apes, including African green monkeys, baboons, chimpanzees, guenons, mangabeys, patas monkeys, rhesus macaques, squirrel monkeys, and vervet monkeys [6, 33, 75, 77, 120]. The route of infection is penetration of the skin by cercariae; ingestion of cercariae in contaminated water is also possible. Infections are typically subclinical. Adults are 1–3 cm long, typically in constant copulation and reside in the mesenteric, pelvic, or portal veins of the definitive host [6]. Eggs are carried in the bloodstream and lodge in the intestine, urinary bladder, liver, pancreas, lung, and various other organs. Grossly, adults often go undetected despite sometimes causing phlebitis and thrombosis. Eggs may cause thickening or stenosis of the intestinal or urinary bladder wall and small nodules in other affected organs. Histologically, adults induce eosinophilic phlebitis with thrombosis. Histologic features of adult trematodes are described above. Heme pigments are present in affected tissues and regional lymph nodes. Eggs, which may be mineralized, are surrounded by granulomatous inflammation/microgranulomas, multinucleated giant cells, and eosinophils. Lymphoplasmacytic inflammation or microabscesses with neutrophils and eosinophils can also occur (Fig. 3.39). Affected livers have fibrosis, particularly around portal vessels [6, 33, 120]. *S. mansoni* eggs are 114–175 μm by 45–68 μm, elongated-ovoid, round at both ends, and have a lateral spine. *S. haematobium* eggs are 112–117 μm by 40–70 μm, elongated-ovoid, rounded at the anterior end, and have a posterior terminal spine [6].

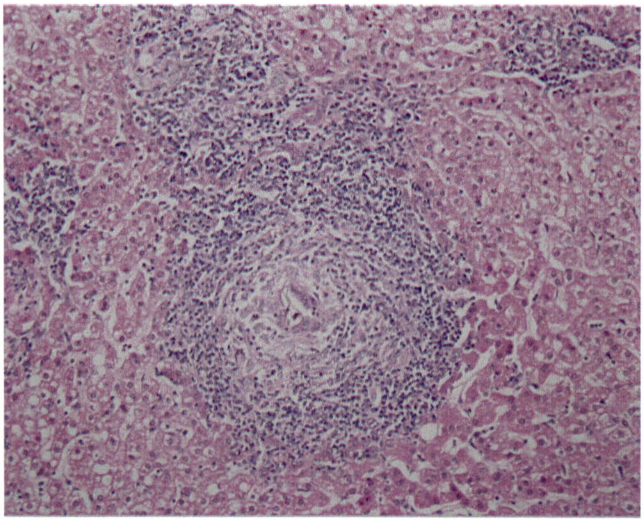

Fig. 3.39 *Schistosoma* sp. Pyogranulomatous inflammation with remnant of a brown refractile schistosome egg present in the center (*M. fascicularis*, liver, H&E stain)

3.2.4 Acanthocephalans (Thorny-Headed Worms)

Prosthenorchis elegans is the most commonly reported acanthocephalan, or thorny-headed worm, of nonhuman primates with fewer reports of *P. spirulina* and *Moniliformis moniliformis*. They have been documented in a wide variety of New and Old World monkeys, apes, and prosimians, including callitrichids, capuchins, chimpanzees, galagos, gibbons, lemurs, macaques, pottos, spider monkeys, and squirrel monkeys [6, 33, 76, 77, 121, 122]. The route of infection is ingestion of an infected arthropod intermediate host [6]. Adults are white to yellow, cylindrical, transversely wrinkled and reside in the ileum, cecum, and colon where they are anchored to the mucosa by a proboscis containing several rows of hooks. A small, white, firm nodule at the site of attachment can sometimes be seen from the serosal side of the intestine. Mucosal hyperemia and regional lymphadenomegaly may occur. [6, 33, 121]. If worms completely penetrate the intestinal wall, there may fibrinosuppurative peritonitis with free worms in the abdominal cavity [122]. Heavy infections can cause obstruction, intussusception, or rectal prolapse. *P. elegans* adults are 20–55 mm long [6]. Histologically, the proboscis is deeply embedded in the mucosa, and sometimes the submucosa and muscularis, and is accompanied by mucosal ulceration, necrosis, granulomatous enteritis or abscess formation, and secondary bacterial infection [6, 33, 121]. Characteristic histologic features that help distinguish acanthocephalans from other metazoan worms include a spined proboscis at the anterior end; a thick hypodermis that has a thinner felted layer beneath the cuticle and deeper, thicker layer of cross fibers with lacunar channels; a pseudocoelom lined by thin muscular layer that has both circular and longitudinal components; and no digestive tract [78, 79].

3.2.5 Arthropods

Arthropods have some combination of the following histologic features: chitinized exoskeleton, jointed appendages, setae or chitinized "hairs," striated skeletal muscle, mouth parts, a body cavity with a simple digestive tract lined by columnar to cuboidal cells, chitinized tracheal ring system, and reproductive structures [78, 79].

Insects

Phthirapterids (Lice)

Anoplurids (Sucking Lice)

Numerous species of sucking lice, including the genera *Pedicinus* and *Pediculus*, have been reported from New and Old World monkeys, apes, and prosimians, including baboons, chimpanzees, colobus monkeys, galagos, gibbons, gorillas, green monkeys, guenons, howler monkeys, langurs, lemurs, macaques, marmosets, tamarins, sakis monkeys, siamangs, spider monkeys, and uakaris [6, 33, 75, 76, 123]. Aside from finding the lice and eggs in the hair of primates, there are minimal lesions. Alopecia, anemia, and pruritic self-mutilation can occur in heavy infestations [6, 33, 75, 76, 123].

Mallophagids (Biting Lice)

Numerous species of biting lice have been reported in New and Old World monkeys and prosimians, including colobus monkeys, howler monkeys, indri, lemurs, lorises, rhesus macaques, owl monkeys, and woolly monkeys [6, 33, 75, 76, 123].

Siphonapterids (Fleas)

Tunga penetrans

Tunga penetrans, also called stick-tight flea, chigoe flea, or jigger, has been reported in wild-caught baboons, gorillas, guenons, and howler monkeys [6, 33, 75, 76, 123]. The female is 1 mm long and brown and invades the epidermis of tail and limbs, especially the ischial callosities. Grossly, they appear as firm, brown nodules that contain friable dark brown material on cut section. Histologically, the flea is embedded within epidermal/dermal cavity lined by stratified squamous epithelium. The anterior end of the flea faces the dermis while the posterior end is oriented toward the skin surface. There is epidermal ulceration, acanthosis, parakeratotic hyperkeratosis, dermal and epidermal necrosis, hemorrhage and eosinophilic and granulomatous dermatitis [6, 33, 77, 123, 124].

Other Fleas

Fleas that are the natural parasites of some domestic animals, including *Ctenocephalides canis*, *C. felis*, and *Echidnophaga gallinacea*, can opportunistically infest nonhuman primates, although they appear to be of minimal importance.

Dipterids (Flies)

The larvae of several species of flies, including *Cuterebra* sp., *Cochliomyia hominivorax*, *Cordylobia anthropophaga*, *Dermatobia hominis*, and *Alouatta* sp., can infest New and Old World monkeys and prosimians [6, 33, 75, 76, 123]. The larvae, or bot, resides within a fistulous swelling containing a central pore in the skin and subcutis, often in the cervical region. Chronic inflammation surrounds the bot and secondary bacterial infection sometimes occurs [6, 33, 76, 77, 123].

Arachnids

Acariformes (Mites)

In addition to the histologic features of arthropods listed above, mites also have brightly eosinophilic yolk material within eggs and the yolk glands of female mites [79].

Cutaneous Mites

Sarcoptiform Mites

Sarcoptes scabiei is the most common cause of sarcoptic mange in nonhuman primates, although many other mites, including *Prosarcoptes pitheci*, *Dunnalges lambrechti*, and *Rosalialges cruciformis*, have been reported. New and Old World monkeys and apes have been found to be affected, including African green monkeys, baboons, cebus monkeys, chimpanzees, drills, gibbons, gorillas, marmosets, orangutans, owl monkeys, and siamangs [6, 33, 75, 77, 123]. Gross lesions include alopecia and thickening and scaling of the skin that most often occurs over the back, neck, and shoulders and less commonly over the lower trunk and extremities. Self-mutilation due to intense pruritus can result in hemorrhage and secondary bacterial infections. Histologic lesions include acanthosis, parakeratotic hyperkeratosis, crusting, and epidermal tunnels containing mites and eggs [6, 33]. Adult mites are broadly oval, 170–270 by 220–380 μm, and translucent white [6].

Trombidiformes Mites

Several species of trombidiformes mites can cause cutaneous lesions similar to sarcoptic mange. Demodectic mange caused by *Demodex* spp. is the most well-known. Rare mites can be found in the hair follicles of clinically normal animals; animals that develop symptoms of demodectic mange are often young, old, or immunocompromised. Gross lesions include alopecia, crusting, erythema, and formation of nodules, pustules, or plaques (Fig. 3.40). Adult mites reside with hair follicles and sebaceous glands and are 100–300 μm long by 30–50 μm in diameter, with a tapered posterior end and four pairs of short legs at the anterior end (Fig. 3.41) [6]. Affected hair follicles have mild lymphoplasmacytic folliculitis with epidermal hyperplasia. When furunculosis occurs, mite fragments and keratin induce granulomatous dermatitis [6, 125, 126].

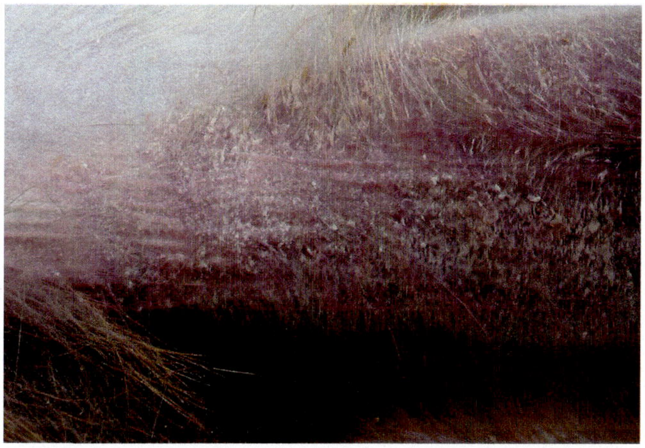

Fig. 3.40 *Demodex* sp. Generalized hyperkeratosis (erythema is representative of sex skin) (*Macaca mulatta*, caudal thigh skin)

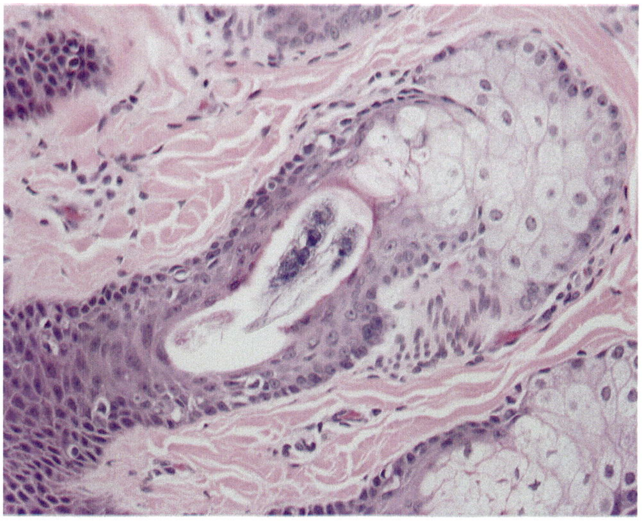

Fig. 3.41 *Demodex* sp. Intrafollicular mite (*Macaca mulatta*, fascial skin)

Pulmonary Mites

Pulmonary acariasis is caused by at least ten species of mites within the genus *Pneumonyssus*, with *Pneumonyssus simicola* being the most commonly encountered, as well as *Pneumonyssoides stammeri*. They have been reported in a wide variety of New and Old World monkeys and apes, including baboons, chimpanzees, colobus monkeys, gorillas, guenons, howler monkeys, langurs, macaques, mangabeys, orangutans, patas monkeys, proboscis monkeys, and woolly monkeys [6, 33, 75, 77, 81, 127]. *P. simicola* is extremely common in imported rhesus macaques. Suggested routes of transmission include inhalation, ingestion during grooming, or through feces [6, 127]. Gross lesions include 1–7 mm diameter, translucent yellow to white to gray, soft, cystic bullae that are throughout the surface and deeper parenchyma (Fig. 3.42). These so-called mite houses contain adult and larval mites and eggs that can be seen under a dissecting scope. Adjacent lesions may become confluent. Those in the periphery of the lung are nodular and elevate the pleura. There may be fine, fibrous adhesions between the visceral and parietal pleura, bullous emphysema, and hemorrhage. There are reports of pneumothorax, pleuritis, pericarditis, and pulmonary arteritis associated with mite infections. Histologically, there is granulomatous bronchiolitis and peribronchiolitis, often with a mixture of neutrophils, eosinophils, lymphocytes, plasma cells, and macrophages that thicken the bronchiolar wall. Bronchiolar epithelial loss, bronchiectasis, bronchiolar smooth muscle hypertrophy, peribronchiolar lymphoid aggregates, interstitial fibrosis, alveolar emphysema, and atelectasis may also be present. Macrophages, areas of peribronchiolar and peribronchial inflammation, and blood vessel walls contain mite excrement with golden brown to black, birefringent, granular pigment, and needle-like crystals [6, 33, 127, 128]. Adult *P. simicola* mites are yellow-white, elongate-ovoid, and 500–850 μm long; eggs are white, spherical, and 250–450 μm in diameter (Fig. 3.43) [6].

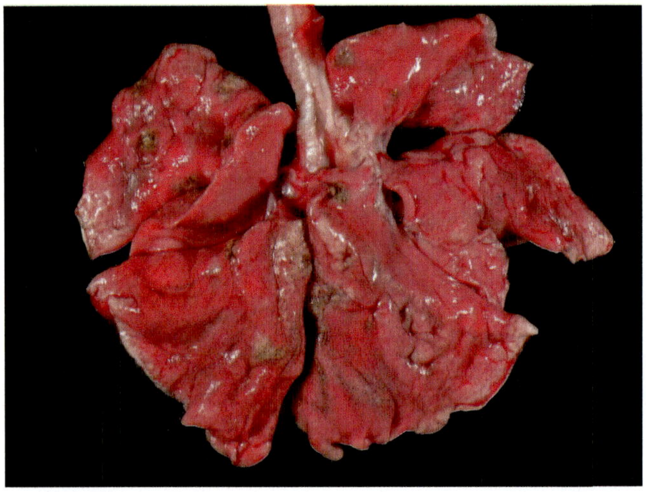

Fig. 3.42 *Pneumonyssus simicola*. Multifocal tan nodules expand the pulmonary parenchyma, often centered on bronchioles (*Macaca mulatta*, lung)

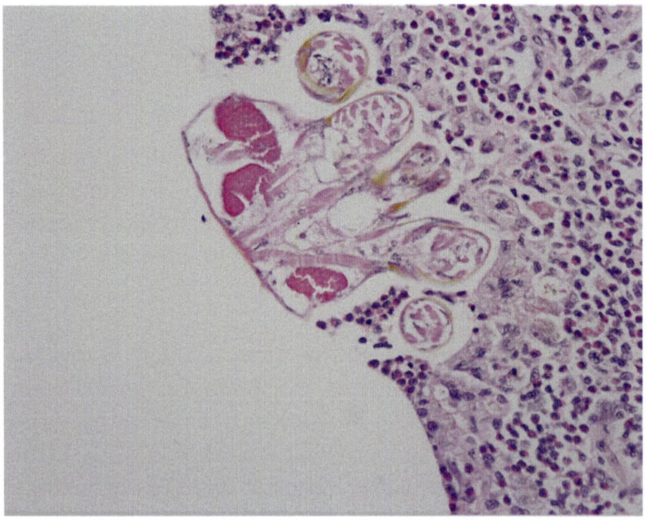

Fig. 3.43 *Pneumonyssus simicola*. Once common, *P. simicola* is now rare with the widespread use of ivermectin. Note striated muscle present within the halarachnid mite (*Macaca mulatta*, lung, H&E stain)

Nasal Mites

Rhinophaga cercopitheci, *R. dinolti*, *R. elongata*, *R. papionis*, and *R. pongicola* have been described within the nasal cavity and sinuses of Old World monkeys and apes, including baboons, bonobos, gorillas, guenons, macaques, and orangutans [6, 33, 75, 127]. The host response to nasal mites is typically limited to a small mucosal polyp or nodule. *R. cercopitheci*, *R. dinolti*, and *R. papionis* have also been found in the lungs where they cause pneumonitis and excessive mucus production [6, 33, 127]. Mites have the following histologic features: chitinized exoskeleton, jointed appendages, striated skeletal muscle, a body cavity with a digestive tract and reproductive structures, and yolk material within eggs and the yolk glands of female mites [78, 79].

Parasitiformes (Ticks)

A variety of hard (ixodid) and soft (argasid) ticks have been reported from numerous nonhuman primates [6, 33, 75–77, 123]. Most tick infestations are asymptomatic as engorged ticks disengage from the host following feeding. Heavy infestations can cause weight loss and anemia. Only a mild inflammatory reaction consisting of hyperemia, edema, and hemorrhage is induced, unless there is secondary bacterial infection [6, 33].

Pentastomids

Armillifer sp., *Linguatula serrata*, and *Porocephalus* sp. are the most commonly reported pentastomes of nonhuman primates. They have been reported in New and Old World monkeys, apes, and prosimians, including baboons, chimpanzees, galagos, guenons, macaques, mangabeys, marmosets, squirrel monkeys, tamarins, and titi monkeys

[6, 33, 77, 129]. Primates act as the intermediate host for the nymphal stage of these highly aberrant arthropods and become infected by ingesting eggs within food or water that is contaminated by the feces or respiratory mucus of the definitive host [6]. Encysted nymphs are typically found incidentally at necropsy in the lung, liver, mesenteric lymph nodes, brain, omentum, and the parietal or visceral pleura and peritoneum (Figs. 3.44 and 3.45) [6, 33, 129]. Nymphs are white, "C"-shaped, flattened (*Linguatula*) or cylindrical (*Porocephalus* and *Armillifer*) with spinous body rings or annulations and two pairs of hooks on the anterior end. *Linguatula serrata* are 4–6 mm long with approximately 90 annulations; *Porocephalus* are 7–15 mm long with 30–45 annulations; and *Armillifer* are 13–22 mm long with 15–22, thick, projecting annulations [6]. Dead nymphs invoke a foreign body response and overwhelming intestinal infections can cause fatal peritonitis [33, 129, 130]. Characteristic histologic features that help distinguish pentastomes from other metazoan worms include a pseudosegmented body (when section longitudinally); a cuticle with sclerotized openings; striated musculature; two pairs of hooks surrounding the mouth; and an intestine that is multicellular and bordered by two acidophilic glands [78, 79].

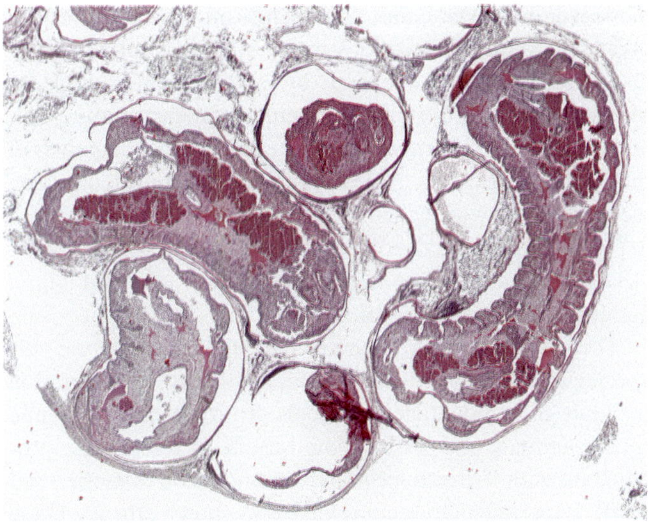

Fig. 3.44 Pentastomiasis (*Armillifer* sp). Large pentastomes present in the serosa demonstrating pseudosegmentation (*Macaca fascicularis*, serosa, H&E stain, case courtesy of Dr. K. Matz-Rensing)

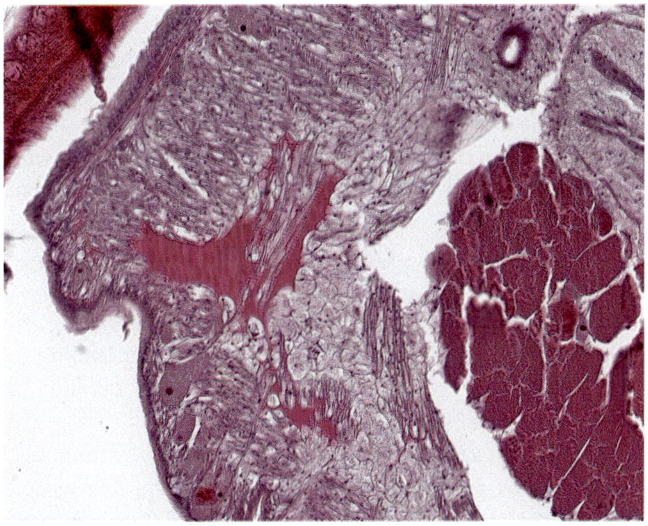

Fig. 3.45 Pentastomiasis (*Armillifer* sp). Pentastome demonstrating cuticle, acidophilic glands, and striated muscle (*Macaca fascicularis*, serosa, H&E stain, case courtesy of Dr. K. Matz-Rensing)

3.2.6 Annelids (Segmented Worms)

Dinobdella ferox

The leech *Dinobdella ferox* has been reported in rhesus and Formosan macaques that have been obtained from the wild [6]. They are acquired when drinking and reside in the nasal cavity and pharynx. The leeches are 3.5–6.0 cm long by 0.5–0.8 cm wide, dorsoventrally flattened, and become dark red when engorged. Infections are typically subclinical; heavy infection can result in anemia, epistaxis and asphyxiation. Histologically, there is localized chronic inflammation at the site of attachment and mucous cell hyperplasia.

3.3 Fungal Infections

3.3.1 *Pneumocystis* Species

Pneumocystis species are extracellular obligate fungal pathogens. The organisms have high host specificity and isolates from individual hosts should be considered distinct species. Nomenclature is in flux and the organism infecting humans has recently been renamed *P. jirovecii*, the organisms infecting rats named *P. carinii* and *P. wakefieldiae*, and the organism infecting macaques *P. macacae*. Multiple genetically distinct strains identified based on mitochondrial sequences may circulate concurrently within populations [131]. Disease has been described in a variety of new world primates (NWP) and old world primates (OWP) but is most frequently observed in immunodeficient macaques [132, 133].

The organism is transmitted by the aerosolized route. Infection of neonatal animals is likely uniform and not associated with overt disease. These infected neonatal animals most commonly clear infection but more infrequently become carriers for variable periods of time serving as reservoirs within a population. The juvenile carrier animal plays an important role in propagation and survival of the organism in colonies. While reactivation of latent infection was believed to play a role, more recent evidence suggests acquisition of the organism during periods of immunosuppression as the most common timing of infection leading to overt disease. An epizootic has been described in a colony of immunocompromised macaques [134].

In debilitated or immunosuppressed animals, *Pneumocystis* may cause pneumonitis (often abbreviated PCP) likely resulting from re-acquisition of the organism from carrier animals [135]. Disease appears to develop slowly allowing time for compensation which may manifest as an increase in blood hemoglobin content or packed cell volume (PCV). Cough is uncommon with the most frequent clinical sign being dyspnea or tachypnea at rest. Grossly lesions vary from a multifocal to coalescing interstitial pneumonitis in milder cases to multifocal gray to white nodules in severe cases (Figs. 3.46 and 3.47) [136].

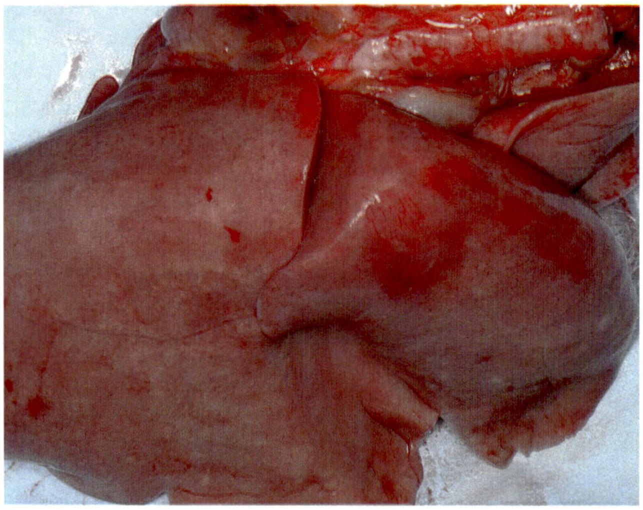

Fig. 3.46 *Pneumocystis macacae*. Diffuse interstitial pneumocystis pneumonitis (PCP) (*Macaca mulatta*, lung)

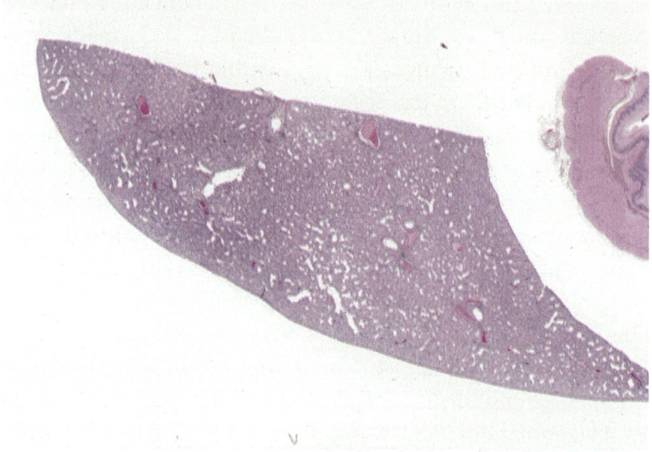

Fig. 3.47 *Pneumocystis macacae*. Diffuse interstitial pneumonitis (*Macaca mulatta*, lung, subgross H&E stain)

Histologically, the organism can be recognized on H&E sections as pale pink foamy material within alveolar spaces (Fig. 3.48) [133, 137, 138]. Inflammatory infiltrates are mild consisting of lymphocytes, plasma cells, and macrophages found within alveoli and the interstitial space. There is often pronounced type II pneumocyte hyperplasia (Fig. 3.49). *Pneumocystis* spores can be demonstrated with Grocott's methenamine silver stains. Infection is often complicated by other opportunistic infections in the simian acquired immunodeficiency syndrome caused by experimental inoculation of macaques with the simian immunodeficiency virus (SIV) and immunohistochemistry may be used to assist diagnosis in mild cases (Fig. 3.50). A second angiocentric form has been recognized in severe cases. In this instance, organisms appear to proliferate within alveolar as well as the periarterial spaces [139]. Very rarely infection may disseminate to region lymph nodes and the thoracic wall.

There is no known zoonotic potential for *Pneumocystis* species infecting OWP. Prevention is difficult but separation of adult immunodeficient animals from potential juvenile carrier animals may be of some benefit. Alternatively, prophylaxis with Trimethoprim-sulfamethoxazole may be used [140]. Experimental vaccines have also shown efficacy in the SIV model [141].

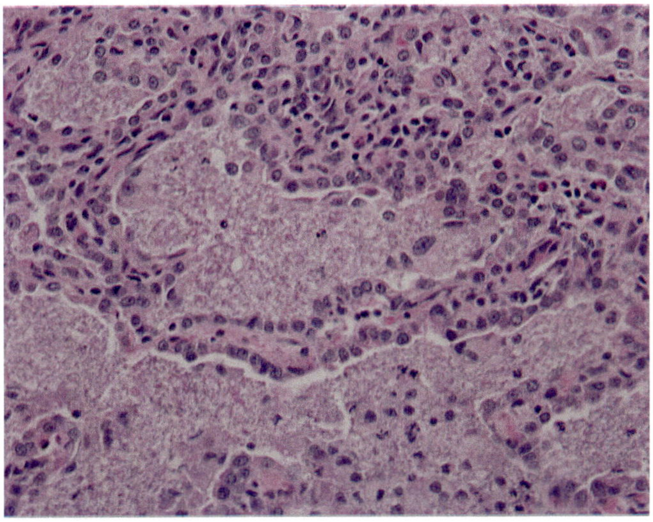

Fig. 3.48 *Pneumocystis macacae*. Septal thickening accompanied by marked type II pneumocyte hyperplasia and mixed inflammatory cell infiltrate (*Macaca mulatta*, lung, H&E stain)

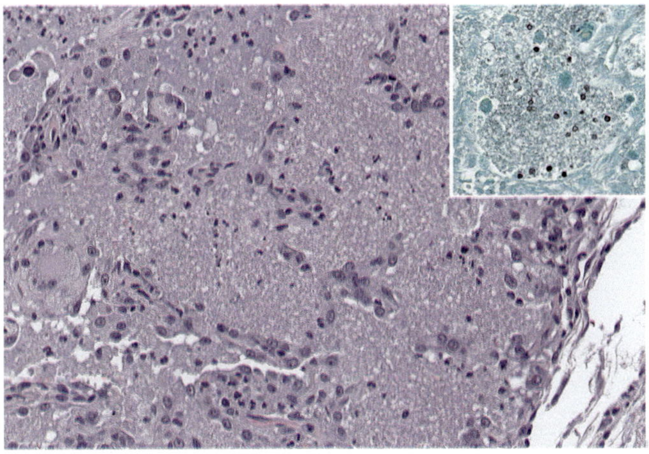

Fig. 3.49 *Pneumocystis macacae*. Characteristic foamy eosinophilic appearance of pneumocystis casts consisting of cysts and trophozoites admixed with cellular exudates (*Macaca mulatta*, lung, H&E stain; insert GMS stain demonstrating cysts)

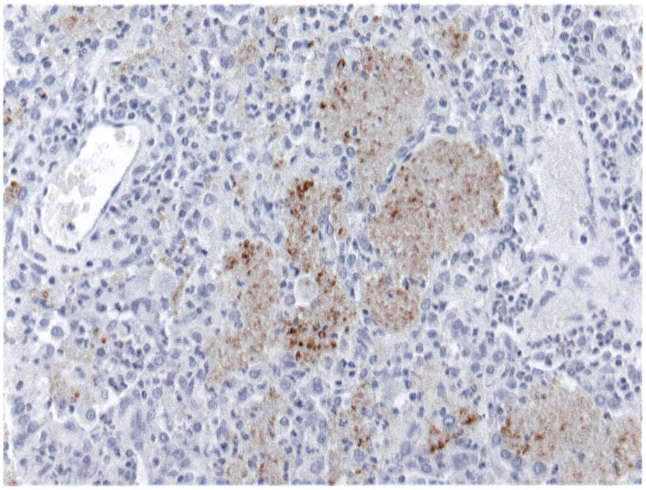

Fig. 3.50 *Pneumocystis macacae*. Immunohistochemistry may be used to detect pneumocystis organisms when typical histological features are missing (*Macaca mulatta*, lung, ABC immunostain technique, Clone 3F6, DAB chromogen)

3.3.2 Dermatophytosis

Dermatophytosis is rarely described in OWPs and NWPs and has been caused by fungi within the genera *Microsporum* and *Trichophyton* [142]. Clinical signs vary from diffuse alopecia to the typical ring-shaped areas of crusts and alopecia. Diagnosis is based on morphologic features of the fungal organisms within and around hair follicles and positive dermatophyte cultures. Neutrophilic folliculitis accompanied by orthokeratosis and parakeratosis is observed histologically.

3.3.3 *Coccidioides immitis*

C. immitis is a saprophytic soil fungus that may opportunistically cause disease in a variety of mammals. Distribution is limited to semiarid areas of the southwestern US and Central and South America. Infection of nonhuman primates is uncommon but has been reported in baboons, chimpanzees, and gorillas housed in the southwest [143, 144]. Transmission occurs through inhalation of arthrospores that contaminate dry environments.

Clinical signs are dependent on the primary system involved and may be related to pulmonary disease or osteomyelitis [145]. Respiratory signs such as cough and dyspnea are commonly described. *C. immitis* infection may result in disseminated disease involving the lungs, bone, spleen, liver, and kidneys (Figs. 3.51 and 3.52). Histologically, pyogranulomatous inflammation with multinucleated giant cells is observed (Table 3.1). The characteristic organism may be visualized in these areas as large 10–60μm diameter thick-walled spherules which contains 2–4μm diameter endospores (Fig. 3.53). The wall appears double-contoured and refractile. No capsule or budding is observed.

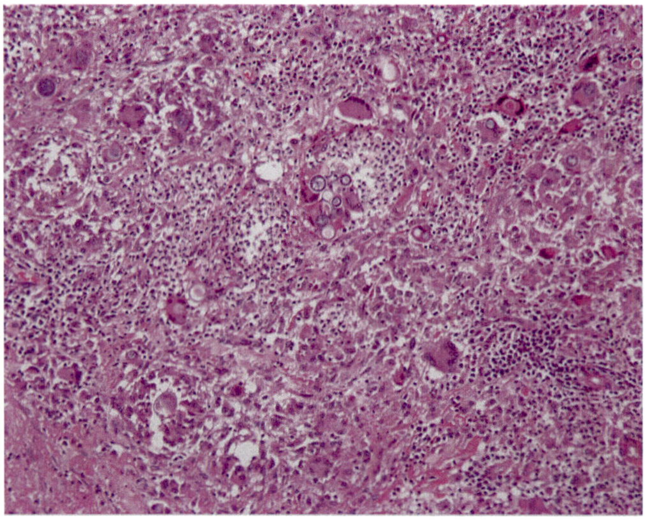

Fig. 3.51 *Coccidioides immitis*. Pyogranulomatous inflammation surrounding spherules (*Macaca fuscata*, lung, H&E stain)

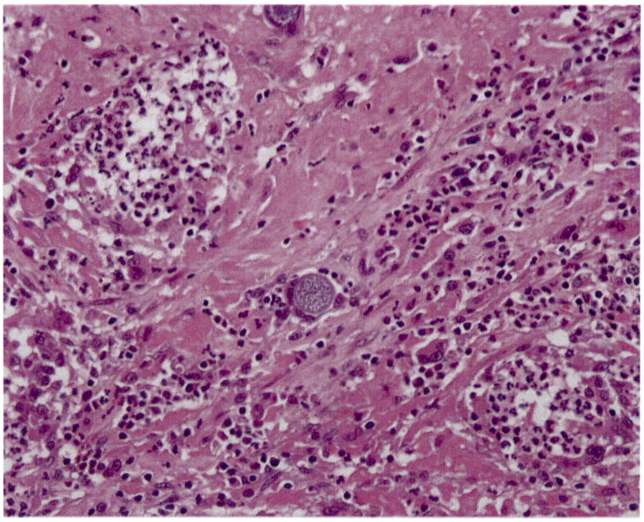

Fig. 3.52 *Coccidioides immitis*. Pyogranulomatous inflammation with thick-walled spherule containing endospores (*Macaca mulatta*, lung, H&E stain)

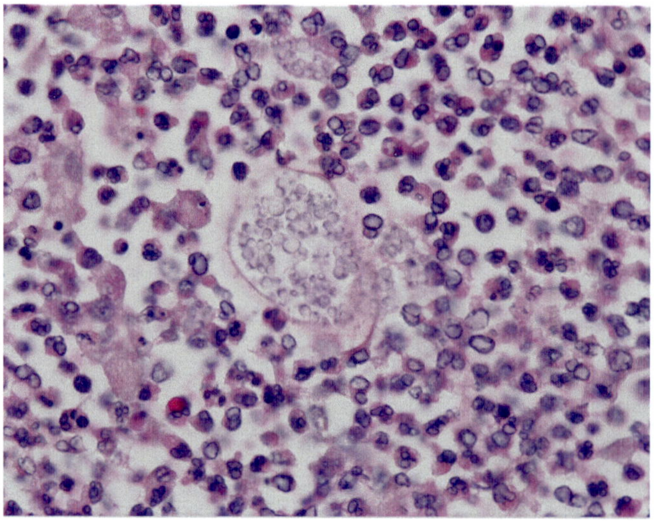

Fig. 3.53 *Coccidioides immitis*. Ruptured spherule releasing endospores (*Macaca mulatta*, lung, H&E stain)

Table 3.1 Systemic mycosis

Agent	Size	Appearance	Budding
Hyphomycetes			
Coccidioides immitis	20–200 μm	Thick-walled (double contoured and refractile) spherule containing 2–4 μm endospores	None; endosporulation
Histoplasma capsulatum var. capsulatum	2–5 μm	Thin-walled yeast; found intracellularly with histiocytes	Single, narrow based
Histoplasma capsulatum var. Duboisii	8–15 μm	Thick-walled yeast; no capsule	Single, narrow based
Sporothrix schenkii	2–6 μm	Oval to cigar shaped; pleomorphic	1–2, narrow based
Paracoccidioides brasiliensis	8–20 μm	Spherical yeast	Multiple peripheral buds
Blastomycetes			
Blastomyces dermatitidis	5–25 μm	Double contour and refractile	Single, broad based
Cryptococcus neoformans	2–10 μm	2–8 μm thick capsule	Narrow-based budding
Candida albicans	3–5um	Yeast, Pseudohyphae, and blastospores	One or more buds

3.3.4 *Candida albicans*

C. albicans is a common saprophytic agent that colonizes mucosal surfaces shortly following birth. In most animals, it exists as a commensal agent and does not produce clinical signs. It is a dimorphic fungus that grows on Sabouraud and blood agar. Localized and systemic forms of disease are recognized in young and immunocompromised or debilitated animals [146, 147]. The localized form also known as thrush occurs along mucous membranes principally the oral mucosa and esophagus (Fig. 3.54) [148]. It is common in young nursery reared animals and SIV-infected macaques [138]. The condition can be painful and stop animals from eating and swallowing normally. Less frequently, overgrowth of Candida may be detected along the course of the gastrointestinal tract often in association with prolonged use of antibiotics and associated with diarrhea.

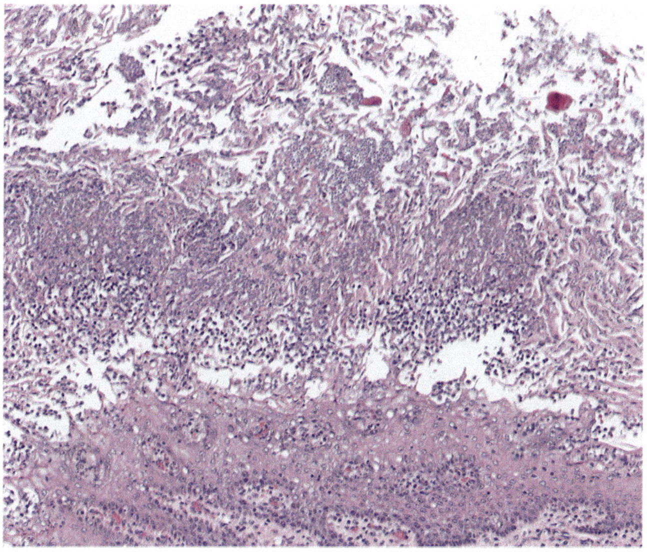

Fig. 3.54 *Candida albicans*. Thick mat of budding yeast admixed with keratin and cellular debris with spongiosis of adjacent epithelium (*Macaca fascicularis*, esophagus, H&E stain)

Systemic infection is rarely observed but can develop in debilitated animals treated with antibiotics and in animals with other current diseases such as diabetes mellitus that may adversely impact the host's immune system. In these cases, the animals present with evidence of severe sepsis (febrile and hypotensive shock). Systemic disease has also been recognized in animals on immunosuppressive therapy.

Thrush appears grossly as pale tan to white plaques that are tightly adhered to the underlying epithelium which may be proliferative and raised. These plaques termed pseudomembranes are composed of masses of pseudohyphae and blastospores mixed with degenerate and sloughed epithelial cells, neutrophils, and proteinaceous exudates (Fig. 3.55) [146]. The organisms stain well with Grocott's methenamine silver (GMS) (Fig. 3.56) or periodic acid-Schiff stains (Figs. 3.57 and 3.58). In disseminated forms, pyogranulomatous inflammation is observed.

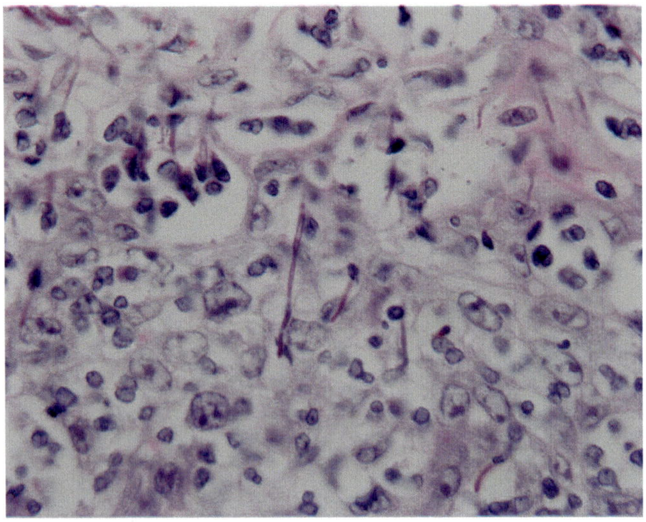

Fig. 3.55 *Candida albicans*. Blastoconidia budding from the pseudohyphae invading the underlying epidermal mucosa (*Macaca mulatta*, oral mucosa, PAS stain)

Fig. 3.56 *Candida albicans*. Blastoconidia budding from the pseudohyphae invading the underlying epidermal mucosa (*Macaca mulatta*, oral mucosa, GMS stain)

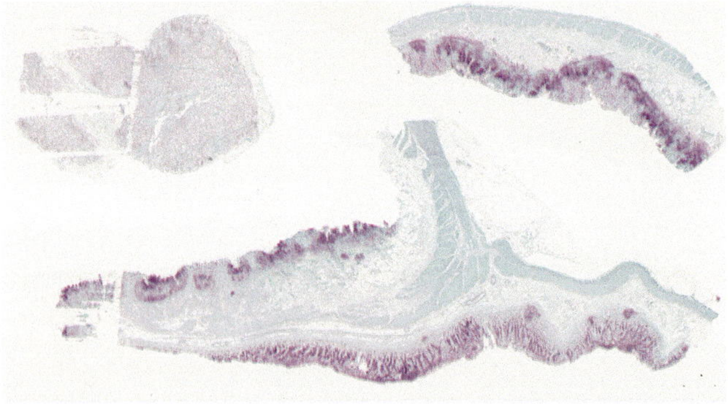

Fig. 3.57 *Candida albicans*. Diffuse candidiasis of the gastrointestinal tract in an immunosuppressed baboon (*Papio anubis*, colon, subgross PAS stain)

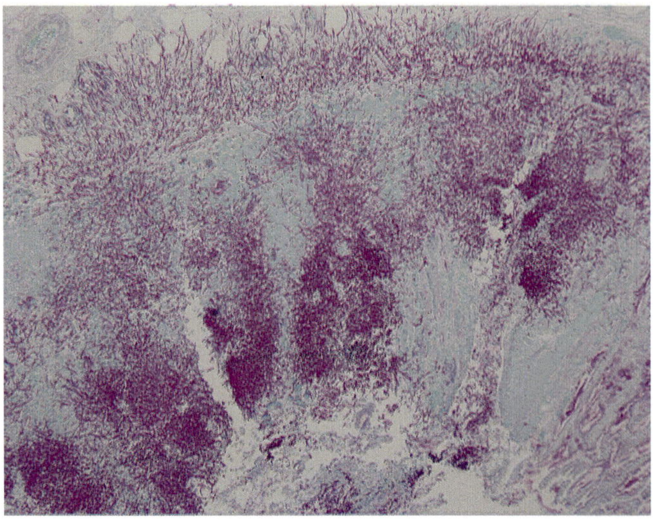

Fig. 3.58 *Candida albicans*. Diffuse candidiasis of the gastrointestinal tract demonstrating pseudohyphae with invasion of the submucosa (*Papio anubis*, colon, subgross PAS stain)

3.3.5 *Histoplasma capsulatum* var. *capsulatum* and *duboisii*

H. capsulatum var. capsulatum is a soil-borne fungus that is found in association with soil contaminated with the feces of bats and birds [149]. While disease has rarely been reported in NHPs, serologic surveys suggest that exposure may be common in NWPs. Disease has been recognized in squirrel monkeys, owl monkeys, and rhesus and may be produced experimentally in macaques [150–153]. Morphologically disseminated microgranulomas are observed in multiple organs and are typical of those seen in other species. These granulomas are composed of large numbers of histiocytes and giant cells that contain the characteristic 2–4μm diameter thin-walled yeast forms (Figs. 3.59 and 3.60). Rarely disease has been recognized in SIV-infected macaques.

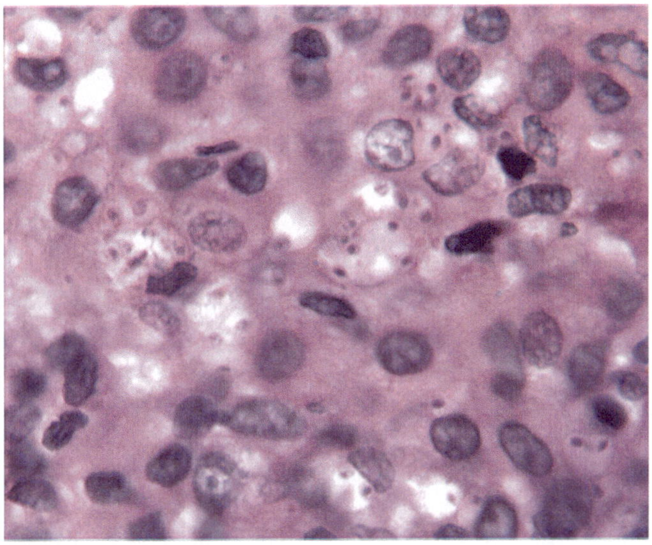

Fig. 3.59 *Histoplasma capsulatum var. capsulatum*. Intracellular yeast with histocytes (*Macaca mulatta*, liver, H&E stain)

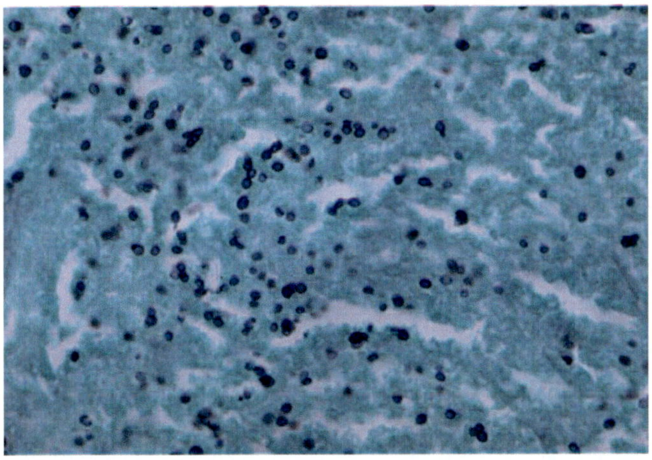

Fig. 3.60 *Histoplasma capsulatum var. capsulatum*. Silver stained used to detect small 2–5 μm diameter yeast forms with narrow-based budding (*Macaca mulatta*, lung, GMS stain)

H. capsulatum var. duboisii is a saprophytic soil fungus originally found naturally in Africa. The organism appears to have been imported with NHPs to Texas [154–156]. All cases to date have been described in baboons of the genus Papio and Cynocephalus. The disease caused by *H. capsulatum var. duboisii* is primarily manifested as multiple cutaneous nodules that appear as papules and pustules that may ulcerate and reach 2–3 cm in diameter. Radiographically involvement of underlying bone may be observed.

Transmission of *H. capsulatum var. duboisii* is through direct contact with contaminated soil or animals. Dermal, GI, and respiratory routes have been observed. The epizootic in Texas was likely propagated through direct contacts between animals. A long incubation period may be possible. Histologically, pyogranulomatous inflammation with large numbers of histiocytes and multinucleated giant cells is observed [154]. Nodules often ulcerate and secondary bacterial infections may be observed. The organism may be found in the areas of inflammation and are thick-walled unicellular yeast forms that reproduce through single narrow based budding (Figs. 3.61 and 3.62). Histiocytes or multinucleated cells may contain large numbers of organisms.

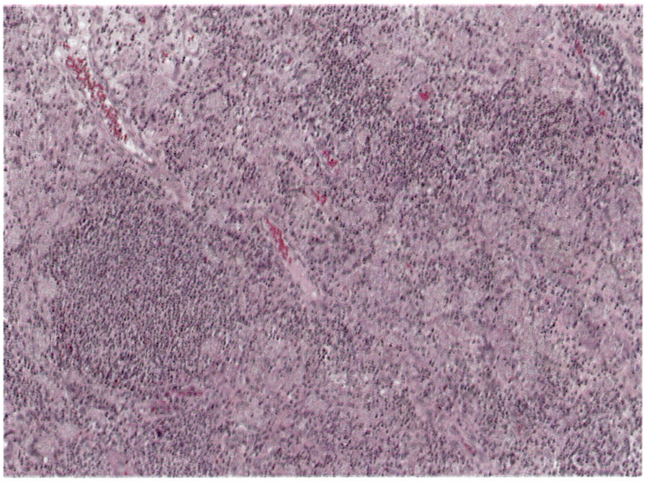

Fig. 3.61 *Histoplasma capsulatum var. duboisii*. Pyogranulomatous inflammation in skin with myriad of thick-walled yeast (*Papio anubis*, skin, H&E stain)

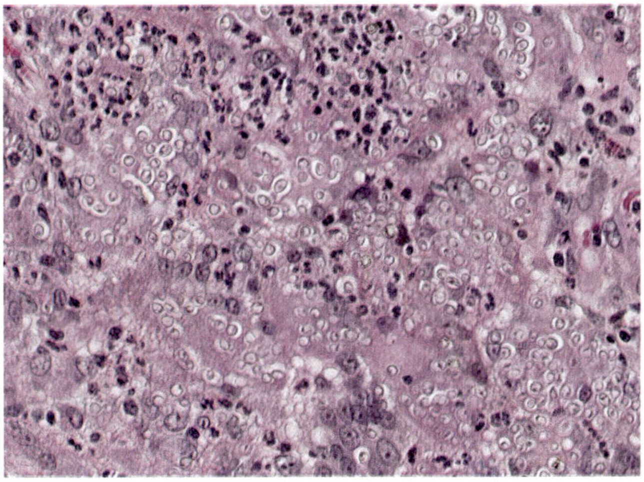

Fig. 3.62 *Histoplasma capsulatum var. duboisii*. Intra- and extracellular thick-walled yeast 8–15 μm in diameter with no capsule (*Papio anubis*, skin, H&E stain)

3.3.6 *Cryptococcus neoformans*

C. neoformans is a saprophytic fungal organism that has world-wide distribution and is commonly found in pigeon droppings [157]. The organism may be transmitted by the aerosolized route or through direct contact. Clinical disease has been recognized in a variety of NWP and OWPs. The organism disseminates to bone and the CNS and clinical signs often relate to neurologic involvement [158, 159].

The organism disseminates and gross lesions may be found in multiple organs. Mass lesions may appear gelatinous due to the organism's capsule. Histologically, the lesion is characterized by masses of yeast organisms that infiltrate and replace normal tissues. There is often remarkably little inflammatory response. The yeast appear as small 2–20 μm diameter forms that reproduce by narrow based budding and are surrounded by a wide polysaccharide capsule (Figs. 3.63 and 3.64). This capsule stains clear with H&E and may be accentuated by carminophilic stains.

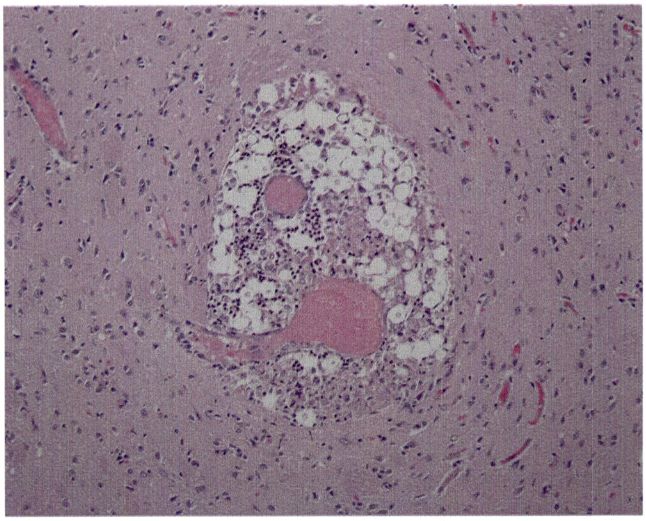

Fig. 3.63 *Cryptococcus neoformans*. Perivascular neutrophilic and histiocytic infiltrate with 2–10 μm diameter yeast forms with thick non-staining capsule (*Lemur catta*, brain, H&E stain)

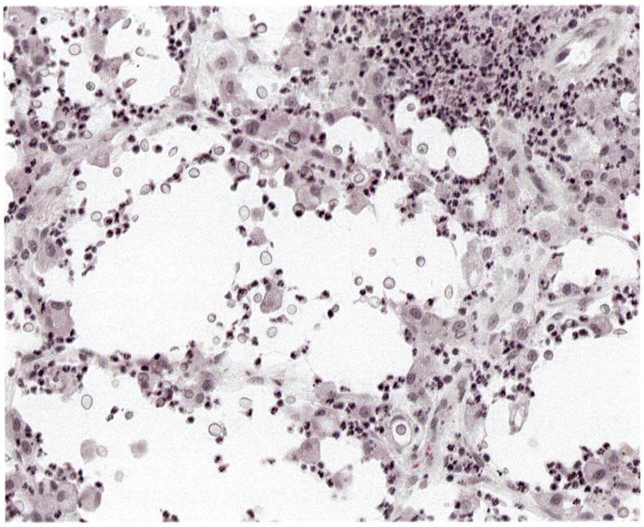

Fig. 3.64 *Cryptococcus neoformans*. Histiocytic infiltrate with 2–10 μm diameter yeast forms with thick non-staining capsule (*Mandrillus sphinx*, H&E stain)

A related organism *Cryptococcus gattii* has recently be recognized in SIV-infected rhesus macaques causing pulmonary lesions [160]. Morphologically similar to *C. neoformans* polymerase chain reaction on formalin fixed paraffin embedded tissue and sequencing was required to differentiate.

3.3.7 *Sporothrix schenckii*

S. schenckii is a saprophytic dimorphic fungus that is found in soil, moss, and tree bark. Disease has been rarely reported in NHP but serological data suggests exposure may be common in NWPs [161]. Initially, lesions arise on distal extremities and appear as multiple small nodules that coalesce and may following draining lymphatics. Infection is initiated through injuries or small punctures induced by thorns or other objects. The organism causes pyogranulomatous inflammation with extensive necrosis in which the pleomorphic yeast may be visualized. These are often ovoid or cigar shaped 2–6 μm in size and reproduce by narrow-based budding.

3.3.8 *Paracoccidioides brasiliensis*

P. brasiliensis is an endemic dimorphic fungus native to South America. It is the most common systemic mycoses in humans in this region of the world but has been rarely described in NHPs [162]. A single case has been reported in a squirrel monkey recently imported from Bolivia [163]. The organism is a large oval yeast 8–50 μm in diameter which reproduces by multiple buddings along the periphery.

3.3.9 Aspergillosis

Aspergillus fumigatus is the most common agent of aspergillosis. While a ubiquitous saprophytic fungi found in decaying vegetation and soil, it has been infrequently reported as a cause of disease in NHPs. A single epizootic has been reported in a zoological collection in association with an outbreak of tuberculosis. We have recognized it as a cause of disseminated disease in chemically immunosuppressed baboons and cynomolgus macaques and in a rhesus macaque with chronic diabetes mellitus. In all cases, the histologic picture was similar consisting of disseminated microgranulomas in the lung, liver, kidneys, and spleen. Clinical signs were nonspecific and included intermittent fever, anorexia, and depression prior to euthanasia. Histologically, hyphae appear as uniform and septate with a diameter of 3–6 μm. Hyphae branch at 45 degree angles.

Table 3.2 Systemic Hyphal Infections

Agent	Appearance	Size
Hyphomycetes		
Aspergillus	Septate hyphae; branching at 45 degree angle	3–6 µm with uniform diameter
Zygomycetes		
Mucorales	Irregularly branching, bulbous, rare septate	4–24 µm diameter; 10–60 µm in length
Entomophthorales	Irregularly branching; thin-walled hyphae; may be associated with Splendore–Hoeppli reaction	6–12 µm diameter

Other systemic mycosis have been recognized in NHPs and include Paecilomyces, Mucorales, and Entomophthorales (Table 3.2). Of these, members of the mucorales grouping have been most commonly observed in NHPs. This includes the genera Absidia, Rhizopus, Rhizomucor, Mucor, and Mortierella. Three clinical forms are observed, including cutaneous, rhinocerebral, and systemic. In all cases, granulomatous inflammation with angioinvasion and thrombosis predominate. Cases are usually observed in animals debilitated by concurrent disease or under immunosuppressive therapy.

References

1. Baker DG. Parasitic diseases. The common marmoset in captivity and biomedical. Research. 2019:289–303. https://doi.org/10.1016/b978-0-12-811829-0.00017-0.
2. Hegner R. Intestinal protozoa of chimpanzees. Am J Epidemiol. 1934;19:480–501.
3. Saleh MN, Thomas JE, Heptinstall JR, Herbein JF, Wolf RF, Reichard MV, et al. Immunologic detection of Giardia duodenalis in a specific pathogen-free captive olive baboon (Papio cynocephalus anubis) colony. J Vet Diagn Investig. 2017;29:916–9.
4. Ye J, Xiao L, Li J, Huang W, Amer SE, Guo Y, et al. Occurrence of human-pathogenic Enterocytozoon bieneusi, Giardia duodenalis and Cryptosporidium genotypes in laboratory macaques in Guangxi, China. Parasitol Int. 2014;63:132–7.
5. Armstrong J, Hertzog RE, Hall RT, Hoff GL. Giardiasis in apes and zoo attendants, Kansas City, Missouri. CDC Vet Public Health Notes. 1979.
6. Flynn RJ. Parasites of Laboratory Animals. Ames, IA: The Iowa State University Press; 1973.
7. Wenrich DH. A species of Hexamita (Protozoa, Flagellata) from the intestine of a monkey (Macacus rhesus). J Parasitol. 1933;19:225–9.
8. da Cunha AM, Muniz J. Nota sobre os parasitas intestinaes do Macacus Rhesus com a descripção de uma nova especie de Octomitus. Mem Inst Oswaldo Cruz. 1929;22:34–7.
9. Taylor MA, Coop RL, Wall RL. Veterinary parasitology. Wiley; 2015.
10. Scimeca JM, Culberson DE, Abee CR, Gardner WA Jr. Intestinal trichomonads (Tritrichomonas mobilensis) in the natural host Saimiri sciureus and Saimiri boliviensis. Vet Pathol. 1989;26:144–7.
11. Brady AG, Pindak FF, Abee CR, Gardner WA Jr. Enteric trichomonads of squirrel monkeys (Saimiri sp): natural infestation and treatment. Am J Primatol. 1988;14:65–71.
12. Kondova I, Simon MA, Klumpp SA, MacKey J, Widmer G, Domingues HG, et al. Trichomonad gastritis in rhesus macaques (Macaca mulatta) infected with simian immunodeficiency virus. Vet Pathol. 2005;42:19–29.
13. Inoue T, Hayashimoto N, Yasuda M, Sasaki E, Itoh T. Pentatrichomonas hominis in laboratory-bred common marmosets. Exp Anim. 2015;64:363–8.
14. Hegner R, Chu HJ, Others. A survey of protozoa parasitic in plants and animals of the Philippine Islands. Philipp J Sci. 1930:43.
15. Poindexter HA. A study of the intestinal parasites of the monkeys of the Santiago Island primate colony. 1942.
16. Jeong E-S, Park J-H, Ryu S-H, Choi S-Y, Lee K-S, Kim J-M, et al. Detection of Chilomastix mesnili in common marmoset (Callithrix jacchus) and treatment with metronidazole. Iran J Parasitol. 2019;14:334–9.
17. Kouassi RYW, McGraw SW, Yao PK, Abou-Bacar A, Brunet J, Pesson B, et al. Diversity and prevalence of gastrointestinal parasites in seven non-human primates of the Taï National Park, Côte d'Ivoire. Parasite. 2015;22:1.
18. Levine DN, Levine DN. Veterinary protozoology. Ames, IA: Iowa State University Press; 1985.
19. Kuntz RE, Myers BJ. Microbiological parameters of the baboon (Papio sp.): parasitology. Baboon Med Res. 1967;2:741–55.
20. Knowles R, Das G, Others. Some observations on the intestinal protozoa of macaques. Indian J Med Res. 1936:24.
21. Mynářová A, Foitová I, Kváč M, Květoňová D, Rost M, Morrogh-Bernard H, et al. Prevalence of Cryptosporidium spp., Enterocytozoon bieneusi, Encephalitozoon spp. and Giardia intestinalis in wild, semi-wild and captive orangutans (Pongo abelii and Pongo pygmaeus) on Sumatra and Borneo, Indonesia. PLoS One. 2016;11:e0152771.
22. Mansfield KG, Carville A, Shvetz D, MacKey J, Tzipori S, Lackner AA. Identification of an Enterocytozoon bieneusi-like microsporidian parasite in simian-immunodeficiency-virus-inoculated macaques with hepatobiliary disease. Am J Pathol. 1997;150:1395–405.
23. Chalifoux LV, MacKey J, Carville A, Shvetz D, Lin KC, Lackner A, et al. Ultrastructural morphology of Enterocytozoon bieneusi in biliary epithelium of rhesus macaques (Macaca mulatta). Vet Pathol. 1998;35:292–6.
24. Green LC, Didier PJ, Bowers LC, Didier ES. Natural and experimental infection of immunocompromised rhesus macaques (Macaca mulatta) with the microsporidian Enterocytozoon bieneusi genotype D. Microbes Infect. 2004;6:996–1002.
25. Dorn PL, Daigle ME, Combe CL, Tate AH, Stevens L, Phillippi-Falkenstein KM. Low prevalence of Chagas parasite infection in a nonhuman primate colony in Louisiana. J Am Assoc Lab Anim Sci. 2012;51:443–7.
26. Bommineni YR, Dick EJ Jr, Estep JS, Van de Berg JL, Hubbard GB. Fatal acute Chagas disease in a chimpanzee. J Med Primatol. 2009;38:247–51.

27. Hall CA, Polizzi C, Yabsley MJ, Norton TM. Trypanosoma cruzi prevalence and epidemiologic trends in lemurs on St. Catherines Island, Georgia. J Parasitol. 2007;93:93–6.
28. Schielke JE, Selvarangan R, Kyes KB, Fritsche TR. Laboratory diagnosis of Trypanosoma cruzi infection in a colony-raised pig-tailed macaque. Contemp Top Lab Anim Sci. 2002;41:42–5.
29. Pung OJ, Spratt J, Clark CG, Norton TM, Carter J. Trypanosoma cruzi infection of free-ranging lion-tailed macaques (Macaca silenus) and ring-tailed lemurs (Lemur catta) on St. Catherine's Island, Georgia, USA. J Zoo Wildl Med. 1998;29:25–30.
30. Gleiser CA, Yaeger RG, Ghidoni JJ. Trypanosoma cruzi infection in a colony-born baboon. J Am Vet Med Assoc. 1986;189:1225–6.
31. Eberhard M, D'Alessandro A. Congenital Trypanosoma cruzi infection in a laboratory-born squirrel monkey, Saimiri sciureus. Am J Trop Med Hyg. 1982;31:931–3.
32. Kasa TJ, Lathrop GD, Dupuy HJ, Bonney CH, Toft JD. An endemic focus of Trypanosoma cruzi infection in a subhuman primate research colony. J Am Vet Med Assoc. 1977;171:850–4.
33. Strait K, Else JG, Eberhard ML. Parasitic diseases of nonhuman primates. In: Abee CR, Mansfield K, Tardif SD, Morris T, editors. Nonhuman primates in biomedical research: disease. Elsevier; 2012. p. 197–297.
34. Teare JA, Loomis MR. Epizootic of balantidiasis in lowland gorillas. J Am Vet Med Assoc. 1982;181:1345–7.
35. Kumar S, Sundararaj P, Kumara HN, Pal A, Santhosh K, Vinoth S. Prevalence of gastrointestinal parasites in bonnet macaque and possible consequences of their unmanaged relocations. PLoS One. 2018;13:e0207495.
36. Kuze N, Kanamori T, Malim TP, Bernard H, Zamma K, Kooriyama T, et al. Parasites found from the feces of Bornean orangutans in Danum Valley, Sabah, Malaysia, with a redescription of Pongobius hugoti and the description of a new species of Pongobius (Nematoda: Oxyuridae). J Parasitol. 2010;96:954–60.
37. Kessler MJ, Yarbrough B, Rawlins RG, Berard J. Intestinal parasites of the free-ranging Cayo Santiago rhesus monkeys (Macaca mulatto). J Med Primatol. 1984:57–66. https://doi.org/10.1111/j.1600-0684.1984.tb00114.x.
38. Nakamura S, Itagaki I, Nakagawa T, Kawamoto I, Asano T, Arakawa D, et al. Paromomycin sulfate is an effective treatment for balantidiasis in captive cynomolgus monkeys. Exp Anim. 2019;68:285–92.
39. Yi-di HOU. Investigation of infected state of Balantidium coli in rhesus monkey groupos. Zool Res. 1989;10:151–5.
40. Kovatch RM, White JD. Cryptosporidiosis in two juvenile rhesus monkeys. Vet Pathol. 1972;9:426–40.
41. Hahn NE, Capuano SV 3rd. Successful treatment of cryptosporidiosis in 2 common marmosets (Callithrix jacchus) by using paromomycin. J Am Assoc Lab Anim Sci. 2010;49:873–5.
42. Wilson DW, Day PA, Brummer ME. Diarrhea associated with Cryptosporidium spp in juvenile macaques. Vet Pathol. 1984;21:447–50.
43. Baskerville A, Ramsay AD, Millward-Sadler GH, Cook RW, Cranage MP, Greenaway PJ. Chronic pancreatitis and biliary fibrosis associated with cryptosporidiosis in simian AIDS. J Comp Pathol. 1991;105:415–21.
44. Yanai T, Chalifoux LV, Mansfield KG, Lackner AA, Simon MA. Pulmonary cryptosporidiosis in simian immunodeficiency virus—infected rhesus macaques. Vet Pathol. 2000;37:472–5.
45. Dietz HH, Henriksen P, Bille-Hansen V, Henriksen SA. Toxoplasmosis in a colony of New World monkeys. Vet Parasitol. 1997;68:299–304.
46. Epiphanio S, Sinhorini IL, Catão-Dias JL. Pathology of toxoplasmosis in captive new world primates. J Comp Pathol. 2003;129:196–204.
47. Levaditi C. Présence d'un toxoplasme dans l'encéphale du "Cynocaphalus babuin". Masson; 1933.
48. Hendricks LD. Host range characteristics of the primate coccidian, Isospora arctopitheci Rodhain 1933 (Protozoa: Eimeriidae). J Parasitol. 1977;63:32–5.
49. Burrows RB. Protozoa of the Intestinal Tract. 1972. pp. 2–28.
50. Lindsay DS, Dubey JP, Blagburn BL. Biology of Isospora spp. from humans, nonhuman primates, and domestic animals. Clin Microbiol Rev. 1997;10:19–34.
51. Arcay L. Coccidiosis en monos y su comparacion con la isosporosis humana, con description de una nueva especie de Isospora en Cacajao rubicundus (Uakari monkey o mono chucuto). Acta Biol Venez. 1967;5:203–22.
52. Duszynski DW, File SK. Structure of the oocyst and excystation of sporozoites of Isospora endocallimici n. sp., from the marmoset Callimico goeldii. Trans Am Microsc Soc. 1974;93:403–8.
53. Marinkelle CJ. Isospora cebi sp. n. aislada de un mico de Colombia (Cebus albifronis). Rev Bras Biol. 1969;29:35–40.
54. McConnell EE, De Vos AJ, Basson PA, De Vos V. Isospora papionis n. sp. (Eimeriidae) of the chacma baboon Papio ursinus (Kerr, 1792). J Protozool. 1971;18:28–32.
55. McConnell EE, Basson PA, Thomas SE, De Vos V. Oocysts of Isospora papionis in the skeletal muscles of chacma baboons. Onderstepoort J Vet Res. 1972;39:113–6.
56. Baker JR. Protozoa of tissues and blood. Jn pathology of simian primates. In: TW-Fiennes RN, editor. Part II: Infectious and Parasitic Diseases. New York: S. Karger; 1972.
57. King NW. Synopsis of the pathology of New World monkeys. First inter-American conference on the conservation and utilization of American nonhuman primates in biomedical research, Washington, DC. 1976.
58. Shadduck JA. Protozoal and metazoal diseases. Pathol Lab Anim. 1978;
59. Yang Z-Q, Wei C-G, Zen J-S, Song J-L, Zuo Y-X, He Y-S, et al. A taxonomic re-appraisal of Sarcocystis nesbitti (Protozoa: Sarcocystidae) from the monkey Macaca fascicularis in Yunnan, PR China. Parasitol Int. 2005;54:75–81.
60. Mehlhorn H, Heydorn AO, Janitschke K. Light and electron microscopical study on sarcocysts from muscles of the rhesus monkey (Macaca mulatta), baboon (Papio cynocephalus) and tamarin (Saguinus (=Oedipomidas) oedipus). Z Parasitenkd. 1977;51:165–78.
61. Terrell TG, Stookey JL. Chronic eosinophilic myositis in a rhesus monkey infected with sarcosporidiosis. Vet Pathol. 1972;9:266–71.
62. Mayer M. Ein neuer, eigenartiger Blutparasit des Affen (Entopolypoides macaci ng et n. sp.). Zentralblatt für Bakteriologie, Originale. 1934;131:132.
63. Hawking F. Entopolypoides macaci, a Babesia-like parasite in Cercopithecus monkeys. Parasitology. 1972;65(1):89–109.
64. Fairbairn H. The occurrence of a piroplasm, Entopolypoides macaci, in East African monkeys. Ann Trop Med Parasitol. 1948;42(1):118.
65. Gleason NN, Wolf RE. Entopolypoides macaci (Babesiidae) in Macaca mulatta. J Parasitol. 1974;60(5):844–7.
66. Moore JA, Kuntz RE. Entopolypoides macaci Mayer, 1934 in the African baboon (Papio cynocephalus L. 1766). J Med Primatol. 1975;4(1):1–7.
67. Garnham PCC. Blood parasites of East African vertebrates, with a brief description of exo-erythrocytic schizogony in Plasmodium pitmani. Parasitology. 1950:328–37. https://doi.org/10.1017/s0031182000018205.
68. Tanguy Y. La piroplasmose du singe. Ann Inst Pasteur. 1937;59:610–23.
69. Kikuth W. Piroplasmose bei Affen. Arch f Schiffs-u Tropen-hyg. 1927;31:37.
70. Schoeb TR. Klossiella sp infection in a galago. J Am Vet Med Assoc. 1984;185:1381–2.

71. Feng M, Cai J, Min X, Fu Y, Xu Q, Tachibana H, et al. Prevalence and genetic diversity of Entamoeba species infecting macaques in southwest China. Parasitol Res. 2013;112:1529–36.
72. Rideout BA, Gardiner CH, Stalis IH, Zuba JR, Hadfield T, Visvesvara GS. Fatal infections with Balamuthia mandrillaris (a free-living amoeba) in gorillas and other Old World primates. Vet Pathol. 1997;34:15–22.
73. Heyneman D, Garcia LS, Bruckner DA. Diagnostic medical parasitology. J Parasitol. 1988:1070. https://doi.org/10.2307/3282241.
74. Orihel TC, Seibold HR. Nematodes of the bowel and tissues. In: T-W-Fiennes RN, editor. Pathology of simian primates: infectious and parasitic diseases. Karger Publishers; 1972. pp. 76–103.
75. Lowenstine LJ, McManamon R, Terio KA. Apes. In: St. Leger J TKMD, editor. Pathology of wildlife and zoo animals. 2018. pp. 375–412.
76. McAloose D, Stalis IH. Prosimians. In: St. Leger J TKMD, editor. Pathology of wildlife and zoo animals. 2018. pp. 323–342.
77. Mätz-Rensing K, Lowenstine LJ. New world and old world monkeys. In: St. Leger J TKMD, editor. Pathology of wildlife and zoo animals. 2018. pp. 343–374.
78. Chitwood M, Lichtenfels JR. Identification of parasitic metazoa in tissue sections. Exp Parasitol. 1972;32:407–519.
79. Gardiner CH, Poynton SL. An atlas of metazoan parasites in animal tissues. Armed Forces Institute of Pathology, American Registry of Pathology; 1999.
80. Stam AB. Fatal ascaridosis in a dwarf chimpanzee. Ann Parasitol Hum Comp. 1960;35:675.
81. Habermann TR. Disease seen at necropsy of 708 Macaca rhesus (Rhesus monkey) and Macaca philippinensis (Cynomolgus monkey). Am J Vet Res. 1957;18:419–26.
82. Ball RL, Dryden M, Wilson S, Veatch J. Cerebrospinal nematodiasis in a white-handed gibbon (Hylobates lar) due to Baylisascaris sp. J Zoo Wildl Med. 1998;29:221–4.
83. Campbell GA, Hoover JP, Russell WC, Breazile JE. Naturally occurring cerebral nematodiasis due to Baylisascaris larval migration in two black-and-white ruffed lemurs (Varecia variegata variegata) and suspected cases in three emus (Dromaius novaehollandiae). J Zoo Wildl Med. 1997;204–207
84. Garlick DS, Marcus LC, Pokras M, Schelling SH. Baylisascaris larva migrans in a spider monkey (Ateles sp.). J Med Primatol. 1996;25:133–6.
85. Hanley CS, Simmons HA, Wallace RS, Clyde VL. Visceral and presumptive neural baylisascariasis in an orangutan (Pongo pygmaeus). J Zoo Wildl Med. 2006;37:553–7.
86. Kazacos KR, Wirtz WL, Burger PP, Christmas CS. Raccoon ascarid larvae as a cause of fatal central nervous system disease in subhuman primates. J Am Vet Med Assoc. 1981;179:1089–94.
87. Kazacos KR. Baylisascaris larva migrans. Abbott RC, van Riper C III, editors. US Geological Survey; 2016. doi:https://doi.org/10.3133/cir1412
88. Shoieb A, Radi ZA. Cerebral Baylisascaris larva migrans in a cynomolgus macaque (Macaca fascicularis). Exp Toxicol Pathol. 2014;66:263–5.
89. Sato H, Une Y, Kawakami S, Saito E, Kamiya H, Akao N, et al. Fatal Baylisascaris larva migrans in a colony of Japanese macaques kept by a safari-style zoo in Japan. J Parasitol. 2005;91:716–9.
90. Jimenez Martinez M-A, Valderrabano Cano E, Rois JL. Baylisascaris procyonis larva migrans in two white-headed lemurs (Eulemur albifrons) in Spain and response to treatment derived from a human pediatric protocol. Vet Parasitol. 2015;210:246–9.
91. Orihel TC. Dirofilaria corynodes (Von Linstow, 1899): morphology and life history. J Parasitol. 1969;55:94.
92. Gamble KC, Fried JJ, Rubin GJ. Presumptive dirofilariasis in a pale-headed saki monkey (Pithecia pithecia). J Zoo Wildl Med. 1998;29:50–4.
93. Ayers KM, Jones SR. The cardiovascular system. Pathol Lab Anim. 1978:1–69. https://doi.org/10.1007/978-1-4612-9942-4_1.
94. Bain O, Mutafchiev Y, Junker K, Guerrero R, Martin C, Lefoulon E, et al. Review of the genus Mansonella Faust, 1929 sensu lato (Nematoda: Onchocercidae), with descriptions of a new subgenus and a new subspecies. Zootaxa. 2015;3918:151–93.
95. Bergner JJE, Jachowski LJA, Naval Medical Research Unit No 2 Manila (Philippines) Manila. The filarial parasite, Macacanema formosana from the Taiwan monkey and its development in various arthropods. Naval Medical Research Unit No 2 Manila (Philippines) Manila; 1968 Jan. Available: https://apps.dtic.mil/sti/citations/AD0669709
96. Orihel TC, Esslinger JH. Meningonema peruzzii gen. et sp. n. (Nematoda: Filarioidea) from the Central Nervous System of African Monkeys. J Parasitol. 1973;59:437.
97. Schmidt RE, Prine JR. Severe Enterobiasis in a Chimpanzee. Pathol Vet. 1970;7:56–9.
98. Murata K, Hasegawa H, Nakano T, Noda A, Yanai T. Fatal infection with human pinworm, Enterobius vermicularis, in a captive chimpanzee. J Med Primatol. 2002;31:104–8.
99. DePaoli A, Johnsen DO. Fatal strongyloidiasis in gibbons (Hylobates lar). Vet Pathol. 1978;15:31–9.
100. Beglinger R, Illgen B, Pfister R, Heider K. The parasite *Trichospirura leptostoma* Associated with wasting disease in a colony of common marmosets, *Callithrix jacchus*. Folia Primatol. 1988;51:45–51.
101. Bleier T, Hetzel U, Bauer C, Behlert O, Burkhardt E. Gongylonema pulchrum infection and esophageal squamous cell carcinoma in a vari (Lemur macaco variegata; Kehr 1792). J Zoo Wildl Med. 2005;36:342–5.
102. Adkesson MJ, Langan JN, Paul A. Evaluation of control and treatment of Gongylonema spp. infections in callitrichids. J Zoo Wildl Med. 2007;38:27–31.
103. Montali RJ, Gardiner CH, Evans RE, Bush M. Pterygodermatites nyctcebi (Nematoda: Spirurida) in Golden Lion Tamarins. Lab Anim Sci. 1983;33:194–7.
104. Machida M, Araki J, Koyama T, Kumada M, Horii Y, Imada I, et al. The life cycle of Streptopharagus pigmentatus (Nematoda, Spiruroidea) from the Japanese monkey. Bull Natl Mus Nat Sci Ser B Bot. 1978;4:1–9.
105. Lumb GD, Beamer PR, Rust JH. Oesophagostomiasis in feral monkeys (Macaca mulatta). Toxicol Pathol. 1985;13:209–14.
106. Brack M, Myers BJ, Kuntz RE. Pathogenic properties of Molineus torulosus in capuchin monkeys, Cebus apella. Lab Anim Sci. 1973;23:360–5.
107. Sly DL, Toft JD, Gardiner CH, London WT. Spontaneous occurrence of Angiostrongylus costaricensis in marmosets (Saguinus mystax). Lab Anim Sci. 1982;32:286–8.
108. Miller CL, Kinsella JM, Garner MM, Evans S, Gullett PA, Schmidt RE. Endemic infections of Parastrongylus (=Angiostrongylus) costaricensis in two species of nonhuman primates, raccoons, and an opossum from Miami, Florida. J Parasitol. 2006;92:406–8.
109. Duffy MS, Miller CL, Kinsella JM, de Lahunta A. Parastrongylus cantonensis in a nonhuman primate, Florida. Emerg Infect Dis. 2004;10:2207–10.
110. Stidworthy MF, Lewis JCM, Masters NJ, Boardman SI, Hopper JS, de Linan FJL, et al. Capillaria hepatica in primates in zoological collections in the British Isles. Vet Rec. 2009;164:66.
111. Graczyk TK, Lowenstine LJ, Cranfield MR. Capillaria hepatica (Nematoda) infections in human-habituated mountain gorillas (Gorilla gorilla beringei) of the Parc National de Volcans, Rwanda. J Parasitol. 1999;85:1168–70.
112. Myers BJ. Echinococcosis, Coenurosis, Cysticercosis, Sparganosis, etc. In: T-W-Fiennes RN, editor. Pathology of Simian primates: infectious and parasitic diseases. Karger Publishers; 1972. p. 124–43.

113. Nobrega-Lee M, Hubbard G, Loverde P, Carvalho-Queiroz C, Conn DB, Rohde K, et al. Sparganosis in wild-caught baboons (Papio cynocephalus anubis). J Med Primatol. 2007;36:47–54.
114. Gardiner CH, Imes GD Jr. Sparganosis in a saddle-back tamarin: another case of viral-induced proliferation? J Wildl Dis. 1986;22: 437–9.
115. Kuntz RE, Myers BJ, Katzberg AA. Sparganosis and "Proliferative" Spargana in Vervets (Cercopithecus aethiops) and Baboons (Papio sp.) from East Africa. J Parasitol. 1970;56:196.
116. Fincham JE, Seier JV, Verster A, Rose AG, Taljaard JJ, Woodroof CW, et al. Pleural Mesocestoides and cardiac shock in an obese vervet monkey (Cercopithecus aethiops). Vet Pathol. 1995;32: 330–3.
117. Bacciarini LN, Gottstein B, Pagan O, Rehmann P, Gröne A. Hepatic alveolar echinococcosis in cynomolgus monkeys (Macaca fascicularis). Vet Pathol. 2004;41:229–34.
118. Rehmann P, Gröne A, Lawrenz A, Pagan O, Gottstein B, Bacciarini LN. Echinococcus multilocularis in Two Lowland Gorillas (Gorilla g. gorilla). J Comp Pathol. 2003;129:85–8.
119. Dyer NW, Greve JH. Severe Cysticercus longicollis cysticercosis in a black lemur (Eulemur macaco macaco). J Vet Diagn Investig. 1998;10:362–4.
120. Kuntz RE. Others. Trematodes of the intestinal tract and biliary passages. In: T-W-Fiennes RN, editor. Pathology of Simian primates: infectious and parasitic diseases. Karger Publishers; 1972. p. 104–23.
121. Schmidt GD. Acanthocephala of captive primates. In: T-W-Fiennes RN, editor. Pathology of Simian primates: infectious and parasitic diseases. Karger Publishers; 1972. p. 144–56.
122. Chen PH, Miller GF, Powell DA. Colitis in a female tamarin (Saguinus mystax). Contemp Top Lab Anim Sci. 2000;39:47–9.
123. T-W-Fiennes RN. Ectoparasites and vectors. In: T-W-Fiennes RN, editor. Pathology of Simian primates: infectious and parasitic diseases. Karger Publishers; 1972. p. 158–76.
124. Schott D, Ribeiro PR, de Souza VK, Surita LE, de Amorim DB, Bianchi MV, et al. Clinical and pathological aspects of first report of Tunga penetrans infestation on southern brown howler monkey (Alouatta guariba clamitans) in Rio Grande do Sul, Brazil. J Med Primatol. 2020;49:315–21.
125. Starost MF, Karjala Z, Brinster LR, Miller G, Eckhaus M, Bryant M, et al. Demodex spp. in the hair follicles of rhesus macaques (Macaca mulatta). J Med Primatol. 2005;34:215–8.
126. James SB, Raphael BL. Demodicosis in red-handed tamarins (Saguinus midas). J Zoo Wildl Med. 2000;31:251–4.
127. Innes JRM, Hull WB. Endoparasites—lung mites. In: T-W-Fiennes RN, editor. Pathology of Simian primates: infectious and parasitic diseases. Karger Publishers; 1972. p. 177–93.
128. Andrade MCR, Marchevsky RS. Histopathologic findings of pulmonary acariasis in a rhesus monkeys breeding unit. Rev Bras Parasitol Vet. 2007;16:229–34.
129. Self JT, Cosgrove GE. Pentastomida. In: T-W-Fiennes RN, editor. Pathology of Simian primates: infectious and parasitic diseases. Karger Publishers; 1972. p. 194–204.
130. Cosgrove GE, Nelson BM, Self JT. The pathology of pentastomid infection in primates. Lab Anim Care. 1970;20:354–60.
131. Demanche C, Wanert F, Barthélemy M, et al. Molecular and serological evidence of Pneumocystis circulation in a social organization of healthy macaques (Macaca fascicularis). Microbiology. 2005;151(Pt 9):3117–25.
132. Demanche C, Berthelemy M, Petit T, et al. Phylogeny of Pneumocystis carinii from 18 primate species confirms host specificity and suggests coevolution. J Clin Microbiol. 2001;39(6): 2126–33.
133. Baskin GB, Murphey-Corb M, Watson EA, Martin LN. Necropsy findings in rhesus monkeys experimentally infected with cultured simian immunodeficiency virus (SIV)/delta. Vet Pathol. 1988;25 (6):456–67.
134. Vogel P, Miller CJ, Lowenstine LL, Lackner AA, Lackner AA. Evidence of horizontal transmission of Pneumocystis carinii pneumonia in simian immunodeficiency virus-infected rhesus macaques. J Infect Dis. 1993;168(4):836–43.
135. Nevez G, Totet A, Matos O, Calderon EJ, Miller RF, le Gal S. It is still PCP that can stand for Pneumocystis pneumonia: Appeal for generalized use of only one acronym. Med Mycol. 2021;59(8): 842–4.
136. Shipley TW, Kling HM, Morris A, et al. Persistent pneumocystis colonization leads to the development of chronic obstructive pulmonary disease in a nonhuman primate model of AIDS. J Infect Dis. 2010;202(2):302–12.
137. Board KF, Patil S, Lebedeva I, et al. Experimental Pneumocystis carinii pneumonia in simian immunodeficiency virus-infected rhesus macaques. J Infect Dis. 2003;187(4):576–88.
138. Bohm RP, Martin LN, Davison-Fairburn B, Baskin GB, Murphey-Corb M. Neonatal disease induced by SIV infection of the rhesus monkey (Macaca mulatta). AIDS Res Hum Retrovir. 1993;9(11): 1131–7.
139. Yanai T, Simon MA, Doddy FD, Mansfield KG, Pauley D, Lackner AA. Nodular Pneumocystis carinii pneumonia in SIV-infected macaques. Vet Pathol; 1999;36(5):471–474.
140. Kling HM, Shipley TW, Guyach S, Tarantelli R, Morris A, Norris KA. Trimethoprim-sulfamethoxazole treatment does not reverse obstructive pulmonary changes in pneumocystis-colonized nonhuman primates with SHIV infection. J Acquir Immune Defic Syndr (1988). 2014;65(4):381–9.
141. Rabacal W, Schweitzer F, Kling HM, Buzzelli L, Rayens E, Norris KA. A therapeutic vaccine strategy to prevent Pneumocystis pneumonia in an immunocompromised host in a non-human primate model of HIV and Pneumocystis co-infection. Front Immunol. 2022;13:7187.
142. Bernstein JA, Didier PJ. Nonhuman primate dermatology: a literature review. Vet Dermatol. 2009;20(4):306. https://doi.org/10.1111/j.1365-3164.2009.00742.x.
143. Lowenstine LJ, Osborn KG. Respiratory system diseases of nonhuman primates. Nonhuman Primates Biomed Res. 2012;2:413.
144. Beaman L, Holmberg C, Henrickson R, Osburn B. The incidence of coccidioidomycosis among nonhuman primates housed outdoors at the California Primate Research Center. J Med Primatol. 1980;9(4):254–61. https://doi.org/10.1159/000460147.
145. Koistinen K, Mullaney L, Bell T, et al. Coccidioidomycosis in nonhuman primates: pathologic and clinical findings. Vet Pathol. 2018;55(6):905–15.
146. Migaki G, Schmidt RE, Toft JD, Kaufmann AF. Mycotic infections of the alimentary tract of nonhuman primates: a review. Vet Pathol. 1982;19:93–103.
147. Pecoraro HL, Berg MR, Dozier BL, et al. Candida albicans-associated sepsis in a pre-term neonatal rhesus macaque (Macaca mulatta). J Med Primatol. 2019;48(3):186–8.
148. Colman K, Andrews RN, Atkins H, et al. International harmonization of nomenclature and diagnostic criteria (INHAND): Non-proliferative and proliferative lesions of the non-human primate (M. fascicularis). J Toxicol Pathol. 2021;34:1–182.
149. Retallack DM, Woods JP. Molecular epidemiology, pathogenesis, and genetics of the dimorphic fungus Histoplasma capsulatum. Microbes Infect. 1999;1(10):817–25.
150. Bauman DS, Chick EW. Acute cavitary histoplasmosis in Rhesus monkeys: influence of immunological status. Infect Immun [Internet]. Infect Immun. 1973;8(2):245–8.
151. Bauman DS, Chick EW. Experimental histoplasmosis in rhesus monkeys. Infectious dose and extrapulmonary dissemination determination. Chest. 1973;63(2):254–8.

152. Miyashiro JE, Jester J, v., Delmage JM, Smith RE. Immunofluorescent detection of histoplasma capsulatum in primate experimental ocular histoplasmosis. Curr Eye Res. 1986;5(11):833–40.
153. Kaplan W, Kaufman L, McClure HM. Pathogenesis and immunological aspects of experimental histoplasmosis in cynomolgus monkeys (Macaca fascicularis). Infect Immun. 1972;5(6):847–53.
154. Baskin GB. Disseminated histoplasmosis in a SIV-infected rhesus monkey. J Med Primatol. 1991;20(5):251–3.
155. Butler TM, Hubbard GB. An epizootic of *histoplasmosis duboisii* (African histoplasmosis) in an American baboon colony. Lab Anim Sci. 1991;41(5):407–10.
156. Migaki G, Hubbard GB, Butler TM. *Histoplasma capsulatum* var. *duboisii* infection, Baboon. Berlin: Springer; 1993. p. 19–23. https://doi.org/10.1007/978-3-642-84924-4_4.
157. Johannigman TA, Gonzalez O, Dutton JW, Kumar S, Dick EJ. Gingival histoplasmosis: An atypical presentation of African histoplasmosis in three baboons (Papio spp). J Med Primatol. 2020;49(1):47–51. https://doi.org/10.1111/jmp.12443.
158. Srikanta D, Santiago-Tirado FH, Doering TL. Cryptococcus neoformans: historical curiosity to modern pathogen. Yeast. 2014;31(2):47.
159. Pal M, Dube GD, Mehrotra BS. Pulmonary cryptococcosis in a rhesus monkey (Macaca Mulatta). Mycoses. 1984;27(6):309–12. https://doi.org/10.1111/j.1439-0507.1984.tb02035.x.
160. Johnson AL, Lewis AD. Two cases of *Cryptococcus gattii* infection in the rhesus macaque (*Macaca Mulatta*). J Med Primatol. 2021;50(1):67–70. https://doi.org/10.1111/jmp.12497.
161. Costa EO, Diniz LSM, Fava Netto C, Arruda C, Dagli MLZ. Epidemiological study of sporotrichosis and histoplasmosis in captive Latin American wild mammals, São Paulo, Brazil. Mycopathologia. 1994;125(1):19–22.
162. Corte AC, Walfrido AE, Svoboda K, et al. Paracoccidioidomycosis in wild monkeys from Paraná State. Brazil. 2007;164(5):225–8. https://doi.org/10.1007/s11046-007-9059-y.
163. Johnson WD, Lang CM. Paracoccidioidomycosis (South American blastomycosis) in a squirrel monkey (Saimiri sciureus). Vet Pathol. 1977;14(4):368–71.

Nutritional, Metabolic, and Toxic Disorders of Nonhuman Primates

4

Lars Mecklenburg and Sarah Beck

Abstract

In humans, a plethora of metabolic diseases exists, i.e. disorders that disrupt the cell's ability to process or transport proteins, carbohydrates, or lipids. It is likely that in nonhuman primates, the same variety of diseases exists as in humans, but much less is known from those species, which are primarily housed in laboratories under controlled conditions and for a limited life span. In striking similarity to humans, the most common metabolic diseases in captive nonhuman primates are obesity and diabetes. These disorders bear striking similarity to the human counterpart and are associated with a high-caloric diet and insufficient exercise. Amyloidosis is a well-known age-related phenomenon in nonhuman primates and is only diagnosed at necropsy unless it has reached a critical stage. Fatty liver syndrome is a syndrome that occurs in overweight animals which rapidly loose body weight for any reason. It is rare today, but has been well described in the literature. Other metabolic diseases of primates, such as metabolic bone diseases, are rarely reported in modern-day humans but can be relatively common in captive nonhuman primates. Metabolic bone disease in common marmosets has proven to be a difficult disease to manage due to differing clinical and pathologic case presentations and undetermined cause, although a metabolic pathogenesis such as vitamin deficiency is suspected.

Keywords

Metabolic disease · Fatty liver · Obesity · Diabetes mellitus · Amyloidosis · Metabolic bone disease

4.1 Fatal Fatty Liver Syndrome

A syndrome in which obese rhesus macaques die after rapidly losing about 30% body weight is known as "fatal fatty liver syndrome" or "fatal fasting syndrome" [1]. This syndrome has mainly been observed in female animals above 8 years of age. The reason for the anorexia can be manifold, e.g. changes in the housing environment or social hierarchy.

Common histological findings in this condition are severe fatty change in the liver and in proximal convoluted renal tubules, occasionally associated with renal tubule atrophy. Other histopathological findings that have been described in this condition are pancreatitis, pancreatic ductal ectasia, and pancreatic necrosis (Fig. 4.1) [2].

L. Mecklenburg
Labcorp Early Development Services GmbH, Münster, Germany
e-mail: lars.mecklenburg@covance.com

S. Beck (✉)
Charles River Laboratories, Veterinary Pathologist II, Columbia, MD, USA
e-mail: sarah.beck2@crl.com

I. Kondova - Perseng et al. (eds.), *Atlas of Diagnostic Pathology in Nonhuman Primates*,
https://doi.org/10.1007/978-3-031-41280-6_4

a
b
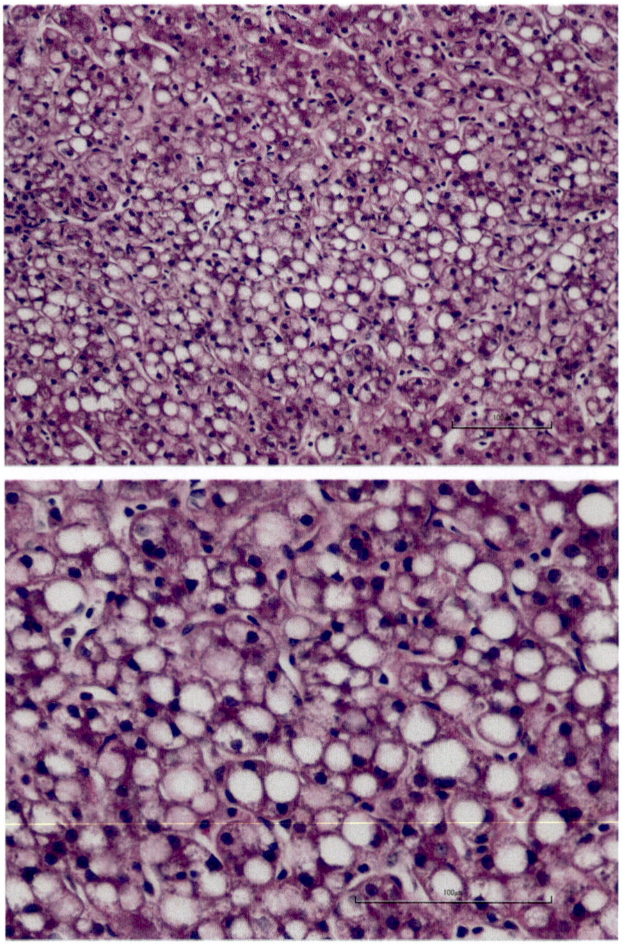

Fig. 4.1 Fatal fatty liver syndrome/fatal fasting syndrome in a 23-year-old female rhesus macaque. **(a, b)** Macrovesicular fatty change in hepatocytes with displacement of nuclei to the periphery

4.2 Amyloidosis

Spontaneous amyloidosis occurs in many nonhuman primate species, particularly associated with advanced age [3]. Amyloid deposits are characterized by deposition of amorphous to fibrillar, acellular, eosinophilic, congophilic material which aggregates in interstitial spaces. Depending on the extent of amyloid deposition, cellular function may become impaired resulting in organ dysfunction.

Visceral amyloidosis is a common finding in baboons [4], where it may be found in various tissues such as intestine, adrenal gland, kidney, prostate, spleen, lymph node, liver, gall bladder, stomach, tongue, urinary bladder, or salivary gland [5]. It is also known in pigtail macaques [6] and in rhesus macaques, where amyloid is predominantly deposited in the small intestine, spleen, large intestine, liver, lymph nodes, stomach, and adrenal gland [7]. Amyloid deposition in the liver typically starts in the space of Disse and expands outwards, eventually compressing and replacing normal hepatic parenchyma (Fig. 4.2). Amyloid in the gastrointestinal tract deposits primarily in the mucosal lamina propria, while adrenal amyloidosis is typically localized to the corticomedullary junction.

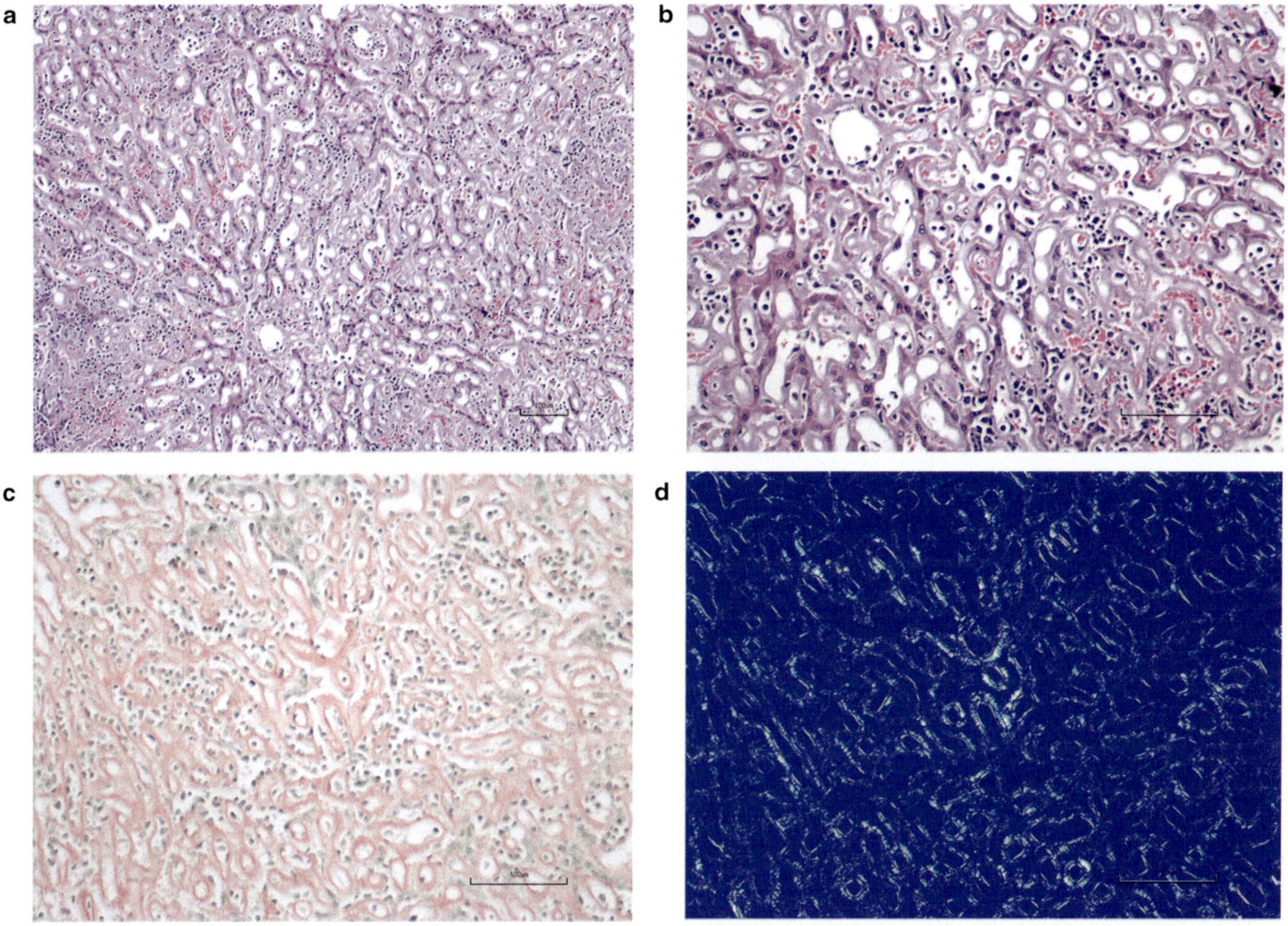

Fig. 4.2 Hepatic amyloidosis in a 24-year-old male rhesus macaque. **(a, b)** Amorphous to fibrillar, extracellular, eosinophilic AA-type amyloid expands the space of Disse in the liver and compresses adjacent hepatocytes, with cord disruption and mononuclear inflammatory cell infiltration; **(c, d)** Amyloid is pinkish-orange and has apple green birefringence under polarized light (Congo Red)

Visceral amyloidosis in primates are most frequently secondary to chronic inflammation, also known as secondary systemic or reactive amyloidosis, and increased serum amyloid A (SAA) can be found circulating in the blood as a precursor to systemic amyloid deposition [8]. Many rhesus macaques with visceral amyloidosis also show histological evidence of colitis, lung mites, or arthritis [9]. Visceral amyloidosis carries a grave prognosis once animals become clinically ill [10]. Subclinical disease may be indicated by elevation of alkaline phosphatase, aspartate aminotransferase, lactate dehydrogenase and decreased concentrations of albumin and total cholesterol [10]. Serum amyloid A can also serve as an effective noninvasive biomarker for amyloidosis, and minimally invasive biopsies from liver or intestine can be used to confirm the diagnosis [8].

Amyloidosis of pancreatic islets is reported in macaques, baboons, and other Old World species where it is strongly associated with diabetes mellitus (see below). Amyloid in pancreatic islets originates from a 37 amino acid polypeptide known as islet associated polypeptide (IAPP).

Cerebral amyloidosis occurs in various nonhuman primate species, particularly in squirrel monkey (*Saimiri sciureus*). It is characterized by beta-amyloidosis mostly located in the basal prefrontal gyrus and the amygdala [11]. Cerebral amyloid plaques may be associated with angiopathy, known as "cerebral amyloid angiopathy." It is independent of visceral amyloidosis, although both disease may concur in the same animal. Cerebral amyloidosis is strongly associated with age [12–14].

4.3 Obesity

Cynomolgus monkeys (*Macaca fascicularis*) and rhesus monkeys (*Macaca mulatta*) exhibit spontaneous onset of obesity in adulthood [15, 16]. Diets containing high amounts of fructose induce rapid weight gain and obesity [17, 18].

Obesity is associated with enhanced synthesis and excretion of leptin in adipocytes, and high levels of hematocytes. Nonhuman primates with obesity show an increased risk to develop multiple metabolic sequelae including type

2 diabetes, cardiovascular disease, or fatty liver disease, and as such represent a valuable and physiologically relevant animal model in metabolic disease research [19].

4.4 Diabetes Mellitus

Naturally occurring non-insulin-dependent (type II) diabetes mellitus (NIDDM) may spontaneously occur in several species of Old and New World nonhuman primates [20, 21]. Consequently, nonhuman primates are considered a valuable model to study NIDDM and therapeutic effects in humans [22]. Diabetes mellitus should be suspected, if a single evaluation of the blood glucose concentration reveals hyperglycemia (>5.0 mmol/L; 123 mg/dl), if blood fructosamine values reveal hyperfructosaminemia, if a glucose tolerance test is positive, or if glucose is excreted in the urine [23].

An evaluation in 440 overweight middle-and old-aged cynomolgus monkeys revealed an incidence of 69% of animals diagnosed with diabetes mellitus [23]. Risk factors that favor development of NIDDM in nonhuman primates comprise single housing (opposed to social housing), insufficient exercise, high-fat diet, middle to old age, and female sex [24].

A typical morphological finding in nonhuman primates with NIDDM is degeneration of pancreatic beta cells, associated with islet amyloid deposits which exhibit typical congophilia, green birefringence, and immunoreactivity for islet amyloid polypeptide [25, 26]. Although islet amyloid can also be found in non-diabetic animals, it is more extensive in diabetics [27]. Nonhuman primates with NIDDM also show typical sequela of diabetes such as microangiopathic retinopathy [28], or cardiomyopathy [29].

NIDDM in nonhuman primates can be controlled by insulin therapy based on fasting glucose levels and glycosylated hemoglobin (HbA1c) values, which are determined periodically by blood sampling, in order to prevent nocturnal hypoglycemia and hypoglycemic seizures [30]. Diabetes mellitus can also be experimentally induced in nonhuman primates either by streptozotocin or by diet [31, 32], and nonhuman primates are commonly used as an animal model for NIDDM [33]. Less common in nonhuman primates is a form of insulin-dependent (type I) diabetes mellitus (Fig. 4.3) [34].

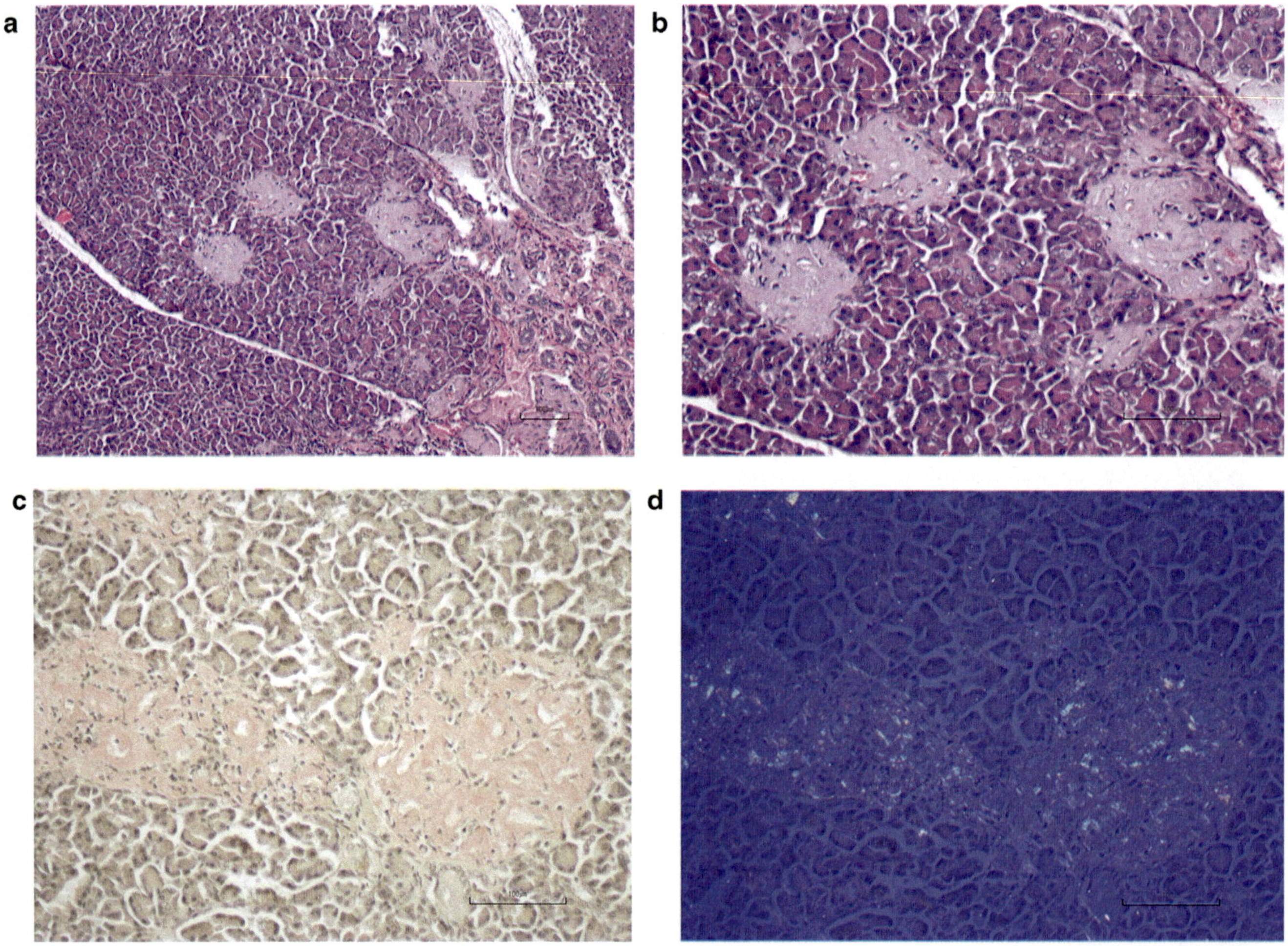

Fig. 4.3 Type II diabetes in a 19-year-old olive baboon with pancreatic islet amyloidosis. **(a, b)** Islet amyloid polypeptide (IAPP)-type amyloidosis replacing degenerate islet amyloid cells; **(c, d)** Amyloid is pinkish-orange and has apple green birefringence under polarized light (Congo Red)

4.5 Metabolic Bone Disease

Common marmosets (*Callithrix jacchus*) have become an increasingly popular experimental animal model over the last few decades but colony management issues have plagued breeding colonies, in part due to their high susceptibility to stress and complex dietary management needs as well as institutional variability in dietary management. Metabolic bone disease is one of the most commonly reported clinical concerns in marmoset colonies, with pathologic fractures as a major concern [35].

Unfortunately, the clinical and pathologic presentation of metabolic bone disease can vary both between institutions and even between animals in the same colony. Depending on the age of presentation and the colony, marmosets have presented with pathologic fractures secondary to bone disease showing features of rickets, fibrous osteodystrophy (FOD), diffuse osteopenia, and/or segmental osteolysis [36]. Young marmosets diagnosed with rickets present with increased physeal thickness due to increased numbers of hypertrophic chondrocytes on histology and variable metaphyseal flaring evident radiographically. In adult marmosets, FOD-like lesions and osteopenia may occur concurrently or separately and may be diffuse or regionally extensive. Commonly affected areas are the mandible and/or maxilla (especially with FOD-like lesions) or long bones adjacent to the stifle and/or tibiotarsal joint. Although metabolic bone disease in marmosets may be challenging to diagnose prior to the development of pathologic fractures without regular radiographic colony screening, progressive weight loss, gastrointestinal disease, and increased serum parathyroid hormone (PTH) are predictive antemortem biomarkers of bone disease [37]. The cause of metabolic bone disease in common marmosets is currently unclear, although insufficient absorption of Vitamin D or calcium has been considered as potential contributing factors although the inconsistent pathologic features do not support one definitive pathogenesis [36].

Most other metabolic bone diseases are relatively rare in captive primate colonies in the modern era. Vitamin C/ascorbic acid deficiency leading to clinical disease (scurvy) in Old and New World primates is well documented and varies in presentation by species [38, 39]. Typical lesions include petechiae and ecchymoses of the subcutis and gingiva as well as hematoma formation due to the essential role that Vitamin C plays in collagen fibril synthesis. Squirrel monkeys (*Saimiri sciureus*) are prone to forming characteristic subperiosteal hematomas (cephalohematomas) with deformation of the cranial bones while rhesus macaques are more susceptible to lameness caused by epiphyseal/metaphyseal fractures of the long bones; however, these presentations are not absolute and squirrel monkeys may also present with lameness. Colony outbreaks of scurvy were often attributed to errors in feed manufacturing or handling, and treatment with ascorbic acid resolves clinical disease in many animals (Figs. 4.4 and 4.5) [38–40].

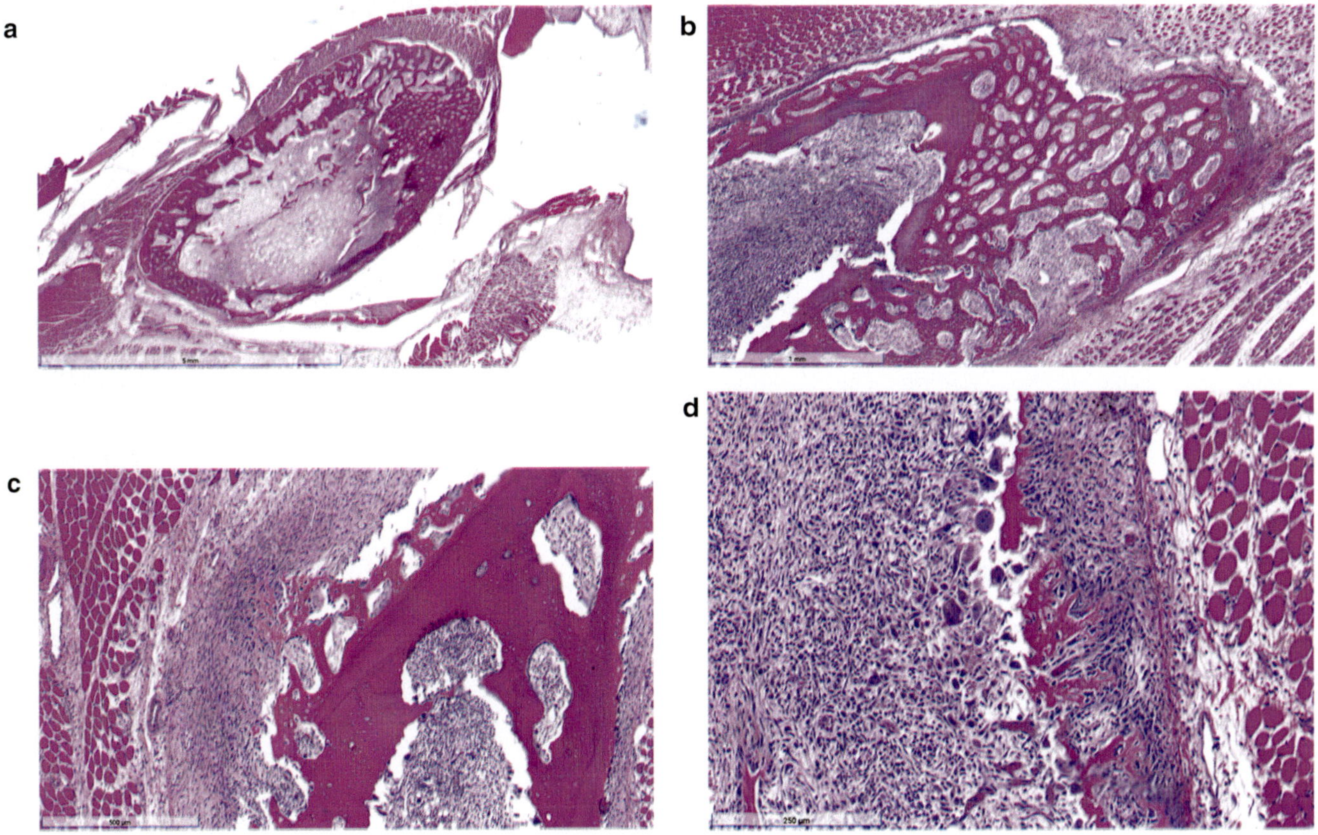

Fig. 4.4 Fibrous osteodystrophy-like metabolic bone disease in common marmosets. **(a, b)** Fibrous osteodystrophy in the ulna **(a)** and radius **(b)**; **(c)** Marked cortical bony remodeling is present; **(d)** Osteolysis with myelofibrosis.

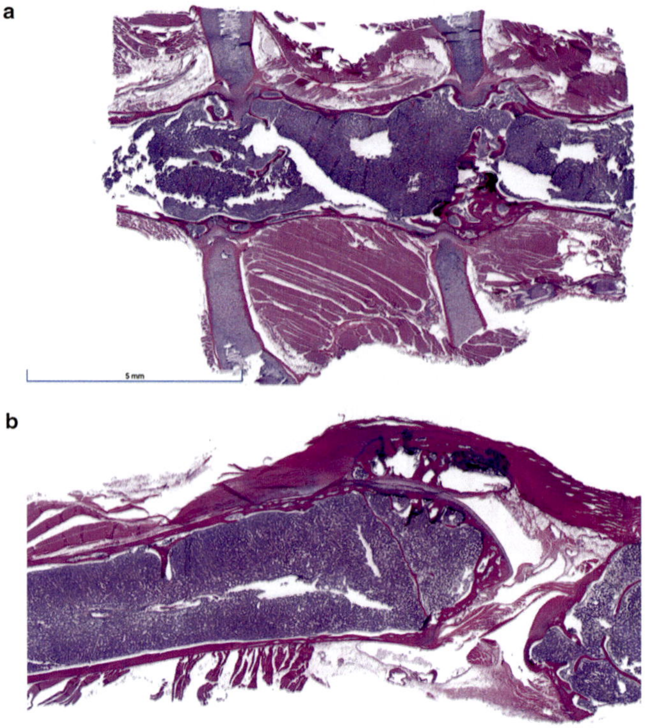

Fig. 4.5 Osteopenia in the common marmoset. **(a, b)** Severe cortical thinning in the sternum **(a)** and femur **(b)**.

References

1. Laber-Laird KE, Jokinen MP, Lehner ND. Fatal fatty liver-kidney syndrome in obese monkeys. Lab Anim Sci. 1987;37(2):205–9.
2. Bronson RT, O'Connell M, Klepper-Kilgore N, et al. Fatal fasting syndrome of obese macaques. Lab Anim Sci. 1982;32 (2):187–92.
3. Simmons HA. Age-associated pathology in rhesus macaques (Macaca Mulatta). Vet Pathol. 2016;53(2):399–416.
4. Dick EJ Jr, Owston MA, David JM, et al. Mortality in captive baboons (Papio Spp.): a-23-year study. J Med Primatol. 2014;43 (3):169–96.
5. Hubbard GB, Steele KE, Davis KJ 3rd, et al. Spontaneous pancreatic islet amyloidosis in 40 baboons. J Med Primatol. 2002;31(2): 84–90.
6. Doepel FM, Glorioso JC, Newcomer CE, et al. Enzyme-linked immunosorbent assay of serum protein SAA in rhesus monkeys with secondary amyloidosis. Lab Investig. 1981;45(1):7–13.
7. Blanchard JL, Baskin GB, Watson EA. Generalized amyloidosis in rhesus monkeys. Vet Pathol. 1986;23(4):425–30.
8. Rice KA, Chen ES, Metcalf Pate KA, et al. Diagnosis of amyloidosis and differentiation from chronic, idiopathic enterocolitis in rhesus (Macaca mulatta) and pig-tailed (M. nemestrina) macaques. Comp Med. 2013;63(3):262–71.
9. Chapman WL Jr, Crowell WA. Amyloidosis in rhesus monkeys with rheumatoid arthritis and enterocolitis. J Am Vet Med Assoc. 1977;171(9):855–8.
10. MacGuire JG, Christe KL, Yee JL, et al. Serologic evaluation of clinical and subclinical secondary hepatic amyloidosis in rhesus macaques (Macaca mulatta). Comp Med. 2009;59(2):168–73.
11. Uno H, Walker LC. The age of biosenescence and the incidence of cerebral beta-amyloidosis in aged captive rhesus monkeys. Ann N Y Acad Sci. 1993;695:232–5.
12. Walker LC. Animal models of cerebral beta-amyloid angiopathy. Brain Res Brain Res Rev. 1997;25(1):70–84.
13. Nakamura S, Nakayama H, Goto N, et al. Histopathological studies of senile plaques and cerebral amyloidosis in cynomolgus monkeys. J Med Primatol. 1998;27(5):244–52.
14. Uno H, Alsum PB, Dong S, et al. Cerebral amyloid angiopathy and plaques, and visceral amyloidosis in aged macaques. Neurobiol Aging. 1996;17(2):275–81.
15. Bauer SA, Arndt TP, Leslie KE, Pearl DL, Turner PV. Obesity in rhesus and cynomolgus macaques: a comparative review of the condition and its implications for research. Comp Med. 2011;61 (6):514–26.
16. Vaughan KL, Mattison JA. Obesity and aging in humans and nonhuman primates: a mini-review. Gerontology. 2016;62(6):611–7.
17. Butler AA, Zhang J, Price CA, et al. Low plasma adropin concentrations increase risks of weight gain and metabolic dysregulation in response to a high-sugar diet in male nonhuman primates. J Biol Chem. 2019;294(25):9706–19.
18. Bremer AA, Stanhope KL, Graham JL, Cummings BP, Wang W, Saville BR, et al. Fructose-fed rhesus monkeys: a nonhuman primate model of insulin resistance, metabolic syndrome, and type 2 diabetes. Clin Transl Sci. 2011;4(4):243–52. https://doi.org/10.1111/j.1752-8062.2011.00298.x.
19. Havel PJ, Kievit P, Comuzzie AG, Bremer AA. Use and importance of nonhuman primates in metabolic disease research: current state of the field. ILAR J. 2017;58(2):251–68.
20. Williams-Fritze MJ, Smith PC, Zelterman D, Scholz JA. Fructosamine reference ranges in rhesus macaques (Macaca mulatta). J Am Assoc Lab Anim Sci. 2011;50(4):462–5.
21. Hansen BC. Investigation and treatment of type 2 diabetes in nonhuman primates. Methods Mol Biol. 2012;933:177–85.
22. Pound LD, Kievit P, Grove KL. The nonhuman primate as a model for type 2 diabetes. Curr Opin Endocrinol Diabetes Obes. 2014;21 (2):89–94.
23. Wan YL, Zhang YC, Peng BL, et al. Screening of spontaneous diabetes mellitus in middle-and old-aged cynomolgus monkey. Dongwuxue Yanjiu. 2011;32(3):307–10.
24. Yue F, Zhang G, Quintero JE, Gash DM, Zhang Z. Role of social interaction, exercise, diet, and age on developing and untreated diabetes in cynomolgus monkeys. Exp Gerontol. 2017;96:82–8.
25. Yasuda M, Takaoka M, Fujiwara T, Mori M. Occurrence of spontaneous diabetes mellitus in a cynomolgus monkey (Macaca fascicularis) and impaired glucose tolerance in its descendants. J Med Primatol. 1988;17(6):319–32.
26. Palotay JL, Howard CF Jr. Insular amyloidosis in spontaneously diabetic nonhuman primates. Vet Pathol Suppl. 1982;7:181–92.
27. O'Brien TD, Wagner JD, Litwak KN, et al. Islet amyloid and islet amyloid polypeptide in cynomolgus macaques (Macaca fascicularis): an animal model of human non-insulin-dependent diabetes mellitus. Vet Pathol. 1996;33(5):479–85.
28. Büchi ER, Kurosawa A, Tso MO. Retinopathy in diabetic hypertensive monkeys: a pathologic study. Graefes Arch Clin Exp Ophthalmol. 1996;234(6):388–98.
29. Gu H, Liu Y, Mei S, et al. Left ventricular diastolic dysfunction in nonhuman primate model of dysmetabolism and diabetes. BMC Cardiovasc Disord. 2015;15:141.
30. Ishizaka T, Sato T, Kato K, Ohba M, Kimotsuki T, Yasuda M. Subcutaneous continuous glucose monitoring and dose adjustment decreases glycosylated hemoglobin in spontaneously diabetic cynomolgus monkeys. Contemp Top Lab Anim Sci. 2003;42(5): 36–40.

31. Tze WJ, Tai J. Xenotransplantation of rat pancreatic endocrine cells in spontaneous and streptozotocin-induced diabetic monkeys. Transplant Proc. 1989;21(1 Pt 3):2736–8.
32. Ji F, Jin LS, Zeng XM, et al. Comparison of gene expression between naturally occurring and diet-induced T2DM in cynomolgus monkeys. Dongwuxue Yanjiu. 2012;33(1):79–84.
33. He S, Wang D, Wei L. Practical and critical instruction for nonhuman primate diabetic models. Transplant Proc. 2013;45(5):1856–65.
34. Thomas FT, Ricordi C, Contreras JL, et al. Reversal of naturally occurring diabetes in primates by unmodified islet xenografts without chronic immunosuppression. Transplantation. 1999;67(6):846–54.
35. Goodroe A, Wachtman L, Benedict W, et al. Current practices in nutrition management and disease incidence of common marmosets (Callithrix jacchus). J Med Primatol. 2021 June;50(3):164–75.
36. Olson EJ, Shaw GC, Hutchinson EK, et al. Bone disease in the common marmoset: radiographic and histological findings. Vet Pathol. 2015;52(5):883–93.
37. Baxter VK, Shaw GC, Sotuyo NP. Serum albumin and body weight as biomarkers for the antemortem identification of bone and gastrointestinal disease in the common marmoset. PLoS One. 2013;8(12):e82747.
38. Eisele PH, Morgan JP, Line AS, et al. Skeletal lesions and anemia associated with ascorbic acid deficiency in juvenile rhesus macaques. Lab Anim Sci. 1992;42(3):245–9.
39. Ratteree MS, Didier PJ, Blanchard JL, et al. Vitamin C deficiency in captive nonhuman primates fed commercial primate diet. Lab Anim Sci. 1990;40(2):165–8.
40. Demaray SY, Altman NH, Ferrell TL. Suspected ascorbic acid deficiency in a colony of squirrel monkeys (Saimiri sciureus). Lab Anim Sci. 1978 Aug;28(4):457–60.

Suggested Reading

Doepel FM, Ringler DH, Petkus AR. Secondary amyloidosis in rhesus monkeys with chronic indwelling venous catheters. Lab Anim Sci. 1984;34(5):494–6.

McClure HM. Nonhuman primate models for human disease. Adv Vet Sci Comp Med. 1984;28:267–304.

Tanaka Y, Ohto H, Kohno M, Cho F, Honjo S. Spontaneous diabetes mellitus in cynomolgus monkeys (Macaca Fascicularis). Jikken Dobutsu. 1986;35(1):11–9.

Tardif SD, Mansfield KG, Ratnam R, et al. The marmoset as a model of aging and age-related. ILAR J. 2011;52(1):54–65.

Congenital Disorders of Nonhuman Primates

5

Anne D. Lewis

Abstract

Congenital anomalies are abnormalities of structure or function that occur before birth although they may be detected prenatally, at birth, or during postnatal life. A number of causes may underlie their development *in utero* including genetic mutations, chromosomal aberrations, teratogenic compounds, maternal factors, and infectious diseases. The increasing availability of genetic sequencing has elucidated the basis of some monogenetic diseases in NHP as well as humans. Diseases caused by single gene mutations may cause local or wide-ranging effects dependent on where the proteins are normally expressed. Defects in genes involved with embryonic patterning and organ development often result in early and severe malformations. Genetic diseases can result from loss of function of proteins such as enzymes, ion channels, and structural elements. Gene mutations can also be grouped by the affected cellular organelles such as ciliopathies and lysosomal storage diseases. The other causes of congenital anomalies are much less commonly identified and the etiology of a vast majority of congenital defects remains unknown. This chapter presents examples of a number of spontaneous congenital defects and genetic diseases in laboratory NHP.

Keywords

Congenital malformation · Congenital anomaly · Hereditary disease · Genetic disease

Congenital anomalies are abnormalities of structure or function that occur *in utero*. They may be detected prenatally, at birth, or during postnatal life. A number of causes may underlie their development *in utero* including genetic mutations, chromosomal aberrations, teratogenic drugs or plants, uterine environment, maternal factors, and infectious diseases. Monogenetic diseases may cause local or wide-ranging effects dependent on where they are expressed. They can result from loss of function of enzymes (e.g., globoid cell leukodystrophy), ion channels (e.g., CLCN2-related leukodystrophy), and metabolic pathways (e.g., hemochromatosis). Genes involved with embryonic patterning and organ development often result in early and severe malformations (e.g., lethal ciliopathies). Genes involved with connective tissue structural elements and adhesion molecules can result in defects in tissue integrity and function (e.g., epidermolysis bullosa simplex). Gene mutations can also be grouped by the affected cellular organelles. A common group of inherited diseases with wide-ranging phenotypes are ciliopathies which may affect either primary immotile or motile cilia. Primary cilia are present on a majority of cells in the body where they function in a number of essential signaling pathways. Genetic defects resulting in disruption of cilia assembly and/or structure are associated with a number of diseases including polycystic kidney diseases, retinopathies, and skeletal malformations.

Several surveys of spontaneous congenital malformations in nonhuman primate (NHP) populations estimate the incidence between 0.3% and 1.7% [1–4]; however, differences in anomaly definition, the number of species assessed, and variable population sizes affect these calculated rates. Regardless, the overall indication is that these are generally rare events in most NHP populations. When compared to causes of congenital malformations in humans and other animal species, NHPs in laboratory setting are less likely to experience exposure to exogenous teratogens and many infectious etiologies. Chromosomal abnormalities have been described in NHP species [5–9]. Several monogenetic diseases are also well described such as globoid cell leukodystrophy [10–12] and achromatopsia [13], both in rhesus macaques (*Macaca mulatta*).

A. D. Lewis (✉)
Pathology Services Unit, Oregon National Primate Research Center, Oregon Health and Science University, Beaverton, OR, USA
e-mail: lewisann@ohsu.edu

I. Kondova - Perseng et al. (eds.), *Atlas of Diagnostic Pathology in Nonhuman Primates*,
https://doi.org/10.1007/978-3-031-41280-6_5

The availability of affordable genetic sequencing of individual animals and access to detailed genomic information from reference resources such as the Macaque Genotype and Phenotype Resource (mGAP) (https://mgap.ohsu.edu/) have led to increased identification of previously unrecognized genetic diseases, several of which are described in this chapter. Many of these genetic diseases may serve as important models of human disease where murine models fail to adequately recapitulate the human condition. However, as in humans and other species, the cause of many of congenital anomalies in individual cases remains undetermined.

The scope of this chapter is presentation of selected examples of spontaneous congenital anomalies and genetic diseases in laboratory nonhuman primates. The chapter is organized by organ system; in many of the cases, multiple organ systems were involved and are noted as such. The references are provided as guides for further information and are not intended to represent a comprehensive review of the literature in NHP.

On a practical note, there are some steps that can be taken at postmortem evaluation that may aid the pathologist in the subsequent characterization and identification of an underlying etiology. In addition to collection of tissues for microscopic evaluation, routine banking of a frozen tissue sample such as liver for later DNA extraction is highly recommended. Careful dissection and observation of complex organs such as heart/great vessels *in situ* may yield important information otherwise missed once the structures are removed. Postmortem radiographs can be an important aid in evaluating the skeletal system. The use of normal age-matched archival images and tissues can be invaluable when reviewing fetal abnormalities to assess delays in development and subtle aberrations from normal morphology. Additionally, as many congenital malformations are incompatible with postnatal life; affected fetuses often die *in utero* and are presented to the pathologist after delivery with significant *in utero* autolysis. Even in these cases, genomic sequencing from liver tissue obtained at necropsy can lead to identification of underlying genetic mutations.

5.1 Defects Associated with Twin Pregnancies

A limited number of congenital defects are specific to monozygotic twin pregnancies. Although twinning is generally rare in Old World monkey species, several of these conditions have been reported in the literature such as conjoined twins in rhesus macaques [2, 14].

Twin reversed arterial perfusion (TRAP) sequence is an extremely rare complication of monochorionic twin pregnancy in which one fetus lacks a functional heart. Circulatory support of the acardiac twin is provided by the other ("pump") twin through arterio-arterial and veno-venous anastomoses in the shared placenta. The acardiac twin receives deoxygenated blood through the umbilical arteries; the reverse blood flow through the umbilical arteries and retrograde up the aorta can be identified *in utero* on Doppler ultrasound [15]. The affected fetus is grossly enlarged and deformed, in part, due to marked subcutaneous edema and intracoelomic fluid (fetal anasarca). In macaques, where the incidence of natural twinning is exceedingly low, only one acardiac twin lacking both heart and cephalic structures (*acardius acephalus*) has been described in the literature in a cynomolgus macaque (*M. fascicularis*) [16]. The case depicted here is a rhesus macaque. It represents the most developed form, *acardius anceps*, with a recognizable body shape and limbs, and poorly developed head and face (Figs. 5.1, 5.2, and 5.3).

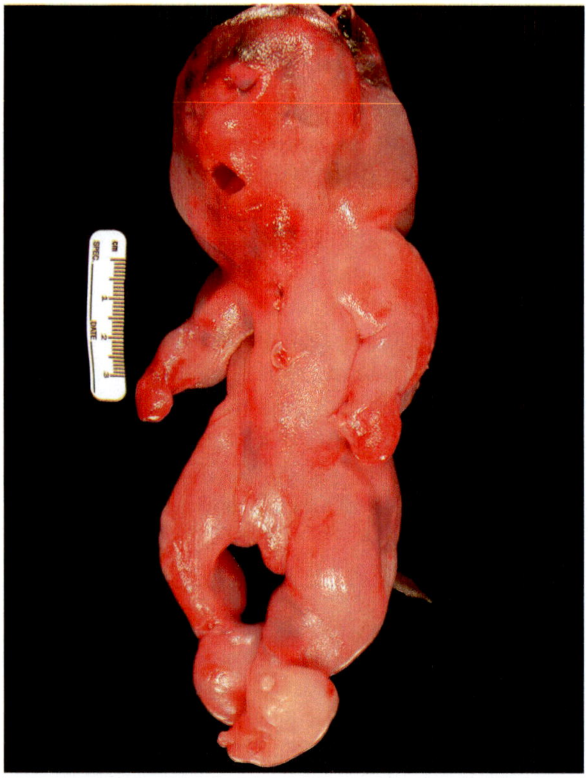

Fig. 5.1 Twin reversed arterial perfusion (TRAP) sequence (acardiac twin), fetal rhesus macaque. Whole body, ventral aspect. There is severe generalized subcutaneous edema (anasarca). The face is flattened with small, poorly formed features. The ears are low set with only vestigial pinnae visible. The digits of the hand are poorly developed and reduced in number

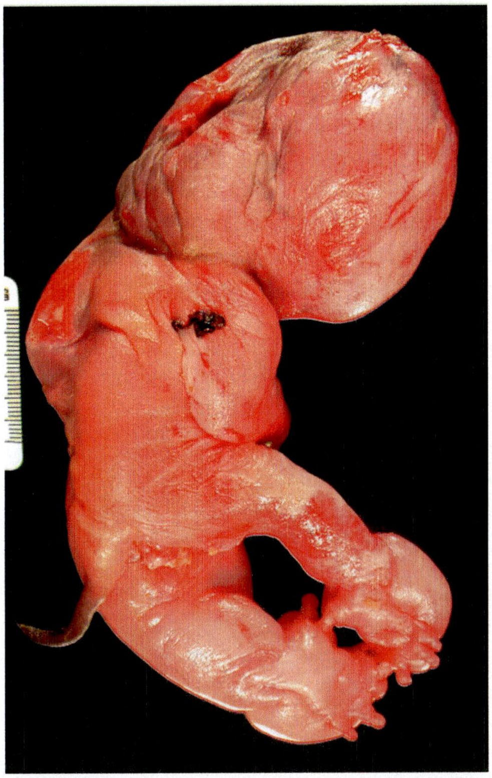

Fig. 5.2 TRAP sequence (acardiac twin), fetal rhesus macaque. Whole body, dorsal aspect. There is diffuse anasarca. Each foot has the normal number of digits in contrast to the hands

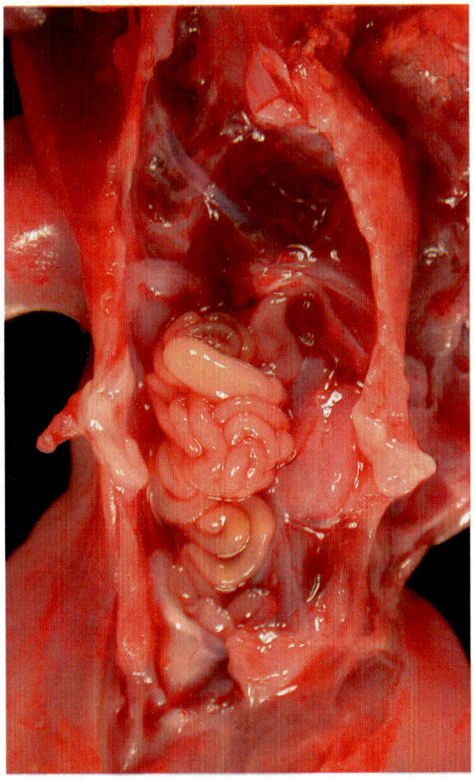

Fig. 5.3 TRAP sequence (acardiac twin), fetal rhesus macaque. Coelomic cavity. The lungs, heart, and thymus are absent. There is no identifiable diaphragm, liver, spleen, or pancreas. The intestines and genitourinary tract appear within normal limits

5.2 Central Nervous System

A variety of congenital central nervous system disorders occur in NHPs; some may be apparent at birth, others may be recognized by the development of neurologic signs with age. The examples described here are classified by the major categories of (a) brain malformations (e.g., neural tube defects, lissencephalies, microcephaly), (b) neurodegenerative diseases (e.g., leukodystrophies, inherited storage disorders), and (c) acquired diseases of the nervous system (e.g., ischemic injury).

5.2.1 Structural Malformations

Neural Tube Defects

Anencephaly, or absence of major portions of the brain and calvaria, is a severe neural tube defect which occurs due to failure of closure of the anterior portion of the tube. The resulting exposure of the developing nervous tissue to amniotic fluid leads to subsequent degradation. The condition is invariably fatal *in utero* or within days of birth. The degree of malformation is variable. The majority of cases in humans involve lack of the neocortex; a similar malformation in a rhesus macaque is shown in Fig. 5.4. Complete agenesis of the majority of cephalic structures can also occur (Fig. 5.5). Cranial meningoencephalocele, the protrusion of brain and meninges through a defect in the calvaria, is another rare form of neural tube defect. In the case depicted here, there is a large defect in the calvaria with protrusion of brain and meninges into the subcutis (Fig. 5.6).

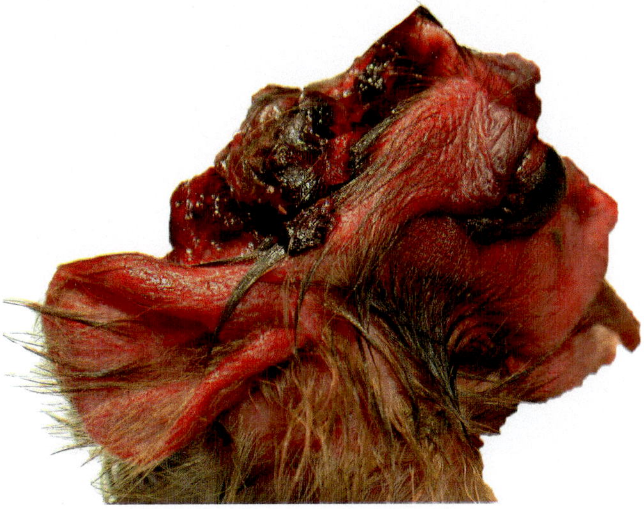

Fig. 5.4 Anencephaly, fetal rhesus macaque. The bones and overlying skin of the skull are absent as are the cerebral hemispheres and cerebellum. The eyes bulge markedly from shallow orbits

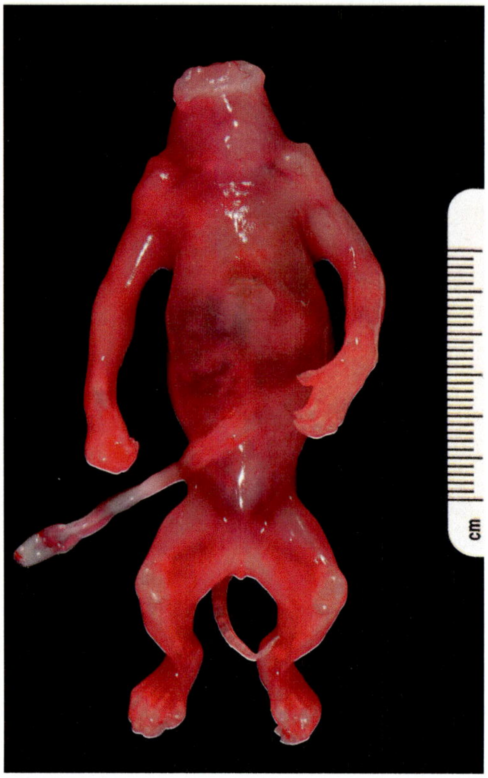

Fig. 5.5 Acephaly, fetal rhesus macaque. Whole body. There is a complete absence of recognizable cephalic structures

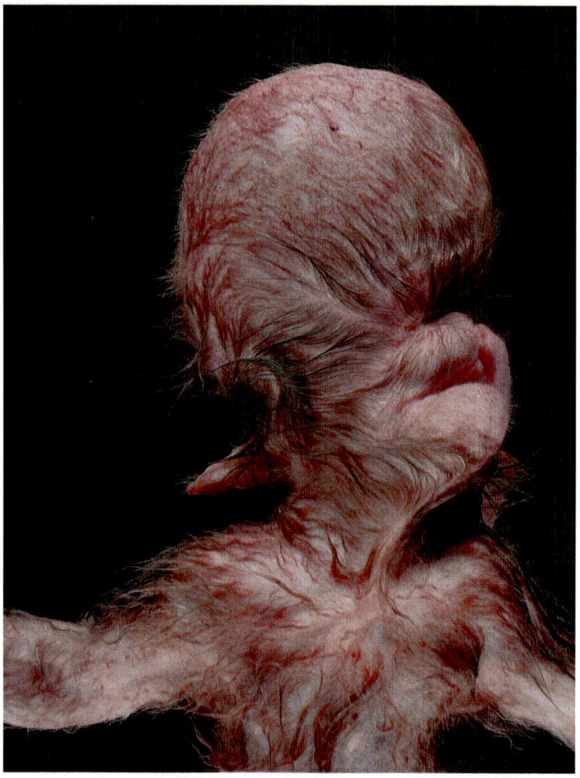

Fig. 5.6 Cranial meningoencephalocele, fetal Japanese macaque. Head. A large defect in the calvaria allowed protrusion of the meninges and cerebrum into the subcutis with subsequent cerebral necrosis. There are additional deformities including low set pinnae, midfacial concavity, cleft lip, hypoplastic orbits, and absent right eye

Holoprosencephaly

Holoprosencephaly is a disorder of early neural development in which the forebrain does not correctly separate to ultimately form two complete cerebral hemispheres. A spectrum of disorders may occur with variable severity. The condition is often accompanied by facial and ocular malformations. Cyclopia represents an extreme end of the phenotypic spectrum (Fig. 5.7).

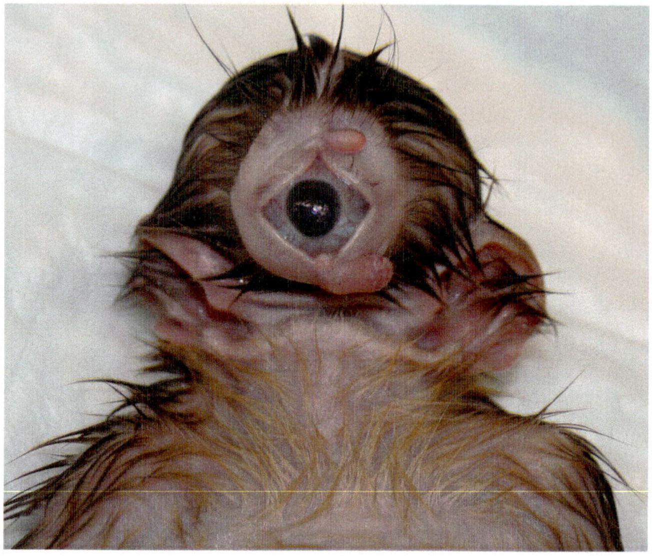

Fig. 5.7 Cyclopia associated with holoprosencephaly, fetal rhesus macaque. Head. Failure of the formation of separate cerebral hemispheres is accompanied by severe facial defects including a single centrally-located eye

Callosal Abnormalities

Agenesis of the corpus callosum is an uncommon defect in most species except specific strains of mice (e.g., BALB/c and 129). The condition may occur as an isolated defect or be associated with multiple congenital abnormalities involving both the brain and other organ systems. In humans, genetic factors are the most commonly identified underlying cause, including chromosomal aberrations, trisomies, and specific gene mutations. Less commonly, it may develop secondary to viral or toxic etiologies. Agenesis may be complete or partial. Causes of corpus callosum agenesis have not been identified in NHPs. Partial agenesis is depicted in Figs. 5.8 and 5.9, in which the caudal portions of the corpus callosum are entirely absent.

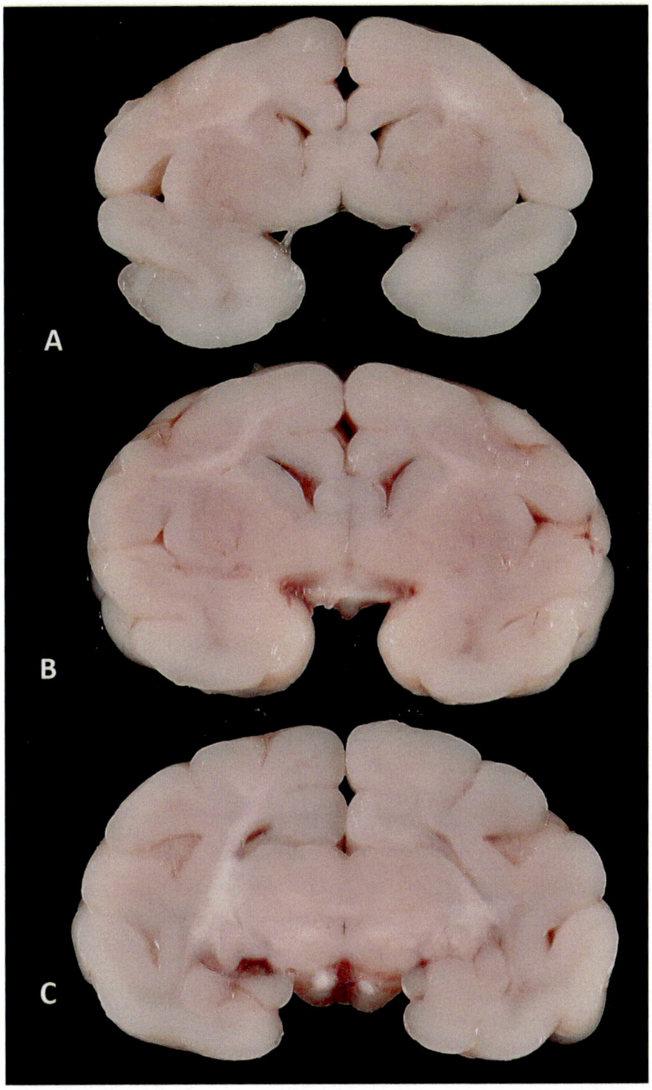

Fig. 5.8 Partial agenesis of the corpus callosum, rhesus macaque. Brain. Coronal sections. The corpus callosum is absent in each section

Malformations of Cortical Development

Malformations of cortical development frequently manifest as changes in the gyration pattern. The brains of Old World primates are lissencephalic in early gestation before maturing to adult gyri and sulci. Some New World primates are naturally lissencephalic (e.g., common marmoset (*Callithrix jacchus*)). In examining fetal specimens, it is often important to have age-matched material for comparison. Sawada et al. [17] provide an excellent review of gyrification in the fetal cynomolgus macaque. Classical lissencephaly or lissencephaly type I features a lack of gyral development due to a failure of normal neuronal migration and may be caused by several known genetic causes in humans. Cobblestone lissencephaly or lissencephaly type II occurs due to defects in the pial–glial interface resulting in overmigration of neurons in the outer cortex [18]. Cobblestone lissencephaly due to a mutation in *POMT1* (encoding for protein O-mannosyltransferase 1) is represented here (Figs. 5.10 and 5.11).

Fig. 5.9 Partial agenesis of the corpus callosum, rhesus macaque. Brain, HE. Coronal sections of cerebrum exhibit a vestigial callosal structure present only in section A

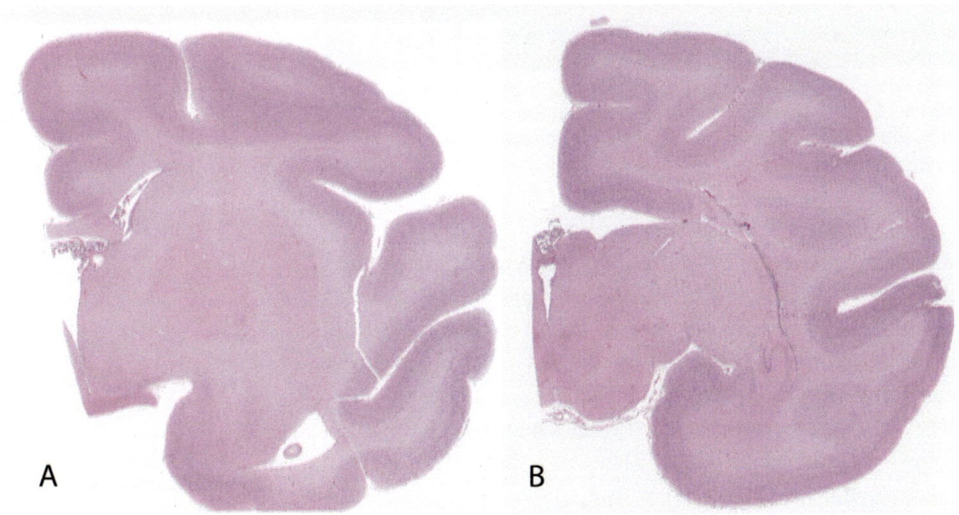

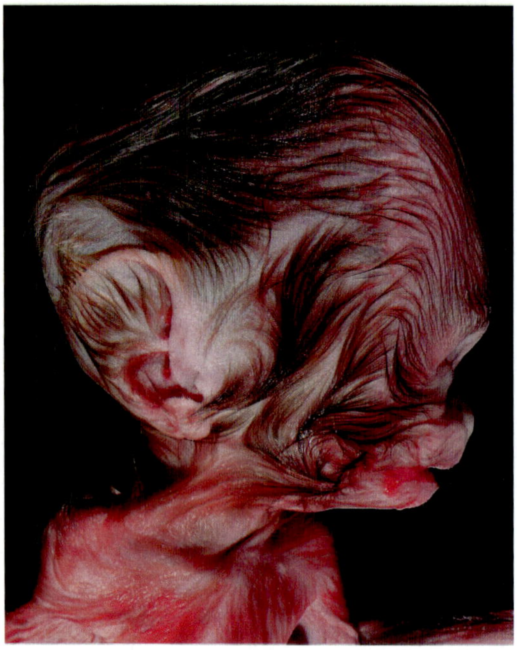

Fig. 5.10 Cobblestone lissencephaly, fetal rhesus macaque. Head. The skull is markedly enlarged and dome-shaped in this fetus homozygous for a mutation in *POMT1*

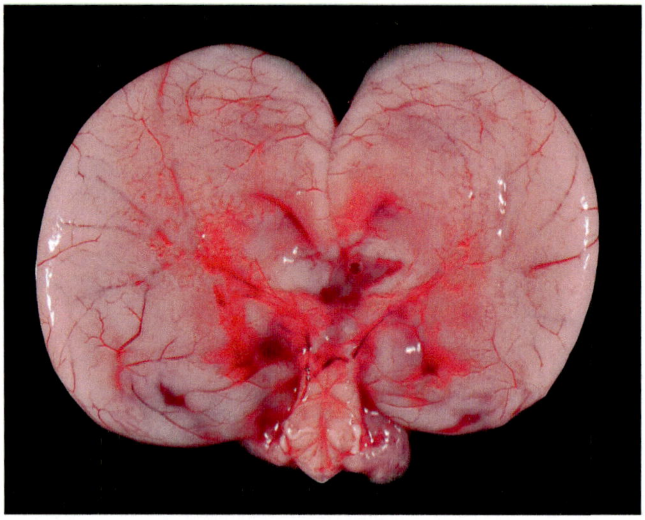

Fig. 5.11 Cobblestone lissencephaly, fetal rhesus macaque. Brain. Fluid filled, flaccid cerebral hemispheres are devoid of normal cortical gyral architecture

Disorders of Brain Growth

Micrencephaly refers to small brain size and weight defined as three standard deviations below norm. It is usually accompanied by microcephaly, a more readily assessable parameter clinically. It may be secondary to extrinsic factors interfering with brain growth such as *in utero* infection (e.g., Zika virus) and hypoxic injury. Primary micrencephaly arises from genetic disorders. Depicted here is severe micrencephaly in an animal with a ciliopathic disorder due to a *TAPT1* (transmembrane anterior posterior transformation 1) mutation resulting in a syndrome analogous to complex lethal osteochondrodysplasia in humans (Fig. 5.12). This syndrome is further discussed in the musculoskeletal system section (see Sect. 5.4.2 and Fig. 5.60).

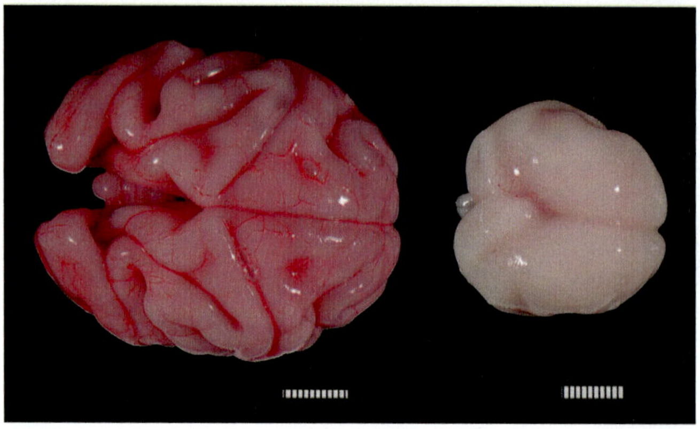

Fig. 5.12 Micrencephaly, rhesus macaque. Brain. On the right is an extremely small and lissencephalic brain from a gestational age 136-day fetus with multiple congenital abnormalities due to a *TAPT1* mutation. On the left is a normal brain from an age-matched control fetus. There are more images of the affected animal in Fig. 5.60

Ventriculomegaly

Ventriculomegaly is defined as abnormal expansion of the ventricles; it includes internal hydrocephalus due to increased pressure and volume of cerebrospinal fluid as well as other causes such as loss of cerebral parenchyma (also termed hydrocephalus *ex vacuo*). The cause of congenital hydrocephalus is often unidentified but is generally thought to be obstructive. Shown here is internal hydrocephalus in a common marmoset (Fig. 5.13). Occipital lobe ventriculomegaly due to loss of parenchyma and expansion of ventricles is seen occasionally as an incidental lesion in rhesus macaques; it is thought to be the result of ischemic injury *in utero* or during parturition (Fig. 5.14).

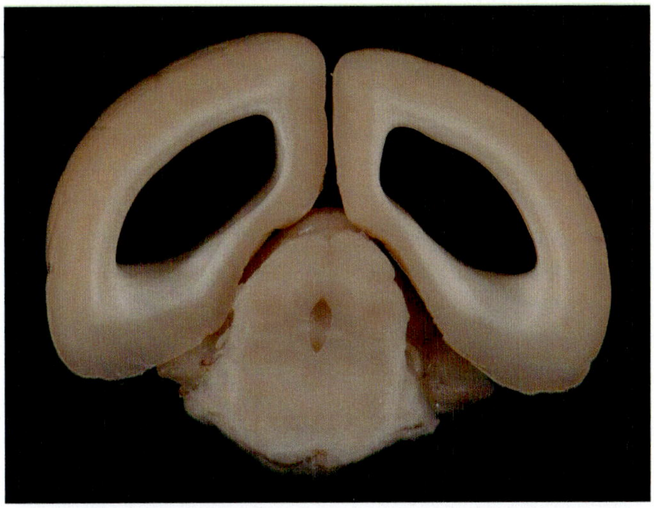

Fig. 5.13 Hydrocephalus, common marmoset. Brain, coronal section. The ventricles are diffusely dilated and there is compensatory loss of cortical tissue

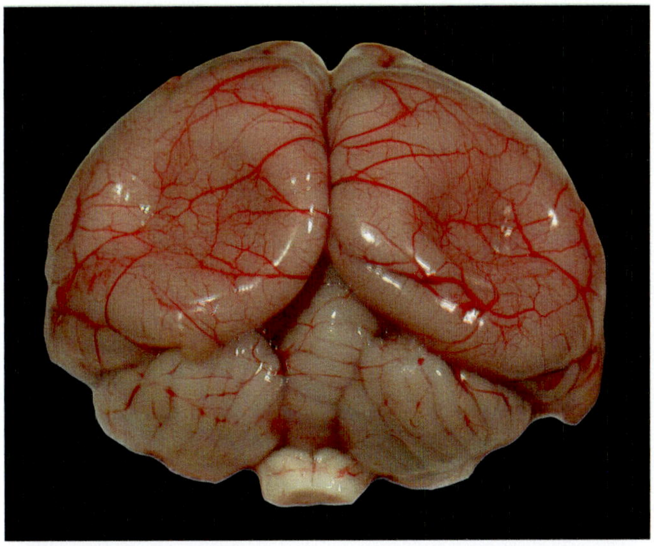

Fig. 5.14 Hydrocephalus, occipital lobes, rhesus macaque. Brain. The occipital lobes are bilaterally collapsed due to hydrocephalus *ex vac*uo affecting the cortical tissue

Intracranial cysts lined by ependyma can occur as ependymal cysts, as part of obstructive hydrocephalus, or associated with specific genetic disorders such as oral–facial–digital syndrome I (OFDS) in humans. Ependymal cyst has been reported in a cynomolgus macaque [19]. Shown here are multiple cerebral cysts of unknown cause in a rhesus macaque [20] (Figs. 5.15 and 5.16).

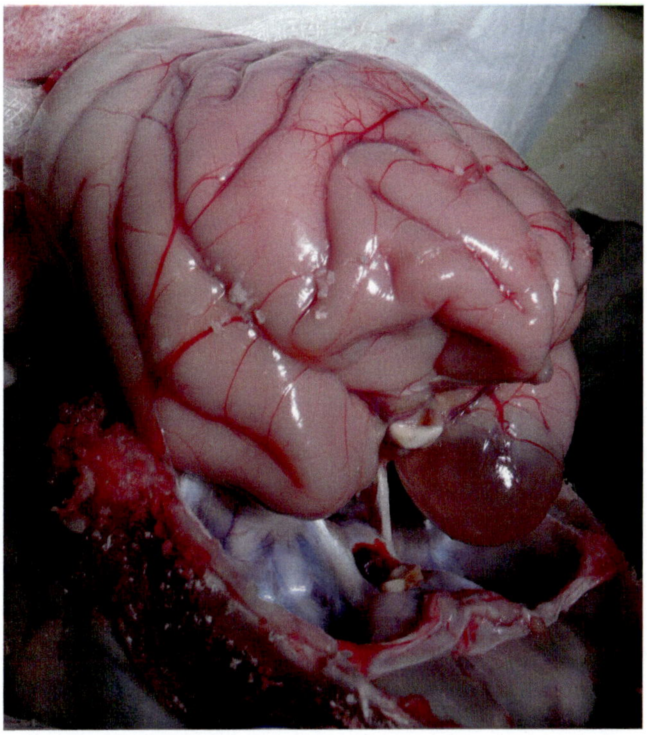

Fig. 5.15 Cerebral cysts, rhesus macaque. Brain. There is a fluid-filled cyst present in the left temporal lobe

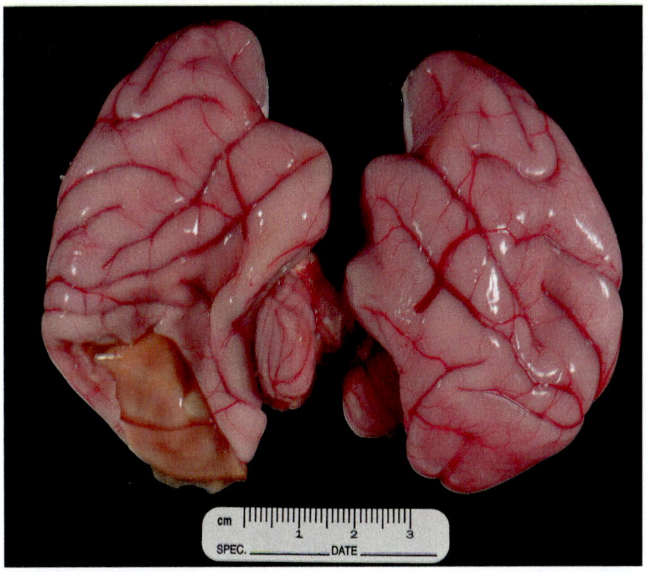

Fig. 5.16 Cerebral cysts, rhesus macaque. Brain. The right occipital lobe is expanded by a large cystic structure. The cyst is collapsed due to loss of fluid upon removal of the brain from the calvaria

5.2.2 Heritable Neurodegenerative Disorders

Disorders Primarily Affecting Gray Matter

Neuronal Ceroid Lipofuscinosis (NCL)

NCLs are a heterogenous group of hereditary neurodegenerative diseases characterized by intracellular accumulation of auto-fluorescent pigment and progressive neuronal degeneration. In humans, 14 forms have been recognized due primarily to defects in genes that code for lysosomal enzymes and lysosomal transport proteins. Historically, NCLs have been diagnosed by age of clinical onset that generally correspond to specific underlying mutations. A spontaneous form of NCLs, CLN7 occurs in Japanese macaques (*M. fuscata*) due to a mutation in *CLN7*, also known as the major facilitator superfamily domain-containing protein 8 (MFSD8) gene [21]. The CLN7 protein is localized principally in the lysosomes, although its *in vivo* functions are not fully understood. Affected macaques are homozygous for a frameshift mutation which is predicted to cause production of a truncated form of the protein. The disease in Japanese macaques is characterized by onset of clinical disease at 4–5 years of age. Neurologic signs include ataxia, incoordination, hypermetria, visual deficits, and intention tremor. At necropsy, the brain is small and discolored (Fig. 5.17). The microscopic findings include severe cerebellar atrophy, cerebral atrophy, and abundant intracytoplasmic accumulation of lightly pigmented, heterogeneously staining, auto-fluorescent material (Figs. 5.18, 5.19, 5.20, 5.21, and 5.22).

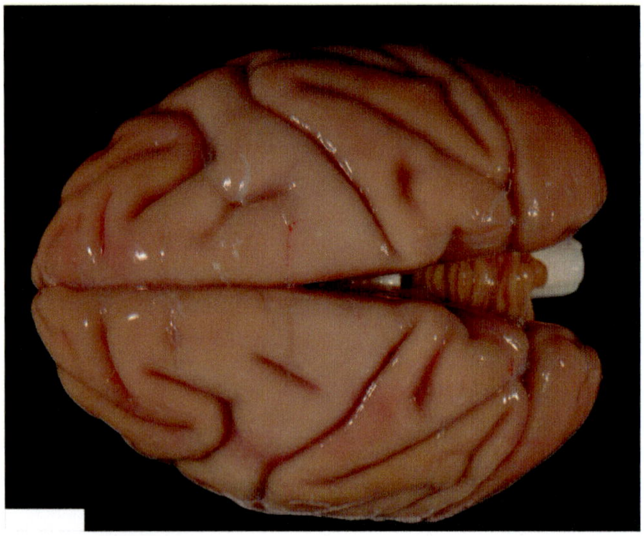

Fig. 5.17 Neuronal ceroid lipofuscinosis, CLN7, Japanese macaque. Brain, saline perfused. The brain of this 5-year-old Japanese macaque, homozygotic for a mutation in *CLN7*, is diffusely atrophic indicated by overall small size, deep cerebral sulci, and narrow gyri. The cerebellum is small and the caudal margin does not extend to the caudal margins of the occipital lobes. Diffuse tan discoloration is due to accumulation of abundant intracellular pigment

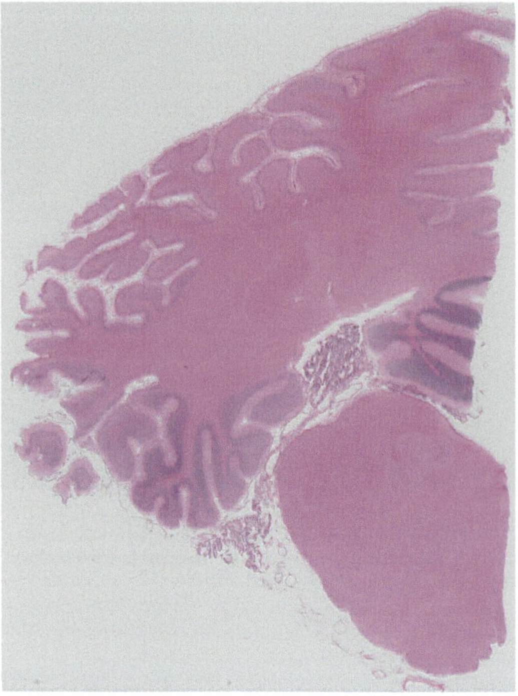

Fig. 5.18 Neuronal ceroid lipofuscinosis, CLN7, Japanese macaque. Cerebellum, HE. There is diffuse atrophy of the cerebellar folia which is most severe in the dorsal aspect

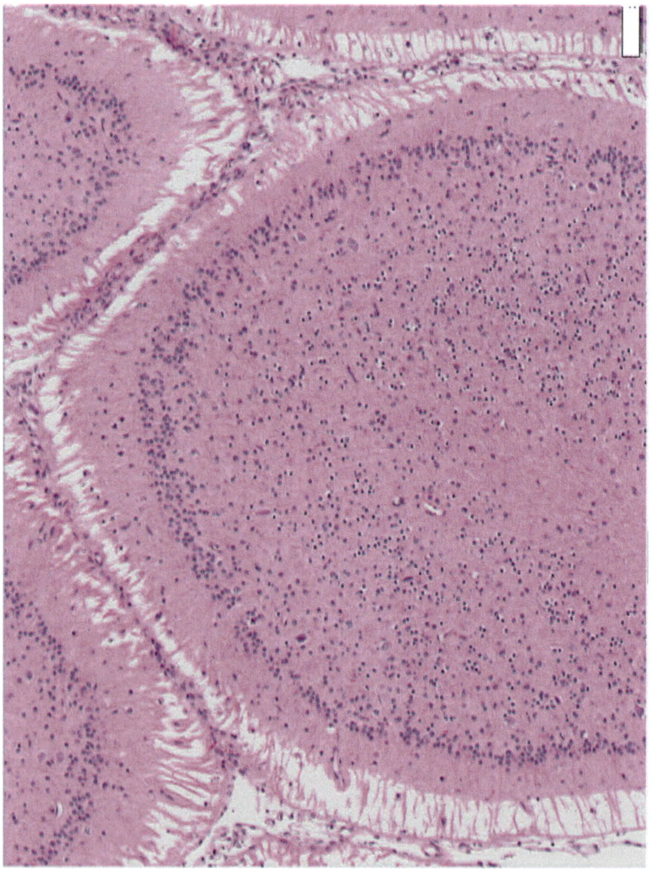

Fig. 5.19 Neuronal ceroid lipofuscinosis, CLN7, Japanese macaque. Cerebellum, HE. Cerebellar atrophy is evidenced by severe thinning of the granular layer, loss of Purkinje cells, and marked hypocellularity of molecular layer

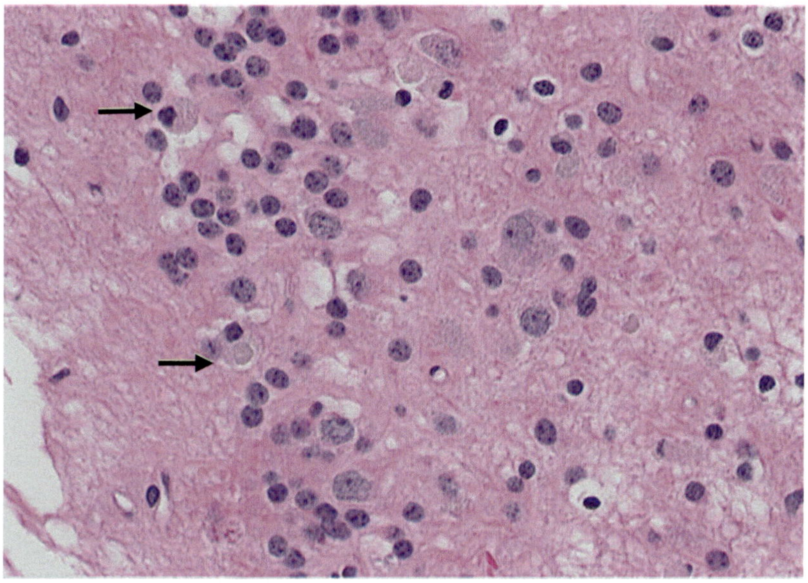

Fig. 5.20 Neuronal ceroid lipofuscinosis, CLN7, Japanese macaque. Cerebellum, HE. Numerous cells (neurons and glia) contain abundant lightly pigmented, intracytoplasmic storage material

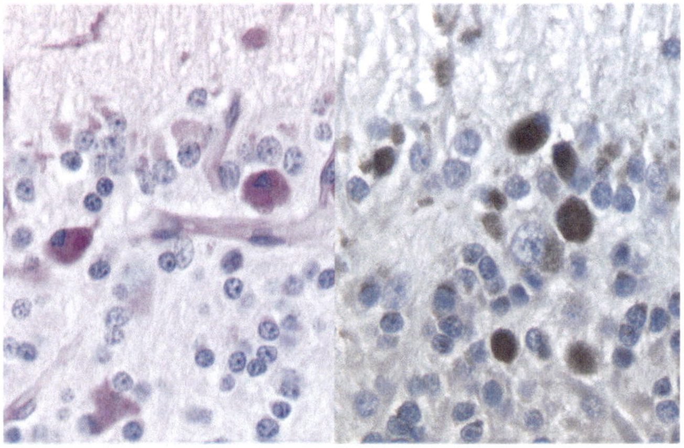

Fig. 5.21 Neuronal ceroid lipofuscinosis, CLN7, Japanese macaque. Cerebellum, periodic acid-Schiff (PAS) (left), Sudan black (right). These histochemical stains reveal the heterogeneous nature of the storage material. PAS highlights glycoproteins (magenta). Sudan black demonstrates lipoprotein components (black)

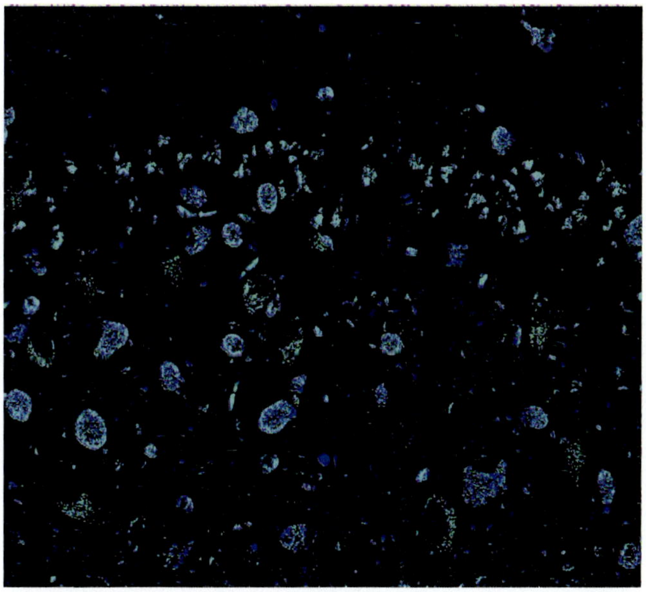

Fig. 5.22 Neuronal ceroid lipofuscinosis, CLN7, Japanese macaque. Cerebellum, unstained tissue. Composite image using DAPI and FITC excitation filters on unstained tissue demonstrates abundant autofluorescence of heterogenous storage material

Disorders Primarily Affecting White Matter

Globoid Cell Leukodystrophy (GCL)

GCL or Krabbe disease is an autosomal recessive heritable lysosomal storage disorder affecting the central and peripheral nervous systems (PNS) that occurs in rhesus macaques as well as in humans, mice, sheep, cats, and dogs [10, 22]. The underlying pathogenesis is deficiency of galactocerebrosidase due to a mutation in the encoding gene, *GALC* [22]. Lack of the enzyme causes disruption of the myelin turnover pathway and buildup of toxic lipid metabolites, most notably psychosine, with toxic effects on oligodendrocytes, leading to loss of myelin and inflammation. The disease has been well described in a colony of rhesus macaques [10]. The gross lesions are primarily evident in the PNS where there is marked enlargement of peripheral nerves and dorsal root ganglia (Figs. 5.23 and 5.24). Microscopically, there is demyelination in both the peripheral and central nerve systems and accumulation of characteristic globoid cells of microglial/monocyte lineage in the white matter (Figs. 5.25 and 5.26).

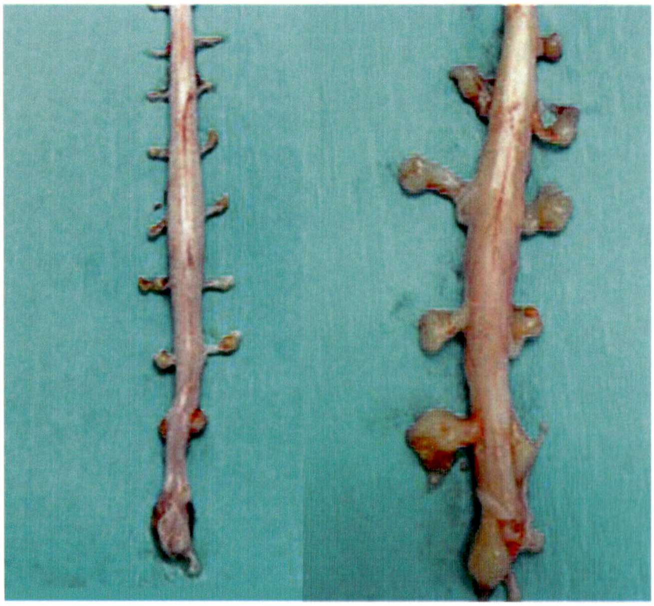

Fig. 5.24 Globoid cell leukodystrophy, rhesus macaque. Spinal cord. On the right is the spinal cord from an affected animal showing the marked enlargement of dorsal root ganglia. On the left is spinal cord from an unaffected animal for comparison

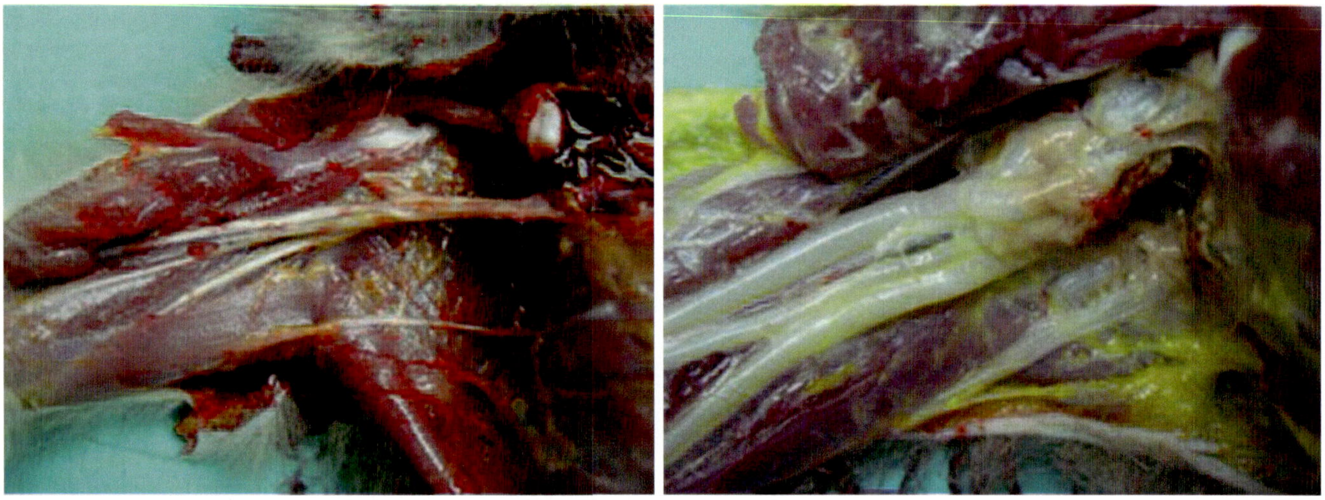

Fig. 5.23 Globoid cell leukodystrophy, rhesus macaque. Brachial plexus. On the right is the brachial plexus from an affected animal showing the marked thickening and enlargement of peripheral nerves. On the left is the brachial plexus of an unaffected animal

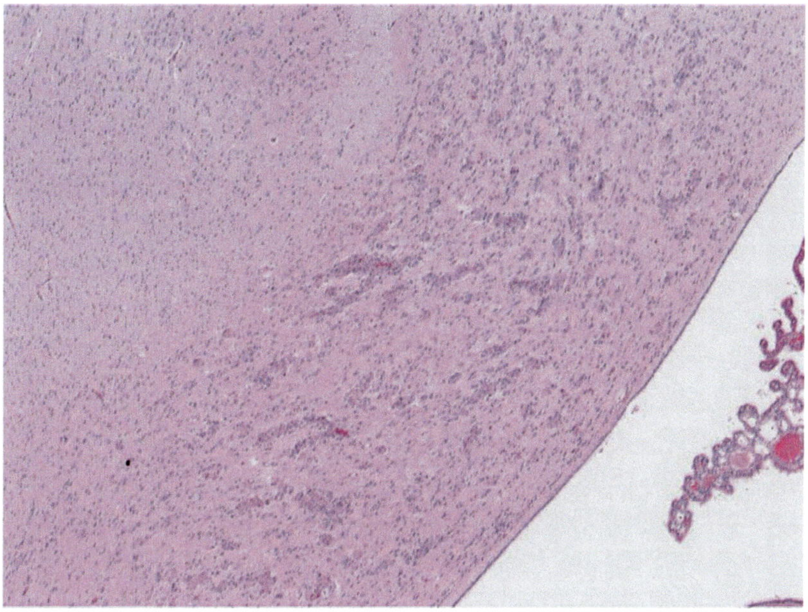

Fig. 5.25 Globoid cell leukodystrophy, rhesus macaque. Cerebrum, HE. The distribution of the lesion is restricted to the white matter

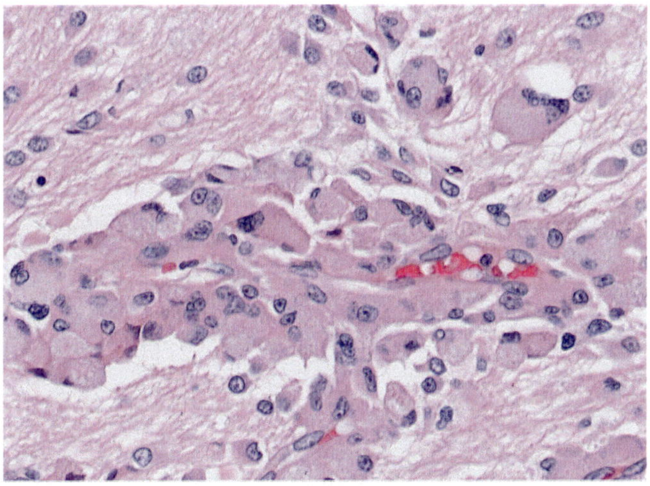

Fig. 5.26 Globoid cell leukodystrophy, rhesus macaque. Cerebrum, HE. Image depicts the predominately perivascular distribution of the characteristic enlarged histiocytic cells (globoid cells) containing storage material

PLP1 Leukodystrophy

A hypomyelinating leukodystrophy resembling Pelizaeus–Merzbacher disease in humans has been identified in infant rhesus macaques [23]. Affected animals have a diffuse paucity of myelin in the central nervous system. Clinical signs include nystagmus, intention tremors, and difficulty nursing. The cause is a missense mutation in the proteolipid protein 1 gene (*PLP1*). This gene is carried on the X chromosome and all affected animals are hemizygous males. Lack of PLP1 protein, a major component of myelin, results in diffuse central nervous system hypomyelination. The gross findings are generally unremarkable and are limited to mild cerebral swelling. Microscopically, there is poor delineation between gray and white matter regions and relative pallor of white matter (Fig. 5.27). In all areas of the CNS white matter, there is moderate gliosis and increased numbers of apoptotic glia (Fig. 5.28). Luxol fast blue staining demonstrates lack of myelin throughout the CNS (Fig. 5.29).

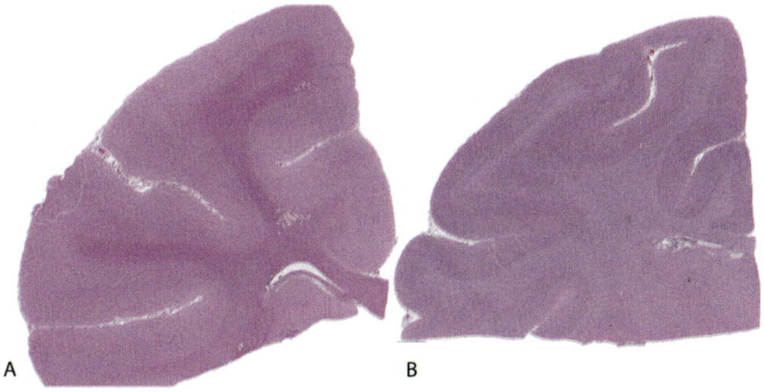

Fig. 5.27 PLP1 leukodystrophy, rhesus macaque. Cerebrum, HE. Image B is tissue from an affected animal showing abnormal staining (pallor) of the cortical white matter. Image A is tissue from an age-matched control

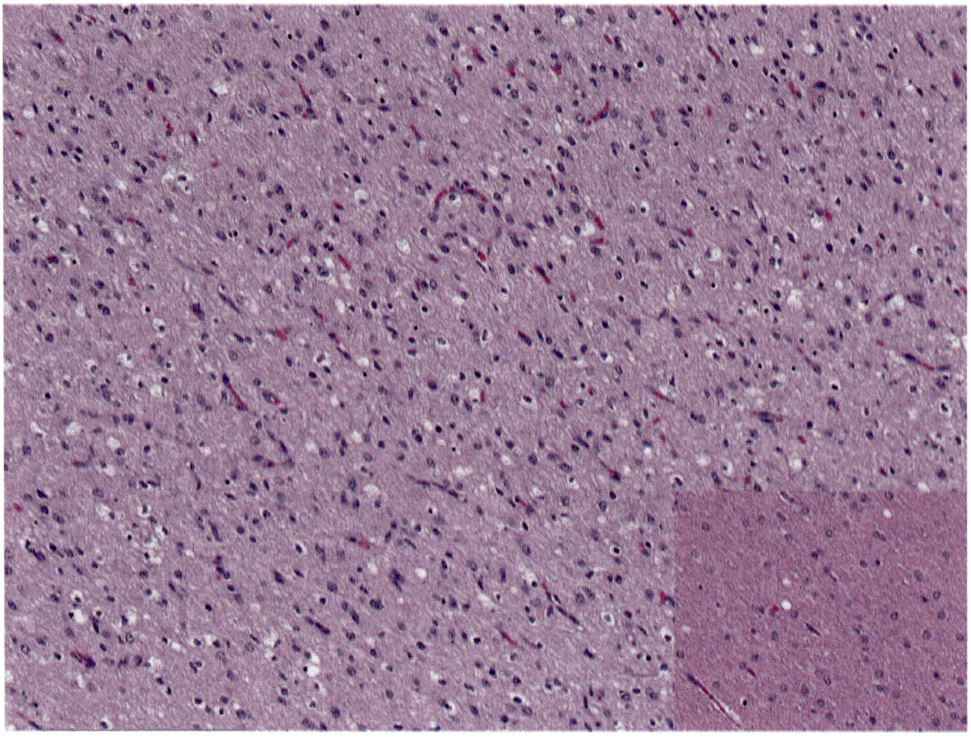

Fig. 5.28 PLP1 leukodystrophy, rhesus macaque. Cerebrum, HE. There is diffuse gliosis with increased numbers of apoptotic cells in the white matter. The inset is from a similar region of cerebrum in an age-matched control

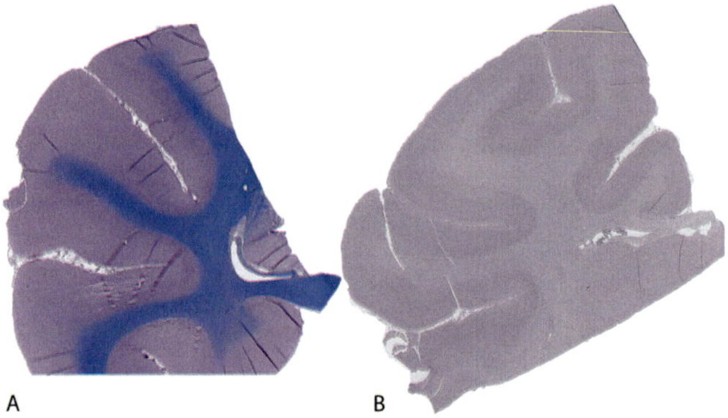

Fig. 5.29 PLP1 leukodystrophy, rhesus macaque. Cerebrum, Luxol fast blue/Periodic acid-Schiff. Image B is from an affected animal exhibiting diffuse absence of myelin as evidenced by lack of staining with Luxol fast blue. Image A is an age-matched control showing normal myelination at this age

CLCN2 Leukodystrophy

A leukodystrophy featuring diffuse white matter vacuolation was identified in a rhesus macaque. The microscopic lesions resemble those of several heritable inborn errors of metabolism such as branched-chain alpha-ketoacid decarboxylase (BCKD) deficiency in Poll Hereford calves [24]. The rhesus macaque depicted here (Figs. 5.30, 5.31, and 5.32) was identified as homozygous for a mutation in *CLCN-2* which is associated with CLCN-2 related leukoencephalopathy, an extremely rare genetic neurologic disorder in humans [25]. *CLCN-2* codes for ClC-2, a chloride channel with ubiquitous expression. In the brain, it is expressed primarily in astrocytes and neurons. ClC-2 regulates ion and water homeostasis. Defective ClC-2 results in intramyelinic edema [26].

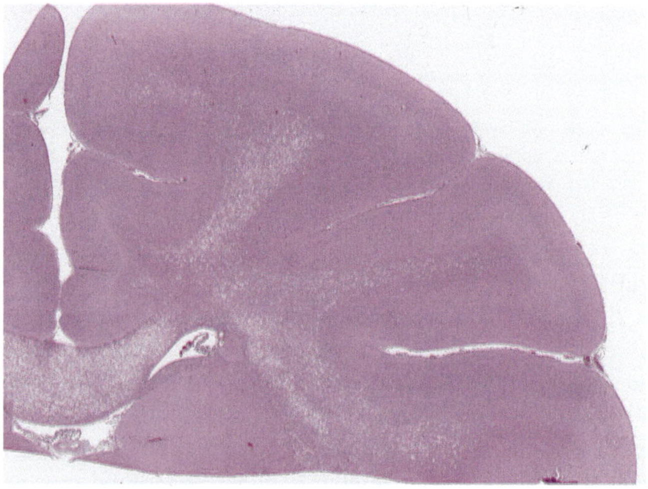

Fig. 5.30 CLCN2 leukodystrophy, rhesus macaque. Cerebrum, HE. There is diffuse pallor of the white matter due to severe vacuolation

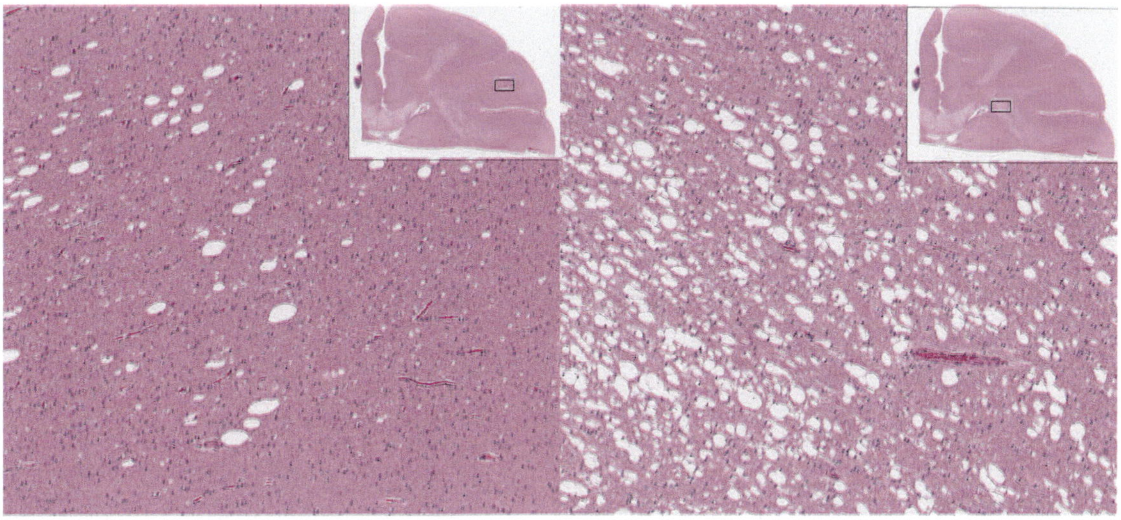

Fig. 5.31 CLCN2 leukodystrophy, rhesus macaque. Cerebrum, HE. Vacuolation is less severe in the peripheral portions of the white matter

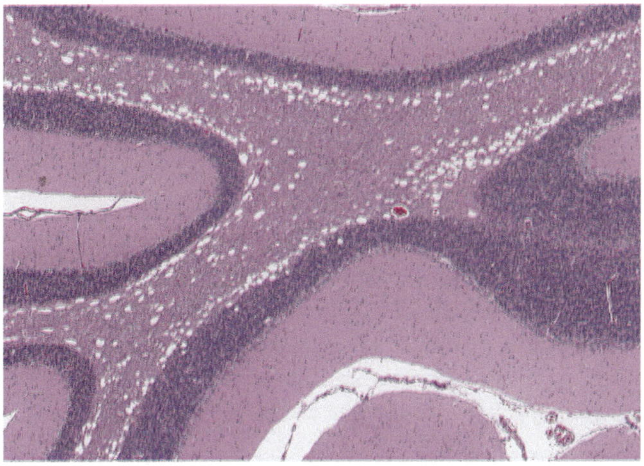

Fig. 5.32 CLCN2 leukodystrophy, rhesus macaque. Cerebellum, HE. Vacuolation is most severe at the interface of white matter and gray matter

5.2.3 Acquired

Hydranencephaly is characterized by near total destruction of the cerebral hemispheres which are replaced by abundant CSF enclosed in a thin walled membrane consisting of leptomeninges and remnants of cortical tissue. It results from insults early in gestation and may occur secondary to a variety of conditions including viral infection, ischemia, and, potentially, specific genetic diseases. The head may be enlarged or of normal size. The calvarium is intact. It may be accompanied by other congenital abnormalities. The case shown here (Figs. 5.33, and 5.34) is from the acardiac twin previously discussed (*see* Figs. 5.1, 5.2, and 5.3). In this case, hydranencephaly is attributed to *in utero* hypoxemia/ ischemia.

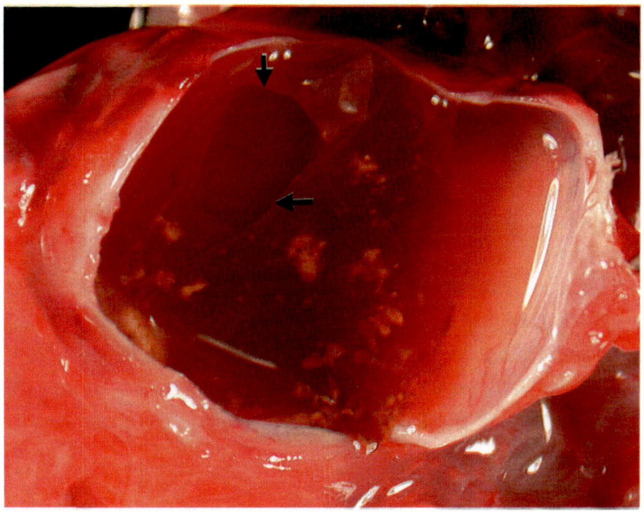

Fig. 5.33 Hydranencephaly, rhesus macaque. Head. Within the calvaria, the meninges form a saccular structure (arrows) containing abundant fluid and flocculent yellow-green material. No organized brain tissue is identified grossly. This is an image from the acardiac fetus depicted in Figs. 5.1–5.3

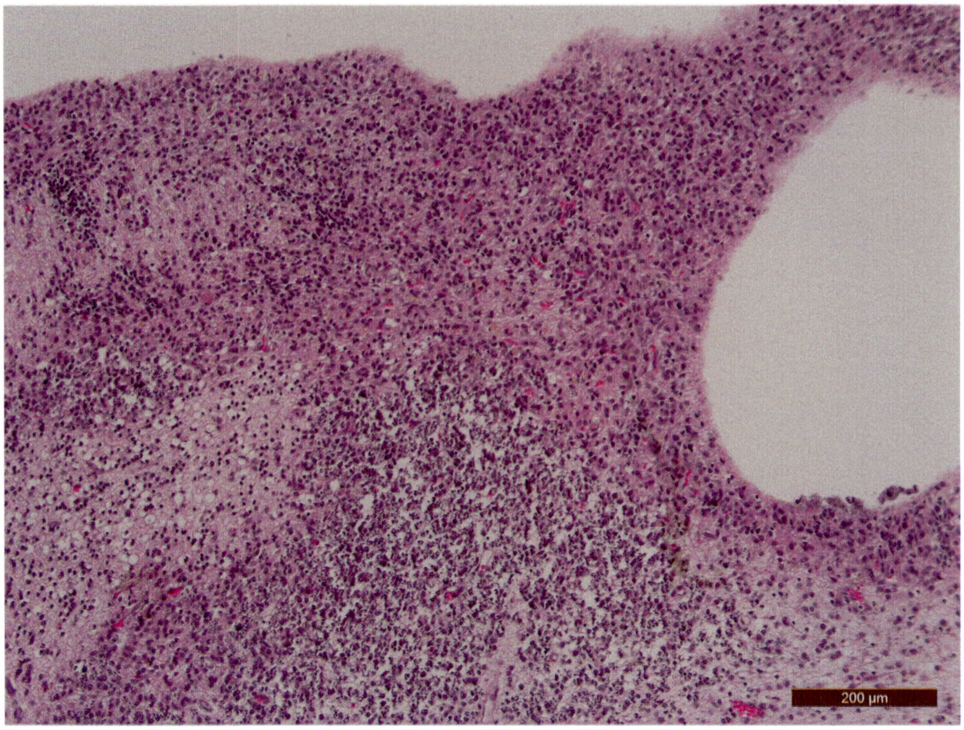

Fig. 5.34 Hydranencephaly, rhesus macaque. Cerebrum, HE. Flocculent material in Fig. 5.33 represents remnants of poorly organized, necrotic cortical tissue

5.3 Special Senses

5.3.1 Eye

Spontaneous defects of eye formation are rarely reported in NHP [3, 27]. Agenesis and microphthalmia may occur alone or be associated with other defects. Agenesis of the right eye was present in the animal with meningoencephalocele (*see* Fig. 5.6).

Heritable causes of retinal degeneration are rare in NHP; however, two forms have recently been described in rhesus macaques. Achromatopsia is a retinal cone disorder which affects the function of all types of cones and results in loss of visual acuity and progressive macular degeneration. In rhesus macaques, it is due to an autosomal recessively inherited mutation in *PDE6C*, which codes for phosphodiesterase 6C, an enzyme necessary for phototransduction in cones [13]. Retinal degeneration due to a heritable ciliopathy has been described in related rhesus macaques [28]. A frameshift mutation in the Bardet–Biedl syndrome 7 gene (*BBS7)* causes a syndrome analogous to Bardet–Biedl syndrome in humans. In affected macaques, there is progressive retinal degeneration with loss of photoreceptors and inner retinal neurons, with thinning and disorganization of all cell layers (Fig. 5.35). In macaques, the mutation is associated with renal disease (*see* Figs. 5.82 and 5.83) and hypogonadism; obesity and polydactyly, seen in human patients, are lacking.

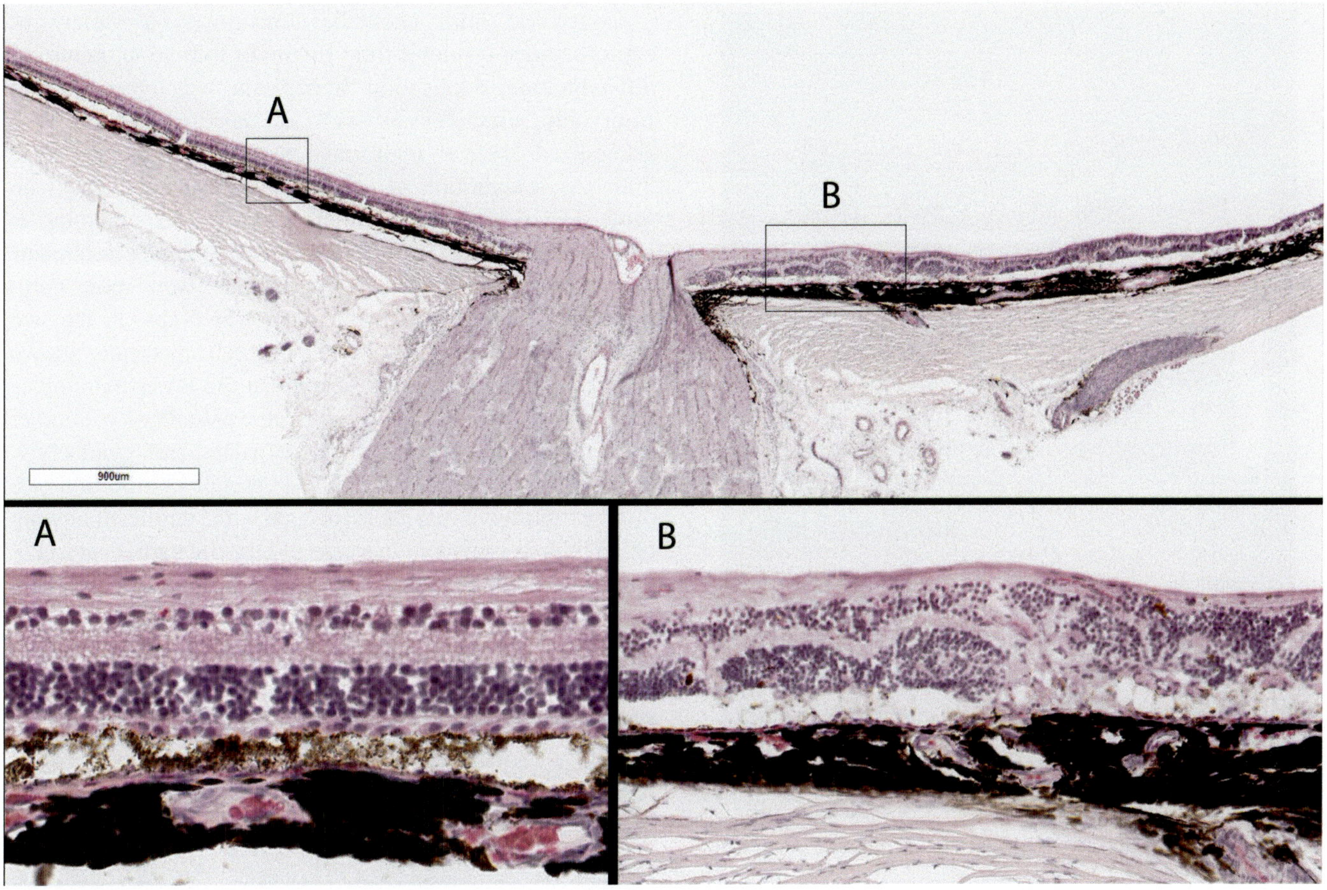

Fig. 5.35 Retinal degeneration, Bardet–Biedl syndrome, rhesus macaque. Retina, HE. Subgross image shows the overall thinning and disorganization of the retinal architecture. Inset A demonstrates the diffuse loss of retinal layers. Inset B shows the disorganization of the retina near the optic nerve

5.4 Musculoskeletal System

5.4.1 Malformations and Limb Deformities

Arthrogryposis is a condition in which there are multiple joint contractures at birth. The underlying common factor is thought to be decreased fetal movement *in utero*. The term is descriptive; the condition can arise from a number of intrinsic and extrinsic causes. Maternal conditions include those resulting in reduced amniotic fluid or constrained uterine size. Fetal conditions associated with arthrogryposis include abnormalities in the central and peripheral nervous systems, muscles, and connective tissues. The condition is associated with a number of genetic and chromosomal defects and syndromes. The cases shown here in rhesus macaques were both associated with central nervous system malformations (Figs. 5.36 and 5.37).

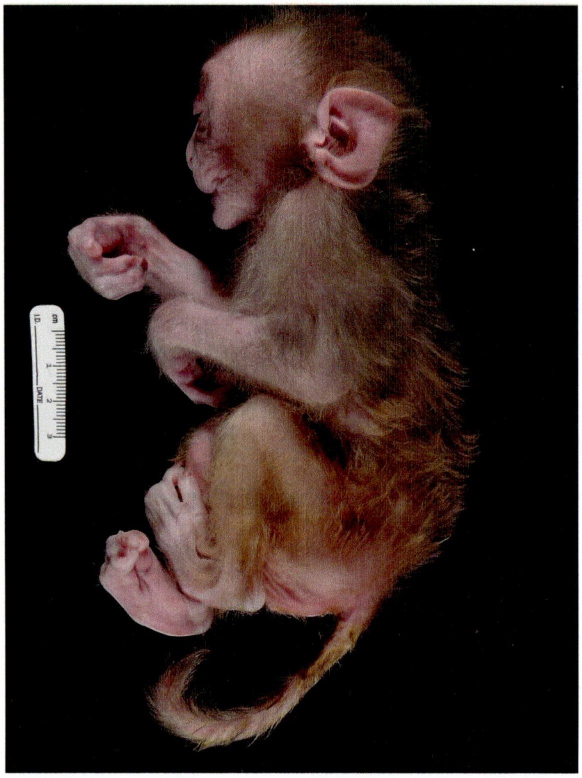

Fig. 5.36 Arthrogryposis, rhesus macaque. Whole body. There are multiple joint contractures as well as facial deformities. The brain of this animal is depicted in Figs. 5.8 and 5.9

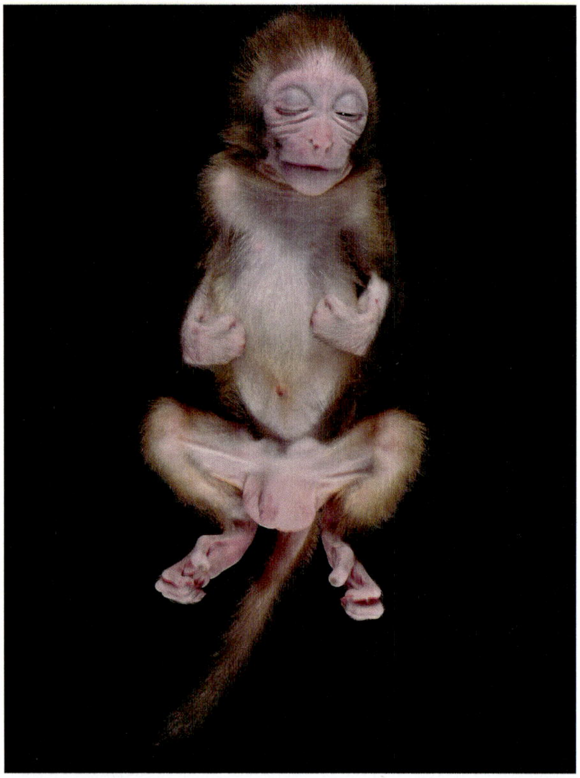

Fig. 5.37 Arthrogryposis, rhesus macaque. Whole body. Multiple joint contractures are present

Congenital limb anomalies encompass a variety of malformations resulting from failure of formation, failure of differentiation, duplication, hypoplasia, and hyperplasia of embryonic structures as well as those associated with generalized skeletal malformations. Phocomelia (or "seal limb") is a condition of abnormally shortened limbs with some preservation of distal extremities. The syndrome is widely known because of the use of thalidomide in humans during pregnancy in which upper limbs were more commonly and severely affected than lower limbs. In the two cases of spontaneous phocomelia in rhesus macaques shown here, both had more severe lesions in the lower extremities (Figs. 5.38, 5.39, and 5.40). Both were associated with other congenital defects such as cleft palate and syndactyly. Hemimelia is a partial to complete absence of a portion of a distal extremity; it may be associated with significant bowing of the limb as is evident in a case of bilateral partial aplasia of the fibula shown in Figs. 5.41 and 5.42. An isolated unilateral malformation of the femoral head is depicted in Fig. 5.43.

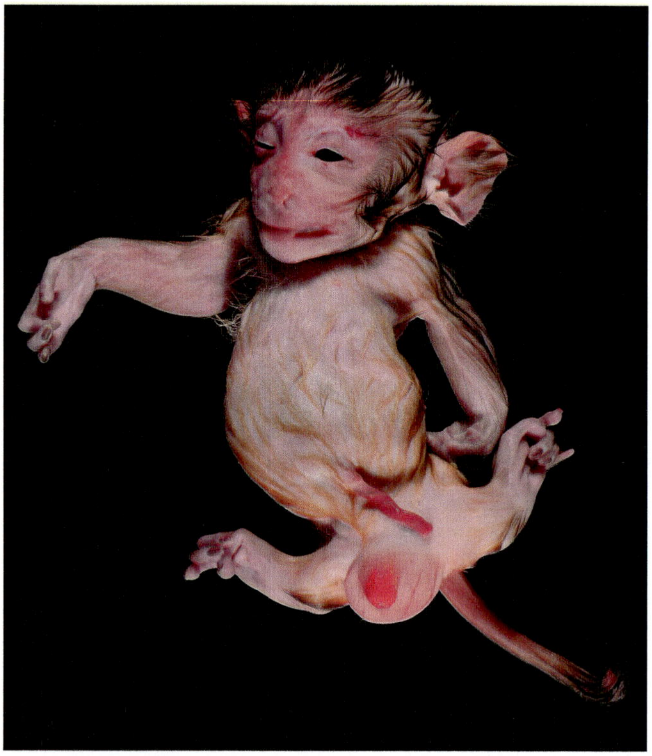

Fig. 5.38 Phocomelia, rhesus macaque. Whole body. In this full-term fetal rhesus macaque, the hind limbs are shortened with "flipper-like" attachment of the feet to the trunk; there is also marked vertebral scoliosis

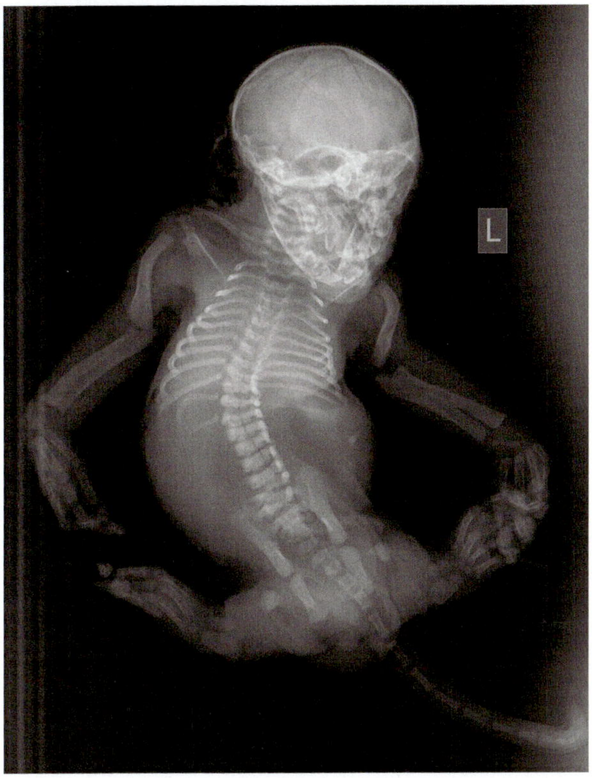

Fig. 5.39 Phocomelia, rhesus macaque. Whole body, radiograph. Radiograph shows skeletal abnormalities of infant in Fig. 5.38. Most notable are the absence/severe hypoplasia of the femurs, tibias, and fibulas and marked shortening and bowing of the humeri

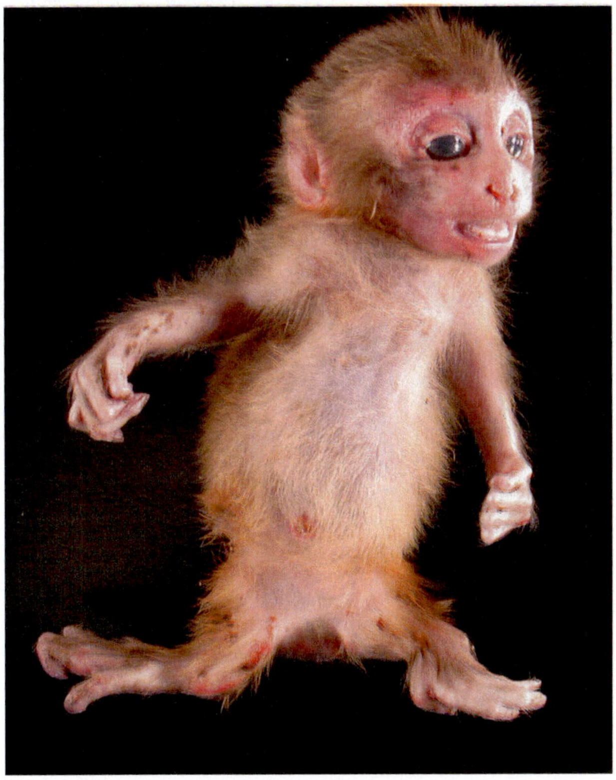

Fig. 5.40 Phocomelia, rhesus macaque. The hind limbs are severely shortened due to hypoplasia/aplasia of the long bones

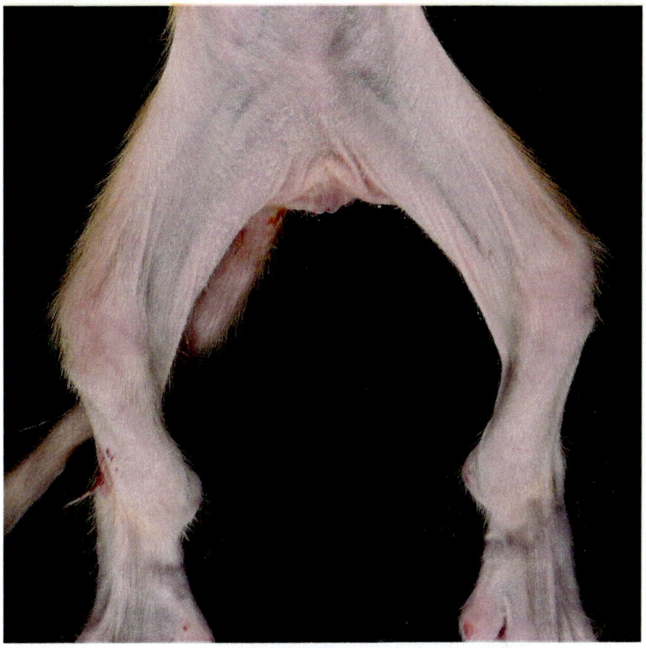

Fig. 5.41 Hemimelia, rhesus macaque. The lower extremities appear abnormally shortened and bowed

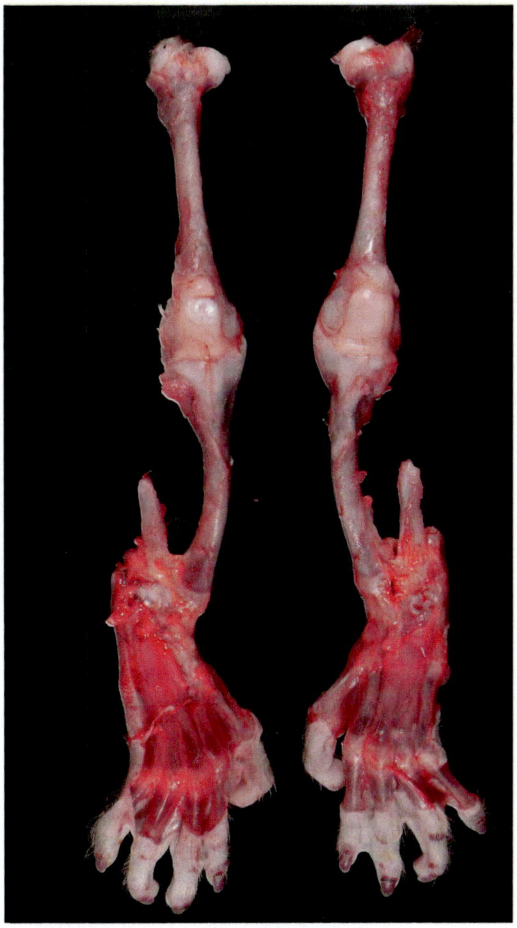

Fig. 5.42 Hemimelia, rhesus macaque. Pelvic limbs. The fibulas are bilaterally hypoplastic consisting only of the distal portion. The tibias are markedly bowed and deviated medially. Note syndactyly of digits 2 and 3 and abnormal curvature of digits

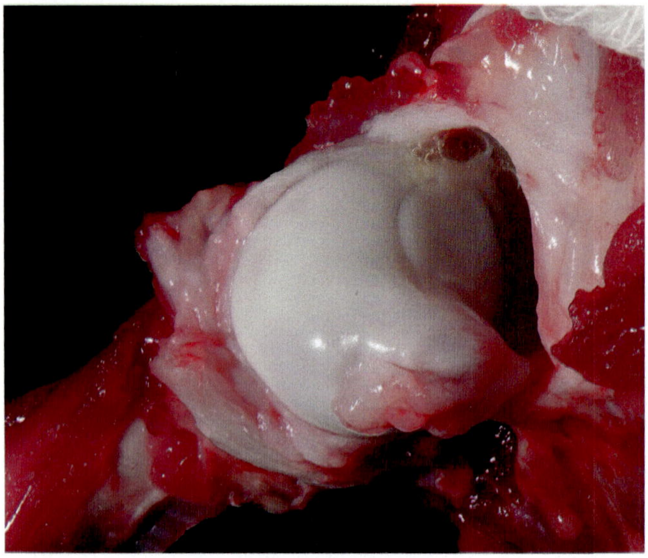

Fig. 5.43 Unilateral malformation of femoral head, rhesus macaque. Coxofemoral joint. The femoral head of this 2-year old macaque is ovoid with a saddle-shaped central depression

Digital malformations can represent a variety of underlying pathologies and may occur in isolation or as part of a group of malformations. Presentations include increased numbers of digits (polydactyly) (Figs. 5.44 and 5.45), decreased numbers of digits (oligodactyly) (Fig. 5.46), and fusion of digits (syndactyly). Syndactyly represents a failure of differentiation resulting in the abnormal interconnection of digits. Simple syndactyly involves soft tissue only. Mild forms are limited to abnormal interdigital webbing (Figs. 5.47 and 5.48). More extensive fusion may involve multiple digits and fusion of bone (synostosis) (Figs. 5.49 and 5.50).

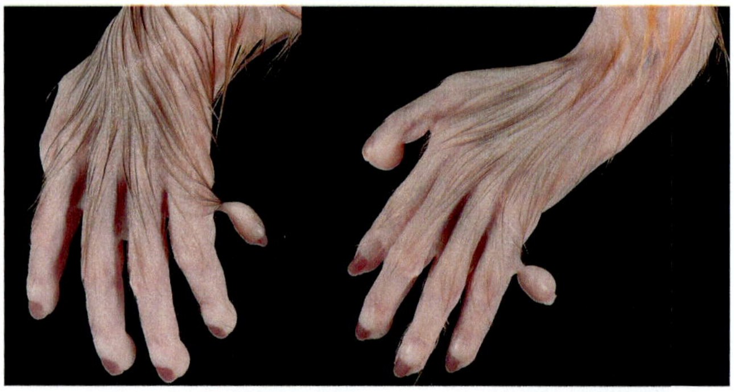

Fig. 5.44 Polydactyly, rhesus macaque. Hand and foot. There is a vestigial digit arising from the fifth digit (post-axial) of the left hand (left) and left foot (right)

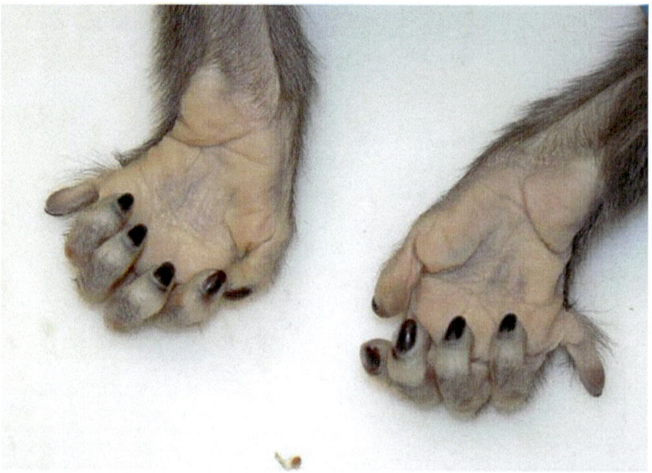

Fig. 5.45 Polydactyly, baboon. Hands. There is an additional digit arising lateral to the fifth digit (post-axial) on both hands

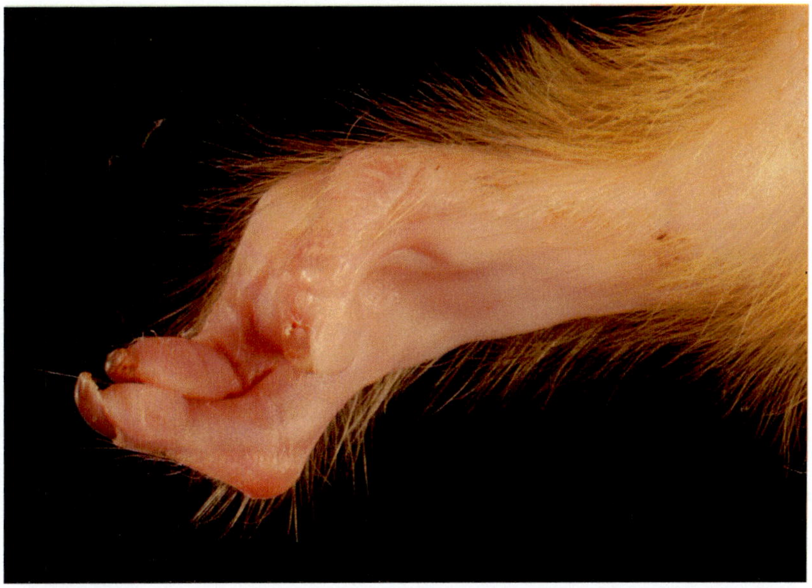

Fig. 5.46 Oligodactyly, rhesus macaque. Foot. Three digits are present on the foot of this animal exhibiting multiple congenital anomalies including phocomelia and cleft palate (Fig. 5.52)

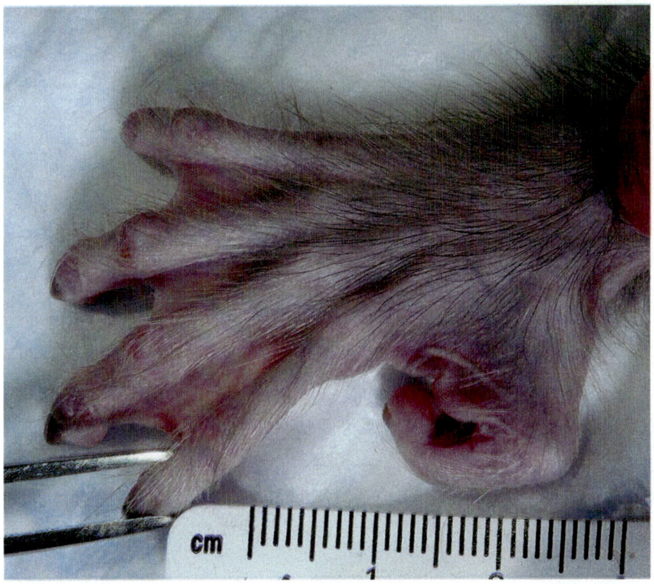

Fig. 5.47 Syndactyly, rhesus macaque. Hand. Cutaneous webs are present between digits 2–5

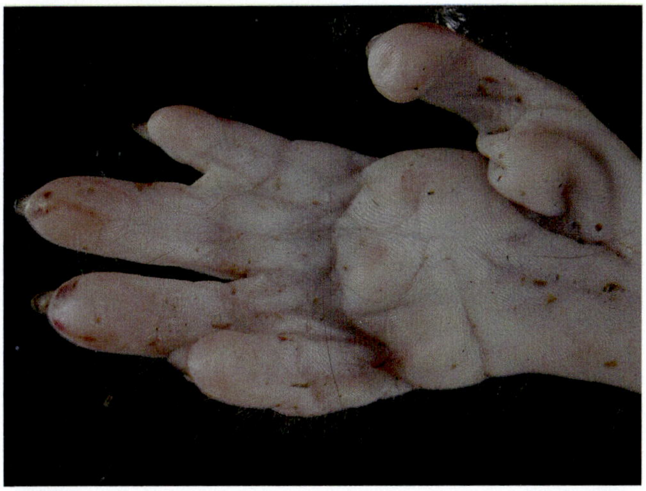

Fig. 5.48 Syndactyly, rhesus macaque. Foot. A cutaneous digital web is present between digits 2 and 3 on the foot

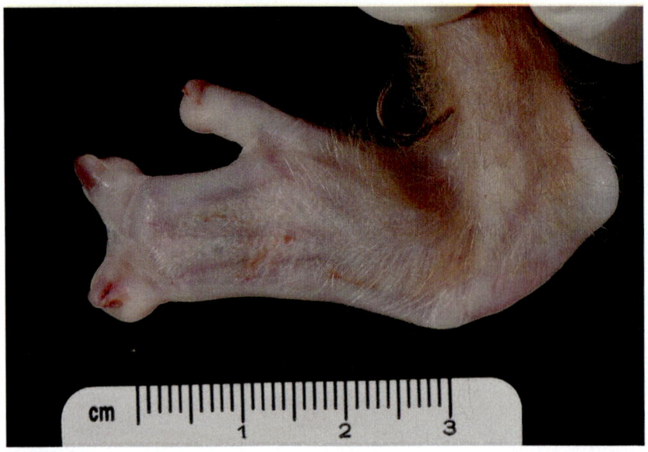

Fig. 5.49 Syndactyly, rhesus macaque. Foot. Marked syndactyly involving fusion of digits 2–5. Note the difference in pattern of syndactyly in the opposite foot (Fig. 5.50) of this animal. This animal had multiple congenital malformations including partial agenesis of the corpus callosum (Figs. 5.8 and 5.9) and arthrogryposis (Fig. 5.39)

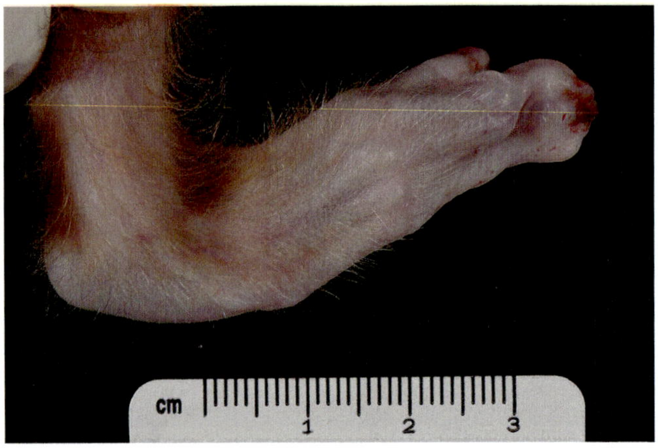

Fig. 5.50 Syndactyly, rhesus macaque. Foot. There is extensive fusion of all digits

Craniofacial malformations include those related to failure of fusion of embryonic components during development. The most common of these in NHPs and other species are cleft lip and cleft palate [2, 4]. The two examples of cleft palate in rhesus macaques shown here (Figs. 5.51 and 5.52) were associated with phocomelia and other skeletal abnormalities. Similar deformities may involve the mandible as shown in this baboon (*Papio* sp.) (Fig. 5.53). Abnormalities of jaw growth can result in mandibular prognathism, i.e. overgrowth of the mandible in relation to the maxilla (Fig. 5.54), and brachygnathism, i.e. abnormally short mandible (Fig. 5.55). In this case, brachygnathism is associated with mandibular hypoplasia and other skeletal and visceral malformations. Macroglossia, enlarged tongue, can be associated with other facial abnormalities as seen here (Fig. 5.56).

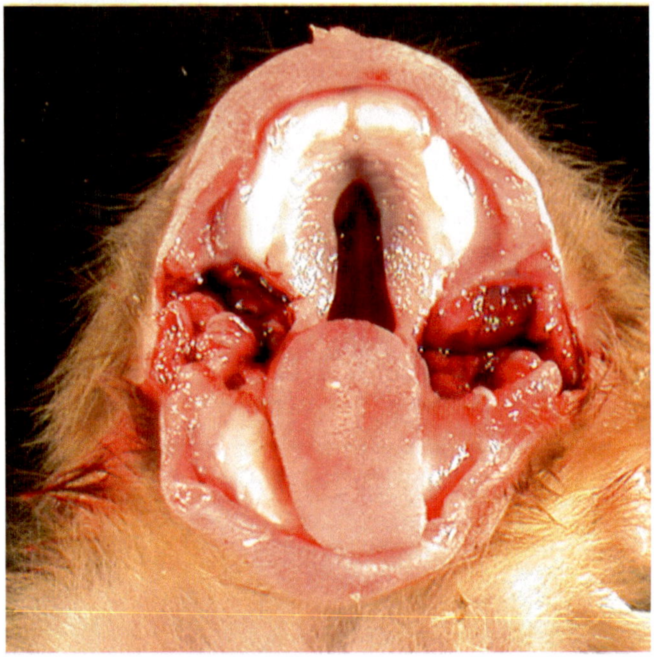

Fig. 5.51 Cleft palate, rhesus macaque. Oral cavity. There is failure of midline closure of both the hard and soft palates with exposure of the nasal cavity. This animal had multiple congenital malformations including phocomelia (Fig. 5.40)

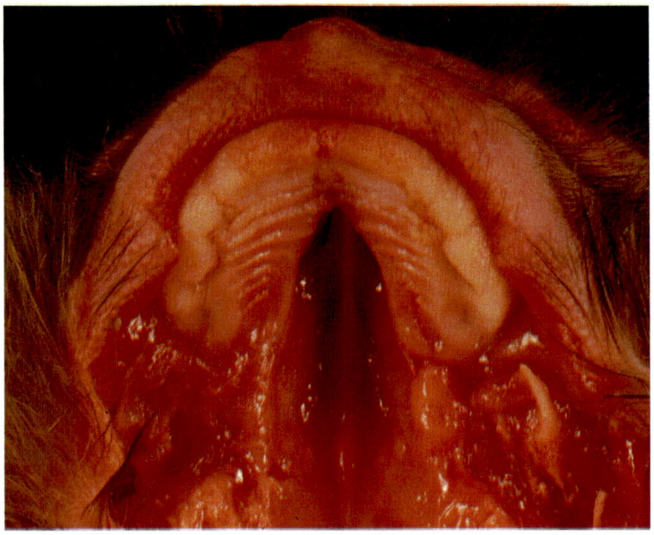

Fig. 5.52 Cleft palate, rhesus macaque. Oral cavity. There is a large defect in the hard and soft palates. This animal had multiple congenital malformations including phocomelia and oligodactyly (Fig. 5.46)

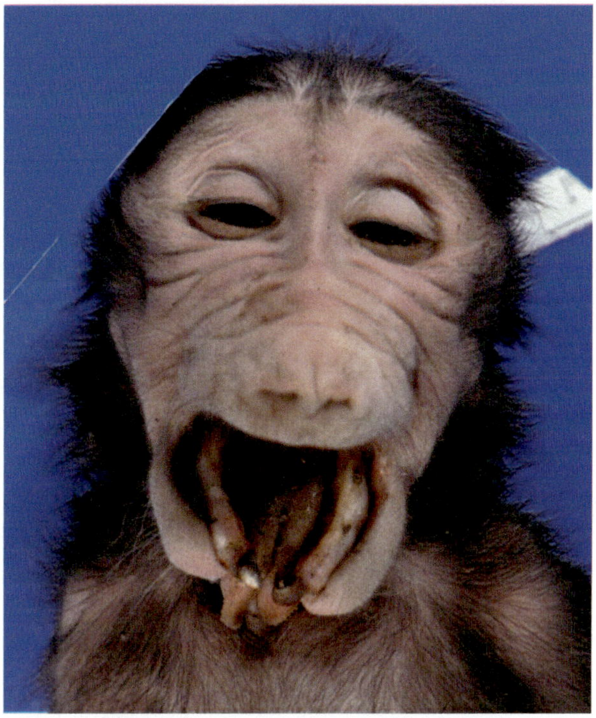

Fig. 5.53 Mandibular schisis, baboon. Head. There is a midline defect of the lower jaw involving the skin, mandible, and tongue

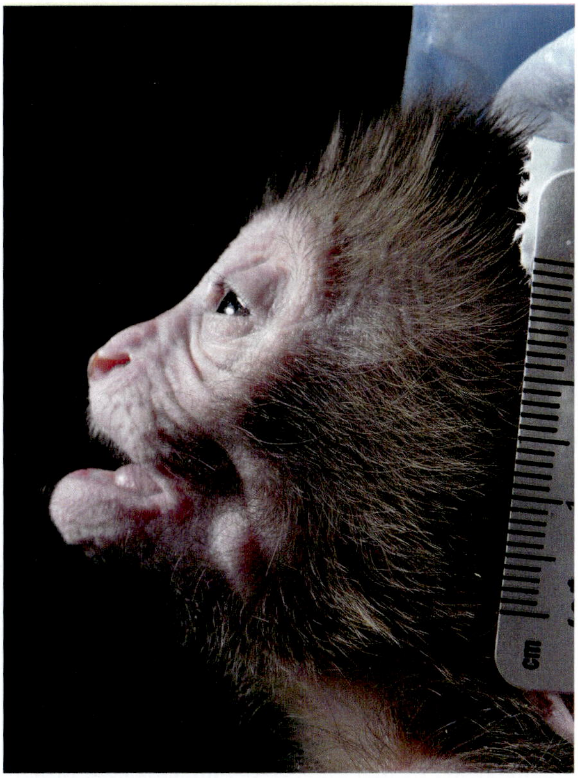

Fig. 5.54 Mandibular prognathism, rhesus macaque. Head. The lower jaw projects rostrally beyond the maxilla

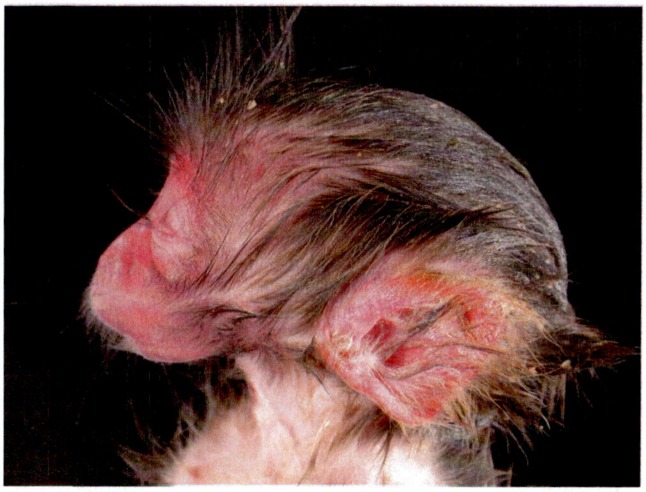

Fig. 5.55 Brachygnathism, rhesus macaque. Multiple facial malformations are present including an abnormally short, underdeveloped mandible

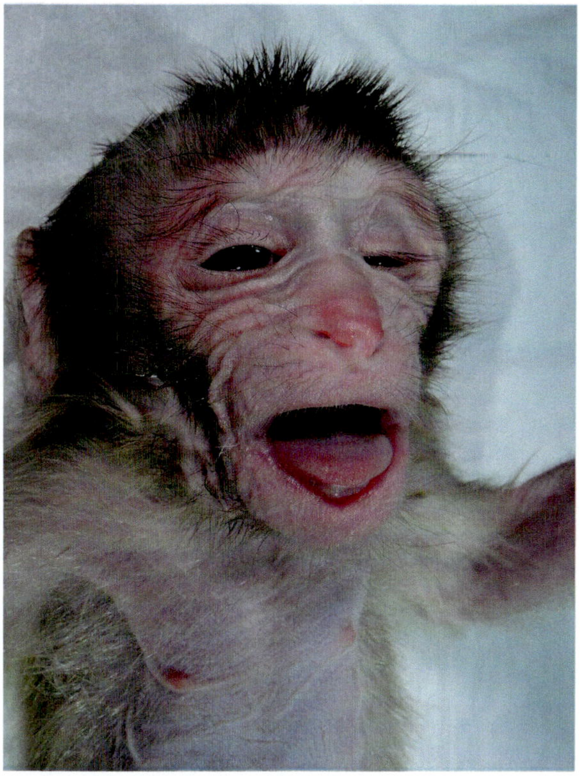

Fig. 5.56 Macroglossia, rhesus macaque. The enlarged tongue protrudes from the mouth

Pectus excavatum is a chest wall deformity in which there is concavity of the sternum ("funnel chest") (Fig. 5.57). Scoliosis may also be present. It is a relatively common skeletal anomaly in NHP [2, 3]. In humans, it may occur in isolation or may be associated with a number of genetic disorders including heritable connective tissue disorders. Similarly, abnormal curvature of the vertebral column (kyphosis, scoliosis, and lordosis) may occur in isolation or associated with other musculoskeletal abnormalities. Shown here are examples of scoliosis in rhesus macaques (Figs. 5.58 and 5.59). Both were associated with additional skeletal abnormalities. Scoliosis is also present in the case of phocomelia depicted in Figs. 5.38 and 5.39.

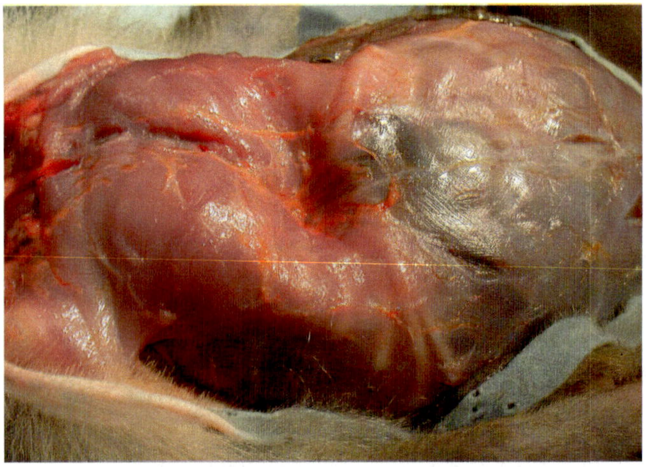

Fig. 5.57 Pectus excavatum, rhesus macaque. There is a concave deformity in the caudal portion of the sternum

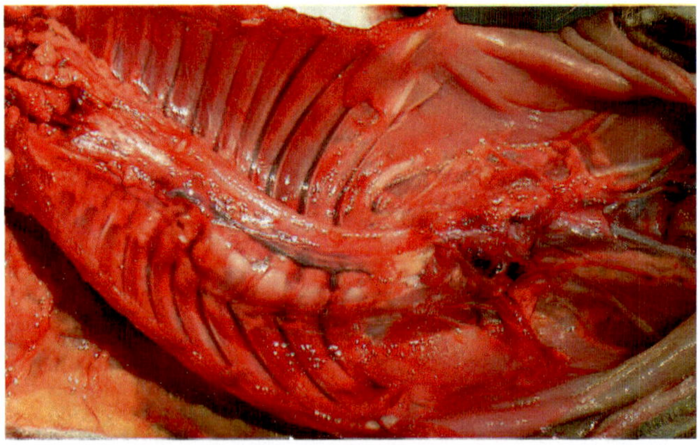

Fig. 5.58 Scoliosis, rhesus macaque. Body cavity showing ventral aspect of spine. This is the thoracic vertebral column of the animal depicted in Fig. 5.57 with pectus excavatum. Marked lateral curvature is most severe in the caudal thoracic vertebral column

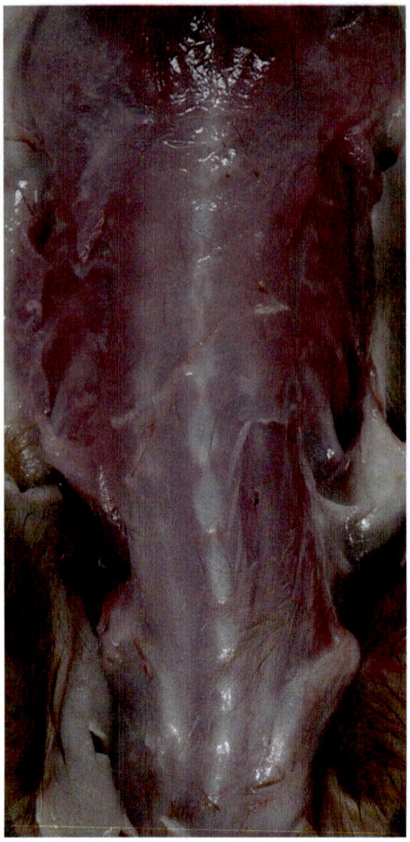

Fig. 5.59 Scoliosis, rhesus macaque. Dorsal aspect of spine. Abnormal curvature of the spine was accompanied in this case by arthrogryposis and syndactyly (Fig. 5.48) as well as severe central nervous system degeneration

5.4.2 Skeletal Dysplasias

Skeletal dysplasias are heritable disorders of bone and cartilage resulting in ongoing abnormal growth and shape of the skeleton. Chondrodysplasia with other skeletal abnormalities of unknown cause has been reported in cotton top tamarins (*Saguinus oedipus*) [29]. A lethal ciliopathy resulting in abnormal bone formation and other congenital deformities has been seen in a rhesus macaque. The animal was homozygous for an early termination variant in the *TAPT1* gene. Skeletal abnormalities included scoliosis, arthrogryposis, micrencephaly (depicted in Fig. 5.12) with abnormal skull shape, and misshapen long bones and thorax (Fig. 5.60). These findings are similar to those in humans with mutations in *TAPT1* which causes complex lethal osteochondrodysplasia, featuring severe congenital abnormalities of multiple organs (flat cerebellum, ventriculomegaly of the brain, hypomineralization of bones, and misshapen thorax) [30].

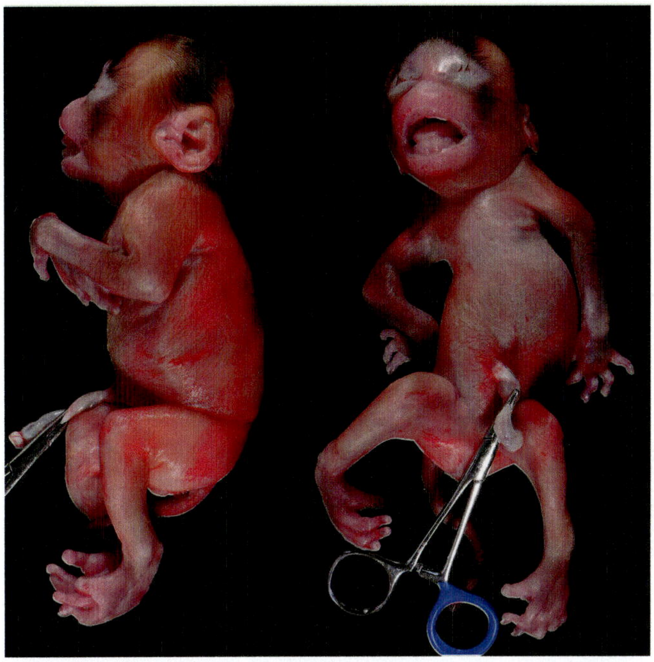

Fig. 5.60 Lethal osteochondrodysplasia-like disease, rhesus macaque. Multiple skeletal abnormalities in a fetus homozygous for a mutation in *TAPT1*. The brain from this animal is shown in Fig. 5.12

Congenital Hyperostosis

Congenital hyperostosis has been seen in a group of related rhesus macaques. All animals presented as third trimester fetuses or neonates and had similar radiographic, gross, and microscopic findings. The primary lesions are marked hyperostosis, periosteal proliferation, and neutrophilic osteitis/periostitis affecting predominately long bones (Figs. 5.61, 5.62, 5.63, and 5.64). No infectious agents have been identified. The lesions have marked similarities to infantile cortical hyperostosis (ICH) or Caffey disease. Caffey disease is a rare autosomal-dominant congenital or neonatal disorder in humans characterized by hyperostosis that targets the mandible, clavicle, ribs, scapulae, and long bones. The bony lesions include periosteal bone proliferation affecting the diaphysis, with an acute inflammatory reaction in the early stage of the disease. The infantile form has been associated with a specific genetic defect (3040C > T) in *Col1A1*, a gene encoding a subunit of type 1 collagen [31, 32]. It is inherited as an autosomal-dominant mutation with incomplete penetrance. A more severe prenatal form with more severe bone deformities, polyhydramnios, and fetal loss has also been reported in humans but is generally not associated with mutations in *Col1A1* [31, 33]. The presentation in rhesus appears intermediate between the prenatal and infantile forms in humans. The genetic basis for this disease in macaques has not been established. Affected animals do not carry deleterious mutations in *COL1A1*. There is a case report describing a condition resembling ICH in a rhesus macaque dam and one stillborn male offspring [34, 35].

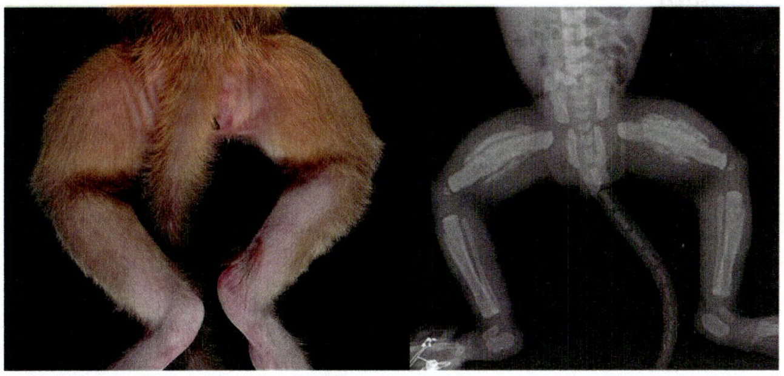

Fig. 5.61 Congenital hyperostosis, rhesus macaque. Hind limbs, gross and radiographic images. The thighs appear enlarged due to the extensive expansion of the underlying femoral bones. Radiographs demonstrate the extensive disruption of cortical architecture and periosteal bony proliferation in the femurs and to a lesser extent the tibias

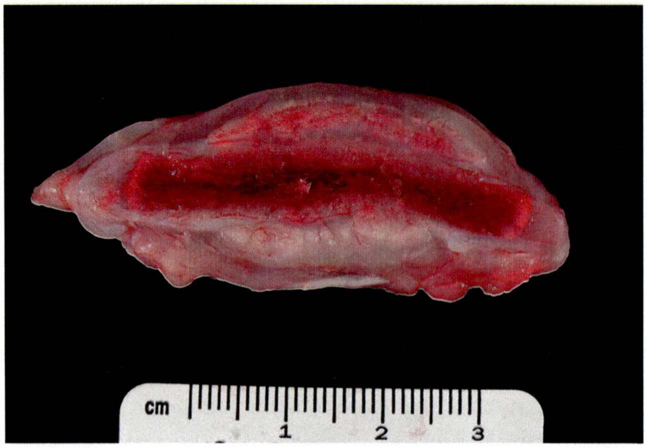

Fig. 5.62 Congenital hyperostosis, rhesus macaque. Femur, cross section. There is marked periosteal new bone formation and disruption of the cortex affecting the diaphysis

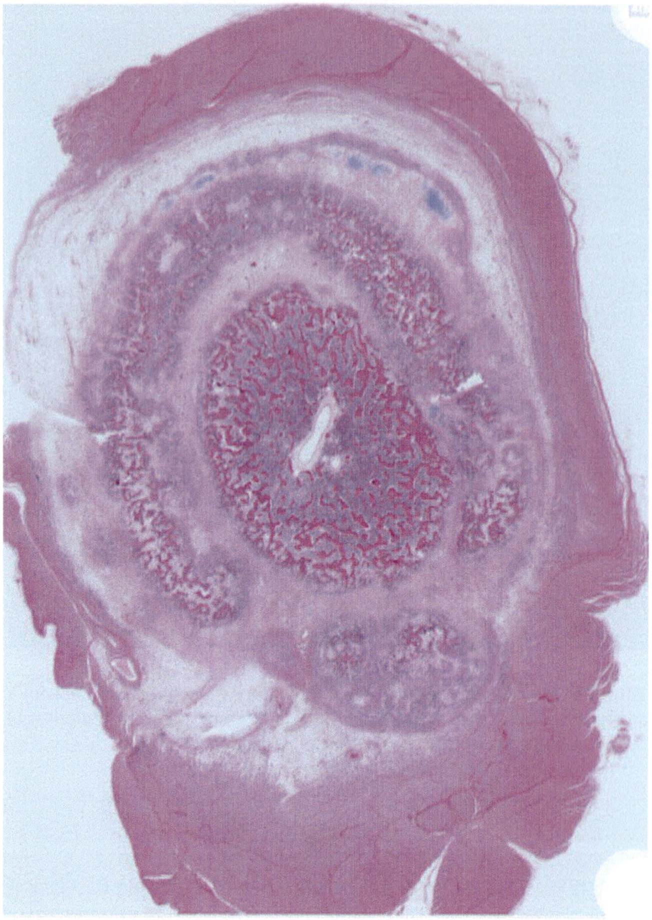

Fig. 5.63 Congenital hyperostosis, rhesus macaque. Femur, cross section, subgross, HE. The diaphyseal cortical bone and periosteum are diffusely, circumferentially, and severely disrupted, thickened up to four times normal width. There is a concentric periosteal proliferation, surrounded by loose connective tissue, interspersed with islands of osteoid and cartilage, and occasional foci of necrotizing inflammation and neutrophilic inflammation. The extra-periosteal connective tissue is markedly edematous, infiltrated by low numbers of lymphocytes, macrophages, neutrophils, and plasma cells. Skeletal myofibers are atrophic and edematous

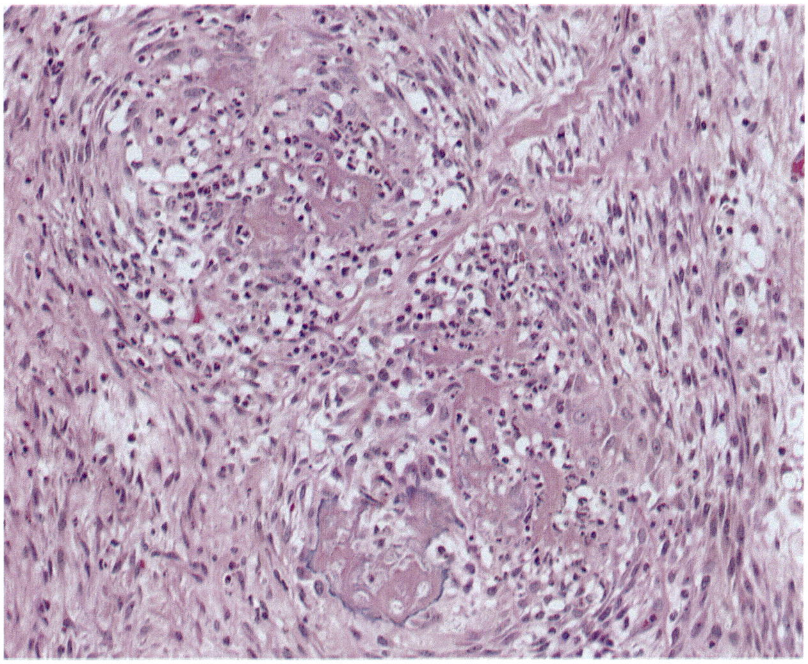

Fig. 5.64 Congenital hyperostosis, rhesus macaque. Femur, HE. Area of necrosuppurative periostitis

5.5 Cardiovascular System

A variety of cardiovascular malformations have been reported in NHP; they represent a relatively common group of congenital defects in several species of NHP. Those causing left-to-right shunts include atrial septal defects (ASD), ventricular septal defects (VSD), and patent ductus arteriosus (PDA). ASDs have been reported in a bonnet macaque (*M. radiata*) [36], baboons [3], cynomolgus macaque [2], stump-tailed macaque (*M. arctoides*) [1], patent foramen ovale in a baboon [3], and ASD with PDA in a cynomolgus and a rhesus macaque [1]. PDA alone has been reported in rhesus [2] and stump-tailed [1] macaques and a baboon [3]. Atrial septal defects in two rhesus macaques are presented in Figs. 5.65 and 5.66. Ventricular septal defects have been reported in rhesus macaques [2, 37], cynomolgus macaques [1, 38], and an African green monkey (*Cercopithecus aethiops*) [1]. Shown here are two cases of VSD in rhesus macaques (Figs. 5.67 and 5.68).

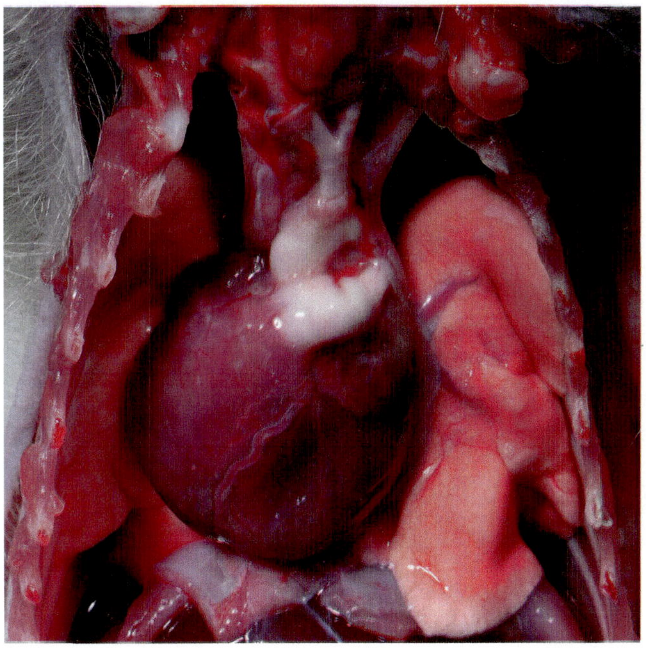

Fig. 5.65 Atrial septal defect, rhesus macaque. There is generalized cardiomegaly and right ventricular hypertrophy due to an atrial septal defect

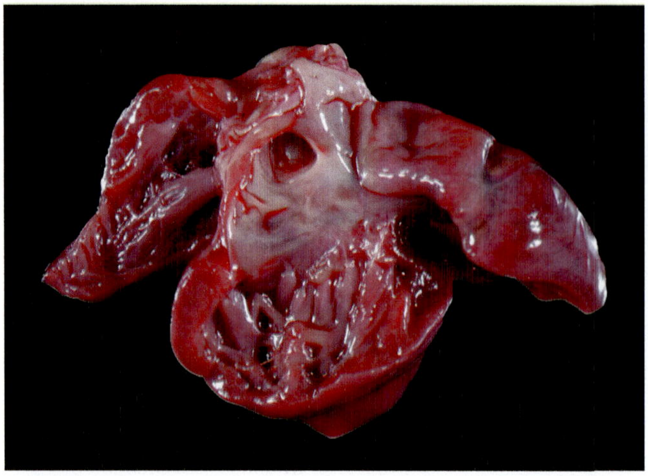

Fig. 5.66 Atrial septal defect, rhesus macaque

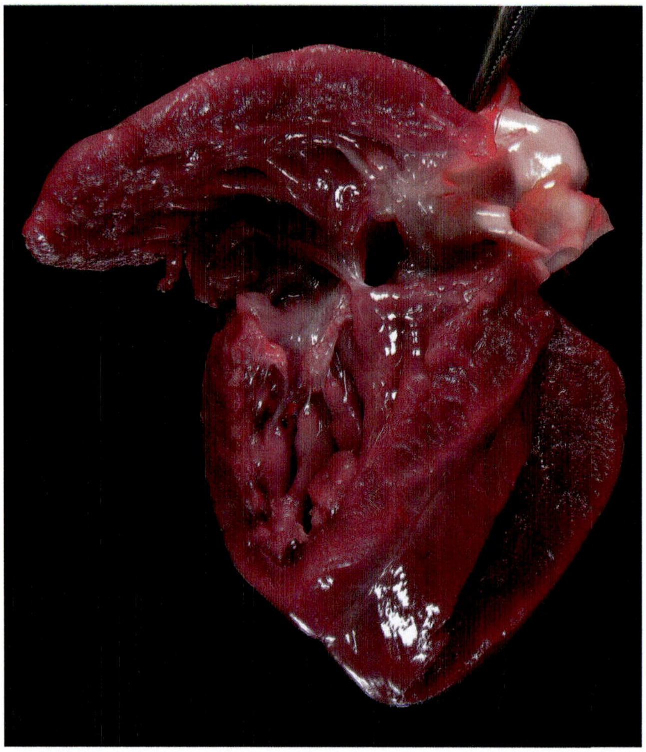

Fig. 5.67 Ventricular septal defect, rhesus macaque. Heart. The left ventricular free wall, interventricular septum, and right ventricular free wall are markedly thickened with hypertrophic papillary muscles. There is a large (0.8 cm × 0.7 cm) defect high in the interventricular septum, beneath the mitral valve

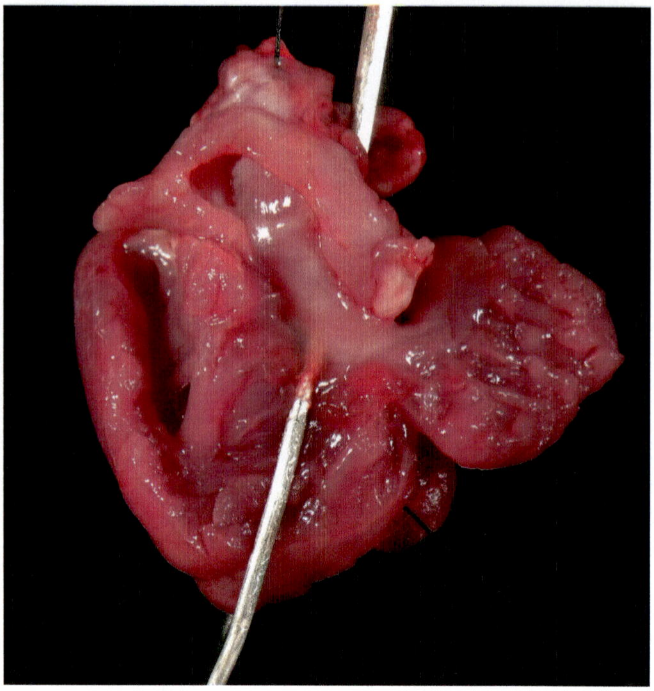

Fig. 5.68 Ventricular septal defect, rhesus macaque. Heart. There is a 0.2 cm ventricular defect high in the interventricular septum shown here beneath the tricuspid valve leaflets

Malformations of cardiac valves are relatively uncommon. Tetralogy of Fallot has been reported in a Japanese macaque [39]. Preductal aortic stenosis has been reported in concert with ASD and PDA in three cynomolgus and one stump-tailed macaque [1]. Bicuspid aortic valve with aortic dilatation in RM shown in Fig. 5.69. Mitral valve malformation with secondary endocardial fibrosis and left ventricular hypertrophy is shown in a rhesus macaque (Fig. 5.70). Various great vessel anomalies and complex cardiac malformations have been reported in rhesus macaques [2, 40, 41]. Several malformations involving the great vessels are illustrated here (Figs. 5.71, 5.72, and 5.73).

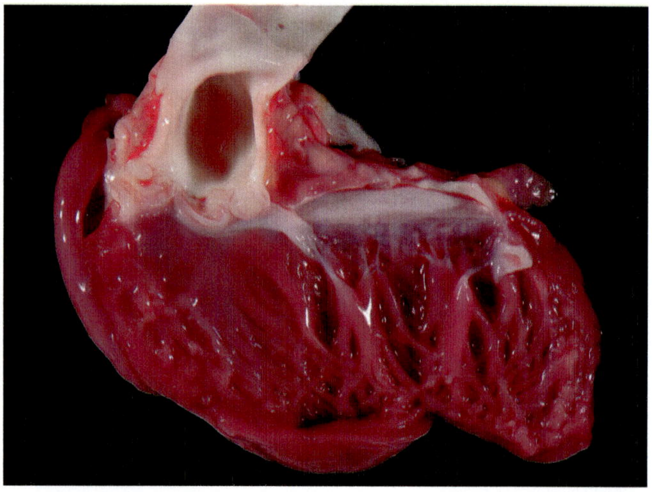

Fig. 5.69 Aortic valve malformation, rhesus macaque. Heart. The aortic valve has only two leaflets (bicuspid) which are thickened and irregular. The base of the aorta is markedly dilated

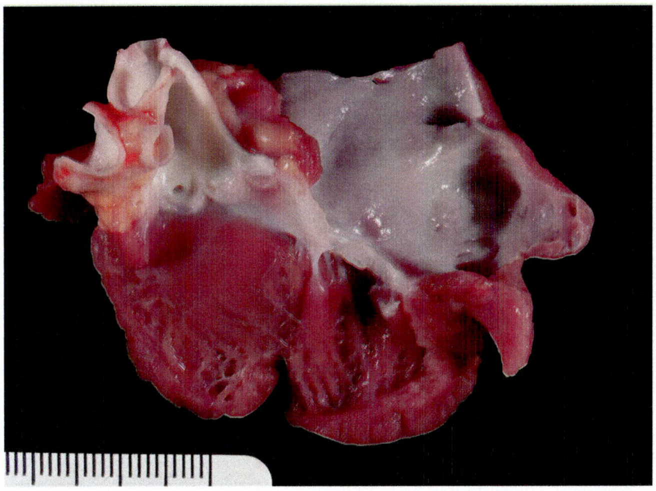

Fig. 5.70 Mitral valve malformation, rhesus macaque, heart. Mitral valve leaflets are hypoplastic and thickened. There is marked left atrial endocardial fibrosis secondary to valvular incompetence

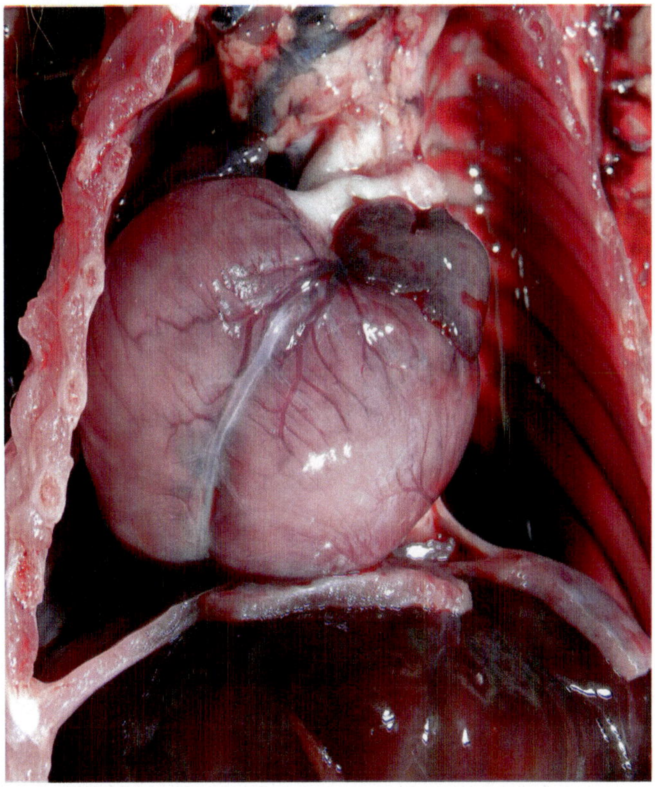

Fig. 5.71 Great vessel anomalies, rhesus macaque. Heart in situ. Marked biventricular hypertrophy is secondary to congenital partial transposition of the great vessels

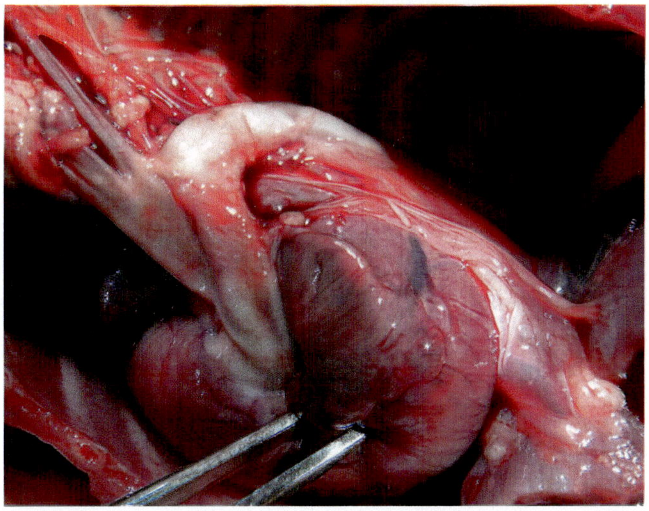

Fig. 5.72 Great vessel anomalies, rhesus macaque. Heart in situ. This is an alternate view of the heart depicted in Fig. 5.71. The aorta/major vessel arising from the left ventricular outflow supplies the coronary arteries and cranial vessels. The pulmonary trunk, arising from the right ventricular outflow tract, supplies the lungs and then continued on to provide the remainder of the aorta. The ductus arteriosus is patent in this one-day-old infant

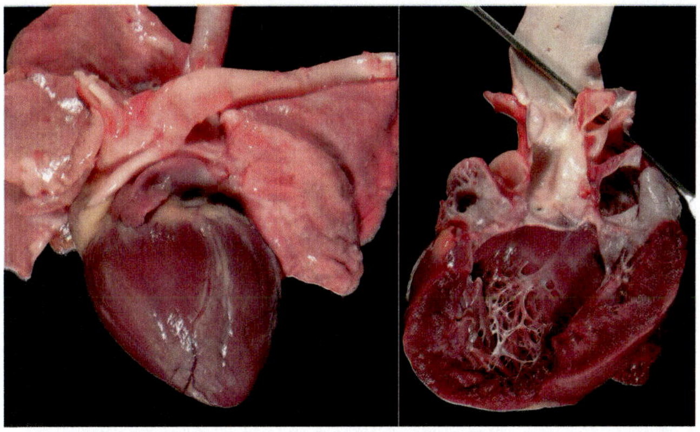

Fig. 5.73 Great vessel anomalies, rhesus macaque. Heart. There are multiple abnormalities including a persistent ductus arteriosus with ductal coarctation of the aorta and preductal aortic dilatation; and bicuspid aortic valve. There is resulting cardiomegaly and right-sided dilatated cardiomyopathy

5.6 Respiratory System

Respiratory tract anomalies are very uncommon in NHPs. Cystic adenomatoid-like malformation [42] and congenital pulmonary dysplasia (Fig. 5.74) have occurred in cynomolgus macaques. Congenital lung lobe fusion in a rhesus macaque is shown in Fig. 5.75.

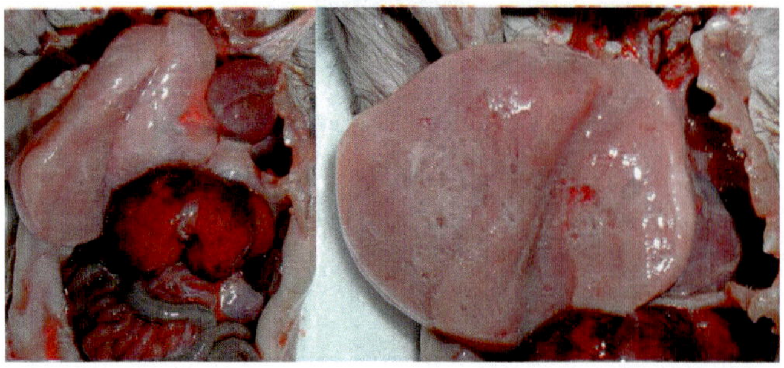

Fig. 5.74 Lung dysplasia, cynomolgus macaque. Thoracic cavity. Normal lung tissue is partially replaced by spongy mass composed of primitive pulmonary tissue

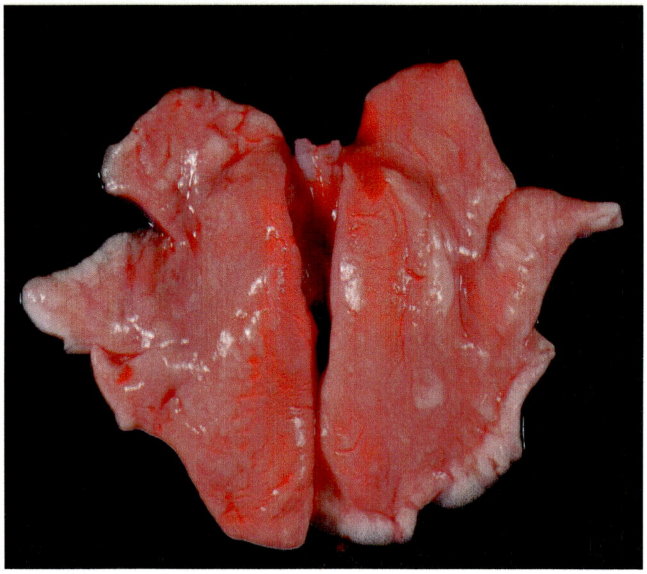

Fig. 5.75 Lung lobe fusion, rhesus macaque. Lung. There is incomplete separation of the lung lobes bilaterally

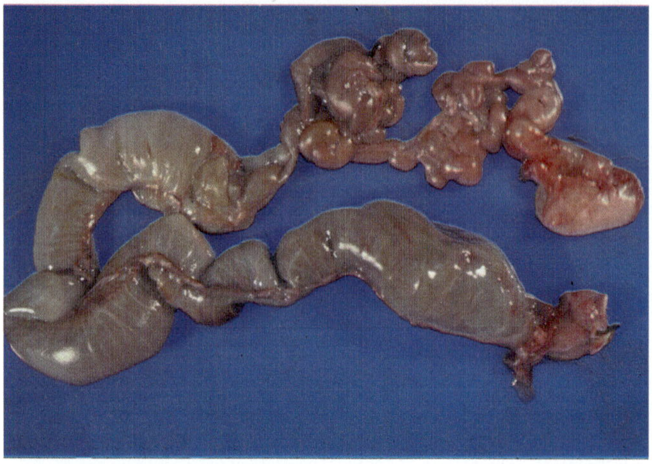

Fig. 5.76 Atresia ani, baboon. Gastrointestinal tract. The anus and distal rectum are nonpatent

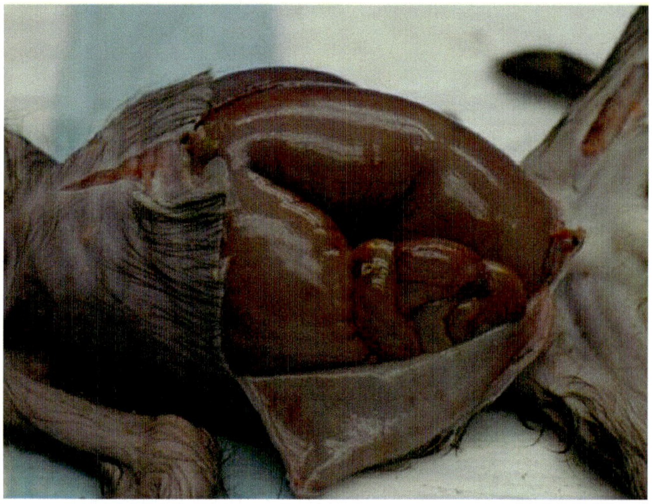

Fig. 5.77 Megacolon, baboon. Abdomen. Failure of formation of a patent distal rectum and anus has resulted in marked dilation of the colon

5.7 Gastrointestinal Tract and Accessory Organs

5.7.1 Alimentary

Relatively few defects of the alimentary tract have been described in NHP except for those involving the distal rectum and anus. Atresia ani can occur in a spectrum of anorectal malformations which may also involve the distal urogenital tract. Atresia ani or imperforate anus has been reported in several NHP species [1–3] (Fig. 5.76). More complex malformation such as cloaca has been reported in the cynomolgus macaque [43] and an imperforate vulva in a stump-tailed macaque fetus, who also had atresia coli, imperforate anus, pulmonary hypoplasia, and right-sided hydronephrosis [44]. Megacolon is a common sequela of anorectal atresia (Fig. 5.77).

5.7.2 Liver

A form of inherited iron storage disease, hemochromatosis type IVB, has been identified in rhesus macaques. It is caused by a mutation in *SLC40A1,* the gene encoding for ferroportin. The pattern of inheritance is autosomal dominant, as is true in humans. It causes a relatively mild phenotype in macaques. Affected animals have mild elevations in circulating liver enzyme, alanine transaminase (ALT). The livers appear grossly normal; the condition is typically an incidental finding during histologic examination of tissues. Iron storage, highlighted by Perl's iron stain, occurs both in the hepatocytes and Kupffer cells (Fig. 5.78).

Duplication of the gallbladder is a common incidental finding at necropsy in rhesus macaques (Fig. 5.79).

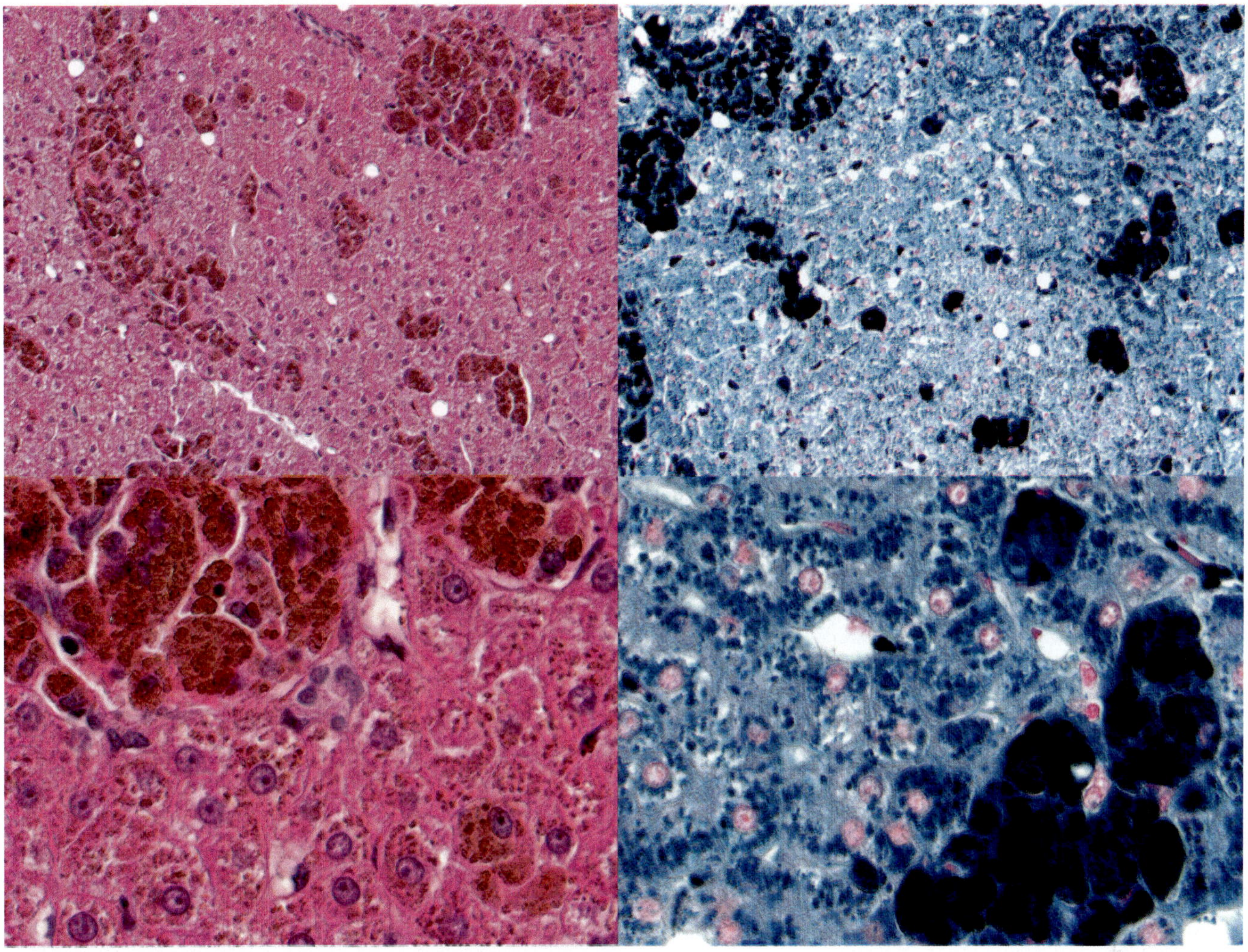

Fig. 5.78 Type IVB hemochromatosis, rhesus macaque. Liver, HE and Perl's iron stain. A mutation in the gene for ferroportin (*SLC40A1*) has resulted in marked iron storage within the liver. HE staining shows abundant brown pigmented material (hemosiderin) in both Kupffer cells and macrophages concentrated in portal regions as well as within individual hepatocytes. Perl's iron stain confirms the iron composition with positive blue stain

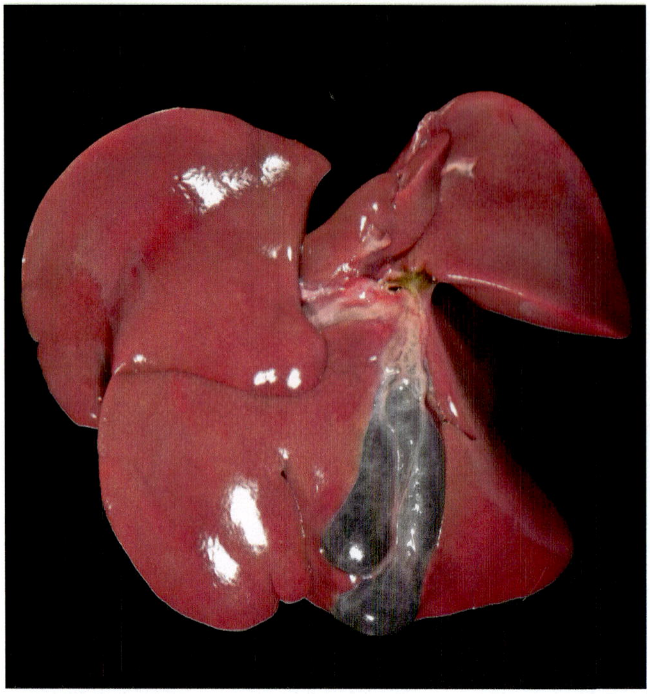

Fig. 5.79 Gallbladder duplication, rhesus macaque. Liver

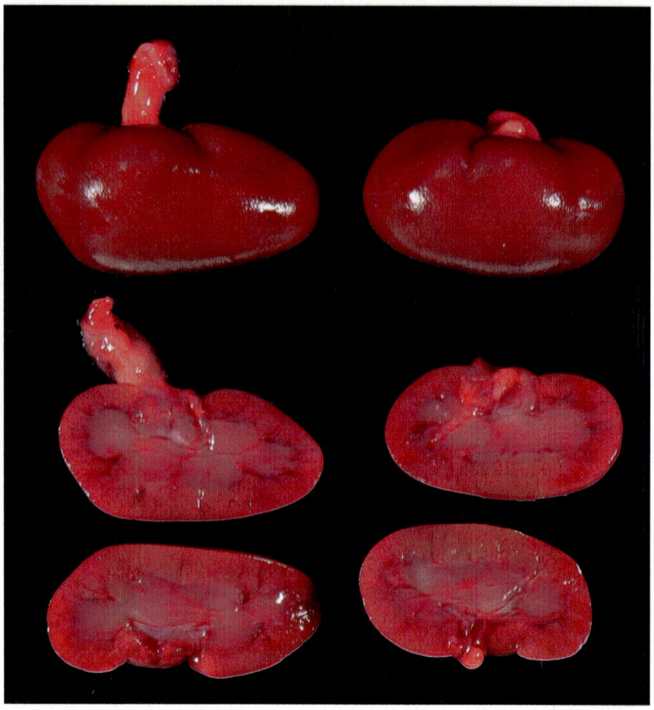

Fig. 5.80 Renal malformation, rhesus macaque. Kidney. Multiple renal calyces create a multilobular appearing hilar region. This animal also had bilateral fibular hypoplasia (Figs. 5.41 and 5.42)

5.8 Genitourinary System

5.8.1 Kidney

Defects of renal formation are generally uncommon. Renal agenesis has been reported in a rhesus macaque [2] and unilateral aplasia in a cynomolgus macaque [45]. Renal ectopia has been described in a squirrel monkey (*Saimiri sciureus)* and an owl monkey (*Aotus trivirgatus)* [46]. Renal malformations involving the presence of multiple papillae have been seen in rhesus macaques (Figs. 5.80 and 5.81). Both affected animals had additional congenital malformations (fibular hemimelia) (Figs. 5.41 and 5.42) and phocomelia, respectively (Figs. 5.38 and 5.39).

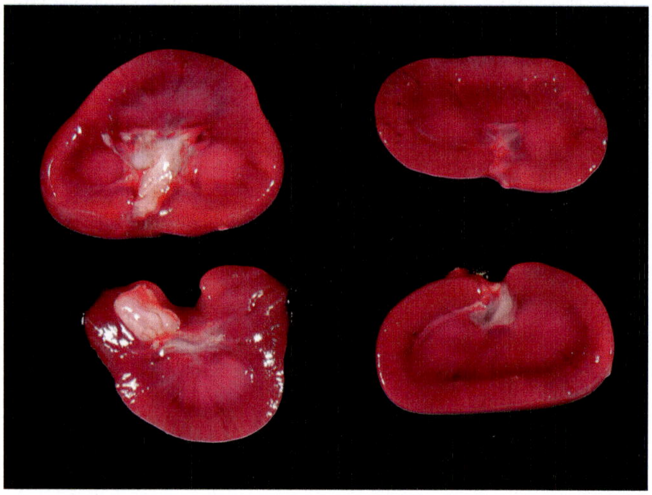

Fig. 5.81 Renal malformation, rhesus macaque. The left kidney has multiple calyces. The right kidney is unaffected. This animal had multiple abnormalities including phocomelia (Figs. 5.38 and 5.39)

A number of the heritable ciliopathies affect the kidney; a classic example is autosomal-dominant polycystic kidney disease (ADPKD) in humans and cats, caused by mutations *PKD1*. The ciliopathic disease, Bardet–Biedl syndrome (BBS7) in rhesus macaques, causes both a retinopathy described earlier (Fig. 5.35) as well as renal disease characterized by renal scarring and cystic tubules lined by irregular hyperplastic epithelium and chronic interstitial inflammation with fibrosis and loss of nephrons (Figs. 5.82 and 5.83). Cystic renal disease of unknown cause has been reported in several NHP species including rhesus [47, 48] and cynomolgus [49] macaques. Polycystic kidney disease is depicted here in a common marmoset (Fig. 5.84). Multiple renal medullary cysts found in an infant rhesus macaque (Figs. 5.85 and 5.86) were accompanied by other congenital abnormalities including ovarian dysplasia (Fig. 5.87); the findings are suggestive of an underlying genetic mutation.

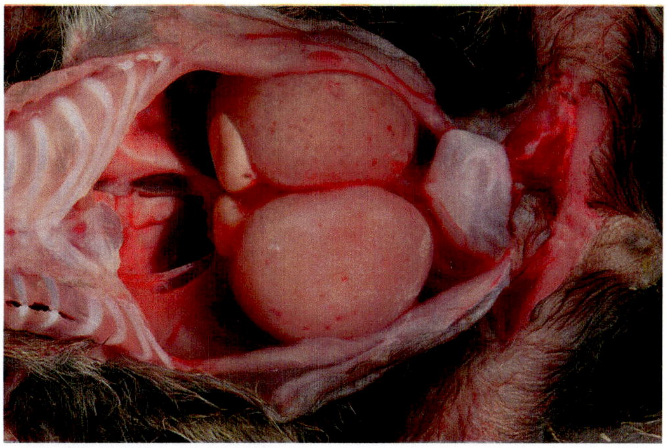

Fig. 5.84 Polycystic kidney, common marmoset. The kidneys are severely enlarged with normal parenchyma effaced by confluent fluid-filled cysts

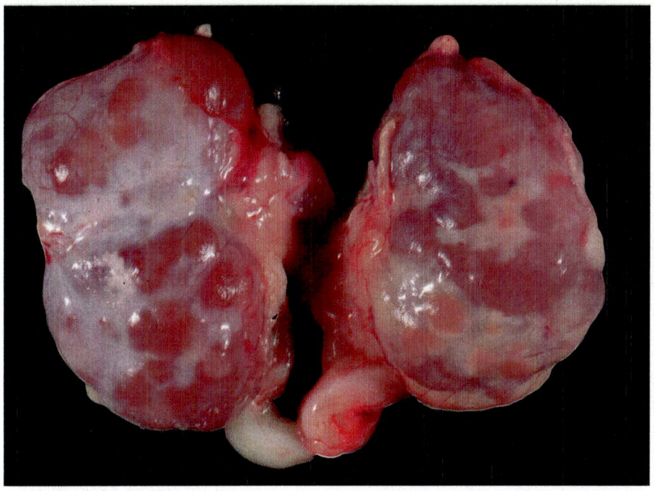

Fig. 5.82 Bardet Biedl syndrome, BBS7, rhesus macaque. Kidneys. The renal capsules are firmly adhered to the cortex and the contours of kidneys are severely and extensively distorted by depressed, fibrotic foci

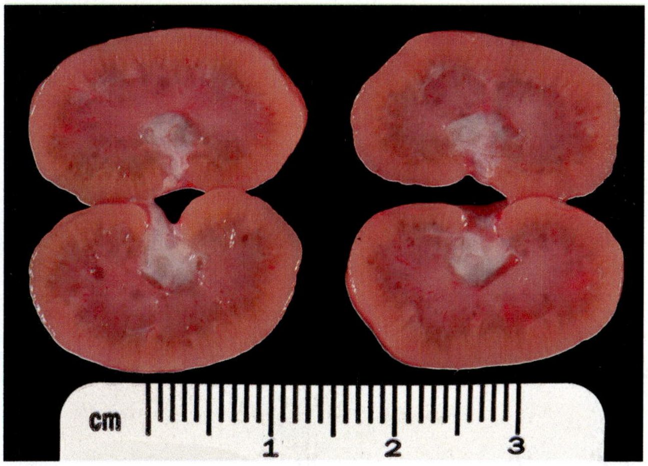

Fig. 5.85 Medullary renal cysts, rhesus macaque. Kidneys. There are numerous ~0.1 cm cysts in the medulla clustered at the corticomedullary junction

Fig. 5.83 Bardet–Biedl syndrome, BBS7, rhesus macaque. Kidney, HE. The renal cortex contains multiple small cysts with are lined by irregularly organized, multi-layered tubular epithelium which occasionally forms polypoid projections

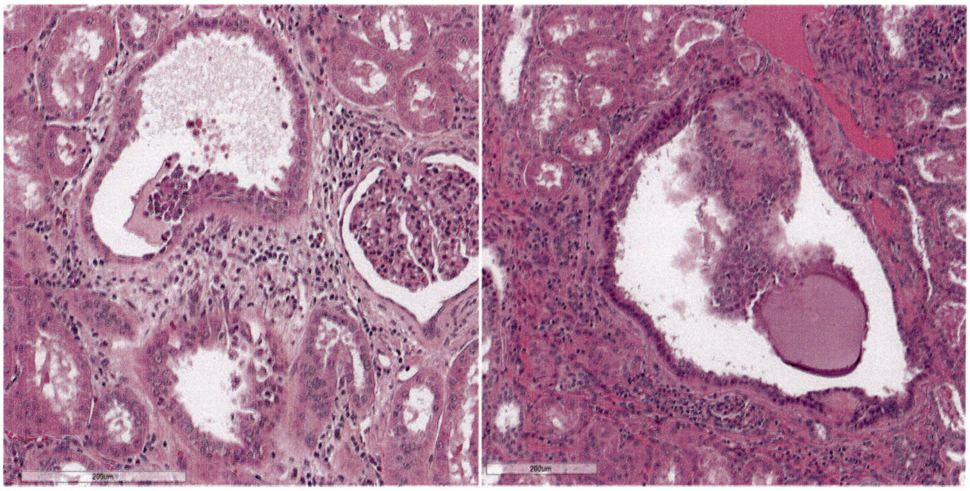

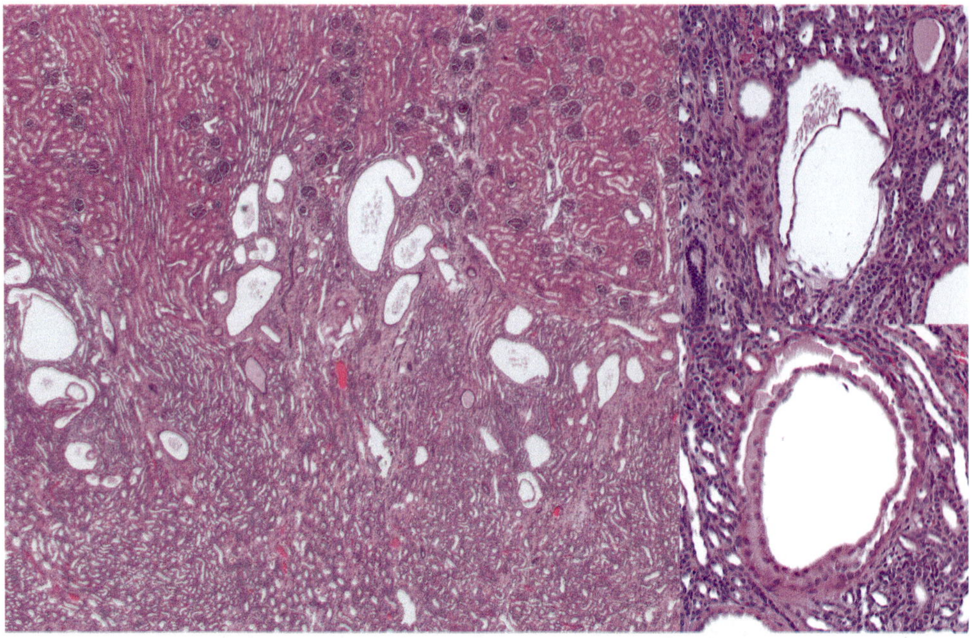

Fig. 5.86 Medullary renal cysts, rhesus macaque. Kidney, HE. Renal medullary cysts are lined by 1–2 layers of cuboidal epithelium

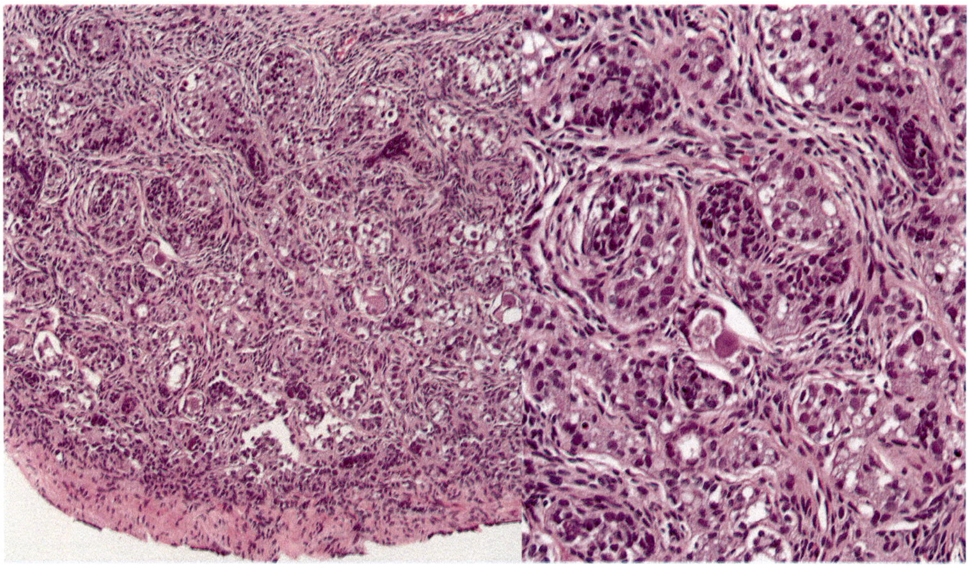

Fig. 5.87 Ovarian dysplasia, rhesus macaque. Ovary, HE. In this 21-day old infant, there is a generalized paucity of oocytes and numerous immature sex cords

5.8.2 Reproductive System

Rare reproductive tract anomalies reported in NHP include uterus didelphys in a rhesus macaque [44] and ovotesticular disorder of sex development in a baboon [50]. Shown here is ovarian dysplasia associated with multiple renal medullary cysts observed in a rhesus macaque (Fig. 5.87).

5.9 Endocrine System

5.9.1 Thyroid Gland

Congenital goiter is a rare finding in nonhuman primates. The general causes of congenital goiter in all species include maternal diets with deficient or excessive iodine, maternal ingestion of goitrogenic compounds, and genetic mutations. Familial goiter has been described in the rhesus macaque; a genetic cause was suspected but not identified [51]. Dyshormonogenetic goiter has been seen in related rhesus macaques which are homozygotic for a mutation in *TG*, the gene encoding for thyroglobulin [52]. Of five cases, three were stillborn or died within 2 weeks of birth. Two animals lived to 1 year of age; one exhibit bone lesions of epiphyseal immaturity and dysplasia, typical of congenital hypothyroidism. In the affected animals, the thyroids glands were enlarged and fleshy (Fig. 5.88). Microscopically, there was diffuse thyroid follicular hyperplasia and hypertrophy with diffuse colloid atrophy (Fig. 5.89).

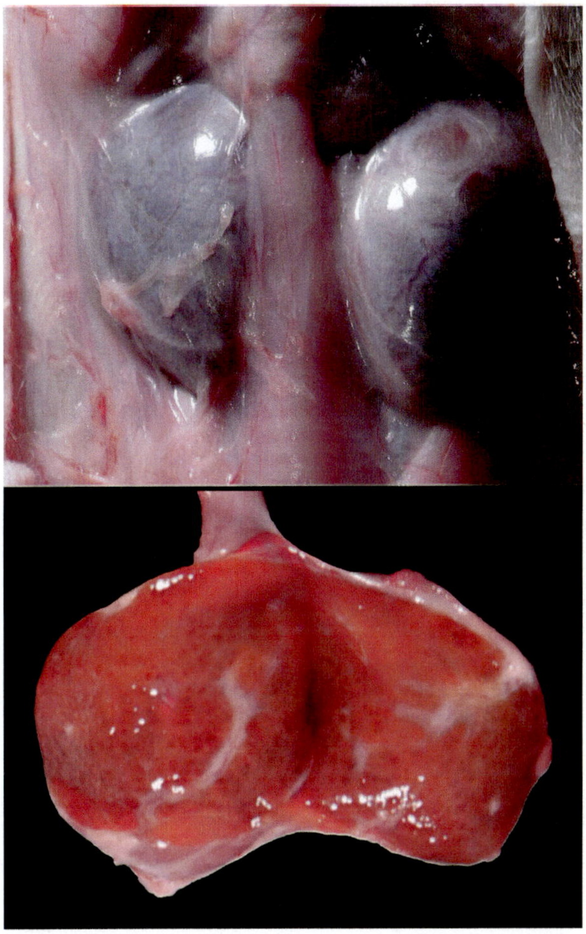

Fig. 5.88 Dyshormonogenetic goiter, rhesus macaque. Thyroid gland. The lobes of the thyroid gland are markedly enlarged in this animal with goiter due to a mutation in *TG*. The parenchyma appears fleshy

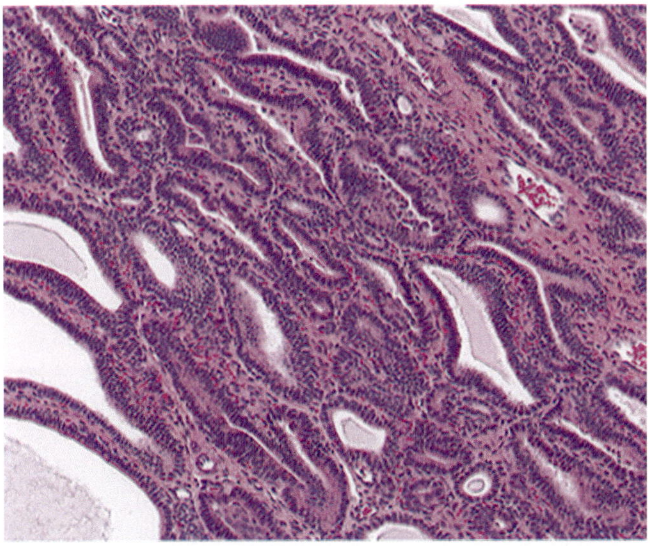

Fig. 5.89 Dyshormonogenetic goiter, rhesus macaque. Thyroid gland, HE. Thyroid follicles are irregular, often angular and lined by diffusely hypertrophic and hyperplastic epithelium. There is diffuse paucity of normal colloid

Thyroid dysplasia is an apparently benign condition seen in related Japanese macaques which features diffuse basilar cytoplasmic vacuolation of thyroid follicular cells (Figs. 5.90 and 5.91). It is similar to a syndrome described as thyroid gland dysplasia in Hannover Wistar GALAS rats [53]. In Japanese macaques, the condition is attributed to a defect in *TG*, as is true for the Wistar rats. A morphologically similar condition has been reported in cynomolgus macaques [54].

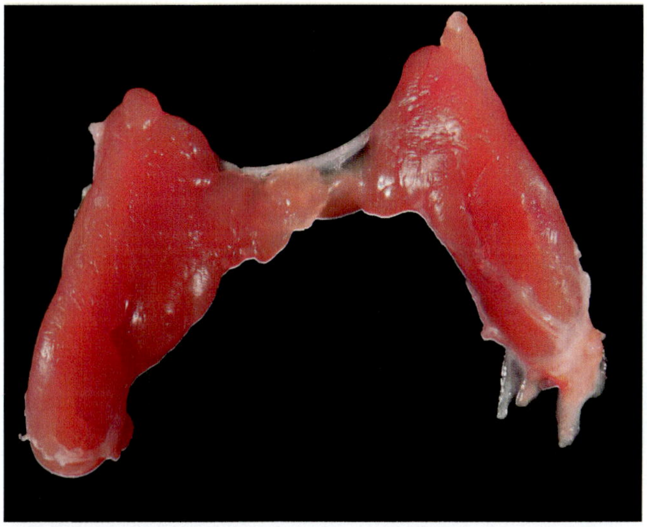

Fig. 5.90 Thyroid dysplasia, Japanese macaque. Thyroid. The gland is mildly enlarged and lobulated

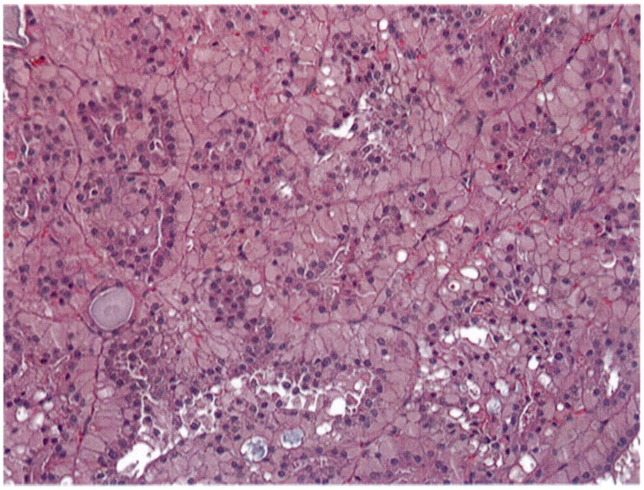

Fig. 5.91 Thyroid dysplasia, Japanese macaque. Thyroid, HE. Thyroid follicles are lined by enlarged columnar follicular epithelium with prominent basilar eosinophilic vacuoles which displace the nucleus apically. There is diffuse lack of colloid

5.10 Integument

Epidermolysis bullosa simplex has been identified in two fetal rhesus macaques [55]. The condition in macaques is due to an insertion variant mutation in *KRT5*. Loss of keratin 5 protein results in cytoskeletal disruption of the basal keratinocytes leading to vesicle formation and epidermal loss. The affected stillborn fetuses were delivered naturally with extensive loss of the epidermis (Fig. 5.92). Microscopically, lesions were characterized by extensive ulceration and vesicle formation due to intraepidermal clefting (Fig. 5.93).

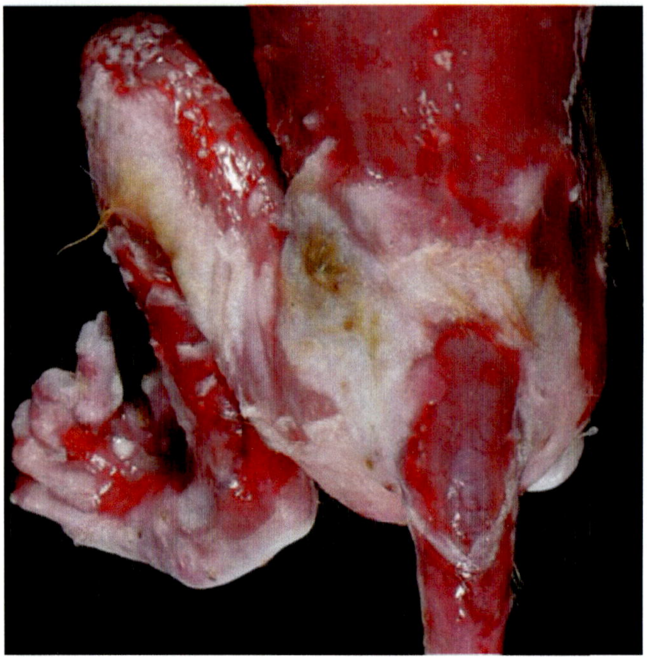

Fig. 5.92 Epidermolysis bullosa simplex due to *KRT 5* mutation, fetal rhesus macaque. There is generalized absence of the epidermis with multiple islands of regeneration, most prominent on the anterior surface of the thigh

Fig. 5.93 Epidermolysis bullosa simplex due to *KRT 5* mutation, fetal rhesus macaque. Skin, HE. There is focally extensive intraepidermal separation transitioning to ulceration

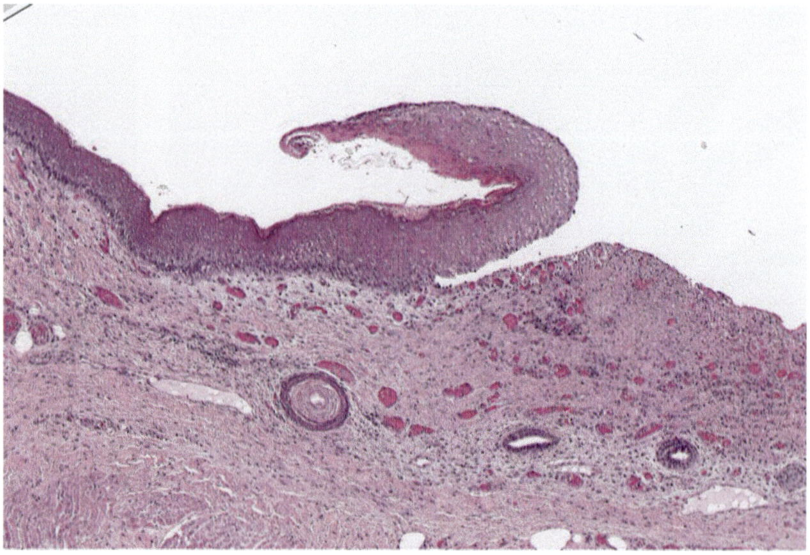

Acknowledgments Lois Colgin for use of case materials and editorial assistance. Samuel Peterson, Betsy Ferguson for consultation and identification of underlying genetic mutations presented in this chapter. Allie Meristem and Wendy Price for technical support. Amanda Johnson, Rebecca Ducore, Heidi Pecoraro, Jonathon Stoddard, Martha Neuringer, Jodi McBride for use of images and case material.

Southwest National Primate Research Center (SNPRC) for use of images (Figs. 5.45, 5.53, 5.74, 5.76, and 5.77).

Lara Doyle-Meyers for use of image (Fig. 5.7); Shelley Falkenstein for use of images and case material (Figs. 5.23, 5.24, 5.25, and 5.26).

Andrew Miller for use of images (Figs. 5.13 and 5.84).

This work was supported in part by NIH Award P51 OD 011092 to the Oregon National Primate Research Center. Digital slide scanning was performed through the Integrated Pathology Core at the Oregon National Primate Research Center by NIH Award 1S10OD025002-01. The Macaque Genotype and Phenotype Resource (mGAP) is supported by NIH R24OD021324.

References

1. Jerome CP. Congenital malformations and twinning in a breeding colony of Old World monkeys. Lab Anim Sci. 1987;37:624–30.
2. Peterson PE, Short JJ, Tarara R, et al. Frequency of spontaneous congenital defects in rhesus and cynomolgus macaques. J Med Primatol. 1997;26:267–75.
3. Fox B, Owston MA, Kumar S, et al. Congenital anomalies in the baboon (Papio spp.). J Med Primatol. 2011;40:357–63.
4. Hoopes CW, Jerome CP. Oral-facial clefts and associated malformations in the squirrel monkey (Saimiri sciureus). J Med Primatol. 1987;16:203–9.
5. Kimura A, Yahashi S, Chatani F, et al. Identification of chromosome 17 trisomy in a cynomolgus monkey (Macaca fascicularis) by multicolor FISH techniques. Cytogenet Genome Res. 2021;161:243–8.
6. Moore CM, Hubbard GB, Dick E, et al. Trisomy 17 in a baboon (Papio hamadryas) with polydactyly, patent foramen ovale and pyelectasis. Am J Primatol. 2007;69:1105–18.
7. Moore CM, McKeand J, Witte SM, et al. Teratoma with trisomy 16 in a baboon (Papio hamadryas). Am J Primatol. 1998;46:323–32.
8. Ruppenthal GC, Caffery SA, Goodlin BL, et al. Pigtailed macaques (Macaca nemestrina) with trisomy X manifest physical and mental retardation. Am J Ment Defic. 1983;87:471–6.
9. Ruppenthal GC, Moore CM, Best RG, et al. Trisomy 16 in a pigtailed macaque (M. nemestrina) with multiple anomalies and developmental delays. Am J Ment Retard. 2004;109:9–20.
10. Baskin GB, Ratterree M, Davison BB, et al. Genetic galactocerebrosidase deficiency (globoid cell leukodystrophy, Krabbe disease) in rhesus monkeys (Macaca mulatta). Lab Anim Sci. 1998;48:476–82.
11. Borda JT, Alvarez X, Mohan M, et al. Clinical and immunopathologic alterations in rhesus macaques affected with globoid cell leukodystrophy. Am J Pathol. 2008;172:98–111.
12. Luzi P, Rafi MA, Victoria T, et al. Characterization of the rhesus monkey galactocerebrosidase (GALC) cDNA and gene and identification of the mutation causing globoid cell leukodystrophy (Krabbe disease) in this primate. Genomics. 1997;42:319–24.
13. Moshiri A, Chen R, Kim S, et al. A nonhuman primate model of inherited retinal disease. J Clin Invest. 2019;129:863–74.
14. Canfield D, Brignolo L, Peterson PE, et al. Conjoined twins in a rhesus monkey (Macaca mulatta). J Med Primatol. 2000;29:427–30.
15. Vitucci A, Fichera A, Fratelli N, et al. Twin reversed arterial perfusion sequence: current treatment options. Int J Women's Health. 2020;12:435–43.
16. Hein PR, van Groeninghen JC, Puts JJ. A case of acardiac anomaly in the cynomolgus monkey (Macaca fascicularis): a complication of monozygotic monochorial twinning. J Med Primatol. 1985;14:133–42.
17. Sawada K, Fukunishi K, Kashima M, et al. Fetal gyrification in cynomolgus monkeys: a concept of developmental stages of gyrification. Anat Rec (Hoboken, NJ: 2007). 2012;295:1065–74.
18. Devisme L, Bouchet C, Gonzales M, et al. Cobblestone lissencephaly: neuropathological subtypes and correlations with genes of dystroglycanopathies. Brain. 2012;135:469–82.
19. Bergin IL, Campbell B, Agnew DW. Ependymal cyst in a cynomolgus macaque (Macaca fascicularis). J Med Primatol. 2008;37:239–44.
20. Pecoraro HL, Haertel AJ, Cullin C, et al. Cerebral cysts of ependymal or ventricular origin in a juvenile rhesus macaque (Macaca mulatta) with neurologic signs. J Med Primatol. 2019;48:378–80.
21. McBride JL, Neuringer M, Ferguson B, et al. Discovery of a CLN7 model of Batten disease in non-human primates. Neurobiol Dis. 2018;119:65–78.
22. Graziano AC, Cardile V. History, genetic, and recent advances on Krabbe disease. Gene. 2015;555:2–13.
23. Sherman LS, Su W, Johnson AL, et al. A novel non-human primate model of Pelizaeus-Merzbacher disease. Neurobiol Dis. 2021;158:105465.
24. Harper PA, Healy PJ, Dennis JA. Ultrastructural findings in maple syrup urine disease in Poll Hereford calves. Acta Neuropathol. 1986;71:316–20.
25. Depienne C, Bugiani M, Dupuits C, et al. Brain white matter oedema due to ClC-2 chloride channel deficiency: an observational analytical study. Lancet Neurol. 2013;12:659–68.
26. Blanz J, Schweizer M, Auberson M, et al. Leukoencephalopathy upon disruption of the chloride channel ClC-2. J Neurosci Off J Soc Neurosci. 2007;27:6581–9.
27. Ribka EP, Dubielzig RR. Multiple ophthalmic abnormalities in an infant rhesus macaque (Macaca mulatta). J Med Primatol. 2008;37 (Suppl 1):16–9.
28. Peterson SM, McGill TJ, Puthussery T, et al. Bardet-Biedl Syndrome in rhesus macaques: A nonhuman primate model of retinitis pigmentosa. Exp Eye Res. 2019;189:107825.
29. Chalifoux LV. Congenital anomalies, tamarins. In: Jones TC, Mohr U, Hunt RD, editors. Nonhuman primates I. Berlin: Springer; 1993.
30. Symoens S, Barnes AM, Gistelinck C, et al. Genetic defects in TAPT1 disrupt ciliogenesis and cause a complex lethal Osteochondrodysplasia. Am J Hum Genet. 2015;97:521–34.
31. Gensure RC, Makitie O, Barclay C, et al. A novel COL1A1 mutation in infantile cortical hyperostosis (Caffey disease) expands the spectrum of collagen-related disorders. J Clin Invest. 2005;115:1250–7.
32. Nistala H, Makitie O, Juppner H. Caffey disease: new perspectives on old questions. Bone. 2014;60:246–51.
33. Kamoun-Goldrat A, Martinovic J, Saada J, et al. Prenatal cortical hyperostosis with COL1A1 gene mutation. Am J Med Genet A. 2008;146A:1820–4.
34. Snook SS. Infantile cortical hyperostosis, rhesus monkey. In: Jones TC, Mohr U, Hunt RD, editors. Nonhuman primates II. Berlin Heidelberg: Springer; 1993.
35. Snook SS, King NW Jr. Familial infantile cortical hyperostosis (Caffey's disease) in rhesus monkeys (Macaca mulatta). Vet Pathol. 1989;26:274–7.
36. Anilkumar TV, Sandhyamani S. Atrial septal defect in a bonnet macaque (Macaca radiata). J Med Primatol. 1990;19:749–52.
37. Swindle MM, Kan JS, Adams RJ, et al. Ventricular septal defect in a rhesus monkey. Lab Anim Sci. 1986;36:693–5.
38. Koie H, Ageyama N, Ono F, et al. Echocardiographic diagnosis of muscular ventricular septal defect in a cynomolgus monkey (Macaca fascicularis). Contemp Top Lab Anim Sci. 2005;44:26–8.
39. Koie H, Abe Y, Sato T, et al. Tetralogy of fallot in a Japanese macaque (Macaca fuscata). J Am Assoc Lab Anim Sci. 2007;46:66–7.
40. Liu DX, Gilbert MH, Kempf DJ, et al. Double-outlet right ventricle and double septal defects in a Rhesus macaque (Macaca mulatta). J Vet Diagn Investig. 2012;24:188–91.
41. Brandt DJ, Canfield DR, Peterson PE, et al. Persistent truncus arteriosus in a rhesus monkey (Macaca mulatta). Comp Med. 2002;52:269–72.
42. Okabayashi S, Ohno C, Kato M, et al. Congenital cystic adenomatoid-like malformation in a cynomolgus monkey (Macaca fascicularis). Vet Pathol. 2008;45:232–5.
43. Lewis RW, Palazzo MC, Kim JC. Persistent cloaca in a cynomolgus monkey (Macaca fascicularis). J Med Primatol. 1978;7:237–41.
44. Kirejczyk S, Pinelli C, Gonzalez O, et al. Urogenital lesions in nonhuman primates at 2 national primate research centers. Vet Pathol. 2021;58:147–60.

45. Obert LA, Suttie A, Abdi M, et al. Congenital unilateral renal aplasia in a cynomolgus monkey (Macaca fascicularis) with investigation into potential pathogenesis. Toxicol Pathol. 2020;48:766–83.
46. Chalifoux LV. Renal ectopia, squirrel monkey and owl monkey. In: Jones TC, Mohr U, Hunt RD, editors. Nonhuman primates II. Berlin: Springer; 1993. p. 147–50.
47. Baskin GB, Roberts JA, McAfee RD. Infantile polycystic renal disease in a rhesus monkey (Macaca mulatta). Lab Anim Sci. 1981;31:181–3.
48. Kessler MJ, Roberts JA, London WT. Adult polycystic kidney disease in a rhesus monkey (Macaca mulatta). J Med Primatol. 1984;13:147–52.
49. Sakakibara I, Honjo S. Spontaneously occurring congenital polycystic kidney in a cynomolgus monkey (Macaca fascicularis). J Med Primatol. 1990;19:501–6.
50. Perminov E, Mangosing S, Confer A, et al. A case report of ovotesticular disorder of sex development (OT-DSD) in a baboon (Papio spp.) and a brief review of the non-human primate literature. J Med Primatol. 2018;47:192–7.
51. Olson LC, Palotay JL, Haines JE, et al. Compensated, goitrous hypothyroidism in rhesus macaques. Lab Anim Sci. 1985;35:629–34.
52. Vallender EJ, Hotchkiss CE, Lewis AD, et al. Nonhuman primate genetic models for the study of rare diseases. Orphanet J Rare Dis. 2023;18:20.
53. Sato A, Abe K, Yuzuriha M, et al. A novel mutation in the thyroglobulin gene that causes goiter and dwarfism in Wistar Hannover GALAS rats. Mutat Res. 2014;762:17–23.
54. Hatakeyama H, Takei Y, Cruz Y, et al. Spontaneous vacuolar degeneration of the thyroid follicular epithelium in cynomolgus monkeys. J Toxicol Pathol. 2011;24:229–32.
55. Johnson AL, Peterson SM, Terry MML, et al. Spontaneous KRT5 gene mutation in rhesus macaques (Macaca mulatta): a novel nonhuman primate model of epidermolysis bullosa simplex. Vet Pathol. 2020;57:344–8.

Age-Related Pathology in Nonhuman Primates

6

Heather A. Simmons

Abstract

This chapter focuses on age-related pathologies in rhesus macaques (*Macaca mulatta*) and common marmosets (*Callithrix jacchus*), two nonhuman primate species often used as models for human disease and for longitudinal studies on aging. Rhesus macaques are well-established models for age-related conditions including hypertension, diabetes, visual accommodation, amyloidosis, osteopenia, osteoporosis, sarcopenia, and frailty. Investigations of age-associated pathologies in the common marmoset include amyloidosis, diabetes, chronic renal disease, osteopenia, vision, and cognitive decline.

Keywords

Macaca mulatta · *Callithrix jacchus* · Common marmoset · Rhesus macaque · Aging · Frailty · Arthritis · Sarcopenia · Diabetes · Amyloid · Diverticulosis · Osteoarthritis · Arteriolosclerosis · Glomerulopathy · Chronic lymphocytic enteritis · Colitis · Cataract

Rhesus macaques are an excellent and well-known model species for aging with a life span of approximately 40 years. Common marmosets have a shorter life span of approximately 16 years and are often considered easier to house and handle due to their smaller body size and the lack of serious zoonotic viruses such as macacine herpesvirus 1. This chapter focuses on the gross and histologic characteristics of the age-related pathologies of rhesus macaques (*Macaca mulatta*) and common marmosets (*Callithrix jacchus*), two nonhuman primate species often used as models for human disease and longitudinal studies on aging.

Frailty is a clinical health assessment which often serves as a 3-year predictor of the incidence of adverse health-related outcomes such as disability or death in humans [1]. In the USA it has been shown that frailty increases with age, affecting 7% of people ≥65 years old and 25–40% of those greater than 80 years of age [2]. The phenotype of frailty is defined as the presence of three or more of the following core elements: poor endurance, low levels of physical activity, tiredness, slow gait speed, weakness, and weight loss [1, 2]. A frail phenotype has been defined in the rhesus macaque [3]. Although frailty is a clinical assessment, the physical pathologies associated with this diagnosis will be described for both the rhesus macaque and common marmoset in this chapter.

H. A. Simmons (✉)
Pathology Services Unit, Wisconsin National Primate Research Center, University of Wisconsin-Madison, Madison, WI, USA
e-mail: hasimmons@wisc.edu

6.1 Musculoskeletal

The presence of apps recording gait and stability on cell phones is an indication of ever-increasing recognition of the serious consequences of diminished mobility due to sarcopenia and arthritis leading to falls that result in osteopenia and osteoporosis associated injuries.

6.1.1 Osteopenia and Osteoporosis

Humans, rhesus macaques, and common marmosets all have increasing bone mineral content between birth and skeletal maturity when there is epiphyseal closure and peak bone mass at the ages of 12–19 years, 7–10 years, and ~1 year, respectively [4–8]. There is subsequent bone loss with advancing age in both sexes of all three species [5, 6, 8]. Osteopenia and osteoporosis are most evident in both sexes of rhesus macaques during gross necropsy when evaluating bone marrow or removing the brain from the calvarium. Bones are noted to be subjectively brittle, less dense, and more easily cut or incised. The female rhesus macaque has been used as a model for pre-, peri-, and post-menopausal osteoporosis due to the marked similarities in

I. Kondova - Perseng et al. (eds.), *Atlas of Diagnostic Pathology in Nonhuman Primates*,
https://doi.org/10.1007/978-3-031-41280-6_6

skeletal response to hormonal changes (diminished estrogen) during menopause [6]. The female marmoset has an estrus cycle rather than a menstrual cycle and thus does not undergo menopause although there is reproductive senescence with diminished ovarian cycles and greater interbirth periods. Interestingly, female marmosets do not develop osteopenia when ovariectomized nor when they have chronic low levels of estrogen, progesterone, and chorionic gonadotropin due to species specific long-term socially induced reproductive suppression [9].

Death caused by the complications of hip fractures is 2–3 times higher in men than women [10]. The orchiectomized male marmoset is a very important model for osteopenia caused by hypogonadism of aging [11]. Young and aged male marmosets lose 5–20% of cancellous density in the tibial metaphysis 6–12 months after castration, while loss of lumbar vertebral density was only observed in the aged marmosets after castration [11, 12].

6.1.2 Osteoarthritis

Primary osteoarthritis otherwise known as degenerative joint disease is strongly associated with increasing age and mechanical wear especially in weight-bearing joints. Primary osteoarthritis consists of articular cartilage degeneration, eburnation of articular bone, and osteophyte formation [13]. Age-related osteoarthritis is documented in both the rhesus macaque and common marmoset and may affect any joint of the axial and abaxial skeleton [5, 6, 14–16]. The most commonly affected joints in the macaque are the stifles, coxofemoral joints, and intervertebral joints of the thoracic spine (Fig. 6.1) [17]. Vertebral osteoarthritis in rhesus macaques and humans is very similar with cartilage degradation, concentric and radial tears in degenerating intervertebral discs, and osteophyte formation on articular facets and vertebral bodies [14]. Aged rhesus macaques can develop spectacular thoracic kyphosis with bridging spondylosis on the anterior aspects of vertebrae due to osteophytosis (Fig. 6.2) [14, 17]. Scoliosis of the vertebral column may occur independently or concurrently with kyphosis in the rhesus with a marked increase in the prevalence of both conditions in aged/geriatric animals >20 years of age (Fig. 6.3) [6, 17].

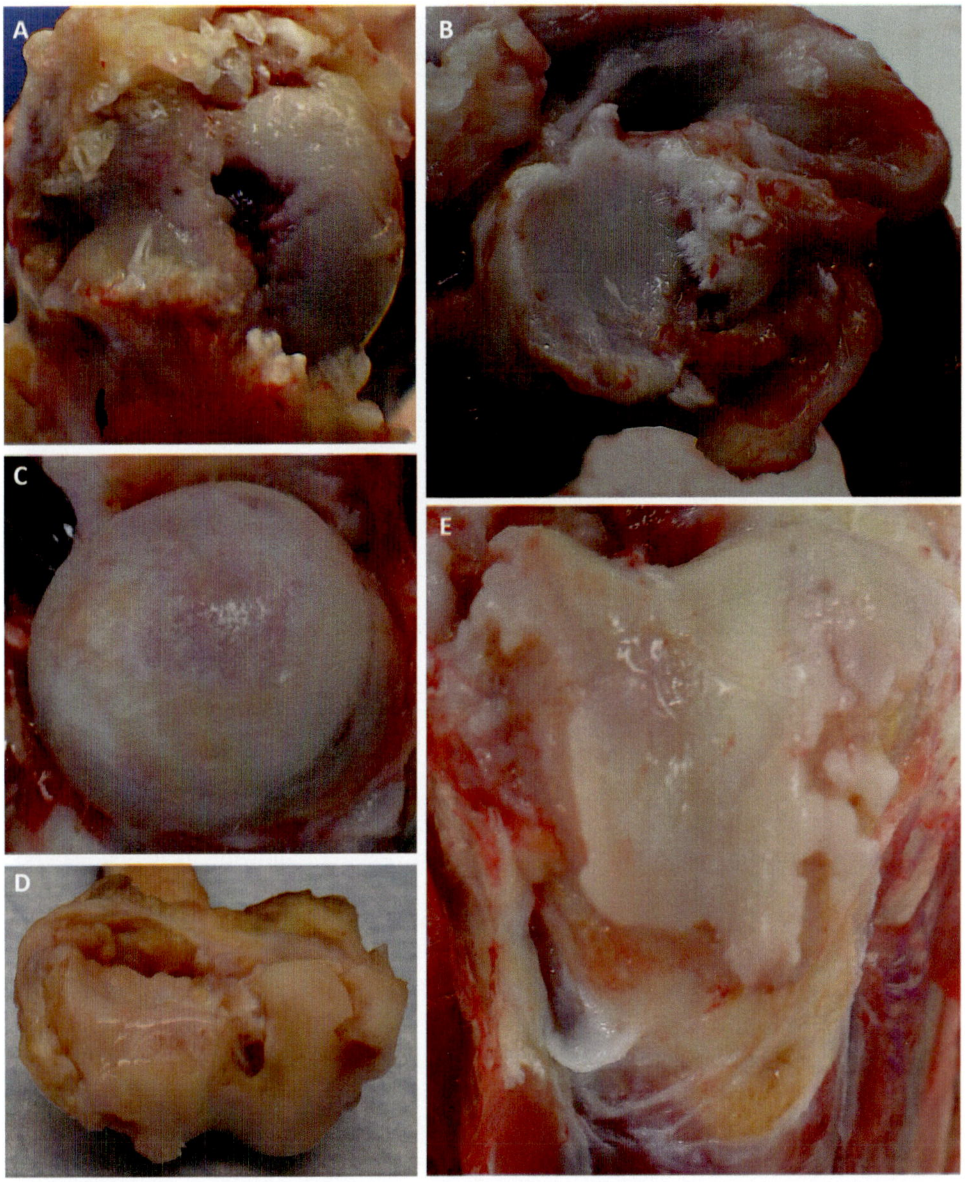

Fig. 6.1 Osteoarthritis in rhesus macaques. (**a**) Osteoarthritis of the proximal humeral head with articular erosion, hemorrhage, osteophyte formation, and nodular thickening of the synovial membranes. (**b**) Scapular glenoid cavity with irregular osteophytes and fibrous thickening of synovial membranes. (**c**) Femoral head eburnation of the articular cartilage. (**d**) Eburnation and of the proximal tibial condyles. (**e**) Erosion, eburnation, and extensive osteophyte formation on the proximal tibia. Images (**c**, **e**) courtesy of Dr. Andres Mejia

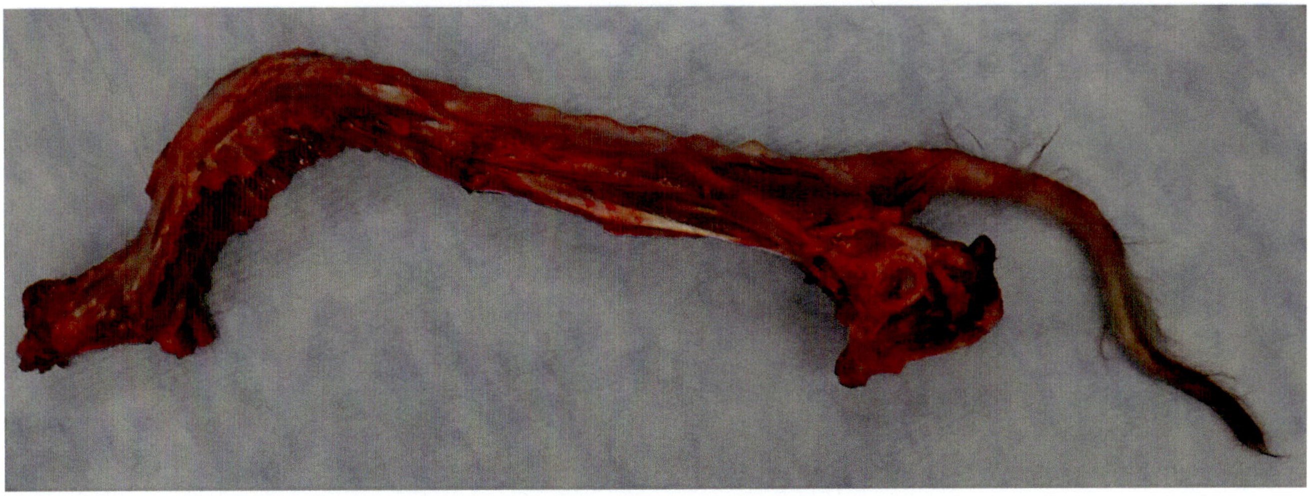

Fig. 6.2 Severe thoracic kyphosis in an aged rhesus macaque. Image courtesy of Abigail Bradford

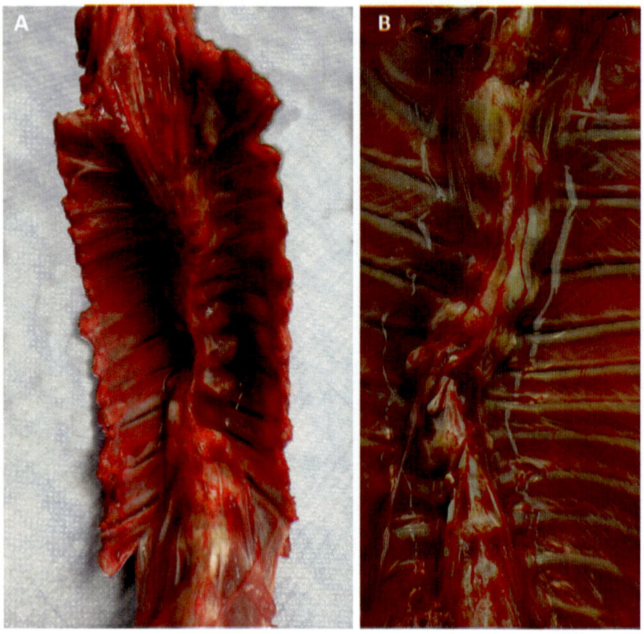

Fig. 6.3 Vertebral osteoarthritis, rhesus macaque. (**a**) Severe kyphosis and scoliosis and ankylosing vertebral spondylosis of the thoracic spine. (**b**) Severe scoliosis with ankylosing vertebral spondylosis characterized by nodular fusion of osteophytes bridging intervertebral joints of the thoracic spine

6.1.3 Sarcopenia

Age-related decrease in muscle mass, sarcopenia, begins in middle-aged humans (50+ years old) and macaques (14–16 years old), and becomes more pronounced over time [12, 18, 19]. The most significant clinical period is in the eighth decade in humas with a 50% loss of leg muscle area in humans and at about 23 years of age in the rhesus with >20% loss of leg muscle are in the legs (Fig. 6.4) [12, 18]. The marmoset differs significantly with humans and macaques by showing increased lean body mass and loss of fat mass with increasing age [20]. This is evident during gross examination with no discernable difference in the quadriceps of older and aged marmosets compared to younger animals. The loss of fat mass in the marmoset is reflected in the omentum, mesentery, and peri-renal adipose. Marmosets have minimal subcutaneous adipose in contrast to humans and macaques, both of which often develop significant subcutaneous adipose in the abdominal regions and within the abdominal cavity with increased body condition scores and obesity.

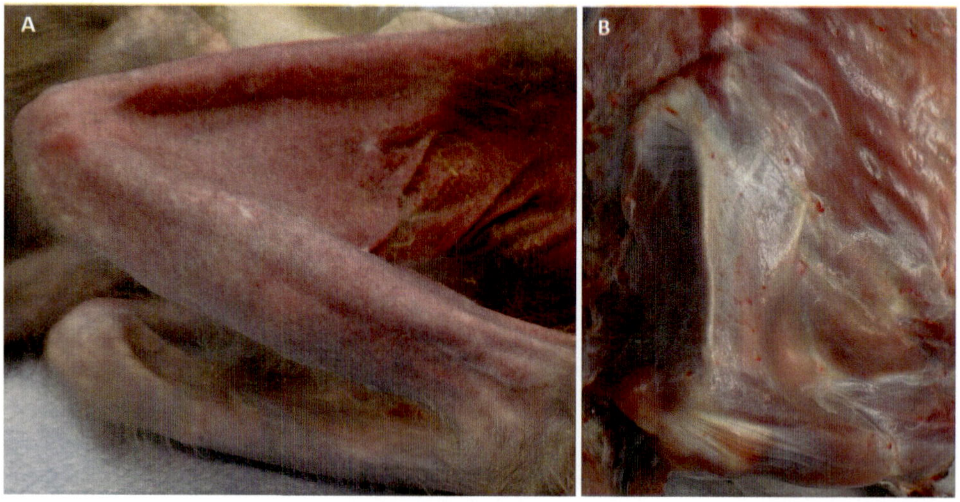

Fig. 6.4 Sarcopenia in an aged rhesus macaque. (**a**) Severe diffuse sarcopenia of the gluteus maximus, biceps femoris, semitendinosus, semimembranosus, and gastrocnemius and anterior tibialis with increased prominence of the underlying femur and tibias. (**b**) Severe diffuse sarcopenia of the deltoid, triceps, supraspinatus, and infraspinatus muscles overlying the scapula

6.2 Cardiovascular

Cardiovascular disease increases with age and is the leading cause of death in humans and great apes, as well as a significant cause of morbidity and mortality in rhesus macaques and common marmosets [17, 21–24]. Common age-associated cardiac changes in rhesus macaques and marmosets include valvular endocardiosis, valvular mineralization, interstitial fibrosis with degeneration and hypertrophy of cardiomyocytes, and myocardial mineralization [15, 25–27]. It should be noted that myocardial mineralization has been noted as an incidental finding in marmosets of all ages, including infants, suggesting that cellular degeneration associated with aging may not be the only mechanism for this change [27].

6.2.1 Valvular Endocardiosis

Valvular endocardiosis is most often noted in the atrioventricular valves and is characterized by chronic nodular thickening of valve leaflets, acid mucopolysaccharide deposition, and in some cases mineralization (Fig. 6.5). This valvular change is occasionally noted at a younger age (~10 years in the rhesus macaque) than other age-associated conditions [17].

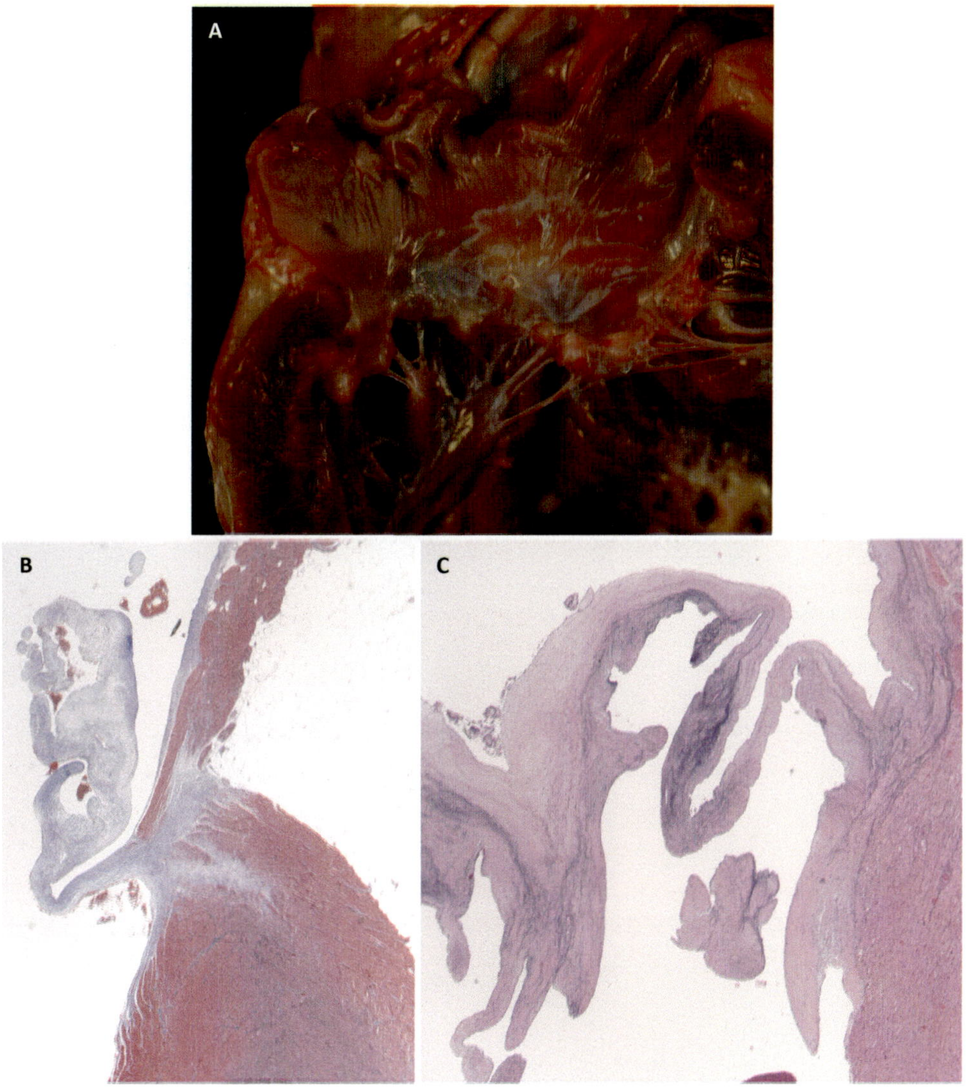

Fig. 6.5 Valvular endocardiosis in the rhesus macaque. (**a**) Mild smooth nodular thickening within the margins of the atrioventricular valve leaflets. (**b**) Photomicrograph of valvular endocardiosis with irregular to nodular thickening of the atrioventricular valve. Stain: Masson's trichrome. (**c**) Photomicrograph of an irregularly thickened and mineralized atrioventricular valve. Stain: HE

6.2.2 Degenerative and Fibrosing Cardiomyopathy

Idiopathic myocardial fibrosis otherwise described as degenerative and fibrosing cardiomyopathy is more prevalent and more extensive in older NHP and great apes with 56% of cases occurring in geriatric and aged rhesus macaques (≥20 years of age) and 80% of cases in geriatric and aged common marmosets (≥8 years of age) in the Wisconsin National Primate Research Center Colony [17, 23, 28]. Degenerative and fibrosing cardiomyopathy is characterized by significant myocardial interstitial fibrosis separating degenerative myocardiocytes that exhibit loss of cytoplasmic cross-striations, sarcoplasmic vacuolization, and variation in nuclear size. Affected regions of myocardium often have individually hypertrophic myocardiocytes and perinuclear accumulations of lipofuscin within cardiomyocytes (Fig. 6.6). MRI studies of common marmosets indicate that, like humans, there is a decrease in ventricular volume with increasing age that is accompanied by an increase in ventricular density, interstitial fibrosis, and no significant change in cardiac weight [24]. Interstitial fibrosis may occur in any region (ventricles, interventricular septum, and/or atria) of the NHP heart. There is a tendency in the marmoset to develop fibrosis in the right ventricular free wall, while fibrosis in the left ventricular free wall is more common in the macaque. There is documentation of more collagen in the right ventricular myocardium than in the myocardium of left ventricle in the marmoset [29].

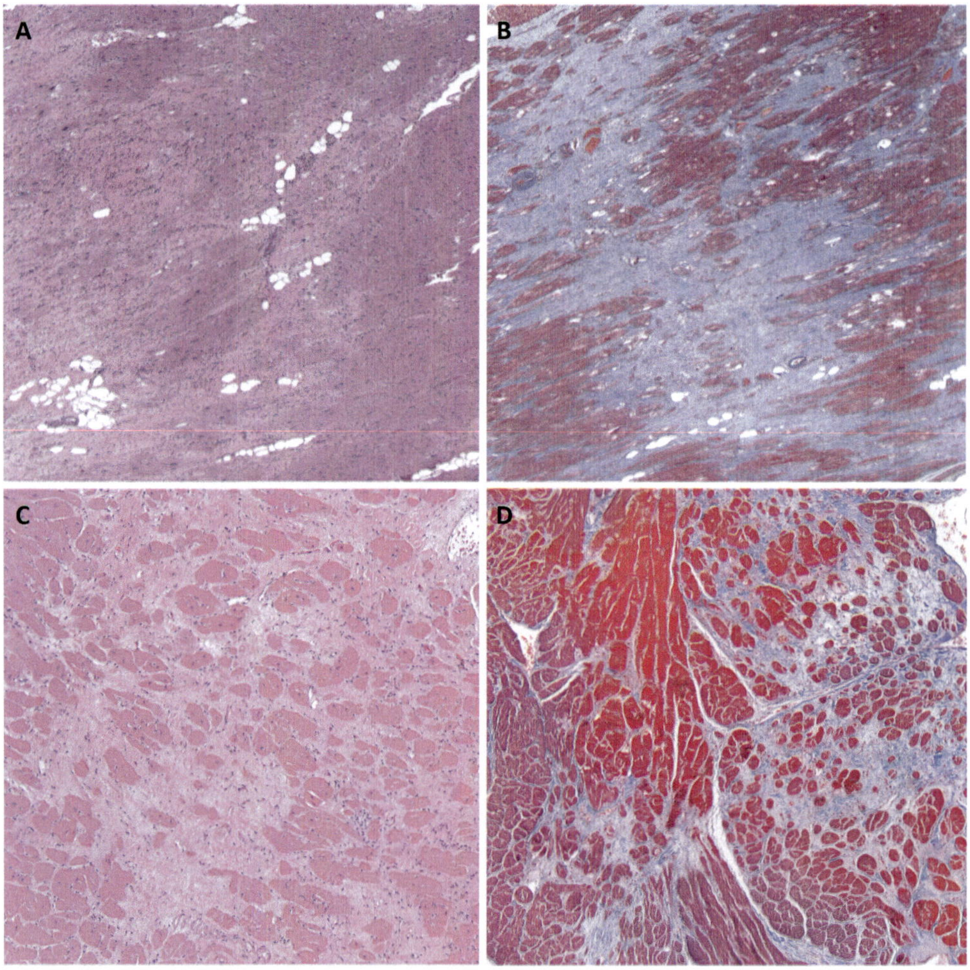

Fig. 6.6 Photomicrographs of degenerative and fibrosing cardiomyopathy. (**a**) Rhesus macaque, left ventricular myocardium with moderate to marked interstitial fibrosis extending between and replacing myocardiocytes with small aggregates of clear adipocytes. Hypertrophic cardiomyocytes have karyomegaly, anisocytosis, and occasionally multiple nuclei. Stain: HE. (**b**) Masson's trichrome stain of the same heart with blue staining of fibrotic regions. (**c**) Common marmoset, left ventricular myocardium with moderate multifocal interstitial fibrosis, myocardiocyte degeneration, and myocardiocyte loss. (**d**) Trichrome stain of the same heart emphasizing the extensive fibrosis (blue color) throughout the myocardium of the left ventricle and interventricular septum

6.2.3 Vascular Fibrosis and Mineralization

Structural changes in large blood vessels such as the aorta are associated with increasing age in humans and include fibrosis of the intima, fibrosis of the media, and calcification of large vascular walls with fragmentation of the intimal elastica [30]. These same changes are noted with increasing frequency in wild-caught and purpose-bred rhesus macaques as they age [15, 17, 31, 32]. Fibrosis of the aorta is characterized by irregular roughening of the intima with patchy pallor and may be associated with age associated valvular pathology (Fig. 6.7a) or along the entire length of the thoracic and abdominal aorta (Fig. 6.7b). Vascular mineralization in the aged marmoset and macaque is quite variable with prominent mineralization in grossly identified segments and very subtle mineralization only detected microscopically in sections collected for routine screening. (Fig. 6.8). These changes may be, but are not always, associated with chronic renal disease in primates. Calcification of the vascular media occurs years earlier in people with chronic kidney disease when compared to general population [33].

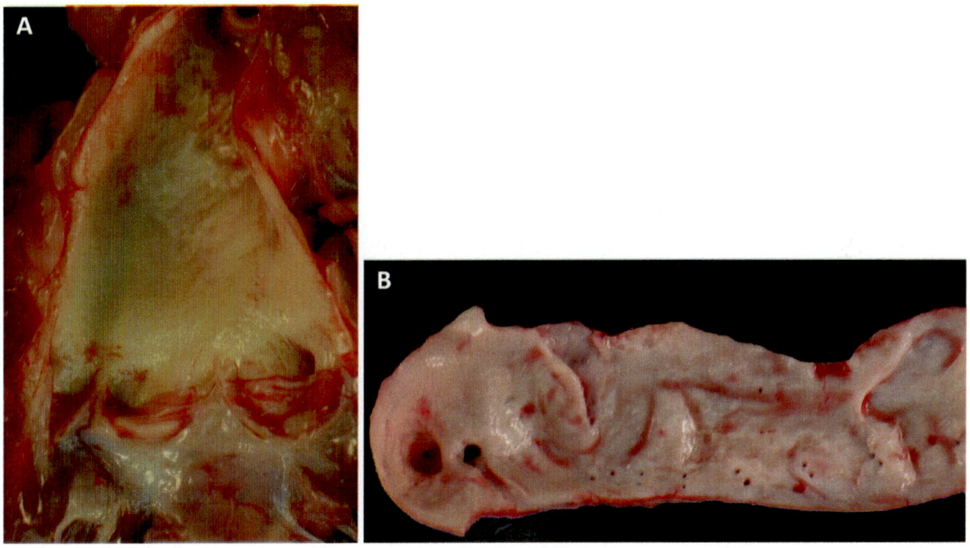

Fig. 6.7 Vascular fibrosis. (**a**) A 29-year-old rhesus macaque with subaortic valve stenosis leading to roughening and fibrosis of the aortic endothelium and intima. (**b**) A 38.6-year-old rhesus macaque with pallor and fibrosis of the endothelium and intima of the thoracic aorta

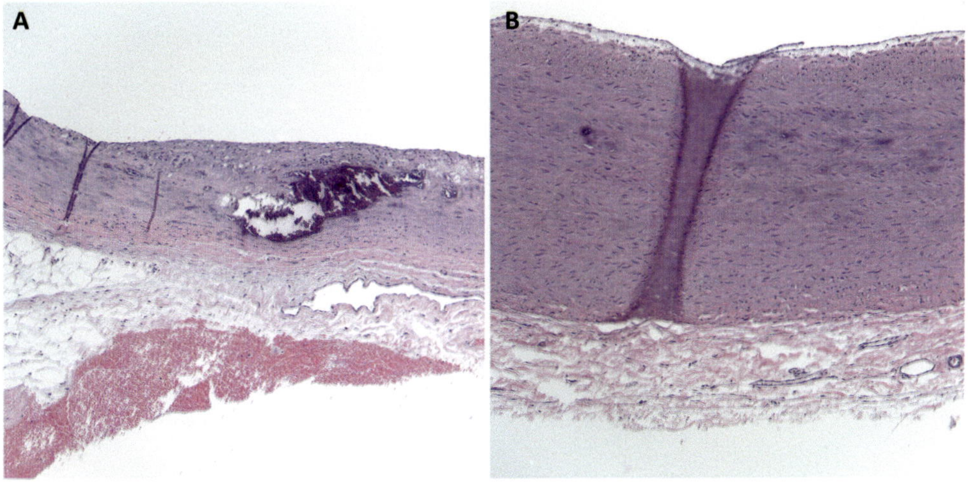

Fig. 6.8 Aortic mineralization. (**a**) Photomicrograph of focally severe mineralization (dark purple) of the aortic wall with vacuolization of cells within the tunica media in a common marmoset. (**b**) Photomicrograph of subtle dark purple mineralization of individual cells within the aortic tunica media of a rhesus macaque

6.2.4 Atherosclerosis and Myocardial Infarction

Free-ranging and laboratory-bred marmosets and macaques do not typically develop the atherosclerotic lesions in the coronary or great vessels, unless they are fed experimental high-fat or Western diets comparable to humans [26, 27, 31, 32, 34–41]. In the macaque, lesions initially develop in the abdominal aorta with subsequent involvement of the thoracic aorta followed by the coronary arteries making them a good experimental model for the human condition [36]. Atherosclerotic lesions in marmosets have been documented in the thoracic aorta, abdominal aorta, and the iliac arteries [41]. Affected vessels may vary from grossly unremarkable with intimal pallor to thickened rigid tubular structures with severe luminal narrowing (Fig. 6.9a, b). Microscopically atherosclerotic vessels may have focal expansion of the subintimal layer with amphophilic material, histiocytic subintimal inflammation, and multifocal mineralization with or without cholesterol clefting (Fig. 6.9c–f). Myocardial infarction (MI) is not a common finding in research colonies of macaques or marmosets. Macaques fed high-fat diets will develop MI and strokes as an experimental model, when fed a high-fat diet [36–38].

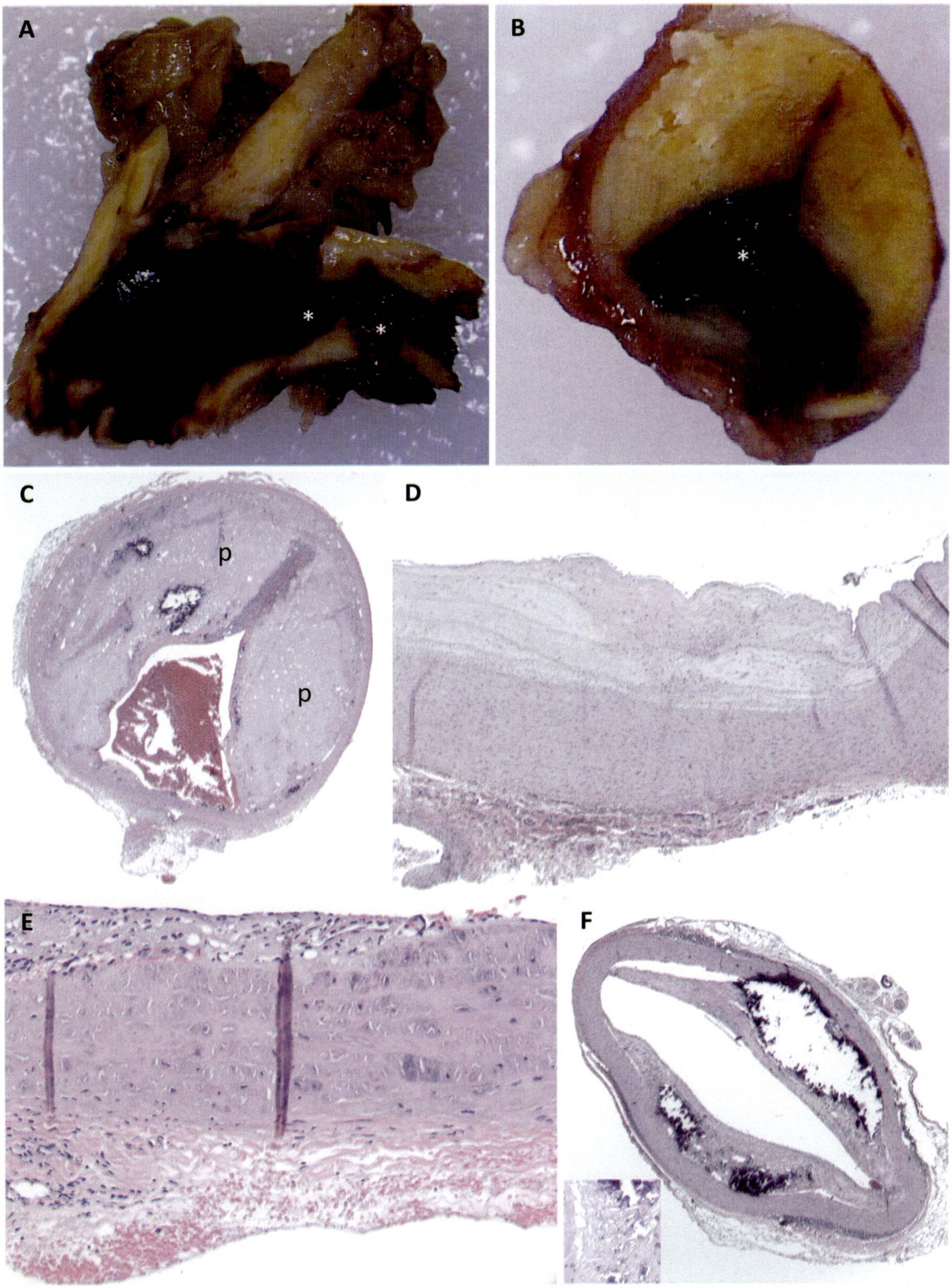

Fig. 6.9 Aortic atherosclerosis. (**a**) The aortic arch, left common carotid artery, and left subclavian artery of an 11-year-old cotton-top tamarin with atherosclerosis significant thickening of the vascular wall and severe luminal narrowing (*). (**b**) Cross-section of the distal thoracic aorta with significant bilateral irregular thickening of the vascular walls and luminal constriction (*). (**c**) Photomicrograph of B with bilateral subintimal plaque formation (p), histiocytic inflammation, and multiple dark purple foci of mineralization. (**d**) Photomicrograph of diffuse disruption and expansion of the subintimal layer of the aorta with abundant hyaline to slightly amphophilic material in a 29-year-old rhesus macaque. (**e**) Photomicrograph of focally extensive expansion of subintimal layer of the aorta with macrophages, lymphocytes, plasma cells, and eosinophilic amorphous material in a 13.3-year-old common marmoset. (**f**) Photomicrograph of a cross-section of the thoracic aorta of a 10.75-year-old common marmoset with histiocytic inflammation, marked multifocal mineralization, and cholesterol clefting (insert) within bilateral atherosclerotic plaques. Case material for (**a–c**) provided courtesy of Dr. Tracey L. Gilbert and the Northeastern Wisconsin Zoo

6.2.5 Arteriosclerosis and Hypertensive Vascular Changes

Arteriosclerosis and arteriolosclerosis both of which are described as vascular thickening with or without luminal narrowing in large and small arteries, respectively, are common spontaneous findings in rhesus macaques and increase with age, similar to humans [15, 31, 34]. The association between hypertension and diabetes, both age-associated and arteriosclerosis associated, is well defined in both the macaque and humans [26, 27, 34, 42, 43]. Rhesus macaques are good model for hypertension investigations and develop hyaline arteriolosclerosis (loss of structural detail and hyaline thickening of vascular walls) and hyperplastic arteriolosclerosis (concentric and/or eccentric lamellar fibrous and muscular thickening of vascular walls with variable luminal narrowing) (Fig. 6.10a) [30, 34, 40, 44, 45]. Although spontaneous hypertension is not well documented in the common marmoset, pulmonary and renal arteriopathy has been noted in several animals, one of which had concurrent hypertrophic cardiomyopathy suggesting pulmonary hypertension (Fig 6.10b) [46]. There is an excellent study of spontaneous hypertension with renal and cardiovascular sequelae in the Western pygmy marmoset (*Cebuella pygmaea*) [47]. Hyaline arteriolosclerosis is prevalent and more severe in diabetics than people without diabetes but is also diagnosed in elderly patients without hypertension or diabetes [34, 48].

6.3 Pulmonary Pathology

6.3.1 Bronchiolar Metaplasia

Aging and age-associated changes in structure and function are considered risk factors for developing chronic respiratory diseases in humans [49]. Alterations in pulmonary responses to injury in the aged lung may lead to aberrant repair with increased interstitial fibrosis and altered populations of alveolar epithelial cells. Epithelial cell senescence is theorized to be the cause of abnormal lung healing in aged individuals with progenitor alveolar epithelial type 2 cells failing to differentiate into type 1 pneumocytes in the lower lung during tissue repair [49]. Aged macaques often have scattered small regions, in the periphery of one or multiple lung lobes, characterized by increased interstitial fibrosis involving terminal bronchioles and adjacent peribronchiolar alveoli lined by low cuboidal bronchiolar epithelium (Fig. 6.11). These are usually considered incidental changes due to the lack of respiratory signs in macaques. In humans, this change is termed bronchiolization, bronchiolar metaplasia, or Lambertosis and may arise in association with small airway dysfunction and diseases such as constrictive bronchiolitis or hypersensitivity pneumonitis [50]. Atypical adenomatous hyperplasia and neoplasia should be ruled out through careful evaluation of cytologic features. Respiratory neoplasia is an uncommon diagnosis in both macaques and marmosets with only one published report of a pulmonary adenocarcinoma in a marmoset and a second case in the Wisconsin National Primate Research Center Colony (Fig. 6.12) [51, 52].

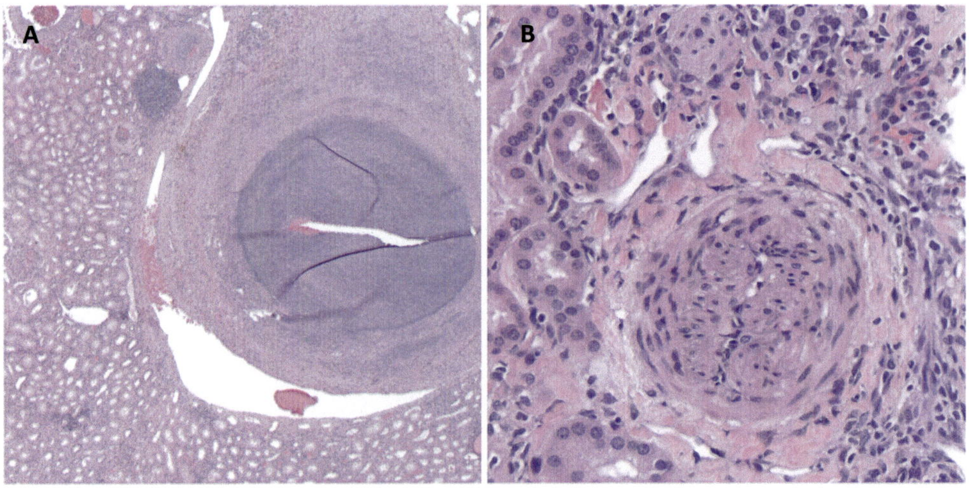

Fig. 6.10 Photomicrographs of hyperplastic arteriolosclerosis of renal arcuate arteries. (**a**) There is severe concentric lamellar fibrous and muscular thickening of the tunica media of a large artery with marked luminal constriction in a rhesus macaque. (**b**) A small renal artery in a common marmoset with moderate hypertrophy of the tunica media and tunica externa with severe luminal constriction

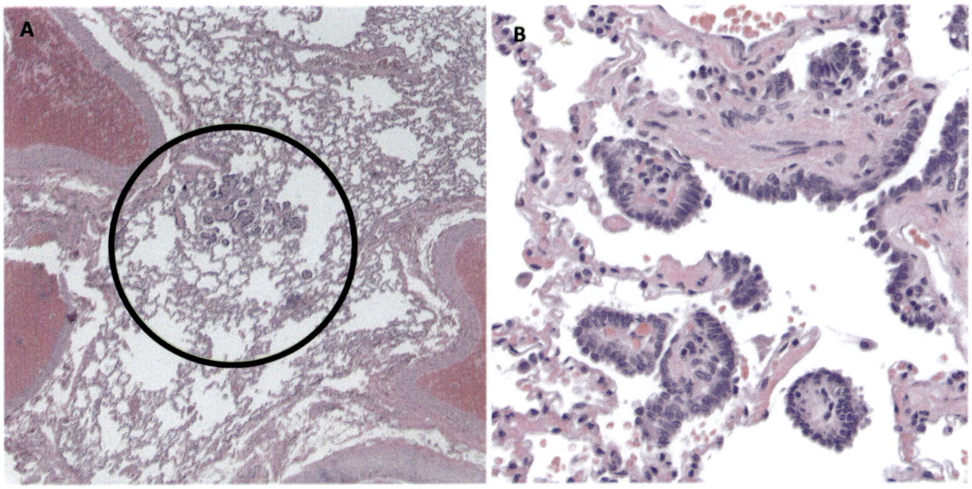

Fig. 6.11 Photomicrographs of bronchiolar metaplasia in a cynomolgus macaque. (**a**) The black circle encloses a region of terminal bronchioles with multiple foci of prominent basophilic epithelium. (**b**) Higher magnification emphasizes the low cuboidal bronchiolar epithelium lining terminal airways with and increased interstitial fibrosis

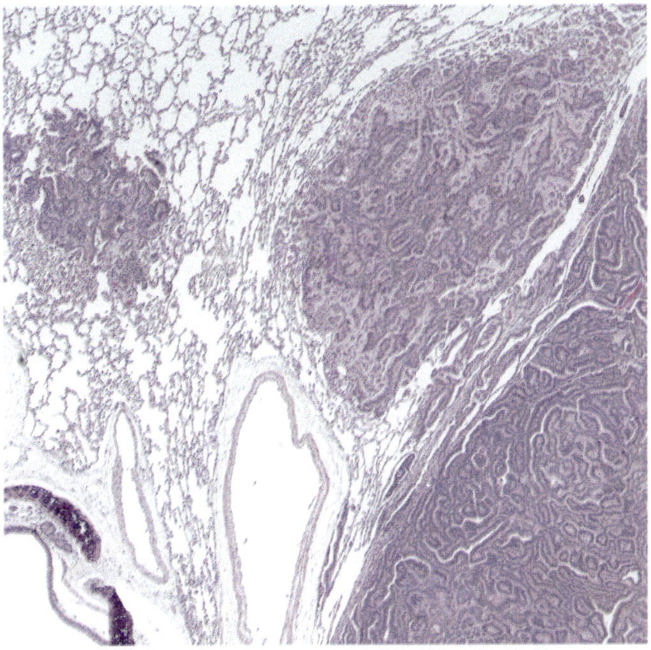

Fig. 6.12 Photomicrograph of a pulmonary adenocarcinoma expanding and replacing the pulmonary parenchyma. Cuboidal to columnar neoplastic cells are arranged in winding cords and papillary projections supported by a fine fibrovascular stroma. Case material courtesy of Dr. Puja Basu

6.3.2 Pleural Fibrosis, Mineralization, and Osseous Metaplasia

Geriatric and aged common marmosets occasionally have areas of pleural pallor on the dorsal surfaces of the lung lobes, most often involving the caudal lobes, that consist of fibrosis and osseous metaplasia (Fig. 6.13a–c) [16, 46]. Pleural fibrosis is noted in the geriatric macaque, but the presence of osseous metaplasia within these regions is not common. Mineralization and osseous metaplasia is commonly noted within cartilaginous tracheal and bronchiolar rings in both the marmoset and macaque (Fig. 6.14).

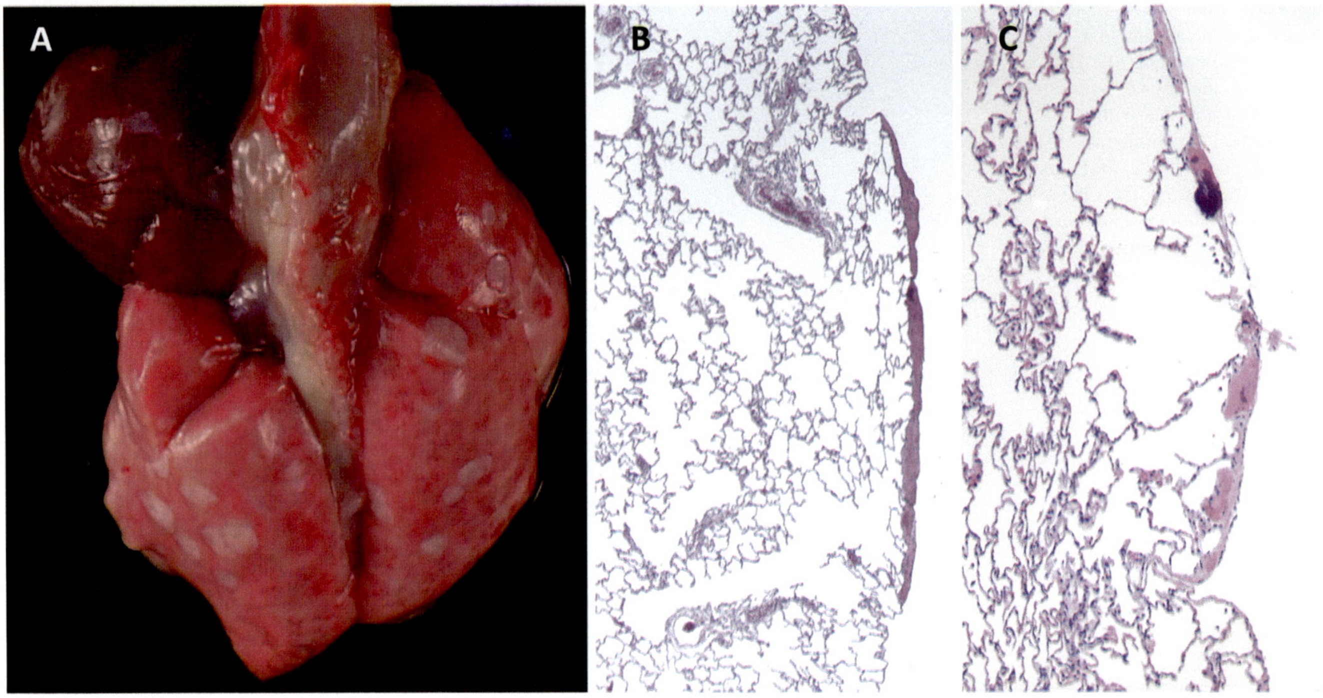

Fig. 6.13 (**a**) Multifocal plural fibrosis (round pale foci) in a 10-year-old common marmoset. (**b**) Photomicrograph of a lung section from (**a**) with a large focus of plural fibrosis. (**c**) Higher magnification of mild multifocal pleural fibrosis with focal mineralization (purple) from a 12-year-old common marmoset

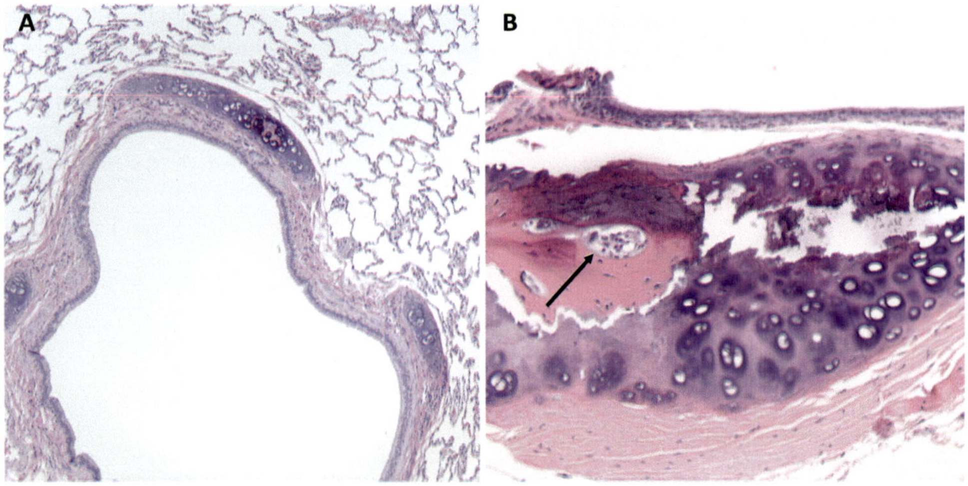

Fig. 6.14 Photomicrographs of osseous metaplasia and mineralization. (**a**) Marmoset lung with osseous metaplasia and mineralization of bronchiolar cartilage. (**b**) Photomicrograph of a common marmoset trachea with osseous metaplasia with scattered osteoclasts and bone marrow formation (arrow)

6.3.3 Alveoli, Pulmonary Emphysema, and Subpleural Bullae

Humans and rhesus macaques both show increases in the number of alveoli between birth and adulthood with the greatest increase in alveolar numbers between birth and 8 years and birth and ~1 year, respectively [53]. Pulmonary aging, especially in postmenopausal women and macaques, is associated with increased alveolar size and declining alveolar numbers [53, 54]. The diminution or loss of estrogen (due to ovariectomy or menopause) in rhesus macaques was shown to be associated with increased alveolar size (Fig. 6.15a, b) [53]. These changes may explain the increase prevalence of subpleural bullae and pleural emphysema in the periphery of lung lobes in geriatric and aged macaques (Fig. 6.15c, d) [15, 17]. Typically, these are noted as incidental findings during necropsy and are not associated with clinical respiratory signs [17]. These changes are not typically noted in the aged marmoset.

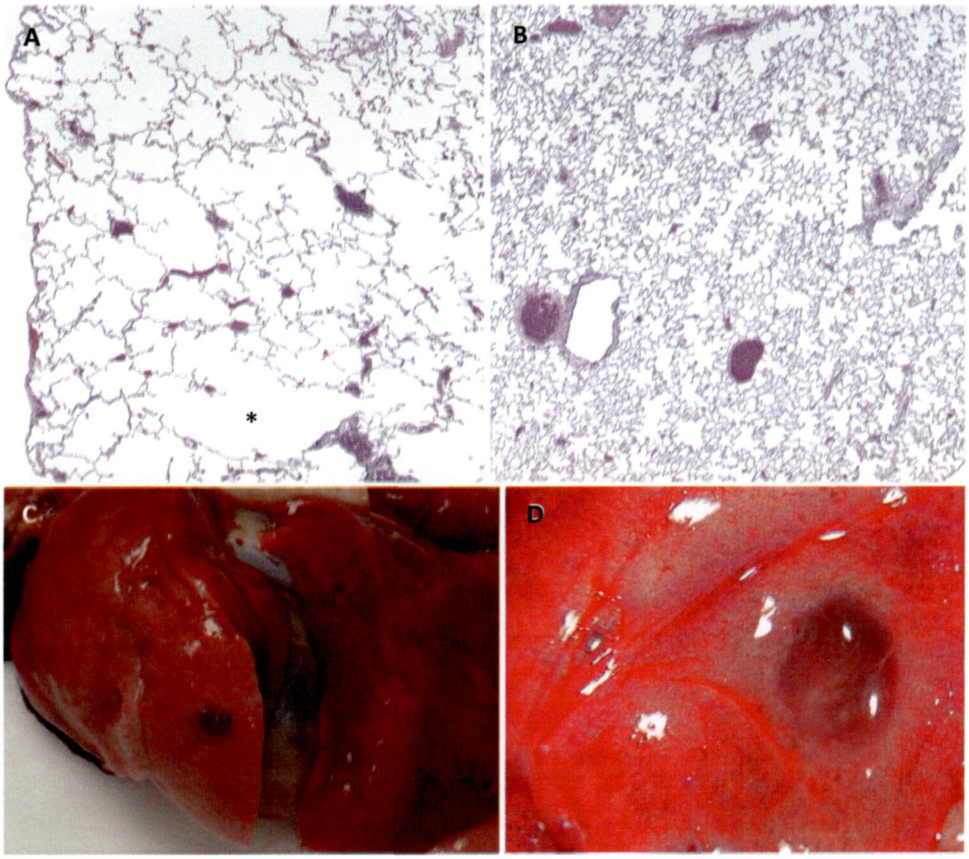

Fig. 6.15 Lung, rhesus macaque. (**a**) Photomicrograph of the lung in an aged rhesus macaque with increased alveolar diameter with microscopic bullous formation (*). (**b**) Photomicrograph of the lung of a young rhesus macaque at the same magnification showing typical alveolar diameters for comparison. (**c**, **d**) Gross images of clear rounded subpleural bullae

6.3.4 Pulmonary Siderophages (Hemosiderin-Laden Macrophages)

Diminished cardiac function, especially of the left ventricle, is associated with the development of pulmonary edema, increased numbers of alveolar macrophages and siderophages (hemosiderin-laden macrophages or "heart failure cells") within the pulmonary alveoli [17, 43]. Subclinical as well as clinically significant left-sided cardiac dysfunction are often confirmed by pathologists with the identification of alveolar siderophages in the macaque and marmoset (Fig. 6.16).

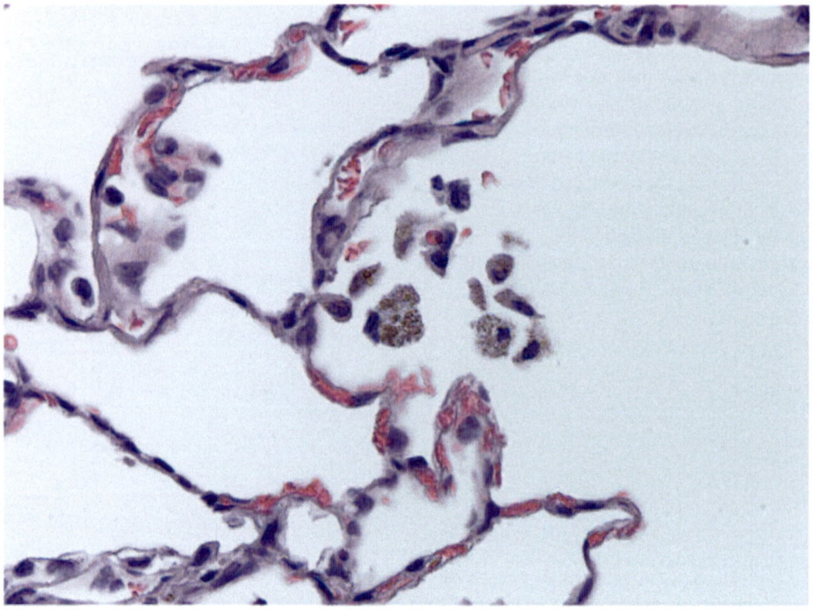

Fig. 6.16 Photomicrograph of pulmonary siderophages (heart failure cells) in the lung of a rhesus macaque characterized by brown-gold pigment within the cytoplasm of alveolar macrophages

6.4 Renal Pathology

6.4.1 Chronic Interstitial Nephritis

The aging phenotype in the kidney is considered to be multifocal nonsuppurative interstitial nephritis with tubulointerstitial fibrosis in most experimental models, including the rhesus macaques and common marmosets, with initial minimal to mild histologic changes noted at any age [42, 55, 56]. The incidence and severity of chronic renal interstitial inflammation with increased interstitial fibrosis (chronic interstitial nephritis) increases in both species with advancing age, as in humans [17, 56]. These cases have grossly pale, firm, kidneys with capsular contracture and often scattered cystic regions within the renal cortex (Fig. 6.17). Histologically there are variable numbers of lymphocytes and plasma cells within the rental interstitium with increased fibrous tissue surrounding and separating renal tubules and glomeruli (Fig. 6.18). Proximal renal tubular atrophy is common within areas of interstitial fibrosis with additional renal tubules exhibiting marked dilation (ectasia) with luminal eosinophilic material (protein). Renal tubules within areas of renal tubular atrophy, degeneration, and loss may have segmental tubular epithelial regeneration with hypertrophy and piling of tubular epithelial cells within lumina. Membranous thickening of Bowman's capsules surrounding glomeruli is also observed.

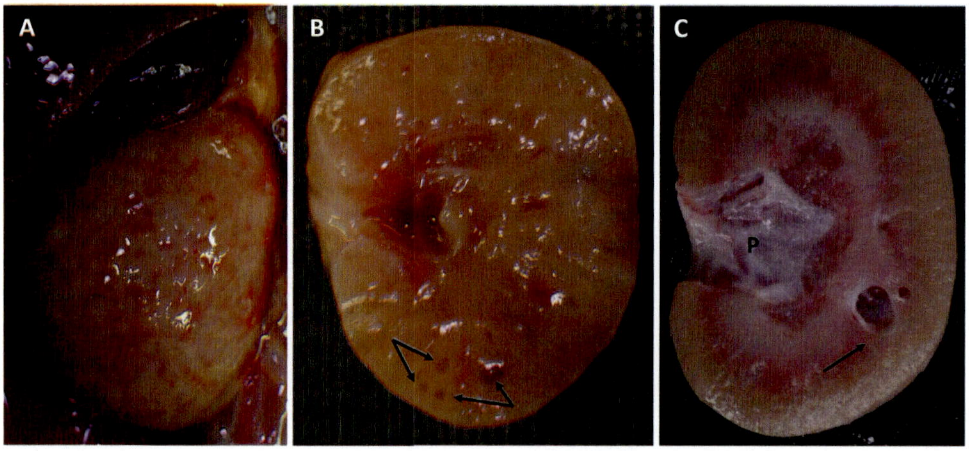

Fig. 6.17 Chronic age-associated renal changes. (**a**) Common marmoset with chronic end-stage kidneys which are pale and firm with roughened surfaces. (**b**) Longitudinal section of a marmoset kidney on cut surface with uniformly pale parenchyma and several small cysts (arrows) within the cortex. (**c**) Longitudinal section of a macaque kidney with a pale cortex, dilated pelvis (P), and large cortical cyst (arrow). Image (**c**) courtesy of Dr. Andres Mejia

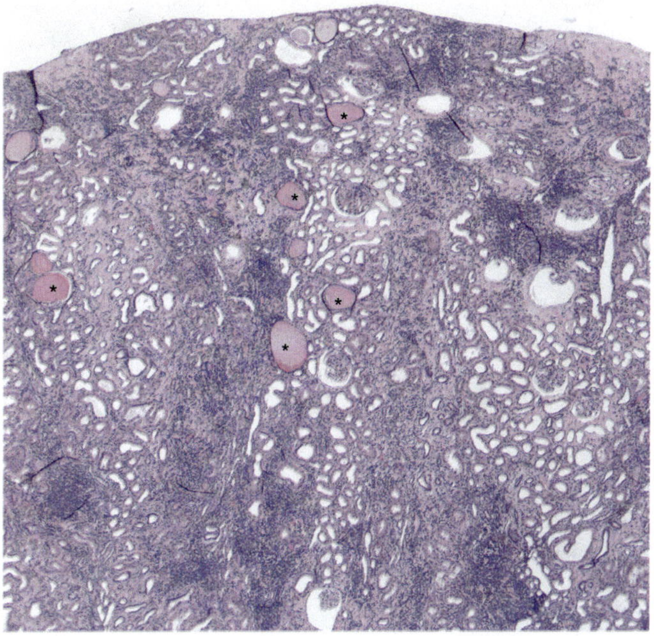

Fig. 6.18 Photomicrograph of chronic interstitial nephritis in a rhesus macaque characterized by dark purple interstitial cellular infiltration, and ectatic (*) proximal renal tubules. Note the capsular undulation of the renal surface due to interstitial fibrosis and tubular atrophy

6.4.2 Membranous Glomerulopathy

Glomerular basement membrane thickness increases with age and plateaus at the age of 18 years in rhesus macaques [57]. It has also been noted that there is glomerular hypertrophy in prediabetic rhesus macaques with hyperinsulinemia [57]. Other causes of thickened glomerular basement membranes are typically inflammatory with fibrin, immunoglobulins, and amyloid causing similar histologic changes (Fig. 6.19) [19]. The incidence of membranous glomerulopathy, membranous thickening of Bowman's capsules, multifocal segmental to global membranous thickening of glomerular capillary loops, synechiation, and glomerulosclerosis occur at a younger age in the rhesus and marmoset than most other age-related pathologies [17]. Glomerular senescence or glomerulosclerosis (severe global fibrosis with synechiation of the glomerulus) (Fig. 6.20) was noted in 64% of geriatric and aged rhesus macaques >20 years of age in one retrospective study [17]. When kidneys are objectively evaluated and scored, there is a significant age-related increase in incidence and severity of renal and more specifically glomerular pathologies [17, 56, 58].

Diabetic nephropathy affects a significant number of both type 2 and type 1 diabetics [57]. The rhesus macaque is a good species for both drug-induced and spontaneous type 2 diabetes [57, 59]. Diabetic nephropathy is histologically characterized by diffuse membranous glomerulopathy, diffuse mesangial sclerosis, and hyaline arteriolosclerosis of both afferent and efferent arterioles [57, 60].

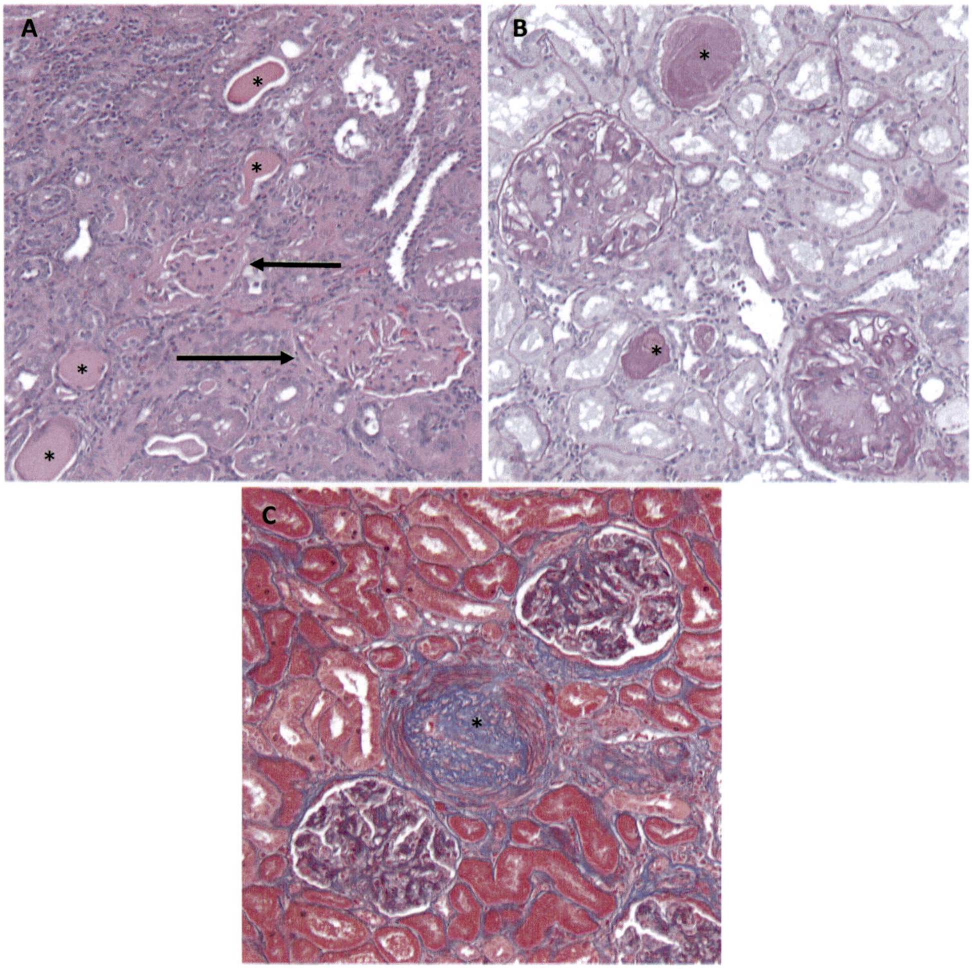

Fig. 6.19 Photomicrographs of age-associated membranous glomerulopathy. (**a**) A macaque kidney with marked membranous thickening of glomerular capillary loops (arrows) and dilated ectatic tubules with luminal protein (*). (**b**) Marmoset kidney sectioned at 3 microns and stained with PAS to emphasize membranous change within glomerular capillary loops. Note ectatic tubules with luminal protein (*). (**c**) Rhesus macaque kidney stained with Masson's trichrome to emphasize periglomerular and interstitial fibrosis as well as hypertrophy of the tunica media of a renal artery with severe luminal constriction (*)

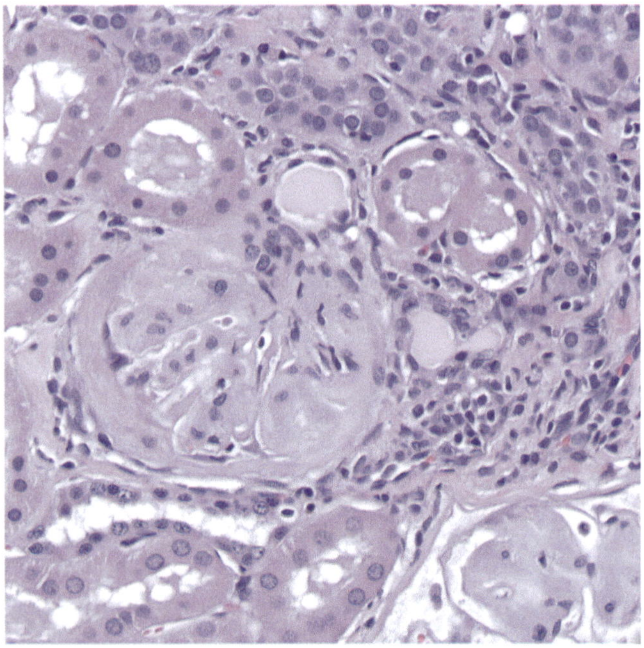

Fig. 6.20 Photomicrograph of membranous glomerulopathy with glomerulosclerosis in a common marmoset

6.5 Alimentary

6.5.1 Dental Disease

A common age-associated change in all species is dental wear, damage, and tooth loss due to mastication, hard-object feeding, foraging, play, and stereotypic behaviors such as gouging (in common marmosets) (Fig. 6.21). Periodontal disease does occur spontaneously in both the macaque and marmoset and can lead to tooth loss as in humans [16, 61]. There may be familial susceptibility to periodontitis in some macaque populations allowing them to serve as excellent models [61, 62].

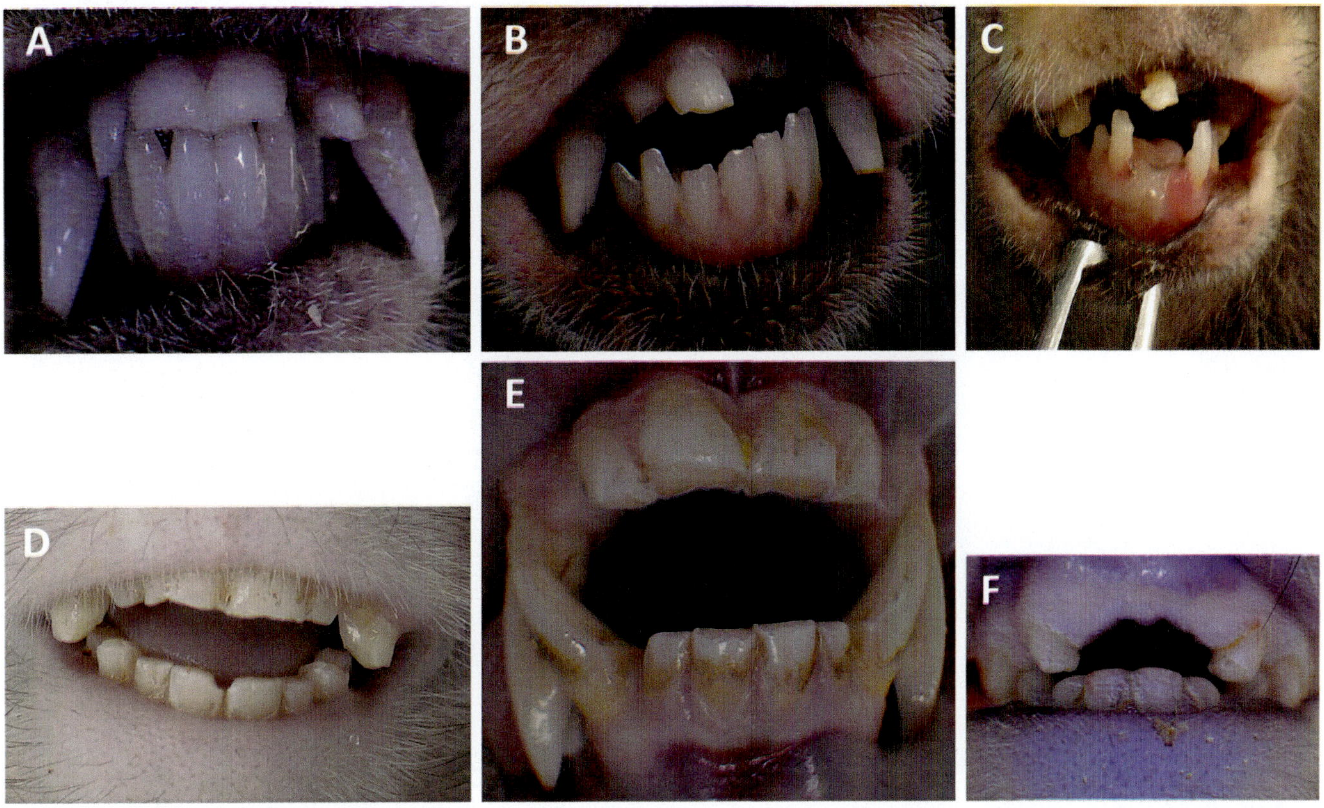

Fig. 6.21 Dentition in young and aged animals. (**a**) Young marmoset dentition. (**b**) Aged marmoset with broken canines and missing upper incisors. (**c**) Aged marmoset with severe tooth wear and loss. (**d**) Aged female macaque with severe tooth wear. (**e**) Aged male macaque with wear on the occlusive surfaces of upper and lower incisors, wear on the anterior surfaces of the upper canines, yellow plaque accumulation, and gingival recession. (**f**) Aged rhesus with occlusive surface tooth wear, loss of upper incisors, and loss of canines

6.5.2 Diverticulosis

Diverticulosis may occur at any level of the gastrointestinal tract in humans and macaques but is not noted in the common marmoset [25, 63, 64]. Diverticula may be focal in one region of the intestine or affect the entire length of the cecum and colon (Fig. 6.22a–b). The gross and histologic appearance are similar in the macaque and human and consist of outpouchings of the mucosa and submucosa through the muscularis which often cause fecal material to become trapped (Fig. 6.22c, d) [25, 63–65]. Entrapped fecal material can cause irritation and secondary diverticulitis. Young macaques and humans (<5 years and <30 years, respectively) have almost no diverticulosis [17, 65]. Diverticulosis increases with age to approximately 50% in people older than 60 years of age in industrialized societies eating "Western-type diets" [66, 67]. Similarly 18% of geriatric macaques and 42% of aged (25+ years) macaques, in the Wisconsin National Primate Research Center population, have diverticulosis of the colon or both the colon and cecum [17].

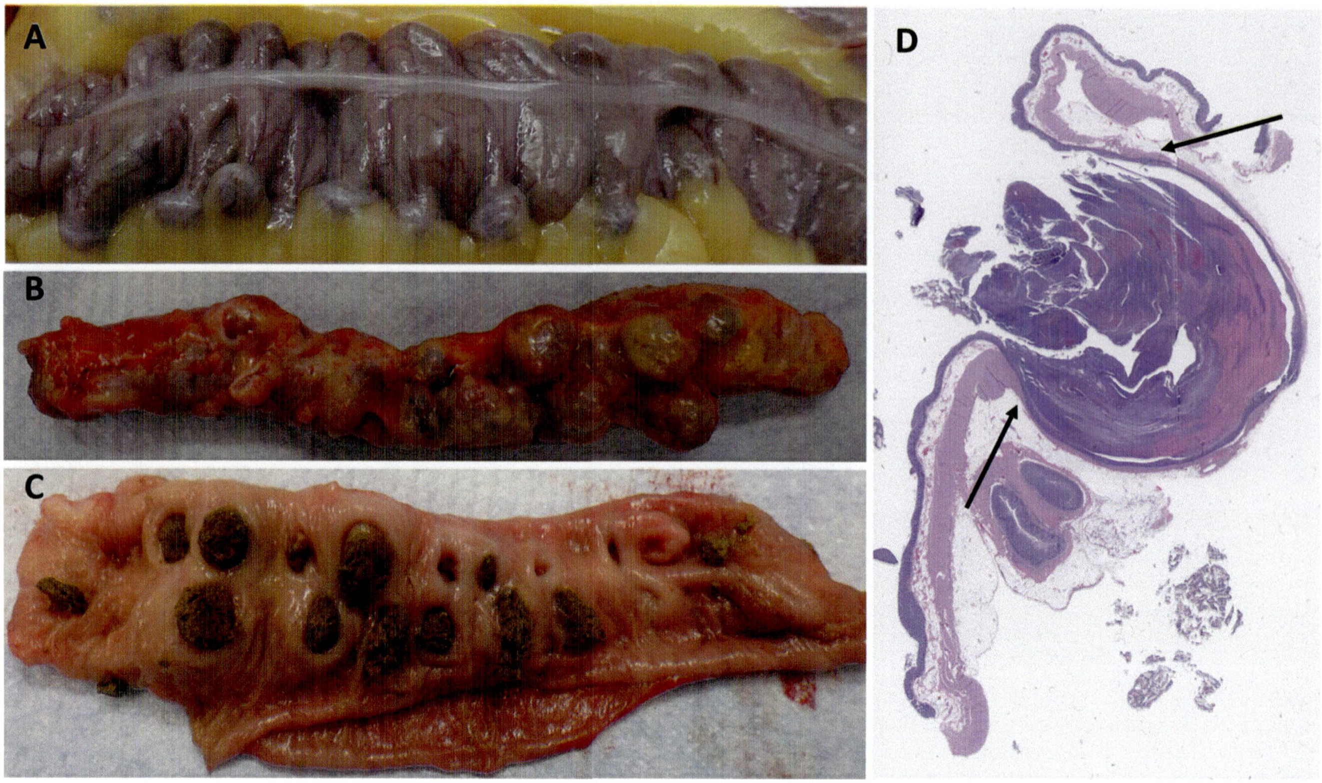

Fig. 6.22 Diverticulosis in rhesus macaques. (**a**) Extensive diverticulosis characterized by numerous outpouchings on either side of the colonic taenia. (**b**) Diverticula after the mesentery has been dissected from a section of the colon. (**c**) Mucosa with numerous diverticula and inspissated fecal material. (**d**) photomicrograph of one section of a colonic diverticula with mucosal outpouching and discontinuity of the inner and outer muscular layers (arrows)

6.5.3 Irritable Bowel Syndrome (IBS), Inflammatory Bowel Disease (IBD), Idiopathic Colitis and Diarrhea (ICD), Chronic Cicatrizing Ulcerative Colitis (CCUC), and Chronic Lymphocytic Enteritis (CLE)

Many enteric disease paradigms involve the recognition that there is a spectrum of mild to severe symptoms and histologic lesions with overlapping and shared characteristics between specific diagnoses. Irritable bowel syndrome (IBS) and inflammatory bowel disease (IBD) are considered by some clinicians to be opposite ends of the same disease spectrum [68]. IBD which includes both Crohn's disease (CD) and ulcerative colitis (CD) is increasingly associated with frailty in humans [69]. Although IBD is typically diagnosed in the third to fourth decade in humans, there is a second peak IBD incidence in the seventh decade [69]. Geriatric IBD patients are living longer irrespective of the age of diagnosis and often experience geriatric syndromes including osteoporosis, fractures, impaired neurocognitive and psychomotor functions, and increased frailty risk scores at earlier ages [69]. As with humans, enteric disease in the macaque manifests through a spectrum of clinical symptoms including recurrent to chronic diarrhea and progressive weight loss. Diarrhea refractory to medical intervention is a major cause of morbidity and mortality in macaque colonies [70]. Idiopathic colitis and diarrhea (ICD) in rhesus macaques also called idiopathic enterocolitis or idiopathic chronic diarrhea is typically noted in juveniles with characteristic dysbiosis and an absence of typical enteric bacterial, viral, and parasitic pathogens [71, 72]. Histologic findings include lymphoplasmacytic colitis with mucosal hypertrophy, diminished numbers of goblet cells, and loss of normal adherent brush border bacteria [71, 73]. Some cases have crypt abscesses and or microscopic colonic ulcers as well [72, 73]. Chronic colitis with intermittent to chronic diarrhea occurs in adult and geriatric macaques at a lower incidence than juveniles leading to euthanasia due to progressive weight loss characterized by sarcopenia, chronic dehydration, and frailty despite ongoing and/or repeated medical interventions and supplemental diets. Although not a distinctly described syndrome in older animals, these cases are grossly and histologically similar to (ICD). Gross findings include mild to marked distention of the colon with fluid to loosely formed feces (Fig. 6.23). Histologically there is prominent lymphoplasmacytic inflammation within the cecal and colonic lamina propria, increased numbers of lymphocytes within in the mucosal epithelium, loss of goblet cells in crypts, displacement and shortening of crypts within the lamina propria, and occasional dilated crypts with commensal protozoa such as *Balantidium* or *Trichomonads* within the lumina (Fig. 6.24).

Gut-associated lymphoid tissues are often hyperplastic and typical bacterial, and parasitic pathogens are not diagnosed. One distinct syndrome in the macaque is chronic cicatrizing ulcerative colitis (CCUC) which occurs in both juvenile and adult macaques and is thought to represent a postinfectious syndrome [73]. CCUC is characterized by discrete constricting annular ulcers in the cecum and proximal colon (Fig. 6.25) [73]. Histologic examination revels chronic ulcers with granulation tissue and superficial neutrophils and hyperplasia to dysplasia of the adjacent intact mucosal epithelium (Fig. 6.26) [73]. Secondary systemic amyloidosis is commonly associated with CCUC as well as other forms of chronic enteric disease in the adult and geriatric macaques [73, 74].

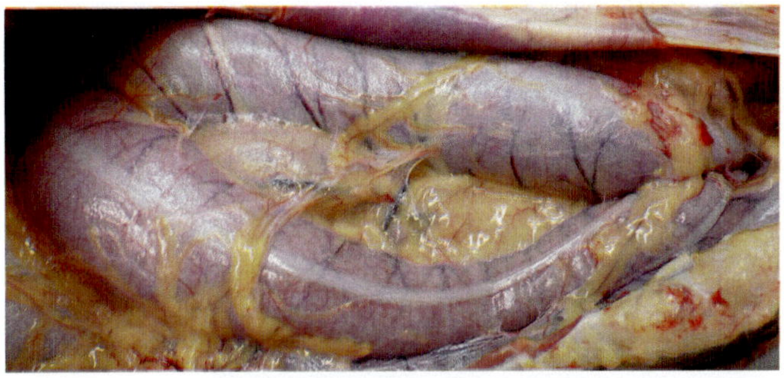

Fig. 6.23 A 17.9-year-old rhesus macaque with chronic colitis. The ascending, transverse, and descending colon are moderately to markedly distended with copious amounts of watery diarrhea. The duodenum, jejunum, and ileum, subjacent to the visible colon, are unaffected

Fig. 6.24 Photomicrographs of chronic colitis in rhesus macaques. (**a**) Severe chronic colitis with marked expansion of the lamina propria with lymphocytes and plasma cells with displacement and shortening of crypts. (**b**) Moderate chronic colitis with loss of clear goblet cells within the crypt epithelium and a crypt abscess (arrow). (**c**) Chronic colitis with prominent lymphoid hyperplasia within the lamina propria (arrows). (**d**) Colonic mucosa with chronic colitis with numerous Balantidium (arrows) within colonic crypts

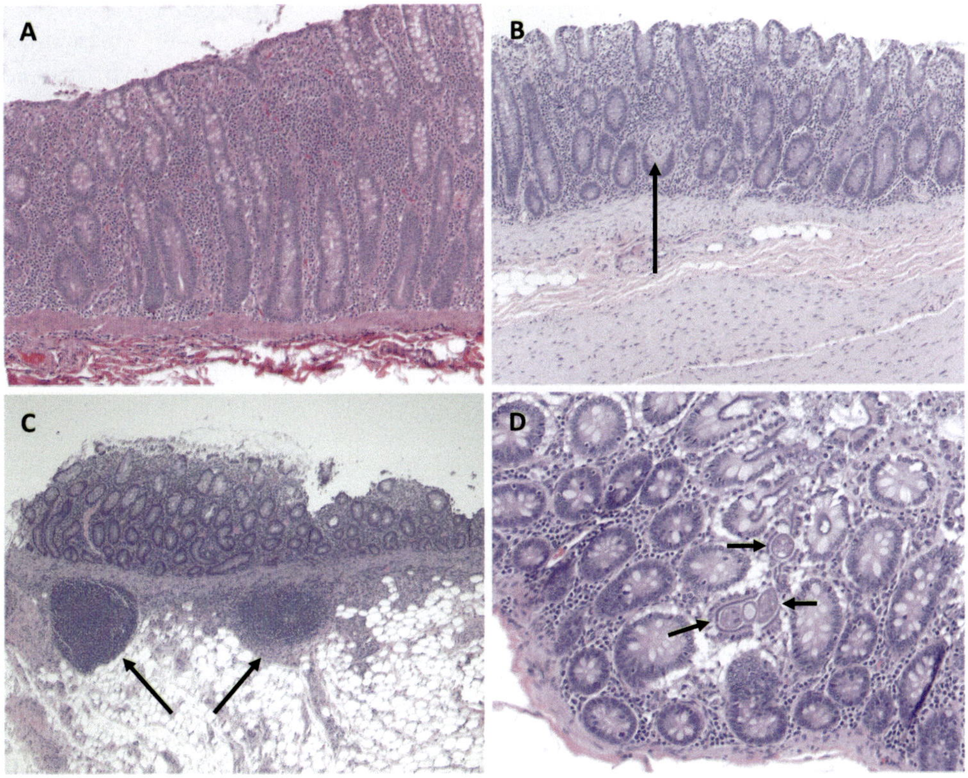

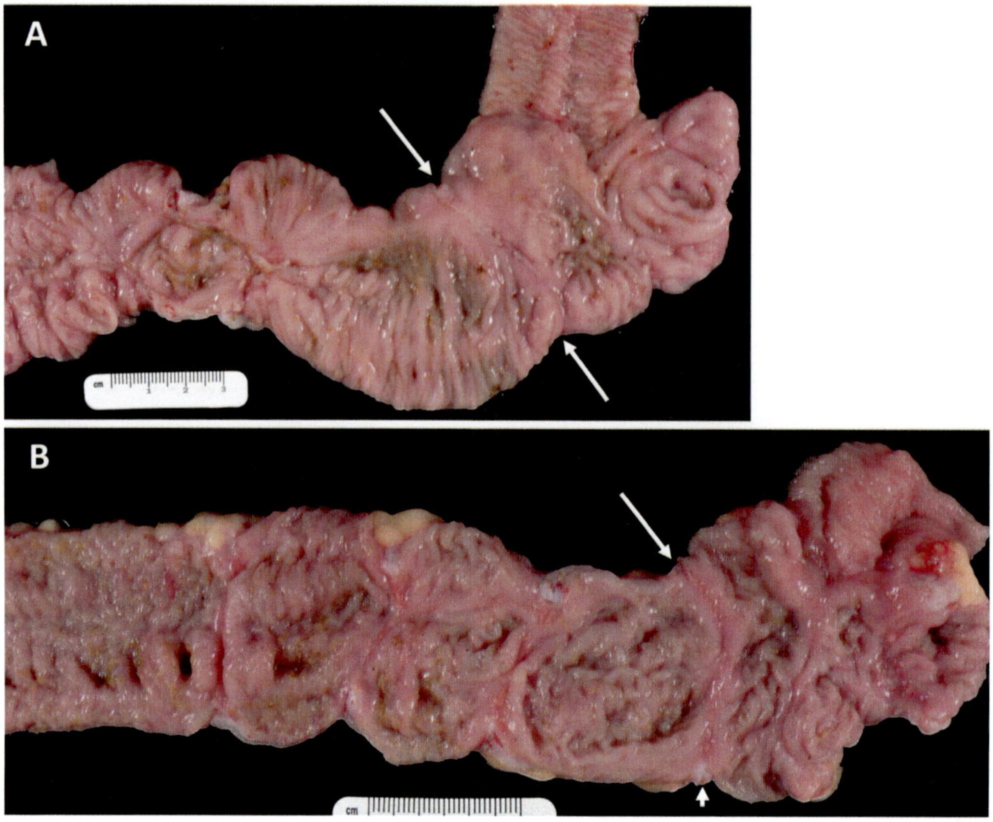

Fig. 6.25 Chronic cicatrizing ulcerative colitis. (**a**) A chronic circumferential ulcer (arrows) 2 cm distal to the ileocecal junction of a macaque. (**b**) An annular ulcer denoted by arrows in the colon of a macaque. Images courtesy of Anne Lewis

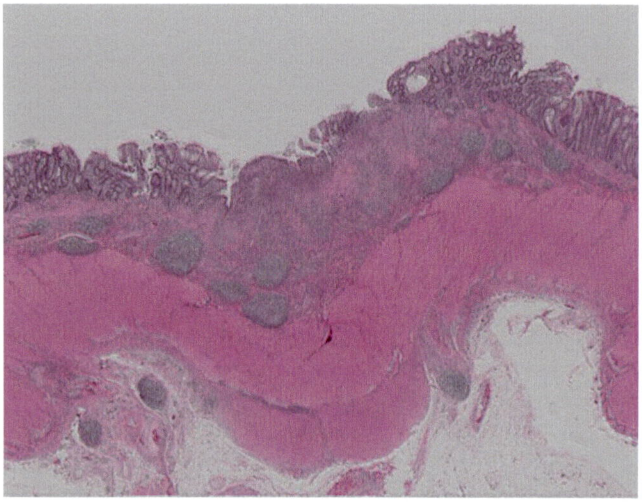

Fig. 6.26 Photomicrograph of chronic cicatrizing ulcerative colitis with chronic mucosal ulceration, granulation tissue, and partial re-epithelialization. There is mild dysplasia of the adjacent mucosa and abundant submucosal fibrosis with numerous lymphoid aggregates. Image and figure legend courtesy of Dr. Anne Lewis

Common marmosets may develop chronic lymphoplasmacytic colitis similar to macaques, but chronic lymphocytic enteritis (CLE) of the small intestine is more commonly diagnosed in this species. CLE is a T-cell-rich enteritis that increases in prevalence with age, although it can occur in animals as young as 2 years of age. Marmosets present with intermittent diarrhea, progressive weight loss, sarcopenia, loss of body condition, and increased frailty. Some colonies report alopecia in CLE cases although this is not a consistent finding in the author's experience [75, 76]. Marmosets with CLE may have hypoproteinemia, elevated liver enzymes, decreased serum calcium, decreased hemoglobin, and anemia with poor clinical outcomes most commonly associated with serum albumin levels of <3.5 g/dl and body weight reduced to <325 g [75–78]. Grossly the duodenum, jejunum, and ileum are segmentally or diffusely thickened with a pale mucosa (Fig. 6.27). The cecum and colon may be completely spared in some cases and moderately to severely involved in other cases. Histologically there is infiltration of the lamina propria with CD3-positive (T) lymphocytes, crypt hyperplasia, and in severe cases villous blunting and fusion. T lymphocytes are noted within the enteric mucosal epithelium and often penetrate into and through the muscularis mucosa (Fig. 6.28). The underlying etiology for this condition has not been definitively defined although associations with enteric parasites, dietary gluten, dysbiosis, stress, pro-inflammatory genes, and other factors have been investigated [75, 78, 79].

Fig. 6.27 Chronic lymphocytic enteritis in the jejunum of a common marmoset characterized by markedly thickened pale mucosa

6.6 Systemic and Metabolic Age-Associated Pathologies

6.6.1 Thymus

The thymus is quite large at the time of birth in most vertebrate species because it is the site of thymus-dependent (T) lymphocyte maturation, positive and negative selection, and export to tissues throughout the body (Fig. 6.29) [80, 81]. Age-related thymic atrophy or involution in humans and NHP is characterized by gross reduction in size with decreased numbers of thymocytes (thymic epithelial cells), loss of cortical and medullary organization within lobules, increased perivascular space, and adipose deposition [80, 81]. Involution in humans begins in childhood with effective cessation of thymic function after the fourth or

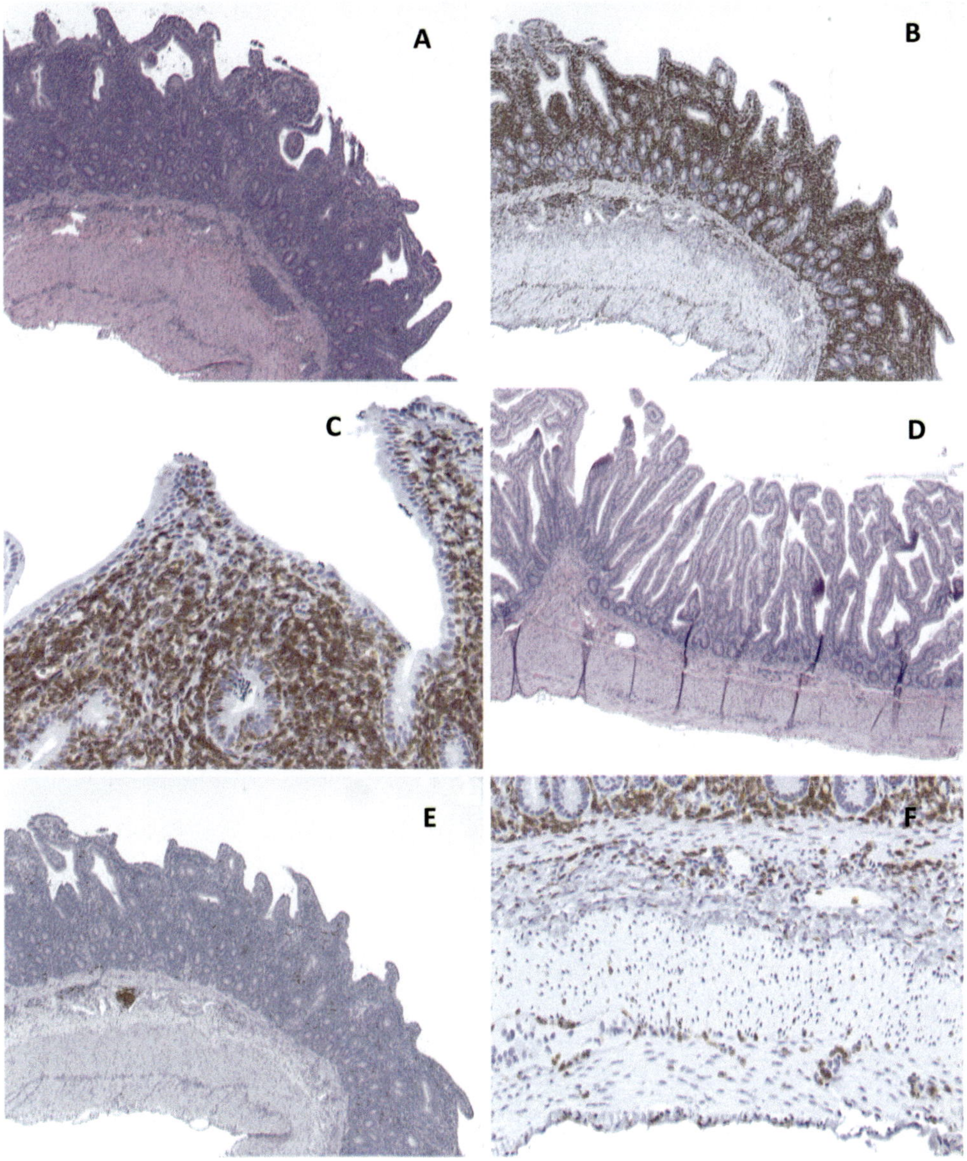

Fig. 6.28 Photomicrographs of chronic lymphocytic enteritis in a common marmoset. (**a**) A section of jejunum with marked lymphoplasmacytic expansion of the lamina propria, villous blunting, and villous fusion. Inflammatory infiltrates are noted within the submucosa at this magnification. (**b**) CD3 immunohistochemistry of the same section illustrating the predominance of T cells. (**c**) Higher magnification of the superficial mucosa demonstrating the infiltration of T cells into the mucosal and glandular epithelium. (**d**) Normal jejunum from a common marmoset for comparison to panel **a**. (**e**) CD20 immunohistochemistry demonstrating the paucity of B cells in the lamina propria. Note the normal distribution of B cells within the Peyer's patch. (**f**) CD3 illustrating the presence of T cells subjacent to jejunal glands, within the muscularis mucosa, submucosa, inner and outer muscular layers, and serosa of affected sections of jejunum

fifth decade [82]. Age-related thymic changes are similar in both macaques and marmosets with visible thymic tissue on the pericardium and within the thoracic mediastinum in infants and young animals and little thymic tissue noted in older and aged animals. The rate of thymic involution was found to be quite variable both for overall size and histologic characteristics in cynomolgus macaques involved in research [83]. Marmosets on study for ~7 years were reported to have no thymic tissue at necropsy [84]. There is documented persistence of residual thymic tissue in some older humans as well as a slower age-associated decline of naïve T cells within lymph nodes compared to other tissue compartments which is theorized to assist with maintenance of immune function during aging (Fig. 6.30) [82].

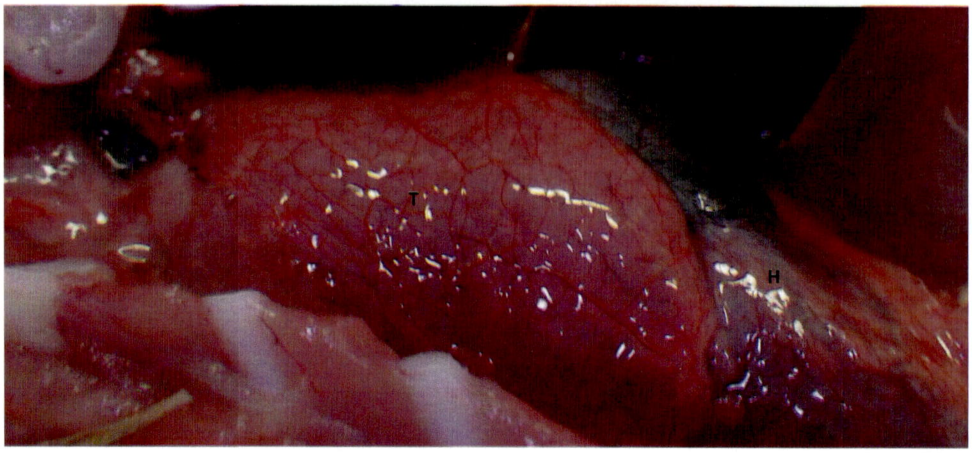

Fig. 6.29 The thymus (T) in an infant rhesus macaque extends from the thoracic inlet to cover the pericardium and majority of the heart (H)

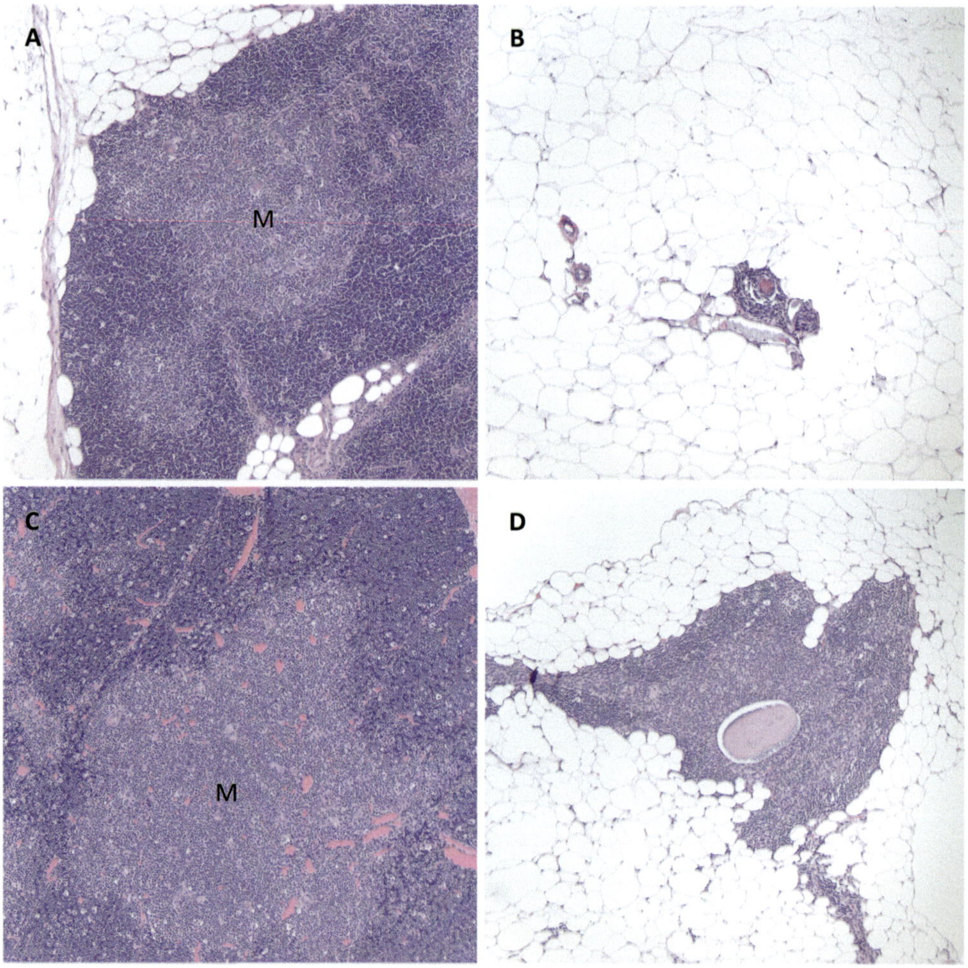

Fig. 6.30 Photomicrographs of thymic tissue. (**a**) Thymic tissue in a 7.9-year-old male macaque with normal cortical and medullary (M) organization within lobules. (**b**) Thymic tissue in an aged 33.9-year-old female macaque with few lymphocytes surrounding a pink Hassal's corpuscle. (**c**) Thymic tissue in a 5-day-old common marmoset with prominent cortical and medullary (M) organization. (**d**) Thymic tissue in an aged 13.5-year-old male common marmoset with loss of defined corticomedullary regions

6.6.2 Amyloidosis

Amyloidosis may occur at any age in the marmoset, macaque, and human, but the incidence of systemic or secondary amyloidosis increases with advancing age [16, 17, 85–88].

Secondary (AA) amyloidosis is associated with generalized persistent inflammation, termed inflammaging in humans, which leads to the production of serum amyloid A by the liver, and the deposition of converted AA amyloid within tissues. Amyloid deposition may occur in a single tissue or multiple locations including the spleen, liver, pancreas, lymph nodes, adrenal glands, kidneys (both glomeruli and renal interstitium), and segmentally within the lamina propria of the stomach, duodenum, jejunum, ileum, cecum (appendix), and/or colon in all three species (Fig. 6.31) [16, 17, 28, 85–87]. The incidence of extensive or severe amyloidosis is much greater in aged marmosets and macaques [17].

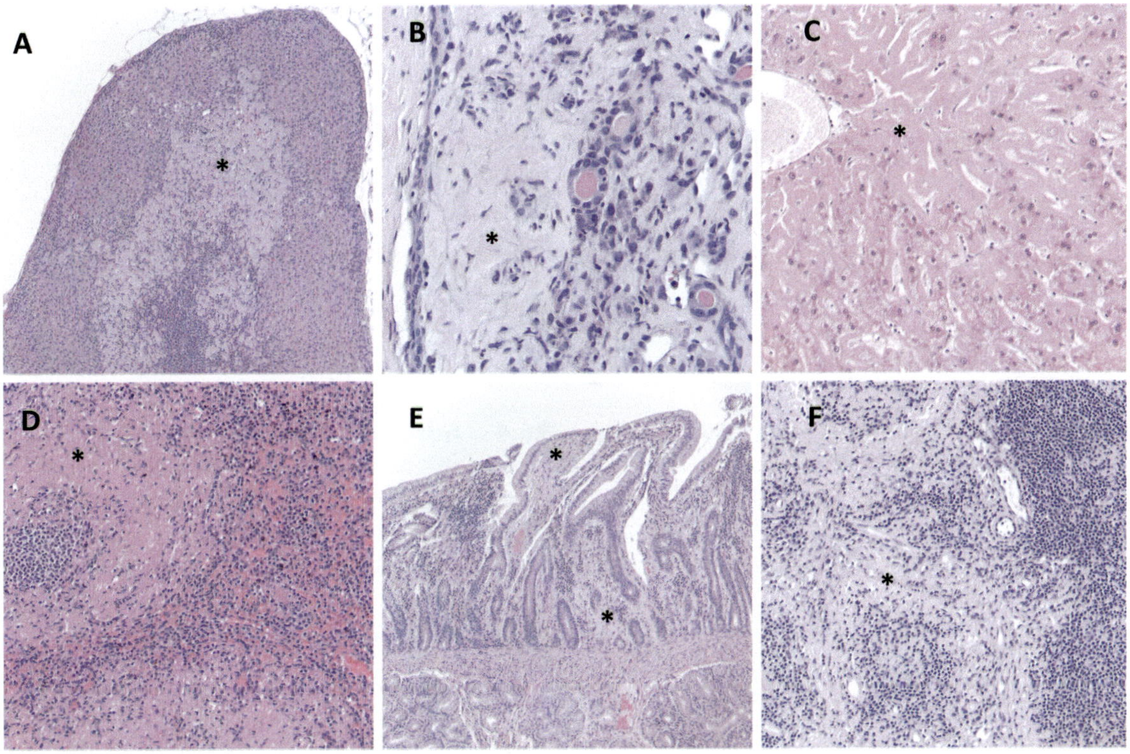

Fig. 6.31 Photomicrographs of systemic amyloidosis. (**a**) Adrenal gland in a common marmoset with amyloidosis (*) of the medulla and extramedullary hematopoiesis. (**b**) Interstitial amyloid surrounding atrophied proximal renal tubules in a common marmoset. (**c**) Hepatic amyloid (*) in a rhesus macaque linearly expanding the space of Disse and causing compression and loss of hepatic cords. (**d**) Splenic amyloid (*) expanding the periarteriolar lymphoid sheath. (**e**) Amyloid in the superficial and deep lamina propria of the duodenum of a rhesus macaque. (**f**) Mesenteric lymph node with moderate amyloid (*) expanding follicular parenchyma

6.6.3 Obesity and Diabetes

Decreased glucoregulatory function is noted with increasing age and obesity in humans, marmosets, and rhesus macaques [40, 48, 57, 89–94]. Disorders associated with diabetes include hypertension, platelet dysfunction, macrovascular disease of the coronary arteries, aorta, and cerebral circulation, myocardial infarction, renal vascular insufficiency, cerebrovascular accidents, diabetic neuropathy, dyslipidemia, microalbuminuria, diabetic retinopathy, glaucoma, cataracts, and enhanced susceptibility to infections [65]. Rhesus macaques are a model for diabetic neuropathy and retinopathy [95]. The distribution of delta and beta cells within pancreatic islets and the ultrastructure secretory granules in the alpha, beta, and delta cells of the common marmoset pancreas have been shown to be very similar to those of humans [96].

Marmosets fed high-fat or high-glucose diets will become obese with dyslipidemia [41, 94]. The high-glucose diet marmoset model causes obesity with dyslipidemia impaired glucose tolerance, increased circulating A1C levels, and pancreatic islet hyperplasia [41].

Amyloid accumulation within pancreatic islets is well described in both diabetic rhesus macaques and approximately 90% of human type 2 diabetics [48, 91, 93, 97]. This deposition is often independent of amyloid deposition in other tissues (Fig. 6.32) [48, 91, 97, 98].

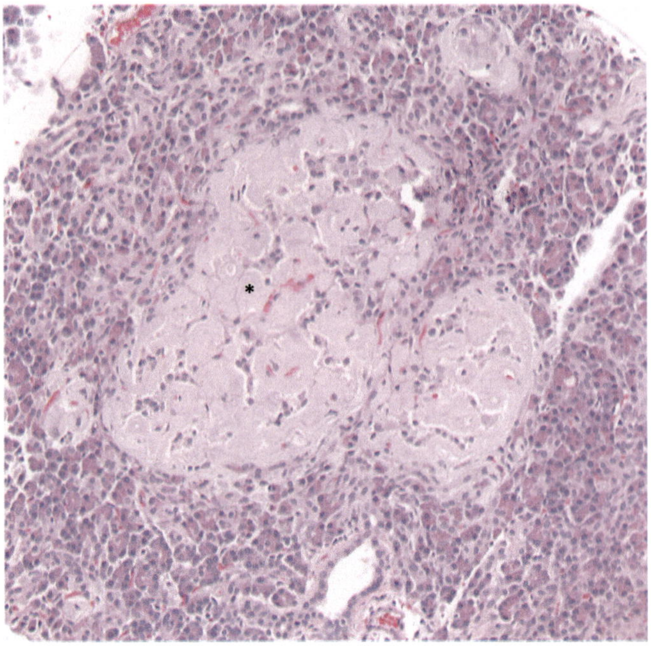

Fig. 6.32 Photomicrograph of amyloid (*) effacing and replacing the majority of endocrine cells within the pancreatic islet in a 24.9-year-old rhesus macaque. The surrounding exocrine pancreas is spared

6.6.4 Adrenal Mineralization

Mineral accumulation within the adrenal gland may occur at any age in both the rhesus macaque and common marmoset (Fig. 6.33). These deposits tend to be larger and more prominent in geriatric and aged animals, although they are not consistently present in all aged animals.

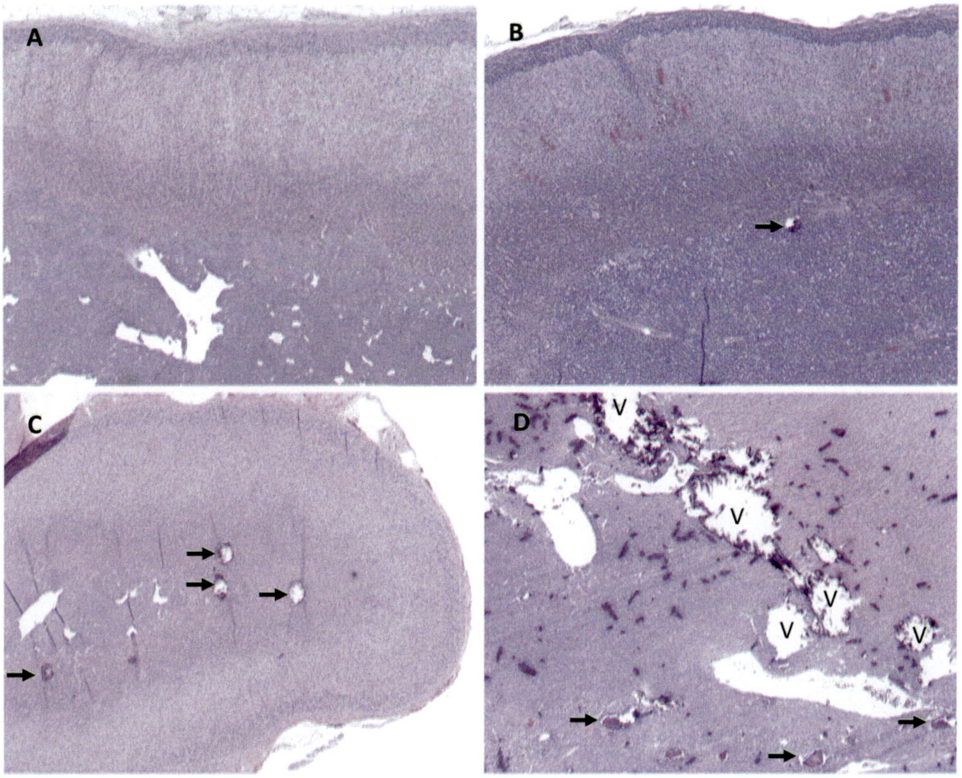

Fig. 6.33 Photomicrographs of adrenal mineralization. (**a**) Normal adrenal gland in a common marmoset. (**b**) Minimal focal mineralization (arrow) in a rhesus macaque. (**c**) Moderate multifocal adrenal mineralization (arrows) in a rhesus macaque. (**d**) Severe multifocal mineralization in the adrenal medulla of an aged rhesus macaque characterized by voids (V) where the shattered mineral material has been displaced as well as intact mineral foci (arrows) are noted

6.6.5 Age-Associated Bone Marrow Changes

The proportions of hematopoietic bone marrow and adipose within the medullary cavities of long bones change with age with hematopoietic cells occupying 90% of the medullary cavity at birth and diminishing to 30% by the seventh decade [99]. In macaques this age-associated change is visible during postmortem examination and corresponds to histologic findings (Fig. 6.34a, d). Obesity appears to influence the rate of adipose acquisition within medullary cavities in the macaque as well. Proportional age-associated changes bone marrow hematopoiesis and adiposity are much more subtle in the common marmoset (Fig. 6.34b, e). Marmosets have propensity to develop EMH throughout the life span which is significantly different from humans and macaques which both shift from EMH in the fetal liver and fetal spleen to intramedullary hematopoiesis at birth (Fig. 6.34c, f) [42, 99].

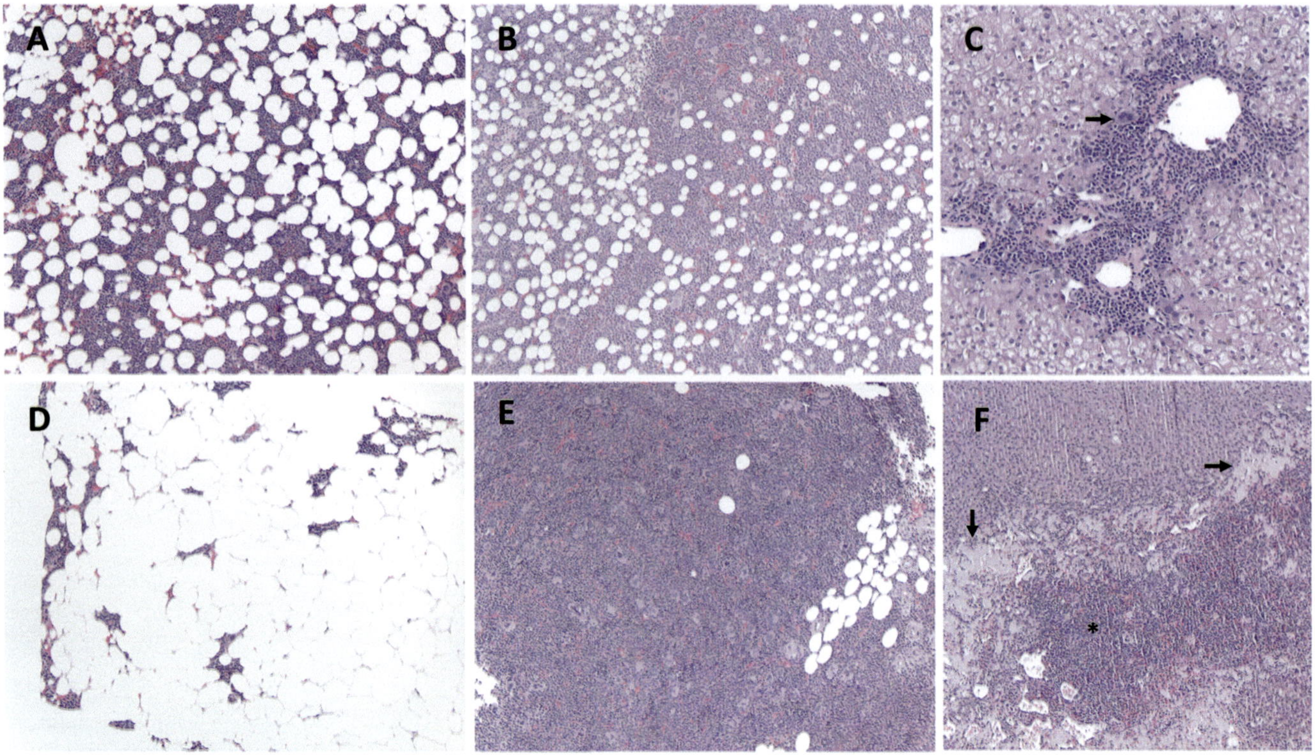

Fig. 6.34 Photomicrographs of age associated bone marrow changes. (**a**) Bone marrow from a 5-year-old rhesus macaque with numerous clear adipocytes. (**b**) Bone marrow from a 1-year-old common marmoset. (**c**) Liver from a common marmoset with periportal extramedullary hematopoiesis (EMH) and a distinct megakaryocyte (arrow). (**d**) Bone marrow from a 20-year-old rhesus macaque with abundant adipose and scant hematopoiesis. (**e**) Bone marrow from a 12-year-old common marmoset with abundant hematopoiesis and scant clear adipocytes. (**f**) Adrenal gland from a common marmoset with abundant EMH (*) within the medulla, mild pale pink amyloid accumulation (arrows) at the corticomedullary junction, and normal cortical parenchyma (**f**)

6.7 Age-Associated Neuropathology

6.7.1 Iron Accumulation

Age-related changes in iron metabolism are very important to brain health due to its high metabolic activity and oxygen consumption [100]. Iron is the most abundant metal within the CNS and is essential for oxidative metabolism, myelination, and the synthesis of neurotransmitters. The CSF has a lower iron-buffering capacity than peripheral serum, and iron deposits are associated with inflammation, neurodegeneration, abnormal protein aggregations, cognitive impairment, and dementia [100]. Iron deposits have been identified using MRI and confirmed histologically in the caudate nucleus, putamen, globus pallidus, and substantia nigra of humans and macaques [15, 100, 101] Additional sites of iron accumulation include the basal nuclei, the red nucleus, and the perirhinal, parietal, and temporal cortices [64, 102]. Macaques fed a calorie-restricted diet exhibit less iron accumulation within the brain than age-matched controls [103]. Evaluation of iron- and ferritin-positive microglia within the brains of common marmosets demonstrated increasing iron content in the hippocampus and cerebral cortices with increased ferritin positive microglia in old marmosets with a decrease in ferritin-positive microglia in very aged >16-year-old marmosets (Fig. 6.35) [104].

Fig. 6.35 Photomicrograph of the cerebrum of a 10-year-old common marmoset stained with Perls Prussian blue histochemical stain to identify iron within the neuropil

6.7.2 Neuromelanin

Neuromelanin is a dark brown pigment that accumulates in the neurons of the catecholamine-producing regions of the human and macaque brain but not in New World monkeys [105–107]. It is a byproduct of catecholamine metabolism, and the loss of this pigment is associated with spontaneous Parkinson's disease in humans and in chemically induced models of Parkinson's in NHP (Fig. 6.36) [105, 108, 109].

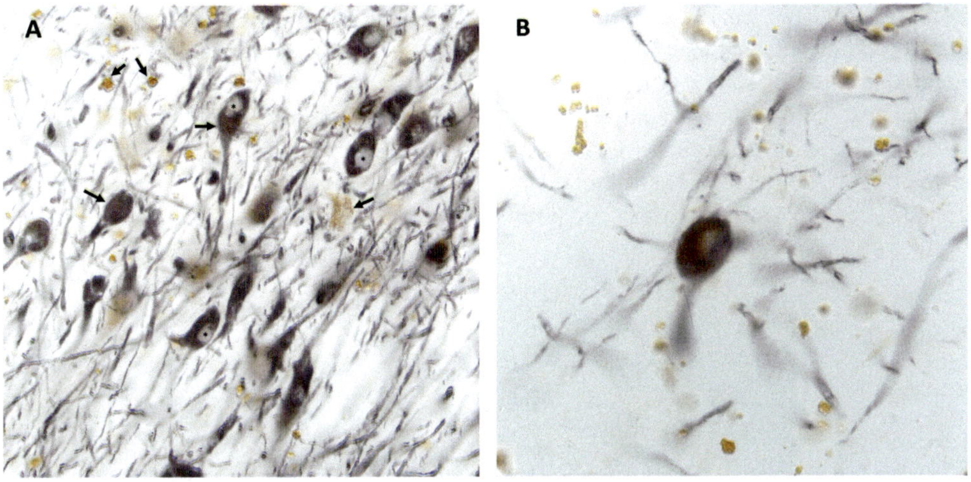

Fig. 6.36 Neuromelanin in the rhesus macaque. (**a**) Photomicrograph of the substantia nigra of an aged (24 years old) rhesus macaque. Brown-gold granular neuromelanin (arrows) is found within the cytoplasm of dopamine producing neurons as well as between axons IHC stained for tyrosine hydroxylase (TH—a marker for dopamine producing neurons and associated axons). The nuclei of neurons (*) do not stain. (**b**) Neuromelanin in the cytoplasm of a dopaminergic neuron of the substantia nigra of an aged rhesus macaque stained with TH. Note the granularity of the brown-gold neuromelanin and pallor of the underling nucleus, 60x magnification. Slides courtesy of Marina Emborg; imaging by Heather Simmons

6.7.3 Cerebral Volume

Cerebral volume diminishes with age in rhesus, human, and marmoset [110–112]. Changes in sulci and gyri are evident grossly in humans and macaques, while changes in the marmoset brain are less obvious due to normal cerebral lissencephaly. The rhesus macaque has diminution of gray matter in the prefrontal, orbitofrontal, and superior temporal sulci with increasing age in MRI studies [113, 114]. Age-associated diminution of both gray and white matter, especially in the corpus callosum, has also been noted in the marmoset, although this is more subtle due to the lissencephalic structure of the marmoset cerebrum (Fig. 6.37) [115, 116].

6.7.4 Cerebral Plaques

Natural or spontaneous accumulation of age-associated cerebral plaques composed of amyloid-beta have been described in humans, rhesus macaques, and common marmosets [117–122]. Plaque accumulation has not been noted in common marmosets <7 years of age and rhesus macaques <19 years of age [119, 121, 122]. Tau is a major component of neurofibrillary tangles (a feature of Alzheimer's disease), when hyperphosphorylated. Although Tau is expressed in both the marmoset and the rhesus macaque brain, neurofibrillary tangles have not been noted in either species. [120, 121, 123]

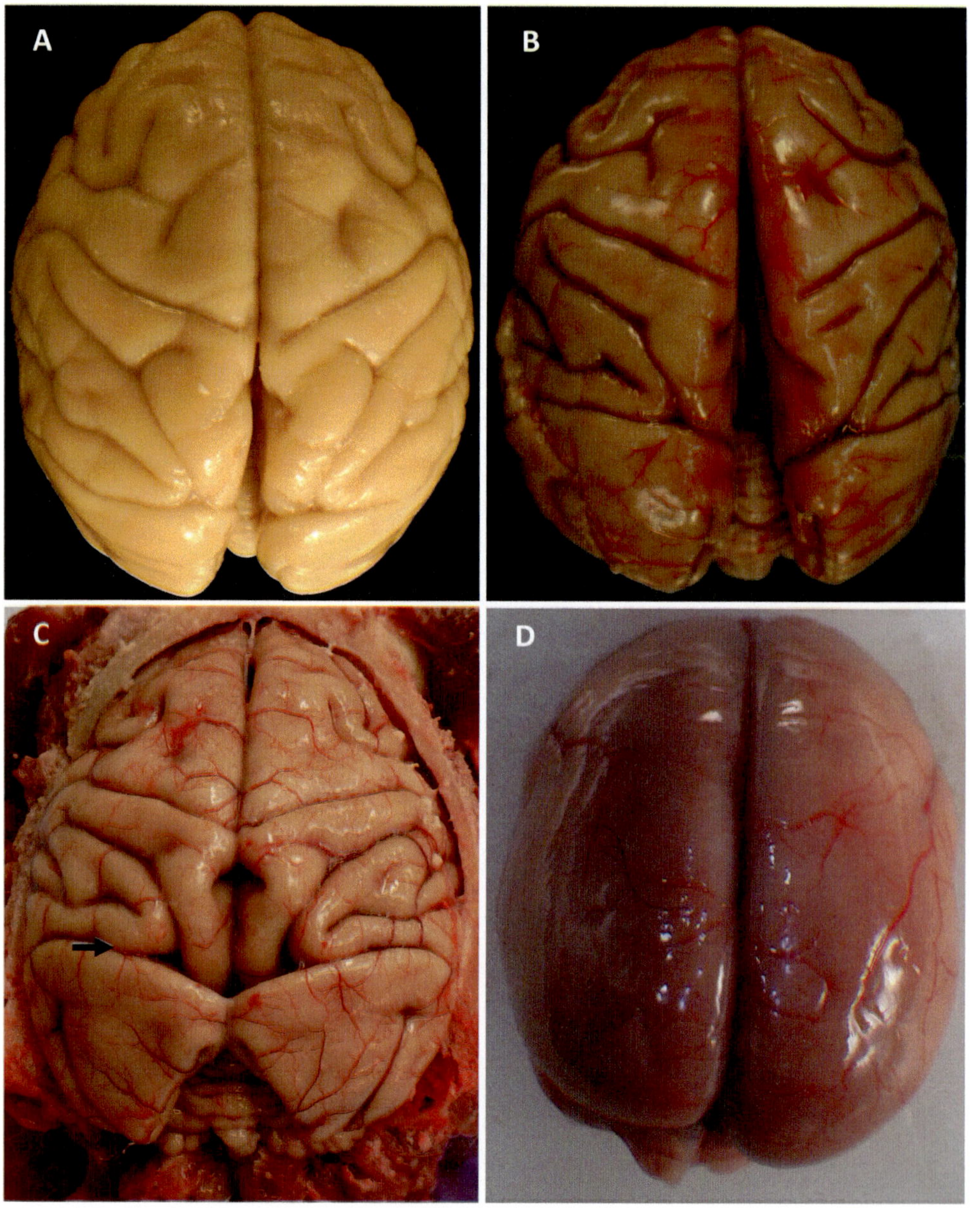

Fig. 6.37 Gross images of changes in cerebral volume. (**a**) A fixed brain of a young 3-year-old rhesus macaque with rounded gyri and narrow sulci. (**b**) A fresh brain of an aged 29-year-old macaque with slightly flattened gyri and widened sulci. (**c**) A fresh brain of a 35-year-old rhesus macaque with atrophy of the superior parietal lobule and medial parietal cortex with marked widening of the parieto-occipital sulcus (arrow). (**d**) Common marmoset brain with normal lissencephalic cerebral hemispheres

6.7.5 Cerebral Infarcts

Spontaneous cerebral infarcts have been demonstrated by MRI (magnetic resonance imaging) in the parietal and temporal lobes of geriatric and aged rhesus macaques [15]. One retrospective study documented gross evidence of cerebral infarction with focal regions of atrophy and loss of the cerebral neuropil in five rhesus macaques (mean age, 25.7 years) at necropsy [17]. Spontaneous infarctions are not reported in common marmosets, but this species has been proven to be a good model for ischemic brain injury [124].

6.7.6 Cerebral Vascular Calcification

Cerebral vascular calcifications are viewed as part of the normal aging process with increasing prevalence in aging humans [125]. Cerebral vascular calcifications may also be considered a predictor of heart attack, stroke, Parkinson's disease, and Alzheimer's disease [125–127]. Calcifications may affect both large arteries and the capillaries that occur throughout the neuroparenchyma [127]. These changes have also been documented in nonhuman primates (Fig. 6.38) [128].

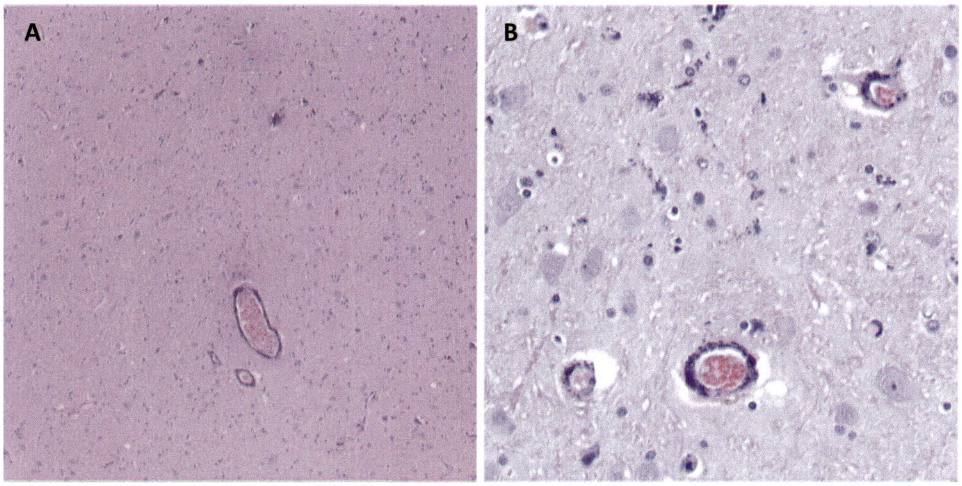

Fig. 6.38 Photomicrographs of vascular calcification. (**a**) There is dark purple mineralization of capillary walls within the temporal lobe of a 10-year-old common marmoset. (**b**) Higher magnification of the eccentric "bead-like" calcification of three cerebral capillaries

6.8 Ocular

6.8.1 Cataracts

Cataracts in humans and macaques are age-related with the majority of cases in older humans and aged and geriatric macaques [15, 17, 129]. Lens opacity generally begins to develop in the rhesus at ~20 years of age (Fig. 6.39) [15]. There are no mentions of cataracts in common marmosets in the literature, nor in the Wisconsin National Primate Research Center records, although there is one case of phacoemulsification of bilateral cataracts in pygmy marmoset (*Callithrix pygmaea*) [130].

6.8.2 Visual Accommodation and Presbyopia

Age-associated changes in visual accommodation and presbyopia are well documented in the rhesus macaque and diminish at a similar rate relative to age as humans [131, 132].

6.9 Reproductive

6.9.1 Prostatic Hyperplasia

Benign basal cell hyperplasia is the most commonly noted change in the aged rhesus macaque prostate gland, although squamous metaplasia, focal glandular pleomorphism, and

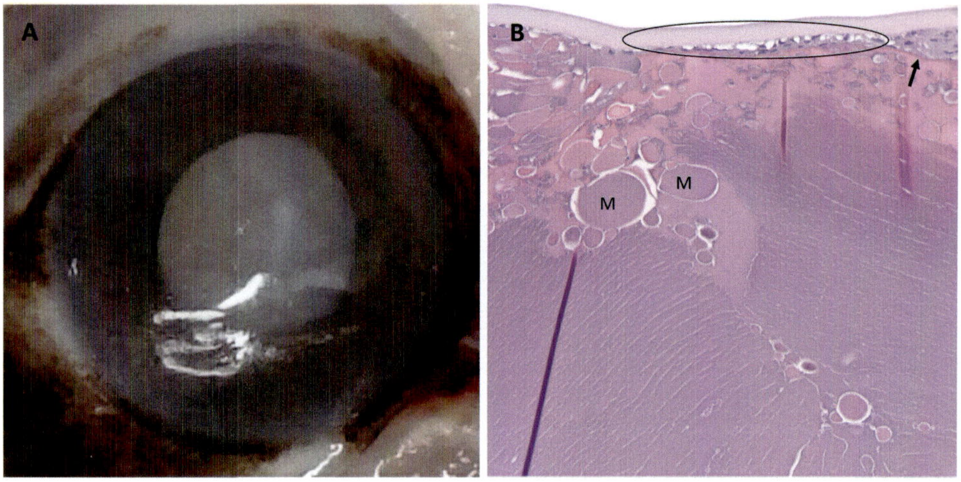

Fig. 6.39 Cataract in a rhesus macaque. (**a**) A hyper-mature cataract in a 37-year-old female with complete opacity of the lens. (**b**) Photomicrograph of a mature cataract with moderate degeneration and fragmentation of lens fibers with spherical protein aggregates (Morgagnian globules) (M). Cuboidal epithelial cells of the lens have cytoplasmic vacuolization (oval) with segmental epithelial hyperplasia (arrow)

stromal hyperplasia have also been noted [133–136]. The rhesus macaque does have the gene to express prostatic-specific antigen (PSA), making it an excellent model for prostatic disease [137]. The common marmoset has not been found to express a PSA that cross-reacts with human assays [137]. Changes in older and aged marmoset prostate glands include interstitial fibrosis, ectasia of glandular acini, cysts, and inspissation of secretory material. Prostatitis may occur in both species at any age.

6.9.2 Testicular Atrophy

Aged humans, macaques, and marmosets may have testicular atrophy characterized by seminiferous tubules with partial or complete loss of mature spermatids, disorganization of germ cells layers, multinucleate germ cells, lipid droplets within Sertoli cells, and maturation arrest leading to diminished or absence of spermatocytes within the epididymis (Fig. 6.40a) [138]. As in humans, productive seminiferous tubules may be noted adjacent to those with hypospermatogenesis or fully sclerotic tubules. In humans, this is postulated to be due to age-associated vascular change, which is supported by the fact that comparable changes are noted after experimental ischemia [138]. A cryopreservation study of sperm collected postmortem from the vas deferens demonstrated that older (19 year macaques) have reduced sperm motility in frozen-thawed samples compared to younger males [139]. Severe testicular atrophy or aspermatogenesis is most often associated with severe chronic systemic disease, unless the animal has a retained (intra-abdominal) testicle (Fig. 6.40b).

6.9.3 Mammary

Mammary (breast) tissue changes associated with aging include diminished glandular tissue due to loss of acinar epithelium (glandular involution), increased adipose, and increased collagenous stroma, often associated with menopause [140]. Age-associated and menopause-associated mammary changes in the rhesus macaque have been well documented due to the many similarities in anatomy and developmental physiology with atrophy of lobules, ductular and lobular hyperplasia, and cystic dilation of ducts, carcinoma in situ of ductus and lobules, and invasive ductal carcinoma [141]. [142] Less information is published on the marmoset mammary gland, although this species has had successful characterization of age-associated changes within mammary stem/progenitor cells [143].

6.9.4 Ovarian Senescence

Ovarian senescence is described in humans and macaques with diminished numbers of oocytes, decreased oocyte quality, and loss ovarian stroma with fibrosis leading to menopause [144, 145]. The marmoset does not undergo menopause although there are age-associated reductions in litter size and ovarian cycles may become intermittent to absent despite the fact that ovaries are shown to be hormonally active [146].

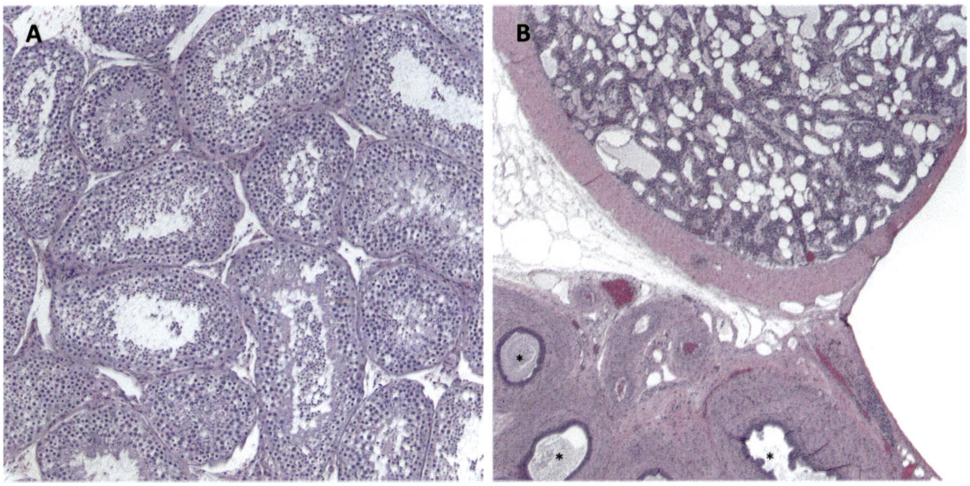

Fig. 6.40 Testicular atrophy in a common marmoset. (**a**) Moderate diffuse testicular atrophy in an aged 12-year-old common marmoset characterized by loss of mature spermatids and disorganization of germ cell layers within seminiferous tubules. (**b**) Severely hypoplastic intra-abdominal testicle in a common marmoset with hypospermatogenesis and sclerotic-dilated tubules separated by interstitial adipose. The adjacent epididymis has no sperm within tubular lumina (*)

6.10 Conclusion

Both macaques and marmosets are excellent nonhuman primate models for studies on aging. Efforts to expand access to research resources are in progress through aged nonhuman primate tissue banks; tissue-sharing programs at the National Primate Research Centers, Primate Aging Databases with physiologic data; and the publication of peer-reviewed research. Typical age-related pathologies in these two model species are reviewed and illustrated with gross and histologic examples in this chapter with the expectation that numerous new discoveries will result from many ongoing studies on aging in nonhuman primates.

Acknowledgments The author gratefully acknowledges the contributions of all previous pathologists of the Wisconsin National Primate Research Center, Dr. Andres Mejia, Dr. Puja Basu, Dr. Anne Lewis, Dr. Tracey L. Gilbert, and the Northeastern Wisconsin Zoo for shared case material and images and Dr. Marina Emborg for providing research materials to be imaged for neuromelanin. Special thanks to Jennifer Hayes, Nicole Oldenberger, Abby Bradford, Alanna Friscino, Chris Huffman, and all previous WNPRC research specialists for able assistance and photography during necropsies.

References

1. De Lepeleire J, Iliffe S, Mann E, et al. Frailty: an emerging concept for general practice. Br J Gen Pract. 2009;59:e177–82. https://doi.org/10.3399/bjgp09X420653.
2. Fried LP, Tangen CM, Walston J, et al. Frailty in older adults: evidence for a phenotype. J Gerontol A Biol Sci Med Sci. 2001;56: M146–56. https://doi.org/10.1093/gerona/56.3.m146.
3. Yamada Y, Kemnitz JW, Weindruch R, et al. Caloric restriction and healthy life span: frail phenotype of nonhuman primates in the Wisconsin National Primate Research Center Caloric Restriction Study. J Gerontol A Biol Sci Med Sci. 2018;73:273–8. https://doi.org/10.1093/gerona/glx059.
4. Olson EJ, Shaw GC, Hutchinson EK, et al. Bone disease in the common marmoset: radiographic and histological findings. Vet Pathol. 2015;52:883–93. https://doi.org/10.1177/0300985815589354.
5. Colman RJ, Lane MA, Binkley N, et al. Skeletal effects of aging in male rhesus monkeys. Bone. 1999;24:17–23. https://doi.org/10.1016/S8756-3282(98)00147-1.
6. Colman RJ, Kemnitz JW, Lane MA, et al. Skeletal effects of aging and menopausal status in female rhesus macaques. J Clin Endocrinol Metab. 1999;84:4144–8. https://doi.org/10.1210/jcem.84.11.6151.
7. Crowder C, Austin D. Age ranges of epiphyseal fusion in the distal tibia and fibula of contemporary males and females. J Forensic Sci. 2005;50:1001–7.
8. Cerroni AM, Tomlinson GA, Turnquist JE, et al. Bone mineral density, osteopenia, and osteoporosis in the rhesus macaques of Cayo Santiago. Am J Phys Anthropol. 2000;113:389–410. https://doi.org/10.1002/1096-8644(200011)113:3<389::Aid-ajpa9>3.0.Co;2-i.
9. Saltzman W, Abbott DH, Binkley N, et al. Maintenance of bone mass despite estrogen depletion in female common marmoset monkeys (Callithrix jacchus). Am J Primatol. 2019;81:e22905. https://doi.org/10.1002/ajp.22905.
10. Center JR, Nguyen TV, Schneider D, et al. Mortality after all major types of osteoporotic fracture in men and women: an observational study. Lancet. 1999;353:878–82. https://doi.org/10.1016/s0140-6736(98)09075-8.
11. Seidlová-Wuttke D, Schlumbohm C, Jarry H, et al. Orchidectomized (orx) marmoset (Callithrix jacchus) as a model to study the development of osteopenia/osteoporosis. Am J Primatol. 2008;70:294–300. https://doi.org/10.1002/ajp.20493.
12. Colman RJ, McKiernan SH, Aiken JM, et al. Muscle mass loss in Rhesus monkeys: age of onset. Exp Gerontol. 2005;40:573–81. https://doi.org/10.1016/j.exger.2005.05.001.
13. Rosenberg AE. Bones, joints, and soft tissue tumors. In: Kumar V, Abbas AK, Fausto N, editors. Robbins and Cotran pathologic basis of disease. 7th ed. Philadelphia: Elsevier/Saunders; 2005. p. 1304–14.
14. Bailey JF, Fields AJ, Liebenberg E, et al. Comparison of vertebral and intervertebral disc lesions in aging humans and rhesus monkeys. Osteoarthritis Cartilage. 2014;22:980–5. https://doi.org/10.1016/j.joca.2014.04.027.
15. Uno H. Age-related pathology and biosenescent markers in captive rhesus macaques. Age. 1997;20:1–13. https://doi.org/10.1007/s11357-997-0001-5.
16. Ross CN, Salmon AB. Aging research using the common marmoset: focus on aging interventions. Nutr Healthy Aging. 2019;5:97–109. https://doi.org/10.3233/nha-180046.
17. Simmons HA. Age-associated pathology in rhesus macaques (Macaca mulatta). Vet Pathol. 2016;53:399–416. https://doi.org/10.1177/0300985815620628.
18. Pugh TD, Conklin MW, Evans TD, et al. A shift in energy metabolism anticipates the onset of sarcopenia in rhesus monkeys. Aging Cell. 2013;12:672–81. https://doi.org/10.1111/acel.12091.
19. Hof PR, Erwin J. Aging in nonhuman primates. Basel: Karger; 2002.
20. Ross CN, Davis K, Dobek G, et al. Aging phenotypes of common marmosets (Callithrix jacchus). J Aging Res. 2012;2012:567143. https://doi.org/10.1155/2012/567143.
21. Murphy SL KK, Xu JQ, Arias E. Mortality in the United States, 2020. In: NCfHS, editor. Hyattsville, MD; 2021.
22. Laurence H, Kumar S, Owston MA, et al. Natural mortality and cause of death analysis of the captive chimpanzee (Pan troglodytes): a 35-year review. J Med Primatol. 2017;46:106–15. https://doi.org/10.1111/jmp.12267.
23. Lowenstine LJ, McManamon R, Terio KA. Comparative pathology of aging great apes: bonobos, chimpanzees, gorillas, and orangutans. Vet Pathol. 2016;53:250–76. https://doi.org/10.1177/0300985815612154.
24. Moussavi A, Mietsch M, Drummer C, et al. Cardiac MRI in common marmosets revealing age-dependency of cardiac function. Sci Rep. 2020;10:10221. https://doi.org/10.1038/s41598-020-67157-5.
25. Bodkin NL, Alexander TM, Ortmeyer HK, et al. Mortality and morbidity in laboratory-maintained Rhesus monkeys and effects of long-term dietary restriction. J Gerontol A Biol Sci Med Sci. 2003;58:212–9.
26. Van Vleet JF, Ferrans VJ. Cardiovascular system. In: McGavin MD, Zachary JF, editors. Pathologic basis of veterinary disease. 4th ed. St. Louis, MO: Mosby Elsevier; 2007. p. 578–9.
27. Chamanza R, Parry NM, Rogerson P, et al. Spontaneous lesions of the cardiovascular system in purpose-bred laboratory nonhuman primates. Toxicol Pathol. 2006;34:357–63. https://doi.org/10.1080/01926230600809737.
28. Ross CN, Adams J, Gonzalez O, et al. Cross-sectional comparison of health-span phenotypes in young versus geriatric marmosets. Am J Primatol. 2019;81:e22952. https://doi.org/10.1002/ajp.22952.

29. Senos R, Gomes Benedicto H, Del Rio M, do Valle C, et al. Collagen quantification in the ventricular walls of the heart of the common marmoset (Callithrix jacchus). Anat Rec (Hoboken). 2021;304:1275–9. https://doi.org/10.1002/ar.24632.
30. Gallagher PJ. Blood vessels. In: Mills SE, Carter D, Greenson JK, Oberman HA, Reuter V, Stoler MH, editors. Sternberg's diagnostic surgical pathology. 4th ed. Philadelphia: Lippincott Williams & Wilkins; 2004. p. 1369–76.
31. Chawla KK, Murthy CD, Chakravarti RN, et al. Arteriosclerosis and thrombosis in wild rhesus monkeys. Am Heart J. 1967;73:85–91.
32. Chakravarti RN, Mohan AP, Komal HS. Atherosclerosis in Macaca mulatta: histopathological, morphometric, and histochemical studies in aorta and coronary arteries of spontaneous and induced atherosclerosis. Exp Mol Pathol. 1976;25:390–401.
33. Palit S, Kendrick J. Vascular calcification in chronic kidney disease: role of disordered mineral metabolism. Curr Pharm Des. 2014;20:5829–33. https://doi.org/10.2174/1381612820666140212194926.
34. Schoen FJ. Bood vessels. In: Kumar V, Abbas AK, Fausto N, editors. Robbins and Cotran pathologic basis of disease. 7th ed. Philadelphia: Elsevier/Saunders; 2005. p. 516–25.
35. Schoen FJ. The heart. In: Kumar V, Abbas AK, Fausto N, editors. Robbins and Cotran pathologic basis of disease. 7th ed. Philadelphia: Elsevier/Saunders; 2005. p. 595–8.
36. Shelton KA, Clarkson TB, Kaplan JR. Nonhuman primate models of atherosclerosis. In: Abee CR, editor. Nonhuman primates in biomedical research. New York: Elsevier; 2012. p. 385–411.
37. Taylor CB, Manalo-Estrella P, Cox GE. Atherosclerosis in Rhesus Monkeys. V. Marked diet-induced hypercholesteremia with xanthomatosis and severe atherosclerosis. Arch Pathol. 1963;76: 239–49.
38. Taylor CB, Patton DE, Cox GE. Atherosclerosis in Rhesus Monkeys. VI. Fatal myocardial infarction in a monkey fed fat and cholesterol. Arch Pathol. 1963;76:404–12.
39. Williams JK, Anthony MS, Clarkson TB. Coronary heart disease in rhesus monkeys with diet-induced coronary artery atherosclerosis. Arch Pathol Lab Med. 1991;115:784–90.
40. Zhang X, Zhang R, Raab S, et al. Rhesus macaques develop metabolic syndrome with reversible vascular dysfunction responsive to pioglitazone. Circulation. 2011;124:77–86. https://doi.org/10.1161/circulationaha.110.990333.
41. Wachtman LM, Kramer JA, Miller AD, et al. Differential contribution of dietary fat and monosaccharide to metabolic syndrome in the common marmoset (Callithrix jacchus). Obesity (Silver Spring). 2011;19:1145–56. https://doi.org/10.1038/oby.2010.303.
42. Chamanza R. Non-human primates: cynomolgus (Macaca fascicularis) and rhesus (Macaca mulatta) macaques and the common marmoset (Callithrix jacchus). In: Mann EFM, editor. Background lesions in laboratory animals. St. Louis: W.B. Saunders; 2012. p. 1–15.
43. Lowenstine LJ, Osborn KG. Respiratory system diseases of nonhuman primates. Nonhuman primates in biomedical research. 2012;2012:413–81.
44. George MP, Champion HC, Simon M, et al. Physiologic changes in a nonhuman primate model of HIV-associated pulmonary arterial hypertension. Am J Respir Cell Mol Biol. 2013;48:374–81. https://doi.org/10.1165/rcmb.2011-0434OC.
45. Sasseville VG, Hotchkiss CE, Levesque PC, et al. Hematopoietic, cardiovascular, lymphoid and mononuclear phagocyte systems of nonhuman primates. In: Abee CR, editor. Nonhuman primates in biomedical research. New York: Elsevier; 2012. p. 357–84.
46. Bleyer M, Kunze M, Gruber-Dujardin E, et al. Spontaneous lung pathology in a captive common marmoset colony (Callithrix jacchus). Primate Biol. 2017;4:17–25. https://doi.org/10.5194/pb-4-17-2017.
47. Cooley AJ, Savage A, Snowdon CT. Vascular, cardiac, and renal lesions attributed to primary systemic hypertension in western pygmy marmosets (Cebuella pygmaea). Vet Pathol. 2022;59: 358–70. https://doi.org/10.1177/03009858211052664.
48. Anirban M, Abbas AK. The endocrine system. In: Kumar V, Abbas AK, Fausto N, editors. Robbins and Cotran pathologic basis of disease. 7th ed. Philadelphia: Elsevier/Saunders; 2005. p. 1189–206.
49. Schuliga M, Read J, Knight DA. Ageing mechanisms that contribute to tissue remodeling in lung disease. Ageing Res Rev. 2021;70: 101405. https://doi.org/10.1016/j.arr.2021.101405.
50. Allen TC. Small airways disease. Surg Pathol Clin. 2010;3:171–86. https://doi.org/10.1016/j.path.2010.04.002.
51. Mineshige T, Inoue T, Kawai K, et al. Spontaneous pulmonary adenocarcinoma in a common marmoset (Callithrix jacchus). J Med Primatol. 2021;50:335–8. https://doi.org/10.1111/jmp.12540.
52. Simmons HA, Mattison JA. The incidence of spontaneous neoplasia in two populations of captive rhesus macaques (Macaca mulatta). Antioxid Redox Signal. 2011;14:221–7. https://doi.org/10.1089/ars.2010.3311.
53. Hyde DM, Blozis SA, Avdalovic MV, et al. Alveoli increase in number but not size from birth to adulthood in rhesus monkeys. Am J Physiol Lung Cell Mol Physiol. 2007;293:L570–9. https://doi.org/10.1152/ajplung.00467.2006.
54. Herring MJ, Avdalovic MV, Quesenberry CL, et al. Accelerated structural decrements in the aging female rhesus macaque lung compared with males. Am J Physiol Lung Cell Mol Physiol. 2013;304:L125–34. https://doi.org/10.1152/ajplung.00226.2012.
55. Cline JM, Brignolo L, Ford EW. Urogenital system. In: Abee CR, editor. Nonhuman primates in biomedical research. New York: Elsevier; 2012. p. 483–562.
56. Lee HJ, Gonzalez O, Dick EJ, et al. Marmoset as a model to study kidney changes associated with aging. J Gerontol A Biol Sci Med Sci. 2019;74:315–24. https://doi.org/10.1093/gerona/gly237.
57. Cusumano AM, Bodkin NL, Hansen BC, et al. Glomerular hypertrophy is associated with hyperinsulinemia and precedes overt diabetes in aging rhesus monkeys. Am J Kid Dis. 2002;40:1075–85. https://doi.org/10.1053/ajkd.2002.36348.
58. Olstad KJ, Imai DM, Keesler RI, et al. Development of a geropathology grading platform for nonhuman primates. Aging Pathobiol Therap. 2020;2:16–9. https://doi.org/10.31491/apt.2020.03.008.
59. Wang D, Liu J, He S, et al. Assessment of early renal damage in diabetic rhesus monkeys. Endocrine. 2014;47:783–92. https://doi.org/10.1007/s12020-014-0211-4.
60. Alpers CE. The kidney. In: Kumar V, Abbas AK, Fausto N, editors. Robbins and Cotran pathologic basis of disease. 7th ed. Philadelphia: Elsevier/Saunders; 2005. p. 966–8.
61. Gonzalez OA, Orraca L, Kensler TB, et al. Familial periodontal disease in the Cayo Santiago rhesus macaques. Am J Primatol. 2016;78:143–51. https://doi.org/10.1002/ajp.22376.
62. Colombo APV, Paster BJ, Grimaldi G, et al. Clinical and microbiological parameters of naturally occurring periodontitis in the non-human primate Macaca mulatta. J Oral Microbiol. 2017;9: 1403843. https://doi.org/10.1080/20002297.2017.1403843.
63. Bunton TE, Bacmeister CX. Diverticulosis and colonic leiomyosarcoma in an aged rhesus macaque. Vet Pathol. 1989;26: 351–2.
64. McClure HM. The Rhesus monkey. New York: Academic Press; 1975.
65. Kumar V, Abbas AK, Ne F. Neoplasia. In: Kumar V, Abbas AK, Ne F, editors. Robbins and Cotran pathologic basis of disease. 7th ed. Philadelphia: Elsevier/Saunders; 2005. p. 1525.
66. Soh YSA, Ooi SQD, Chan YH, et al. Rising prevalence of colonic diverticulosis in a westernized multi-ethnic Asian community. J

Gastroenterol Hepatol. 2021;36:413–20. https://doi.org/10.1111/jgh.15165.
67. Razik R, Nguyen GC. Diverticular disease: changing epidemiology and management. Drugs Aging. 2015;32:349–60. https://doi.org/10.1007/s40266-015-0260-2.
68. Abdul Rani R, Raja Ali RA, Lee YY. Irritable bowel syndrome and inflammatory bowel disease overlap syndrome: pieces of the puzzle are falling into place. Intest Res. 2016;14:297–304. https://doi.org/10.5217/ir.2016.14.4.297.
69. Kochar B, Orkaby AR, Ananthakrishnan AN, et al. Frailty in inflammatory bowel diseases: an emerging concept. Therap Adv Gastroenterol. 2021;14:17562848211025474. https://doi.org/10.1177/17562848211025474.
70. Holmberg CA, Leininger R, Wheeldon E, et al. Clinicopathological studies of gastrointestinal disease in macaques. Vet Pathol Suppl. 1982;19(Suppl 7):163–70.
71. Laing ST, Merriam D, Shock BC, et al. Idiopathic Colitis in Rhesus Macaques is associated with dysbiosis, abundant enterochromaffin cells and altered T-cell cytokine expression. Vet Pathol. 2018;55:741–52. https://doi.org/10.1177/0300985818780449.
72. Delwart E, Tisza MJ, Altan E, et al. Idiopathic chronic diarrhea in rhesus macaques is not associated with enteric viral infections. Viruses. 2021;13(12):2503. https://doi.org/10.3390/v13122503.
73. Johnson AL, Keesler RI, Lewis AD, et al. Common and not-so-common pathologic findings of the gastrointestinal tract of rhesus and cynomolgus macaques. Toxicol Pathol. 2022;50:638–59. https://doi.org/10.1177/01926233221084634.
74. Rice KA, Chen ES, Metcalf Pate KA, et al. Diagnosis of amyloidosis and differentiation from chronic, idiopathic enterocolitis in rhesus (Macaca mulatta) and pig-tailed (M. nemestrina) macaques. Comp Med. 2013;63:262–71.
75. Kramer JA. The common marmoset in captivity and biomedical research. In: Fox JGMR, Wachtman LM, Tardif SD, Mansfield K, editors. American College of Laboratory Animal Medicine series. London: Academic Press, an imprint of Elsevier; 2019. p. 1. online resource.
76. Logan AC, Khan KN. Clinical pathologic changes in two marmosets with wasting syndrome. Toxicol Pathol. 1996;24:707–9. https://doi.org/10.1177/019262339602400605.
77. Baxter VK, Shaw GC, Sotuyo NP, et al. Serum albumin and body weight as biomarkers for the antemortem identification of bone and gastrointestinal disease in the common marmoset. PLoS One. 2013;8:e82747. https://doi.org/10.1371/journal.pone.0082747.
78. Sheh A, Artim SC, Burns MA, et al. Analysis of gut microbiome profiles in common marmosets (Callithrix jacchus) in health and intestinal disease. Sci Rep. 2022;12:4430. https://doi.org/10.1038/s41598-022-08255-4.
79. Kuehnel F, Mietsch M, Buettner T, et al. The influence of gluten on clinical and immunological status of common marmosets (Callithrix jacchus). J Med Primatol. 2013;42:300–9. https://doi.org/10.1111/jmp.12055.
80. Kumar BV, Connors TJ, Farber DL. Human T cell development, localization, and function throughout life. Immunity. 2018;48:202–13. https://doi.org/10.1016/j.immuni.2018.01.007.
81. Thomas R, Su D. Age-related thymic atrophy: mechanisms and outcomes. London: IntechOpen; 2019.
82. Johnsen DO, Johnson DK, Whitney RA. History of the use of nonhuman primates in biomedical research. In: Nonhuman primates in biomedical research. New York: Elsevier; 2012. p. 1–33.
83. Snyder PW, Everds NE, Craven WA, et al. Maturity-related variability of the thymus in cynomolgus monkeys (Macaca fascicularis). Toxicol Pathol. 2016;44:874–91. https://doi.org/10.1177/0192623316649258.
84. Tucker MJ. A survey of the pathology of marmosets (Callithrix jacchus) under experiment. Lab Anim. 1984;18:351–8. https://doi.org/10.1258/002367784780865397.
85. Blanchard JL, Baskin GB, Watson EA. Generalized amyloidosis in rhesus monkeys. Vet Pathol. 1986;23:425–30.
86. Cornwell GG 3rd, Johnson KH, Westermark P. The age related amyloids: a growing family of unique biochemical substances. J Clin Pathol. 1995;48:984–9. https://doi.org/10.1136/jcp.48.11.984.
87. Vaxman I, Dispenzieri A, Muchtar E, et al. New developments in diagnosis, risk assessment and management in systemic amyloidosis. Blood Rev. 2020;40:100636. https://doi.org/10.1016/j.blre.2019.100636.
88. Ludlage E, Murphy CL, Davern SM, et al. Systemic AA amyloidosis in the common marmoset. Vet Pathol. 2005;42:117–24. https://doi.org/10.1354/vp.42-2-117.
89. Colman RJ, Anderson RM, Johnson SC, et al. Caloric restriction delays disease onset and mortality in rhesus monkeys. Science. 2009;325:201–4. https://doi.org/10.1126/science.1173635.
90. Kemnitz JWHK, Colman RJ. Nutrition, aging and reproduction in rhesus monkeys. In: Nutrition and reproduction. Baton Rouge, LA: Louisiana State University Press; 1998.
91. Palotay JL, Howard CF. Insular amyloidosis in spontaneously diabetic nonhuman primates. Vet Pathol. 1982;19:181–92. https://doi.org/10.1177/030098588201907s14.
92. Roth GS, Mattison JA, Ottinger MA, et al. Aging in rhesus monkeys: relevance to human health interventions. Science. 2004;305:1423–6. https://doi.org/10.1126/science.1102541.
93. Wagner JD, Cann JA, Zhang L, et al. Diabetes and obesity research using nonhuman primates. In: Abee CR, editor. Nonhuman primates in biomedical research. New York: Springer; 2012. p. 699–732.
94. Tardif SD, Power ML, Ross CN, et al. Characterization of obese phenotypes in a small nonhuman primate, the common marmoset (Callithrix jacchus). Obesity (Silver Spring, MD). 2009;17:1499–505. https://doi.org/10.1038/oby.2009.77.
95. Islam MS. Animal models of diabetic neuropathy: progress since 1960s. J Diabetes Res. 2013;2013:149452. https://doi.org/10.1155/2013/149452.
96. Plentz RR, Palagani V, Wiedemann A, et al. Islet microarchitecture and glucose transporter expression of the pancreas of the marmoset monkey display similarities to the human. Islets. 2012;4:123–9. https://doi.org/10.4161/isl.19254.
97. Davis KJ, Bell RC, Wilhelmsen CL, et al. Immunohistochemical analysis of spontaneous pancreatic islet amyloid deposits in nonhuman primates. Vet Pathol. 1994;31:479–80. https://doi.org/10.1177/030098589403100414.
98. Wagner JE, Kavanagh K, Ward GM, et al. Old world nonhuman primate models of type 2 diabetes mellitus. ILAR J. 2006;47:259–71. https://doi.org/10.1093/ilar.47.3.259.
99. Bone marrow, thymus and blood: changes across the lifespan. Aging Health. 2009;5:385–393. https://doi.org/10.2217/ahe.09.31.
100. Grubić Kezele T, Ćurko-Cofek B. Age-related changes and sex-related differences in brain iron metabolism. Nutrients. 2020;12:2601. https://doi.org/10.3390/nu12092601.
101. Hardy PA, Gash D, Yokel R, et al. Correlation of R2 with total iron concentration in the brains of rhesus monkeys. J Magn Reson Imaging. 2005;21:118–27. https://doi.org/10.1002/jmri.20244.
102. Kastman EK, Willette AA, Coe CL, et al. A calorie-restricted diet decreases brain iron accumulation and preserves motor performance in old rhesus monkeys. J Neurosci Off J Soc Neurosci. 2010;30:7940–7. https://doi.org/10.1523/JNEUROSCI.0835-10.2010.
103. Kastman EK, Willette AA, Coe CL, et al. A calorie-restricted diet decreases brain iron accumulation and preserves motor

performance in old rhesus monkeys. J Neurosci. 2012;32:11897–904. https://doi.org/10.1523/jneurosci.2553-12.2012.
104. Rodríguez-Callejas JD, Cuervo-Zanatta D, Rosas-Arellano A, et al. Loss of ferritin-positive microglia relates to increased iron, RNA oxidation, and dystrophic microglia in the brains of aged male marmosets. Am J Primatol. 2019;81:e22956. https://doi.org/10.1002/ajp.22956.
105. Vila M. Neuromelanin, aging, and neuronal vulnerability in Parkinson's disease. Mov Disord. 2019;34:1440–51. https://doi.org/10.1002/mds.27776.
106. Herrero MT, Hirsch EC, Kastner A, et al. Neuromelanin accumulation with age in catecholaminergic neurons from Macaca fascicularis brainstem. Dev Neurosci. 1993;15:37–48. https://doi.org/10.1159/000111315.
107. Rose S, Nomoto M, Jackson EA, et al. Age-related effects of 1-methyl-4-phenyl-1,2,3,6-tetrahydropyridine treatment of common marmosets. Eur J Pharmacol. 1993;230:177–85. https://doi.org/10.1016/0014-2999(93)90800-w.
108. Ovadia A, Zhang Z, Gash DM. Increased susceptibility to MPTP toxicity in middle-aged rhesus monkeys. Neurobiol Aging. 1995;16:931–7. https://doi.org/10.1016/0197-4580(95)02012-8.
109. Emborg ME, Moirano J, Raschke J, et al. Response of aged parkinsonian monkeys to in vivo gene transfer of GDNF. Neurobiol Dis. 2009;36:303–11. https://doi.org/10.1016/j.nbd.2009.07.022.
110. Raz N, Lindenberger U, Rodrigue KM, et al. Regional brain changes in aging healthy adults: general trends, individual differences and modifiers. Cerebral Cortex. 2005;15:1676–89. https://doi.org/10.1093/cercor/bhi044.
111. Salat DH, Kaye JA, Janowsky JS. Prefrontal gray and white matter volumes in healthy aging and Alzheimer disease. Arch Neurol. 1999;56:338–44. https://doi.org/10.1001/archneur.56.3.338.
112. Lacreuse A, Raz N, Schmidtke D, et al. Age-related decline in executive function as a hallmark of cognitive ageing in primates: an overview of cognitive and neurobiological studies. Philos Trans R Soc Lond B Biol Sci. 2020;375:20190618. https://doi.org/10.1098/rstb.2019.0618.
113. Alexander GE, Chen K, Aschenbrenner M, et al. Age-related regional network of magnetic resonance imaging gray matter in the rhesus macaque. J Neurosci. 2008;28:2710–8. https://doi.org/10.1523/JNEUROSCI.1852 07.2008.
114. Sridharan A, Willette AA, Bendlin BB, et al. Brain volumetric and microstructural correlates of executive and motor performance in aged rhesus monkeys. Front Aging Neurosci. 2012;4:31. https://doi.org/10.3389/fnagi.2012.00031.
115. Liu JV, Bock NA, Silva AC. Rapid high-resolution three-dimensional mapping of T1 and age-dependent variations in the non-human primate brain using magnetization-prepared rapid gradient-echo (MPRAGE) sequence. Neuroimage. 2011;56:1154–63. https://doi.org/10.1016/j.neuroimage.2011.02.075.
116. Phillips KA, Watson CM, Bearman A, et al. Age-related changes in myelin of axons of the corpus callosum and cognitive decline in common marmosets. Am J Primatol. 2019;81:e22949. https://doi.org/10.1002/ajp.22949.
117. Philippens IH, Ormel PR, Baarends G, et al. Acceleration of amyloidosis by inflammation in the amyloid-Beta marmoset monkey model of Alzheimer's disease. J Alzheimers Dis. 2017;55:101–13. https://doi.org/10.3233/JAD-160673.
118. Ridley RM, Baker HF, Windle CP, et al. Very long term studies of the seeding of β-amyloidosis in primates. J Neural Transm. 2005;113:1243. https://doi.org/10.1007/s00702-005-0385-2.
119. Geula C, Nagykery N, Wu CK. Amyloid-beta deposits in the cerebral cortex of the aged common marmoset (Callithrix jacchus): incidence and chemical composition. Acta Neuropathol. 2002;103:48–58. https://doi.org/10.1007/s004010100429.
120. Shah P, Lal N, Leung E, et al. Neuronal and axonal loss are selectively linked to Fibrillar amyloid-β within plaques of the aged primate cerebral cortex. Am J Pathol. 2010;177:325–33. https://doi.org/10.2353/ajpath.2010.090937.
121. Sridharan A, Pehar M, Salamat MS, et al. Calorie restriction attenuates astrogliosis but not amyloid plaque load in aged rhesus macaques: a preliminary quantitative imaging study. Brain Res. 2013;1508:1–8. https://doi.org/10.1016/j.brainres.2013.02.046.
122. Uno H. The incidence of senile plaques and multiple infarction in aged macaque brain. Neurobiol Aging. 1993;14:673–4.
123. Sharma G, Huo A, Kimura T, et al. Tau isoform expression and phosphorylation in marmoset brains. J Biol Chem. 2019;294:11433–44. https://doi.org/10.1074/jbc.RA119.008415.
124. Teo L, Bourne JA. A reproducible and translatable model of focal ischemia in the visual cortex of infant and adult marmoset monkeys. Brain Pathol. 2014;24:459–74. https://doi.org/10.1111/bpa.12129.
125. Fujita D, Terada S, Ishizu H, et al. Immunohistochemical examination on intracranial calcification in neurodegenerative diseases. Acta Neuropathol. 2003;105:259–64. https://doi.org/10.1007/s00401-002-0640-7.
126. Rennenberg RJMW, Kessels AGH, Schurgers LJ, et al. Vascular calcifications as a marker of increased cardiovascular risk: a meta-analysis. Vasc Health Risk Manag. 2009;5:185–97. https://doi.org/10.2147/vhrm.s4822.
127. Maheshwari U, Huang SF, Sridhar S, et al. The interplay between brain vascular calcification and microglia. Front Aging Neurosci. 2022;14:848495. https://doi.org/10.3389/fnagi.2022.848495.
128. Youssef SA, Capucchio MT, Rofina JE, et al. Pathology of the aging brain in domestic and laboratory animals, and animal models of human neurodegenerative diseases. Vet Pathol. 2016;53:327–48. https://doi.org/10.1177/0300985815623997.
129. National Research Council. Committee on animal models for research on aging. Mammalian models for research on aging. Washington, DC: National Academy Press; 1981.
130. Leiva M, Peña T, Bayón A, et al. Phacoemulsification considerations in nonhuman primates. J Med Primatol. 2012;41:317–24. https://doi.org/10.1111/j.1600-0684.2012.00562.x.
131. Croft MA, McDonald JP, Katz A, et al. Extralenticular and lenticular aspects of accommodation and presbyopia in human versus monkey eyes. Invest Ophthalmol Vis Sci. 2013;54:5035–48. https://doi.org/10.1167/iovs.12-10846.
132. Neider MW, Crawford K, Kaufman PL, et al. In vivo videography of the rhesus monkey accommodative apparatus: age-related loss of ciliary muscle response to central stimulation. Arch Ophthalmol. 1990;108:69–74. https://doi.org/10.1001/archopht.1990.01070030075032.
133. Baskerville A, Cook RW, Dennis MJ, et al. Pathological changes in the reproductive tract of male rhesus monkeys associated with age and simian AIDS. J Comp Pathol. 1992;107:49–57.
134. Jeyaraj DA, Udayakumar TS, Rajalakshmi M, et al. Effects of long-term administration of androgens and estrogen on rhesus monkey prostate: possible induction of benign prostatic hyperplasia. J Androl. 2000;21:833–41.
135. Lewis RW, Kim JC, Irani D, et al. The prostate of the nonhuman primate: normal anatomy and pathology. Prostate. 1981;2:51–70.
136. McEntee MF, Epstein JI, Syring R, et al. Characterization of prostatic basal cell hyperplasia and neoplasia in aged macaques: comparative pathology in human and nonhuman primates. Prostate. 1996;29:51–9. https://doi.org/10.1002/(sici)1097-0045(199607)29:1<51::aid-pros8>3.0.co;2-l.
137. Mubiru JN, Hubbard GB, Dick EJ Jr, et al. Nonhuman primates as models for studies of prostate specific antigen and prostatic diseases. Prostate. 2008;68:1546–54. https://doi.org/10.1002/pros.20814.

138. Paniagua R, Nistal M, Sáez FJ, et al. Ultrastructure of the aging human testis. J Electron Microsc Tech. 1991;19:241–60. https://doi.org/10.1002/jemt.1060190209.
139. Goff K, Liukkonen J, Kubisch HM. Postmortem recovery and cryopreservation of spermatozoa from the vas deferens of rhesus macaques (Macaca mulatta). Theriogenology. 2009;72:834–40. https://doi.org/10.1016/j.theriogenology.2009.06.002.
140. Benz CC. Impact of aging on the biology of breast cancer. Crit Rev Oncol Hematol. 2008;66:65–74. https://doi.org/10.1016/j.critrevonc.2007.09.001.
141. Dewi FN, Cline JM. Nonhuman primate model in mammary gland biology and neoplasia research. Lab Anim Res. 2021;37:3. https://doi.org/10.1186/s42826-020-00053-1.
142. Wood CE, Usborne AL, Starost MF, et al. Hyperplastic and neoplastic lesions of the mammary gland in macaques. Vet Pathol. 2006;43:471–83. https://doi.org/10.1354/vp.43-4-471.
143. Wu A, Dong Q, Gao H, et al. Characterization of mammary epithelial stem/progenitor cells and their changes with aging in common marmosets. Sci Rep. 2016;6:32190. https://doi.org/10.1038/srep32190.
144. Nichols SM, Bavister BD, Brenner CA, et al. Ovarian senescence in the rhesus monkey (Macaca mulatta). Hum Reprod. 2005;20:79–83. https://doi.org/10.1093/humrep/deh576.
145. Park SU, Walsh L, Berkowitz KM. Mechanisms of ovarian aging. Reproduction. 2021;162:R19–33. https://doi.org/10.1530/REP-21-0022.
146. Abbott DH, Barnett DK, Colman RJ, et al. Aspects of common marmoset basic biology and life history important for biomedical research. Comp Med. 2003;53:339–50.

Other Noninfectious Conditions (Inflammatory/Degenerative/Proliferative, Immune-Mediated/Idiopathic/Unknown) in Nonhuman Primates

7

Katherine J. Olstad and Martina Bleyer

Abstract

There are several disease entities in nonhuman primates that primarily manifest as inflammatory, degenerative, and/or proliferative processes and are not assumed or currently known to be caused by infectious agents. Despite uncertainties in etiology and pathogenesis, a number of these entities such as endometriosis, marmoset wasting syndrome, and chronic segmental lymphocytic enteritis are prominent within colonies with a high degree of morbidity and mortality. Endometriosis is a proliferative disease of unidentified etiology that affects several Old World monkey species with a menstrual cycle, while gluten-sensitive enteropathy of macaques is an inflammatory disease, for which an immune-mediated pathogenesis has been established. Given the similarities of these entities in humans, they have been established as an important translational models of disease. Other entities such as diffuse idiopathic skeletal hyperostosis (DISH) and noninfectious alopecia are less debilitating but remain important entities to further try and understand. DISH presents as a chronic degenerative disorder affecting both Old and New World monkeys, and noninfectious alopecia is a recognized health issue in macaque colonies with poorly understood etiology.

Keywords

Chronic glomerulonephritis of marmosets · Endometriosis · Gluten-sensitive enteropathy of macaques · Marmoset wasting syndrome · Chronic lymphocytic enteritis (CLE) in common marmosets · Diffuse idiopathic skeletal hyperostosis (DISH) · Alopecia · Noninfectious

K. J. Olstad (✉)
California National Primate Research Center, University of California, Davis, Davis, CA, USA
e-mail: kjolstad@ucdavis.edu

M. Bleyer
German Primate Center, Pathology Unit, Göttingen, Germany
e-mail: mbleyer@dpz.eu

There are several disease entities in nonhuman primates that primarily manifest as inflammatory, degenerative, and/or proliferative processes and are not assumed or currently known to be caused by infectious agents. Despite uncertainties in etiology and pathogenesis, a number of these entities such as endometriosis, marmoset wasting syndrome, and chronic segmental lymphocytic enteritis are prominent within colonies with a high degree of morbidity and mortality. Endometriosis is a proliferative disease of unidentified etiology that affects several Old World monkey species with a menstrual cycle, while gluten-sensitive enteropathy of macaques is an inflammatory disease, for which an immune-mediated pathogenesis has been established. Given the similarities of these entities in humans, they have been established as an important translational models of disease. Other entities such as diffuse idiopathic skeletal hyperostosis (DISH) and noninfectious alopecia are less debilitating but remain important entities to further try and understand. DISH presents as a chronic degenerative disorder affecting both Old and New World monkeys, and noninfectious alopecia is a recognized health issue in macaque colonies with poorly understood etiology.

7.1 Chronic Glomerulonephritis of Marmosets

Chronic glomerulonephritis, also known as spontaneous progressive glomerulonephropathy, is frequently observed in male and female common marmosets (*Callithrix jacchus*) of all age groups [1]. It is characterized by mesangial expansion, interstitial inflammation, regenerative tubules, and intratubular proteinaceous casts [2, 3]. Clinical symptoms and macroscopic changes are typically not evident until disease has reached the advanced stages. In such advanced cases, the kidneys are pale pink and firm and have a coarsely granular surface with an associated proteinuria (Fig. 7.1) [1, 4, 5].

I. Kondova - Perseng et al. (eds.), *Atlas of Diagnostic Pathology in Nonhuman Primates*,
https://doi.org/10.1007/978-3-031-41280-6_7

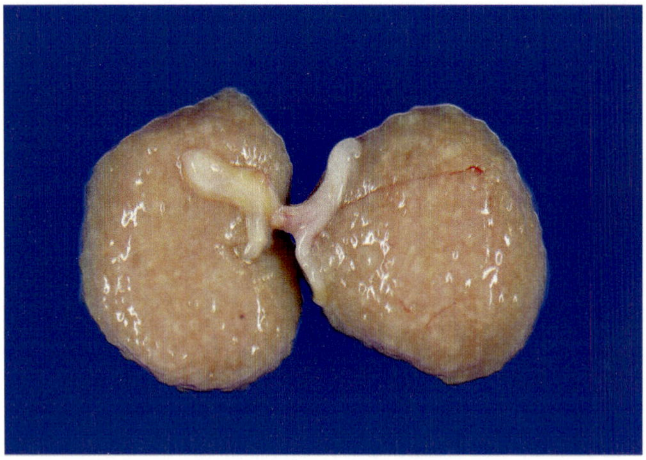

Fig. 7.1 Chronic glomerulonephritis in a 6-year-old male common marmoset (*Callithrix jacchus*). Both kidneys are slightly enlarged and have a pale tan parenchyma and an irregular granular surface. Macroscopic observation of chronic glomerulonephritis is only possible in late-stage disease, when there is prominent tubulointerstitial inflammation and fibrosis and clinically apparent renal insufficiency

For the histological evaluation of renal lesions in marmosets, Yamada et al. (2013) suggested a scoring system with four grades of increasing severity. Grade 1 consist of tubulointerstitial lesions including hyaline cast, regenerative tubules, scattered perivascular foci of interstitial lymphocytic infiltration, and mild interstitial fibrosis. In grade 2, advanced tubulointerstitial changes as previously described are accompanied by glomerular changes characterized by a focal or multifocal increase of mesangial matrix, predominantly in the hilar area, and/or segmental mesangial cell hyperplasia. In grade 3, the glomerular changes involving the mesangial matrix and cell proliferation is global and diffuse, and tubulointerstitial lesions are more pronounced. In grade 4, there is more severe matrix proliferation, collapse of capillaries, dilation of the Bowman's capsule with thickening of its basement membrane, and widespread interstitial fibrosis, and atypical tubular regeneration. With progressing glomerular disease, adhesions develop between Bowman's capsules and glomerular tufts (synechiae), and glomeruli may become sclerotic (Fig. 7.2) [1, 3, 6–8]. In

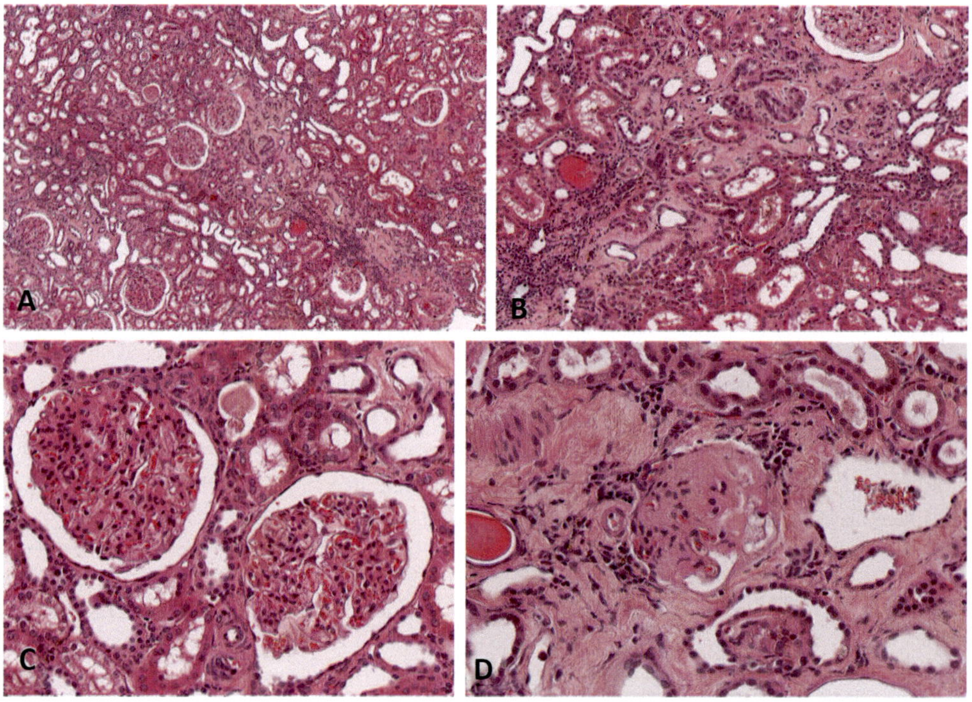

Fig. 7.2 Chronic glomerulonephritis in a 14-year-old male common marmoset (*Callithrix jacchus*) with both glomerular and tubulointerstitial lesions (**a**). The interstitium reveals scattered lymphocytic inflammatory cell foci and moderate fibrosis. There is tubular atrophy and regeneration and occasional hyaline casts (**b**). Glomerular changes are characterized by mesangial expansion affecting almost all glomeruli (diffuse) (**c**). Some glomeruli have global sclerosis (**d**). Paraffin-embedded H&E-stained sections

addition to a hematoxylin and eosin (H&E) stain, special stains like methenamine-silver impregnations and periodic acid-Schiff (PAS) reaction are diagnostically useful to visualize and classify structural changes of the mesangium and the basement membranes (Fig. 7.3) [8]. Electron-dense deposits in the mesangial matrix on transmission electron microscopy is a hallmark lesion (Fig. 7.4) [9]. Immunohistochemical investigations have shown that the electron-dense mesangial particles correspond to accumulations of both IgM and IgA and that there is concurrent glomerular complement factor 3 (C3) deposition indicating activation of the complement pathway (Fig. 7.5). Marmoset glomerulonephritis is, therefore, suspected to have an immune-mediated pathogenesis associated with circulating immune complexes. Specific antigens responsible for immune complex formation have not yet been conclusively identified, but diet components are suspected of being involved [1, 7, 10]. Although glomerular and tubulointerstitial alterations are also regularly observed in kidneys of marmosets clinically diagnosed with "marmoset wasting syndrome," chronic glomerulonephritis of marmosets represents an unrelated spontaneous disease entity in this nonhuman primate species [5].

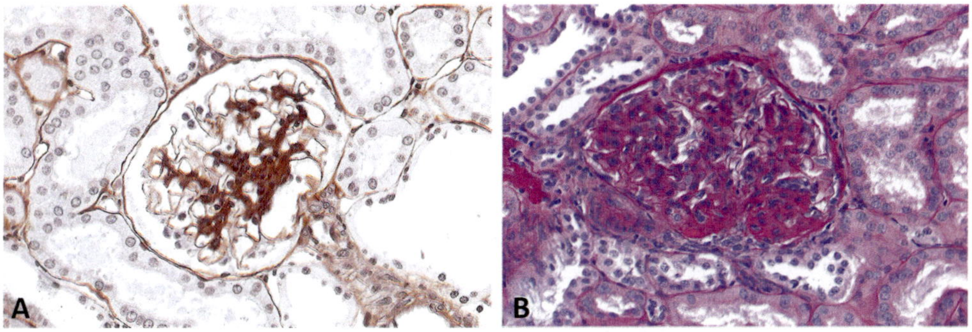

Fig. 7.3 Chronic glomerulonephritis of marmosets in a 4-year-old female common marmoset (Callithrix jacchus). (**a**) Methenamine silver stain highlights segmental increase of mesangial matrix and mesangial cell hyperplasia. Plastic-embedded modified Movat-stained section. (**b**) PAS reaction visualizes thickening of the Bowman's capsule and diffuse synechiae. The distinct increase of mesangial matrix is accentuated in the hilar region. Paraffin-embedded PAS-stained section

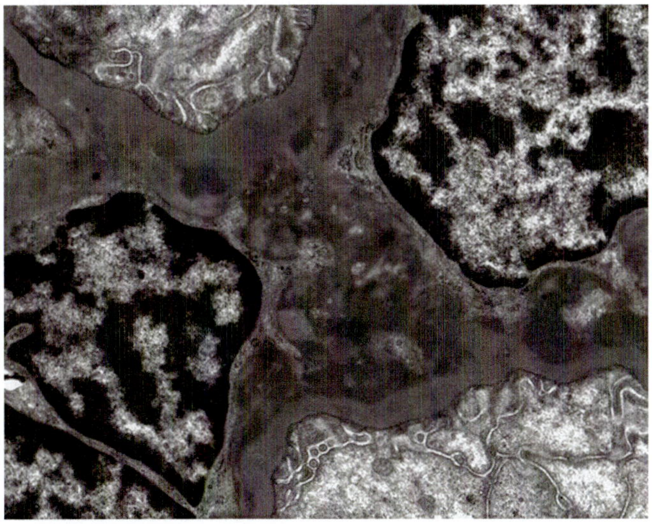

Fig. 7.4 Chronic glomerulonephritis in a 4-year-old male common marmoset (*Callithrix jacchus*). The main ultrastructural feature of marmoset glomerulonephritis is cloudy electron dense material in the mesangial matrix that corresponds to immunoglobulin depositions. Additionally, there is a noticeable overall increase in mesangial matrix. Transmission electron microscopy with uranyl acetate and lead citrate

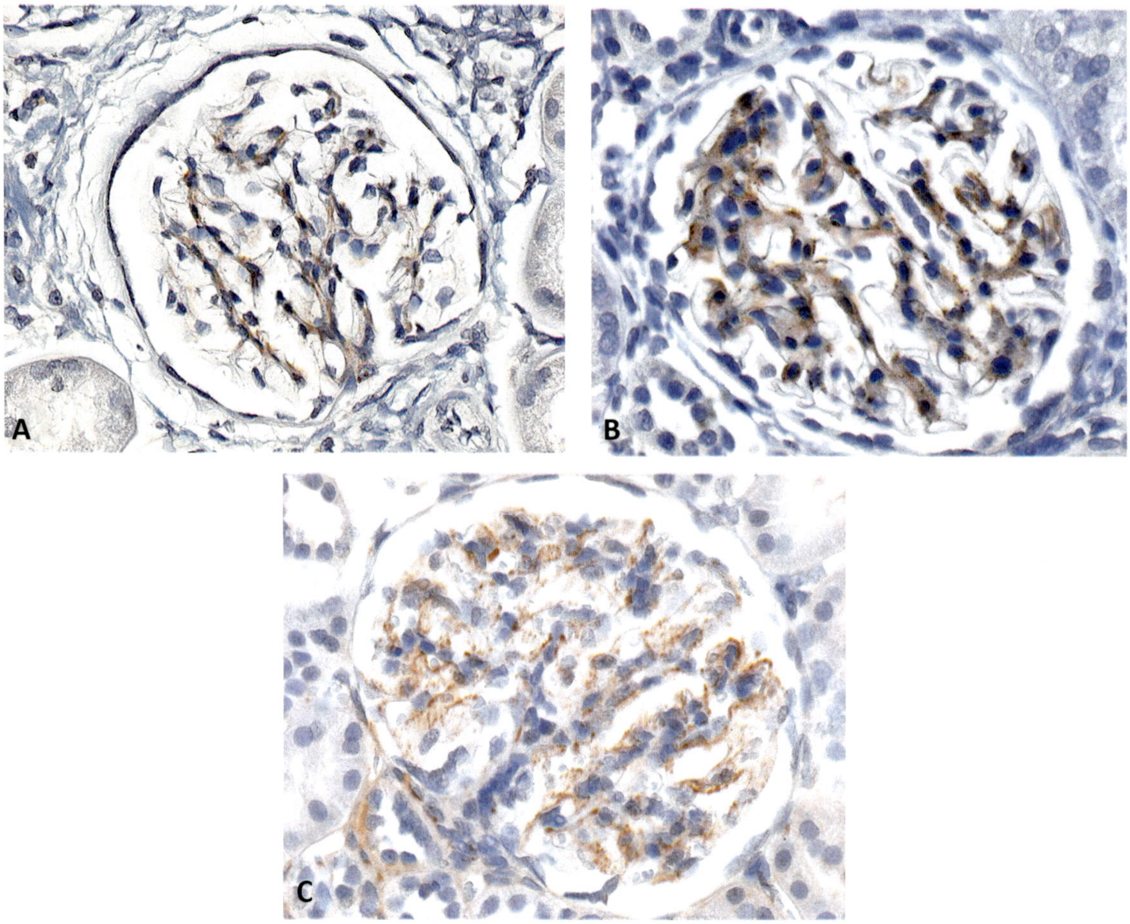

Fig. 7.5 Chronic glomerulonephritis of marmosets in a 7-year-old female (**a, b**) and in a 3-year-old female (**c**) common marmoset (Callithrix jacchus). (**a**) By immunohistochemistry, immunoglobulin A accumulations are demonstrated in the mesangial matrix. (**b**) By immunohistochemistry, immunoglobulin M accumulations are demonstrated in the mesangial matrix. (**c**) Immune complex precipitations are regularly accompanied by complement factor 3 (C3) depositions.

7.2 Endometriosis

Endometriosis occurs in several Old World monkey species with a menstrual cycle and is defined by the presence of ectopic endometrial glandular and stromal tissue outside the uterine cavity. Endometriosis has principally been studied in rhesus macaques and baboons, but spontaneous disease has also been reported in cynomolgus and pig-tailed macaques, African green monkeys, a DeBrazza's monkey, a gray-cheeked mangabey, and a gorilla [11–18]. In captive colonies of rhesus macaques, the prevalence may be as high as 30% in sexually mature females [19]. Clinical symptoms vary with the location of ectopic endometrial tissue and include dysmenorrhea, pelvic pain, abdominal distension, depression, and infertility [15, 20, 21]. Macroscopically, endometriosis presents as cysts or cystic masses on the ovary and/or attached to the uterus as well as fibrous peritoneal adhesions or peritoneal plaques affecting the uterus, ovaries, urinary bladder, and terminal parts of the colon. As cysts often contain a sero-hemorrhagic to bloody fluid, they are also referred to as "chocolate cysts" (Fig. 7.6) [12, 15, 17, 22, 23]. Endometriosis is most frequently observed in the pelvic cavity but may also extend to the serosal surfaces of abdominal organs, like the small intestine, diaphragm, and, rarely, liver, spleen, and stomach [22]. Complications of endometriosis include malignant transformation, hemoperitoneum, suppurative peritonitis, septicemia, intestinal perforation, and hydronephrosis/hydroureter [22, 23] Histologically, endometriotic lesions can be grouped into four different types depending on the degree of ectopic endometrial gland and stromal differentiation: well differentiated, purely stromal, mixed differentiation, and poorly differentiated. In well-differentiated endometriosis, the glandular epithelium is simple to pseudostratified and columnar resembling the epithelium of normal endometrium, while in poorly differentiated endometriosis, the glandular epithelium is simple and cuboidal to flattened. A mixed differentiation refers to an epithelial component composed of both endometrial-like cells and undifferentiated cells. The stromal component in well-differentiated forms is typically highly cellular and matrix-poor, sometimes accompanied by distinct spiral arterioles, but other stromal types of differentiation, e.g., fibrotic, myxoid, fibroangioblastic, or myogenic phenotypes, can be identified in all four histological types [23–25]. Occasionally, endometriotic stroma undergoes decidualization [26]. Endometriotic lesions may also contain cystic structures, either in the form of dilated glands with serous or mucinous secretions or as blood-filled cavities with desquamation of glandular epithelium, stromal inflammatory cell infiltrates, hemorrhages, and necrotic debris in terms of menstrual activity (Figs. 7.7 and 7.8) [23]. Immunohistochemical studies have demonstrated steroid receptor expression in both normal and ectopic endometrium (Fig. 7.8) but have also indicated a decline of estrogen and progesterone receptor expression in less differentiated endometriotic tissue [12, 23, 27]. Although several theories have been proposed to explain the etiopathogenesis of endometriosis, including retrograde menstruation, coelomic metaplasia, and embryonic rest theory, the exact pathogenetic pathways of initial development, growth, and maintenance of endometriotic lesions are still unclear [28–31].

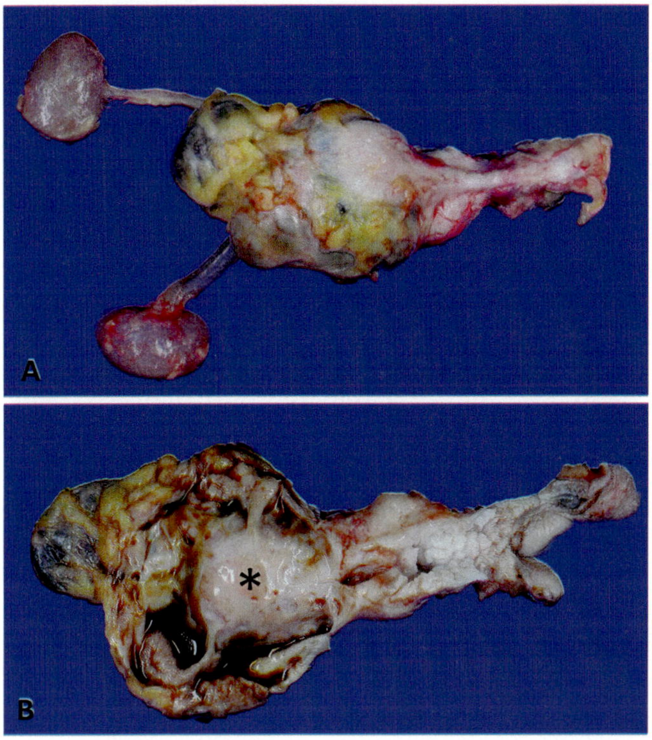

Fig. 7.6 Endometriosis in a 10-year-old female rhesus macaque (Macaca mulatta). (**a**) The uterus is surrounded by a multicystic mass encompassing the ovaries. Serosal fibrous adhesions are present between the mass, the urinary bladder, and the greater omentum. There is unilateral hydroureter. (**b**) The multicystic mass surrounding the uterus* contains a brown fluid (chocolate cyst).

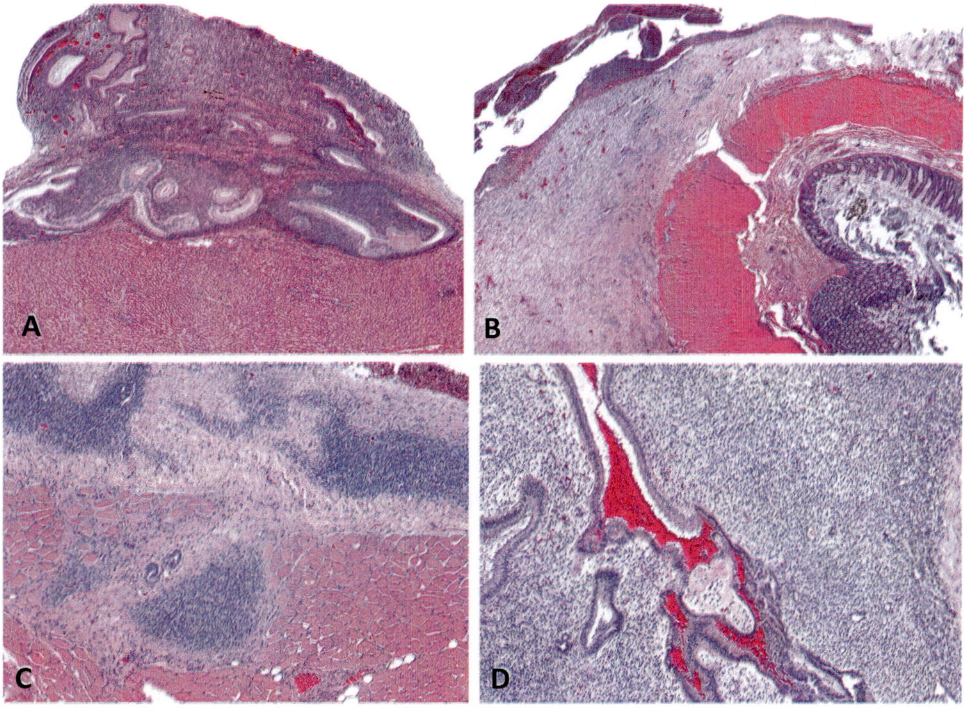

Fig. 7.7 Endometriosis in a 19-year-old female rhesus macaque (Macaca mulatta). (**a**) The subserosal endometriotic lesion consists of a highly cellular and vascularized stroma and well-differentiated glands. (**b**) The colonic serosa is expanded by a moderately cellular endometriotic stroma covered by a fibrinohemorrhagic exudate. (**c**) This predominantly stromal endometriotic lesion is attached to the diaphragm and reveals an infiltrative growth pattern. (**d**) Glands may contain erythrocytes and proteinaceous fluid as well as desquamated epithelial cells reflecting secretory and menstrual activity of the endometriotic tissue

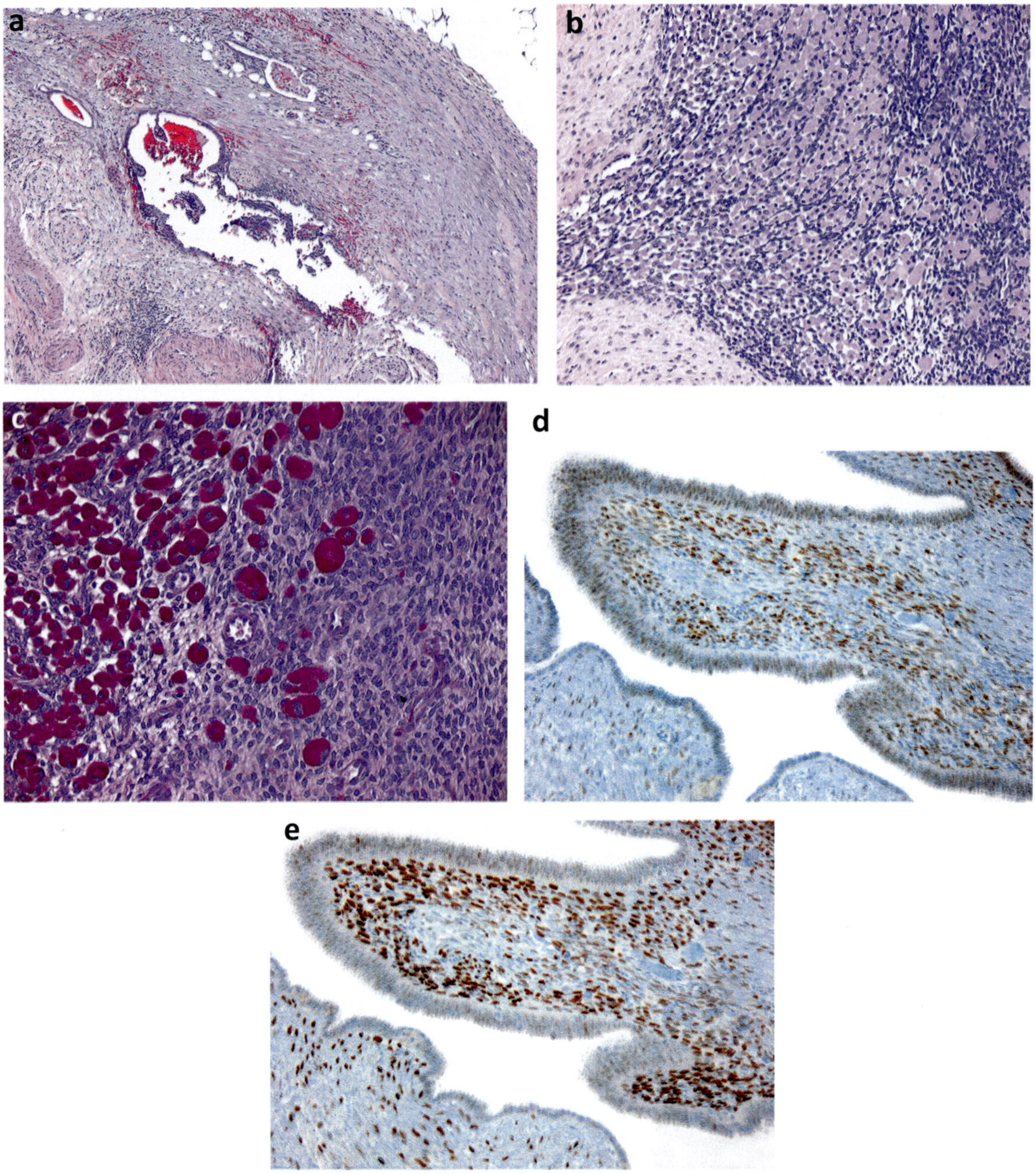

Fig. 7.8 Endometriosis in a 10-year-old female rhesus macaque (Macaca mulatta). (**a**) In poorly differentiated endometriosis, the glandular epithelium is cuboidal to flat. The stroma is loosely arranged, characterized by a myxoid differentiation and reveals multifocal hemorrhages. (**b**) Numerous decidual cells ("deciduosis") of epithelioid shape are present in an endometriotic lesion in the serosa of the urinary bladder. (**c**) The cytoplasm of decidual cells can be highlighted with PAS reaction. (**d**) Estrogen receptor labeling in stromal and epithelial cells in the endometriotic tissue. (**e**) Progesterone labeling in the stromal and epithelial cells in the endometriotic tissue.

7.3 Gluten Sensitive-Like Enteropathy in Macaques

Celiac disease (CD) in humans, also known as celiac sprue, is defined as an inherited enteropathy induced by the intolerance to dietary gluten from wheat (gliadin), rye (secalins), and barley (hordeins) [32, 33]. The first known documented nonhuman primate (NHP) case of a celiac-like enteropathy based on clinical signs, characteristic histology, and response to a gluten free diet (GFD) was reported by the Wake Forest School of Medicine in 1988 in a cynomolgus macaque received from a commercial importer [34]. Since then, a NHP model of naturally occurring gluten sensitive-like enteropathy (GSE) has been established in a predominantly juvenile (≤4 years old) population of Indian rhesus macaques at the Tulane National Primate Research Center (TNPRC). Although this condition has been well characterized at the TNPRC, cases resembling this GSE are rarely identified at other primate centers and often lack diagnostics other than histology. The animals investigated at TNPRC with GSE present with chronic diarrhea, weight loss, dehydration, and abdominal distension [32]. Enteropathy is the most common presentation, but disease can be systemic due to transglutaminase (TG) isotopes within other tissues [35]. One example of a rare concurrent manifestation in humans is a pruritic vesicular skin rash known as dermatitis herpetiformis (DH) [32, 36]. Although rare, DH-like skin lesions have been documented in two rhesus macaques in a single case report from the GSE cohort at the TNPRC [37].

Histology has been considered the "gold standard" for diagnosis, but histologic features are nonspecific and shared with a variety of intestinal disorders [38, 39]. Therefore, for an accurate diagnosis, biopsy should be paired with serology for anti-TG2 or anti-gliadins, and lesions should resolve in the absence of dietary gluten. Most severe lesions are within the proximal jejunum, but the distal duodenum is considered the most practical tissue to evaluate based on the accessibility by endoscope. Lesions can be patchy, so multiple biopsies are recommended. Similar to CD, lesions can consist of intraepithelial lymphocytes, mononuclear inflammation within the lamina propria, shortening of the villi, and crypt hyperplasia [40]. Proper tissue orientation is crucial to properly evaluate the villous height (Fig. 7.9).

It is important to differentiate GSE from idiopathic chronic diarrhea (ICD), which is a common aliment affecting

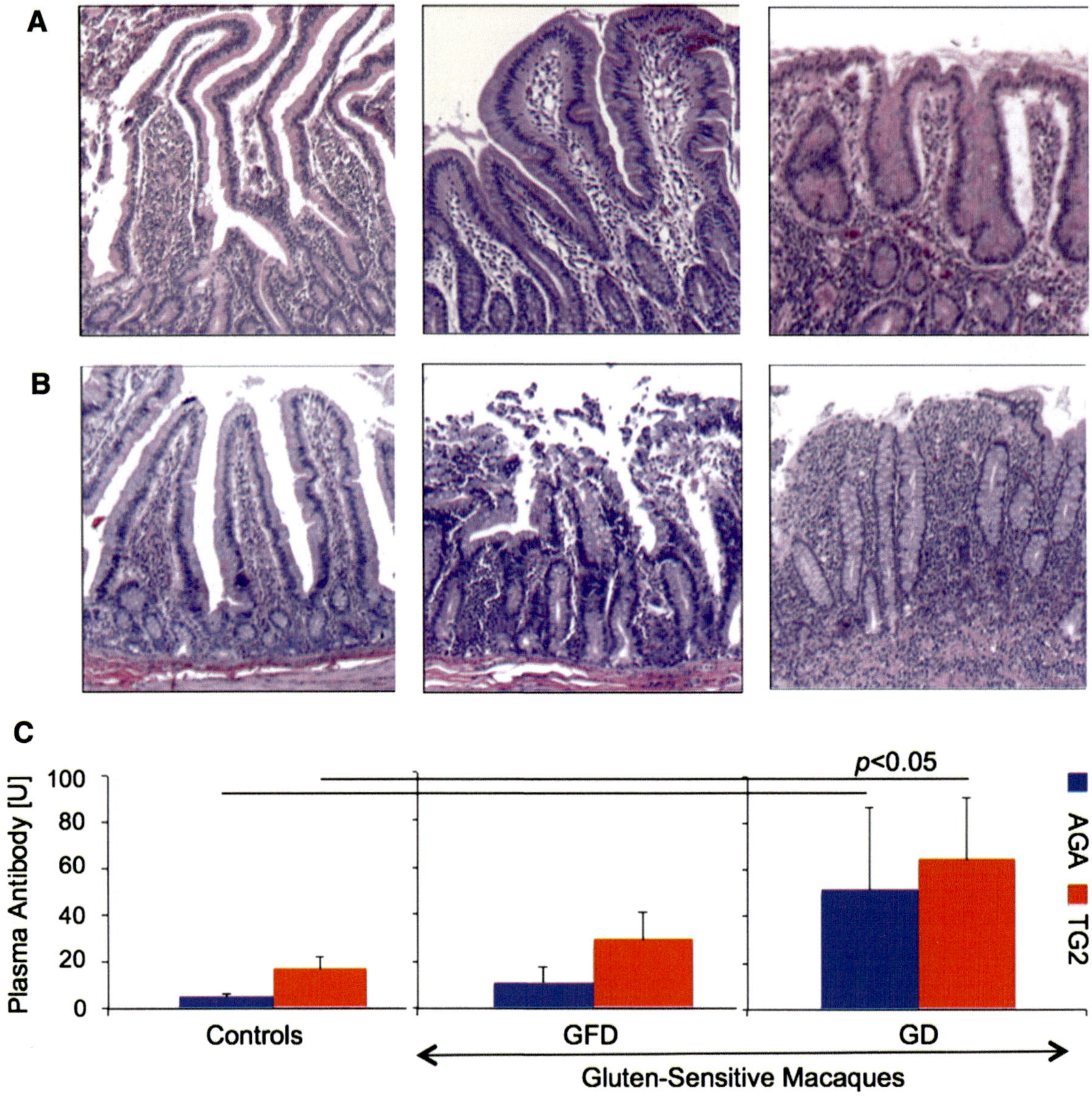

Fig. 7.9 Histopathological evaluation of H&E-stained duodenum (**a**) and jejunum (**b**) biopsy samples from control (left column) and gluten-sensitive rhesus macaques (two right columns). (**a**) Duodenum from a representative healthy control animal did not show any histopathology while on gluten diet (GD). Duodenum from a representative gluten-sensitive macaque on GFD showed, despite being in clinical and immunological remission, some lymphocytic and plasmacytic infiltration of lamina propria. Moderate enteropathy with shortened, fused, and blunted villi appeared within 3 weeks after introduction of GD in the same animal. (**b**) Jejunum from the control animal did not show any histopathological responses to GD. Jejunum from this gluten-sensitive animal shows mild villous atrophy (VA) despite of being on GFD for almost a year. Upon introduction of dietary gluten in the same animal, rapid progression toward severe GSE characterized by disrupted epithelial layer, marked VA, and lymphoplasmacytic enteritis took place. (**c**) AGA and TG2 plasma antibody levels in different cohorts are shown. In contrast to the control group, introduction of dietary gluten to gluten-sensitive group was associated with rise ($p < 0.05$) of AGA and TG2 plasma antibodies. (*From* Xu et al. [41]; *with permission*)

captive rhesus macaques at multiple primate centers. ICD is also a diagnosis of exclusion based on persistence of clinical signs despite supportive therapy and failure to identify infectious agents. In contrast to GSE, ICD clinical signs and lesions should not resolve on a gluten-free diet, and lesions are primarily within the large intestine, although there can also be a concurrent increase in intraepithelial lymphocytes within the small intestine and sometimes villus blunting [42].

7.4 Marmoset Wasting Syndrome

Marmoset wasting syndrome (MWS) is a term used to describe a clinical presentation of varied clinical signs, which most consistently involve weight loss accompanied by decreased muscle mass (Fig. 7.10) and often subsequent death in colonies of captive common marmosets (*Callithrix jacchus*) and cotton-top tamarins (*Saguinus oedipus*). Other reported clinical signs include diarrhea, hindlimb paresis/paralysis [7, 43], tail base alopecia, and anemia [44]. Lesions associated with MWS are numerous and connected with multiple etiologies; therefore, the term MWS is discouraged as it encompasses more than one entity. Associated pathologic lesions include nonspecific lymphoplasmacytic tubulointerstitial nephritis (Fig. 7.11), systemic AA amyloid deposition that can often occur in the renal glomeruli and interstitium, liver space of Disse, and lamina propria of the intestinal tract, pancreatitis, parasitism, enteropathogenic *E. coli*, *Campylobacter* infections, and some forms of inflammatory bowel disease (IBD) [44–46]. Inflammatory bowel disease (IBD) is a consistent finding among marmosets diagnosed with marmoset wasting syndrome [45, 47–50]. In retrospective studies done at the New England Primate Research Center and Southwest National Primate Research Center, approximately 60.5% and 31–44% of marmosets, respectively, had some degree of IBD associated with MWS [49, 51]. Specific entities, such as lymphocytic enteritis, of MWS will be discussed later in this chapter.

A

B

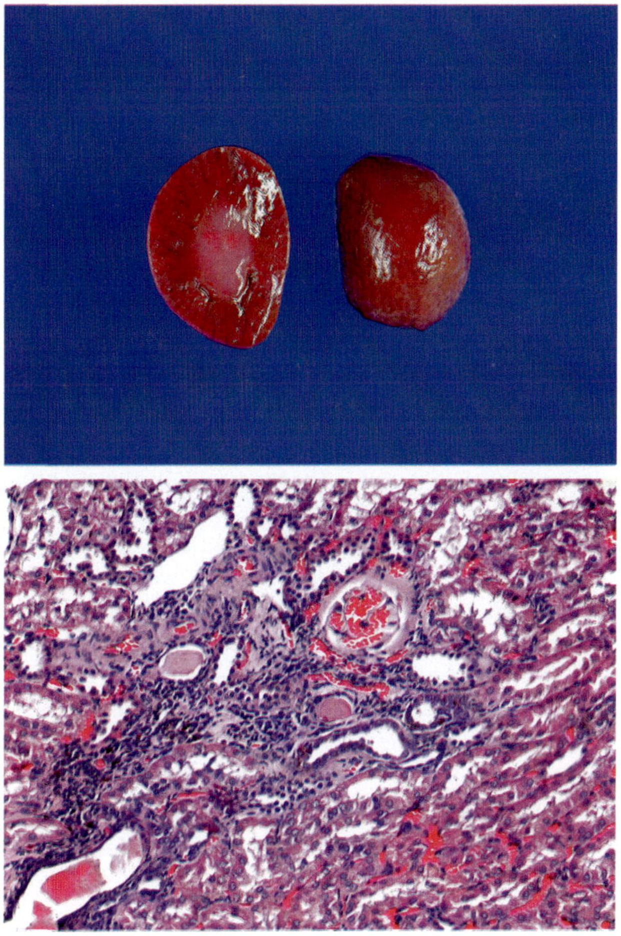

Fig. 7.11 Nonspecific lymphoplasmacytic interstitial nephritis. (**a**) The capsular surface is irregular, and the cortex is diffusely mottled pale tan to red. (**b**) Histologically there is an interstitial nephritis characterized by lymphocytes and plasma cells in the interstitium accompanied by small amounts of fibrosis. Tubules are variably dilated lined by attenuated epithelial cells and contain intraluminal eosinophilic material.

Fig. 7.10 Marmoset diagnosed with marmoset wasting syndrome. There is generalized muscle atrophy, poor haircoat, and alopecia of the tail

7.5 Chronic Segmental Lymphocytic Enteritis (CLE)

Inflammatory bowel disease (IBD) is one of the most common causes of death in captive marmosets and is often diagnosed clinically as marmoset wasting syndrome. However, IBD in marmosets exist as a spectrum of disorders [52]. In particular, a distinct form of IBD diagnosed in marmosets is further classified as chronic segmental lymphocytic enteritis (CLE) which is often associated with low serum albumin [47]. Affected animals clinically present with weight loss in adults, and failure to thrive in juveniles [49]. Diarrhea is only occasionally observed. The pathogenesis of CLE is currently unknown, but there is speculation in an association with viruses, gluten sensitivity, dietary protein deficiency, and pancreatic duct parasitism [49, 53, 54].

A definitive diagnosis is made in combination of clinical signs, low albumin, and biopsy results. Lesions have been described in the small intestine as segmental to diffuse consisting of thickening intestinal wall (Fig. 7.12) [49, 54]. Histologic lesions are characterized by villous blunting, crypt hyperplasia, increase in lymphocytes and plasma cells in the lamina propria, and occasional intraepithelial CD3- and CD8-positive lymphocytes (Fig. 7.13) [49]. In severe cases there can be complete loss of villus architecture. These lesions are nonspecific but does resemble gluten-sensitive enteropathy in macaques [49].

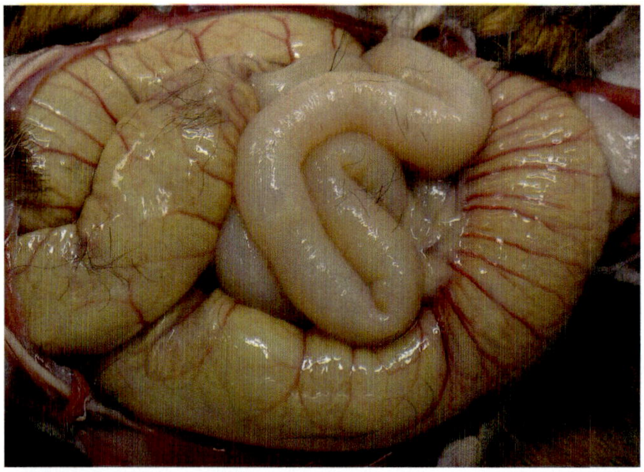

Fig. 7.12 Chronic lymphocytic enteritis in a marmoset. The small intestinal loops are pale tan and dramatically dilated and thickened with a sharp demarcation from the more normal sections of large intestine. Image courtesy of Heather Simmons

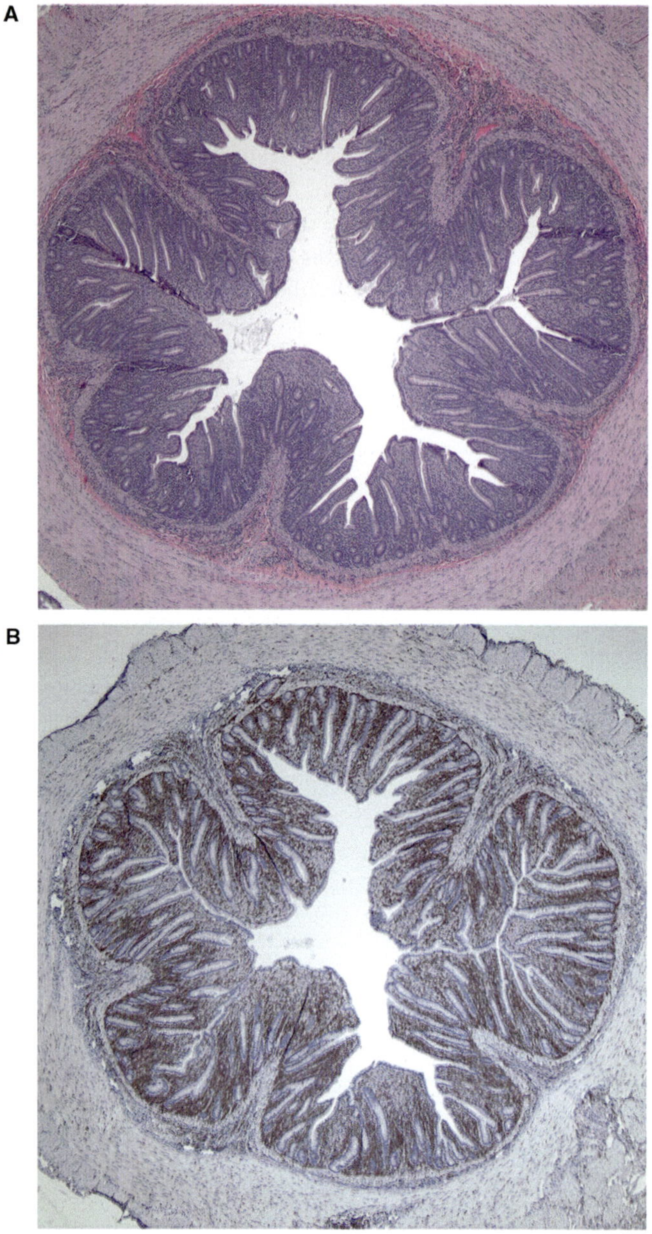

Fig. 7.13 Chronic lymphocytic enteritis (CLE) in common marmosets. (**a**) A section of jejunum with villous blunting and fusion. The lamina propria is expanded by an increased number of lymphocytes and plasma cells that separate and raise the crypts. (**b**) Inflammatory cells in the lamina propria are predominantly CD3 labeled T lymphocytes. Image courtesy of Dr. Heather Simmons.

7.6 Diffuse Idiopathic Skeletal Hyperostosis (DISH)

Diffuse idiopathic skeletal hyperostosis (DISH) is a systemic disease that affects the axial as well as the appendicular skeleton and is characterized by new bone formation at entheseal sites [55]. Although the cause is unknown, DISH in humans has been shown to have an increased prevalence associated with diabetes mellitus, hyperuricemia, dyslipidemia, and glucose intolerance [43] or a complication of long-term treatment with vitamin A analogs such as retinoids [56]. The clinical presentation occurs in older individuals and is often mild considering the severity of the bony lesions. Clinical signs can include tendonitis, back pain, spinal stiffness, and dysphagia secondary to cervical osteophytes [57–59]. Cases of vertebral hyperostosis meeting the criteria of DISH have infrequently been reported in gorillas, rhesus macaques, barbary macaques, and spider monkeys [60–64].

Criteria needed to make a diagnosis requires imaging which reveals the characteristic hyperostotic spondylosis involving at least four adjacent vertebral bodies. The thoracic spine is the most common region affected, but lesions can also occur in other spinal segments as well as in peripheral joints. Excessive calcification and ossification mainly involve the entheses, i.e., the bony attachments of tendons, ligaments, and joint capsules, resulting in anterolateral bridging ("flowing") enthesophytes (Fig. 7.14) in the absence of intervertebral disc abnormalities, apophyseal joint degeneration, or sacroiliac inflammatory changes [55, 65, 66]. Extraspinal entheseal hyperostosis commonly affects periarticular ligaments and tendons of the tibiofibular and metacarpophalangeal joints, shoulder, elbow, knee, and pelvis [61, 67, 68].

Histopathology is nonspecific and largely depends on the age of the lesion. Enthesophytes consist of fibrocartilage, woven bone, or mature lamellar bone with variable degrees of calcification and vascularization (Fig. 7.15), and sometimes inflammatory cell infiltration and degeneration of the peripheral annulus fibrosus [63, 69, 70].

The main differential diagnoses include spinal osteoarthritis and inflammatory spondyloarthropathy (primarily ankylosing spondylitis) [68]. In contrast to DISH, major characteristic findings of spinal osteoarthritis (spondylosis deformans) include the primary involvement of the lumbosacral vertebrae, circumferential osteophytes of the vertebral articular plate, and bone sclerosis across articular surfaces with reduced disc height [55, 68]. In spondyloarthropathy, there typically is early and major sacroiliac joint involvement, lateral osteophytosis of vertebrae progressing to vertical oriented ankylosis, accompanied by erosive arthritis [63, 68, 71]. Differentiation between these disease entities is impeded by the observation that they may coexist, exhibit overlaps, or have specific variations in their anatomical distribution [58, 71].

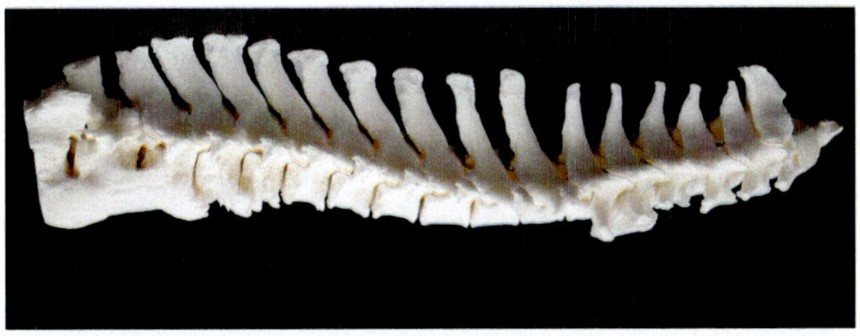

Fig. 7.14 Diffuse skeletal idiopathic hyperostosis (DISH) in a 27-year-old male collared mangabey (*Cercocebus torquatus*). There are several anterolateral osteophytes and anteroventral "flowing" bony spurs throughout the thoracic spine. Severe bridging hyperostosis involving the anterior longitudinal ligament affects several lower thoracic vertebrae

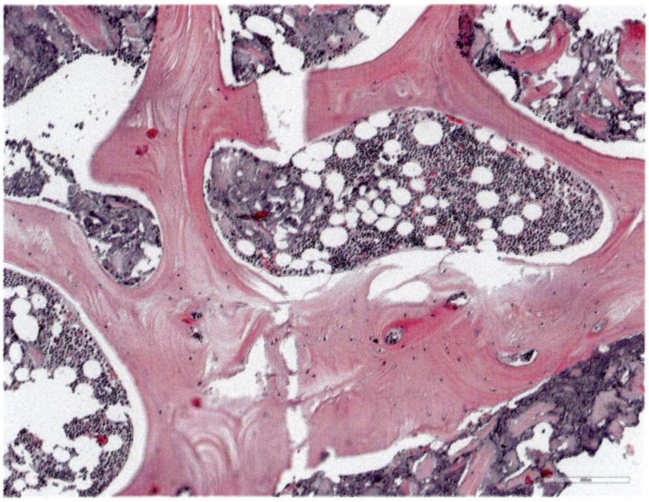

Fig. 7.15 Histology of DISH in the lower spinal cord of a 27-year-old male collared mangabey (*Cercocebus torquatus*). There is noninflammatory metaplastic ossification with presence of hematopoietic bone marrow in the anterior longitudinal ligament. Decalcified, paraffin-embedded H&E-stained section

7.7 Alopecia

Alopecia is a common problem in captive nonhuman primate populations. Reported incidences of hair loss in rhesus macaque colonies range from less than 30% to more than 80% [72–76]. Alopecia is regarded as a multifactorial condition with many contributing extrinsic and intrinsic factors including seasonal variations, aging, rank, sex, housing condition, reproductive state, skin disorders, nutritional deficiencies, behavioral abnormalities, and stress [72, 74, 75, 77, 78]. As part of the health and welfare management for laboratory-housed primate colonies, different scoring systems have been established to monitor hair coat quality; to differentiate between pathological, physiological, and psychological conditions; and to identify animals in need of further diagnosis or treatment [73, 79]. The extent of hair loss can range from small focal spots to large areas of missing hair, often in a bilateral symmetric pattern [72, 76, 79]. All body parts can be affected, but lesions are most common on the back, followed by the extremities (Fig. 7.16) [78]. Skin

Fig. 7.16 Alopecia in a nursing female outdoor-housed rhesus macaque (*Macaca mulatta*). Hair coat thinning and circumscribed total hair loss is present on the dorsum, the lateral abdomen, and the thigh

underlying the alopecic areas do generally not reveal distinct pathological changes, expect for mild lichenification, erythema, and scaling in some cases [74, 80]. Histological changes consist of mild epidermal orthokeratotic hyperkeratosis and mild lymphocytic or lymphohistiocytic perivascular dermatitis that may be accompanied by follicular keratosis, acanthosis, and dermal edema (Fig. 7.17) [74, 78, 81]. There are no differences in the number of anagen, catagen, and telogen follicles [80].

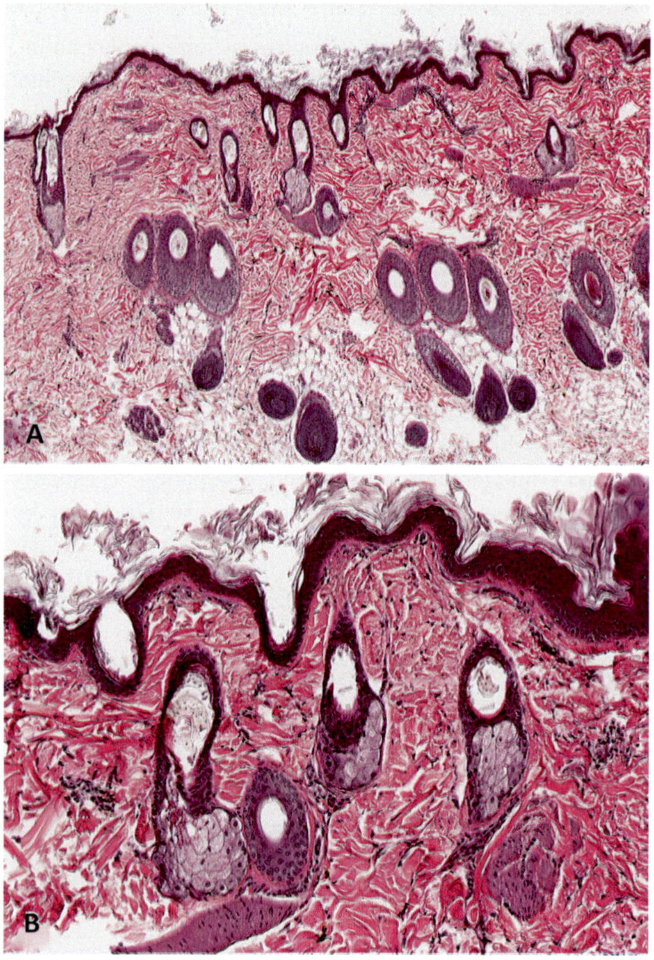

Fig. 7.17 Alopecia in a 4-year-old male rhesus macaque (Macaca mulatta). (**a**) Skin biopsy reveals mild lamellar orthokeratotic hyperkeratosis and infundibular keratosis. (**b**) Minimal perivascular lymphocytic dermatitis and superficial dermal edema is present.

7.8 Idiopathic Chronic Diarrhea

Idiopathic chronic diarrhea (ICD) is an apparent noninfectious diarrhea syndrome that is a major cause of morbidity and mortality of rhesus macaques. ICD most often affects juvenile macaques aged 2–4 years old. At the California National Primate Research Center (CNPRC), up to ~5% of annual diarrhea cases are diagnosed as ICD [82]. The diagnosis is by exclusion based on the lack of identifiable infectious diarrheagenic agents [42]. Clinical signs consist of a reoccurring watery diarrhea often linked to weight loss and dehydration. ICD is a large intestinal disease primarily involving the cecum and proximal ascending colon [42, 82].

The pathogenesis is unknown. Although several viruses and bacterial agents have been associated, there is currently no evidence of causation [82]. Given the overrepresentation of nonsignificant *Campylobacter* species isolated from ICD animals, increased numbers of commensal parasite burdens, and loss of surface associated *Helicobacter macacae* (seen as a blue brush border), dysbiosis likely plays a significant role in the pathogenesis [42].

Gross lesions consist of watery non-bloody diarrhea with a mild to moderately thickened and nonulcerated mucosa (Fig. 7.18). Colonic lymph nodes are often enlarged. Histologic lesions are nonspecific and variable primarily involving the large intestine. Lesions can include crypt hyperplasia, crypt abscesses with mucus plugs or neutrophil infiltrations, loss of goblet cells, increased lymphocytes, and plasma cells in the lamina propria or sometimes intraepithelial, superficial mucosal attenuation or tufting, and a loss of the surface associated blue brush border (Fig. 7.19). Paneth cell metaplasia in the colon can also be rarely seen. There is often an associated increase in abundance of trichomonad-like and *Balantidium* parasites within the lumen and sometimes crypts. Although this is a disease of the large intestine, there can be very mild subtle changes in the small intestine consisting of rare intraepithelial lymphocytes and rare villus blunting. Fibrosis is not a feature of this entity [42, 82].

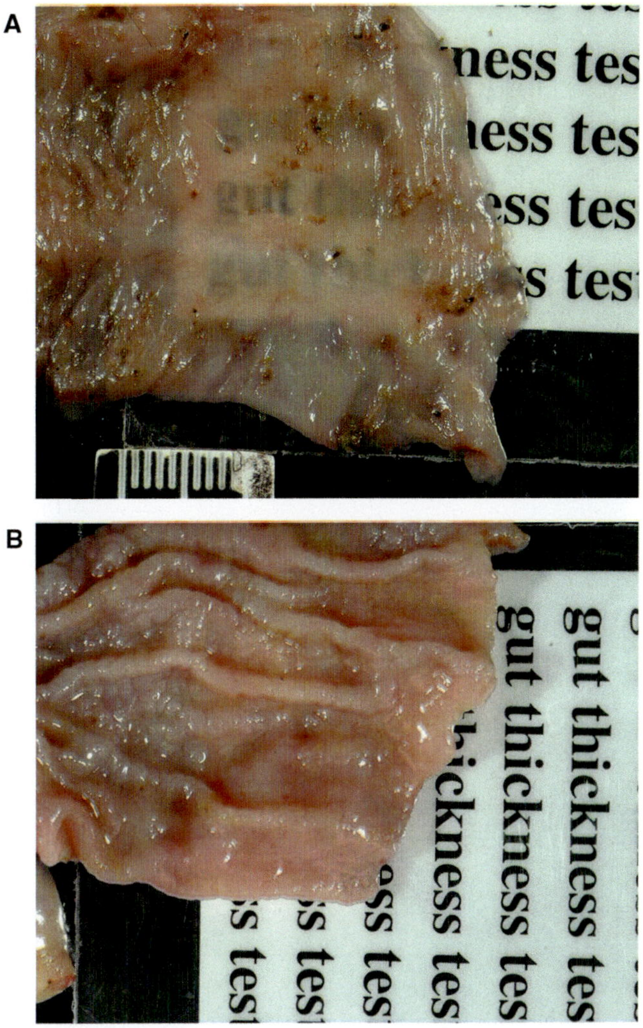

Fig. 7.18 Cecum and colon of a rhesus macaque (Macaca mulatta). (**a**) Normal large intestine should have a thin mucosa in which you are able to see through. (**b**) Chronic idiopathic colitis present with a thick, rugose mucosa that is not see-through. There should be no blood or areas of ulceration

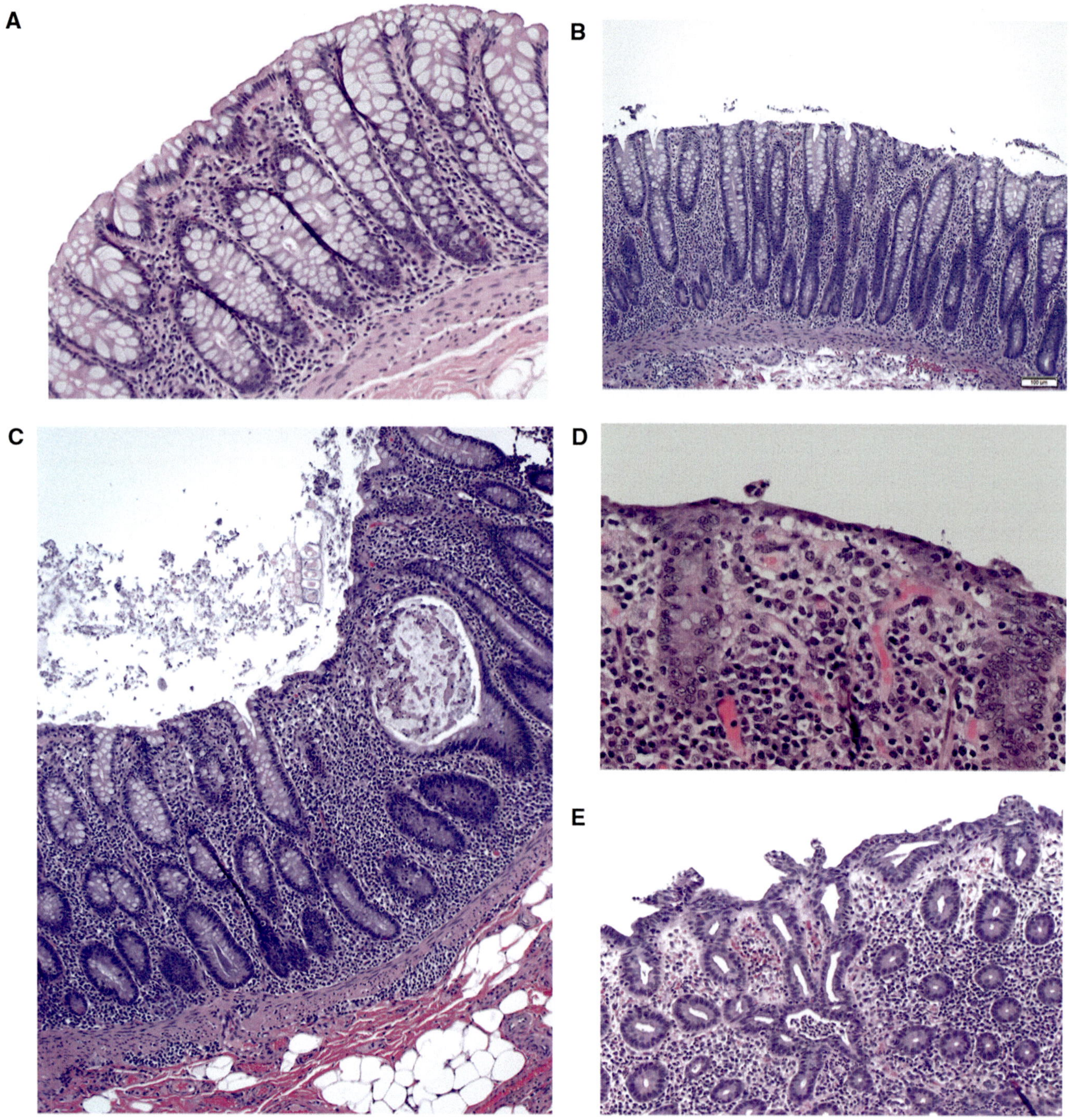

Fig. 7.19 Chronic idiopathic colitis in rhesus macaques. (**a**) In normal sections of cecum and colon goblet cells are prominent, crypt abscesses are absent, and their height is less than a 40× field as a general marker. The lamina propria has few lymphocytes and plasma cells, but the crypts are fairly close together. The mucosal surface is lined by intact columnar enterocytes intermittently lined by a healthy blue brush border. (**b**) Animals with chronic idiopathic colitis have decreased numbers of goblet cells; the crypts are hyperplastic with increased numbers of mitotic figures and increased crypt height that often exceeds a 40× field. (**c**) Crypt abscesses are common. (**d**) The mucosal surface lacks a blue brush border can be attenuated. (**e**) The mucosa can be arranged in a lattice work of tufted surface epithelium in more chronic changes.

References

1. Marini RP. Diseases of the urogenital system. In: Marini RP, Wachtman LM, Tardif SD, Masfield K, Fox JG, editors. The common marmoset in captivity and biomedical research. London: Academic Press; 2019. p. 195–212.
2. Isobe K, Adachi K, Hayashi S, Ito T, Miyoshi A, Kato A, Suzuki M. Spontaneous glomerular and tubulointerstitial lesions in common marmosets (*Callithrix jacchus*). Vet Pathol. 2012;49:839–45. https://doi.org/10.1177/0300985811427151.
3. Yamada N, Sato J, Kanno T, Wako Y, Tsuchitani M. Morphological study of progressive glomerulonephropathy in common marmosets (*Callithrix jacchus*). Toxicol Pathol. 2013;41:1106–15. https://doi.org/10.1177/0192623313478206.
4. Brack M, Rothe H. Chronic tubulointerstitial nephritis and wasting disease in marmosets (*Callithrix jacchus*). Vet Pathol. 1981;18:45–54. https://doi.org/10.1177/030098588101 8s0605.
5. Collins MG, Rogers NM, Jesudason S, Kireta S, Brealey J, Coates PT. Spontaneous glomerular mesangial lesions in common marmoset monkeys (*Callithrix jacchus*): a benign non-progressive glomerulopathy. J Med Primatol. 2014;43:477–87. https://doi.org/10.1111/jmp.12134.
6. Brack M. IgM-mesangial nephropathy in callitrichids. Vet Pathol. 1988;25:270–6. https://doi.org/10.1177/030098588802500404.
7. Brack M, Schroeder C, Fooke M, Schlumberger W. IgM/IgA nephropathy in callitrichids: antigen studies. Nephron. 1999;82:221–31. https://doi.org/10.1159/000045406.
8. Winkelmann, J.M. Kidney alterations in common marmosets (*Callithrix jacchus*) with wasting marmoset syndrome (German). https://elib.tiho-hannover.de/receive/etd_mods_00001143.
9. Brack M, Weber M. Ultrastructural and histochemical mesangial alterations in callitrichid IgM nephropathy (primates: platyrrhina). Nephron. 1995;69:286–92. https://doi.org/10.1159/000188472.
10. Schroeder C, Osman AA, Roggenbuck D, Mothes T. IgA-gliadin antibodies, IgA-containing circulating immune complexes, and IgA glomerular deposits in wasting marmoset syndrome. Nephrol Dial Transplant. 1999;14:1875–80. https://doi.org/10.1093/ndt/14.8.1875.
11. Ami Y, Suzaki Y, Goto N. Endometriosis in cynomolgus monkeys retired from breeding. J Vet Med Sci. 1993;55:7–11. https://doi.org/10.1292/jvms.55.7.
12. Assaf BT, Miller AD. Pleural endometriosis in an aged rhesus macaque (*Macaca mulatta*): a histopathologic and immunohistochemical study. Vet Pathol. 2012;49:636–41. https://doi.org/10.1177/0300985811406890.
13. Binhazim AA, Tarara RP, Suleman MA. Spontaneous external endometriosis in a De Brazza's monkey. J Comp Pathol. 1989;101:471–4. https://doi.org/10.1016/0021-9975(89)90031-5.
14. Dick EJ Jr, Hubbard GB, Martin LJ, Leland MM. Record review of baboons with histologically confirmed endometriosis in a large established colony. J Med Primatol. 2003;32:39–47. https://doi.org/10.1034/j.1600-0684.2003.00008.x.
15. Mattison JA, Ottinger MA, Powell D, Longo DL, Ingram DK. Endometriosis: clinical monitoring and treatment procedures in rhesus monkeys. J Med Primatol. 2007;36:391–8. https://doi.org/10.1111/j.1600-0684.2006.00208.x.
16. Nishimoto-Kakiuchi A, Netsu S, Okabayashi S, Taniguchi K, Tanimura H, Kato A, Suzuki M, Sankai T, Konno R. Spontaneous endometriosis in cynomolgus monkeys as a clinically relevant experimental model. Hum Reprod. 2018;33:1228–36. https://doi.org/10.1093/humrep/dey095.
17. Story L, Kennedy S. Animal studies in endometriosis: a review. ILAR J. 2004;45:132–8. https://doi.org/10.1093/ilar.45.2.132.
18. Waterton JC, Miller D, Morrell JS, Dukes M, West CD, Wadsworth PF. A case of endometriosis in the macaque diagnosed by nuclear magnetic resonance imaging. Lab Anim. 1992;26:59–64. https://doi.org/10.1258/002367792780809066.
19. Zondervan KT, Weeks DE, Colman R, Cardon LR, Hadfield R, Schleffler J, Trainor AG, Coe CL, Kemnitz JW, Kennedy SH. Familial aggregation of endometriosis in a large pedigree of rhesus macaques. Hum Reprod. 2004;19:448–55. https://doi.org/10.1093/humrep/deh052.
20. Miller AD. Neoplasia and proliferative disorders of nonhuman primates. In: Abee CR, Mansfield K, Tardif S, Morris T, editors. Nonhuman primates in biomedical research: diseases. London: Academic Press; 2012. p. 325–56.
21. Mounsey AL, Wilgus A, Slawson DC. Diagnosis and management of endometriosis. Am Fam Physician. 2006;74:594–600.
22. Cline JM, Brignolo L, Ford EW. Urogential system. In: Abee CR, Mansfield K, Tardif S, Morris T, editors. Nonhuman primates in biomedical research: diseases. London: Academic Press; 2012. p. 484–562.
23. Gruber-Dujardin E, Bleyer M, Mätz-Rensing K. Morphological and immunohistochemical characterization of spontaneous endometriosis in rhesus macaques (*Macaca mulatta*). Primate Biol. 2017;4:77–91. https://doi.org/10.5194/pb-4-77-2017.
24. Abrao MS, Neme RM, Carvalho FM, Aldrighi JM, Pinotti JA. Histological classification of endometriosis as a predictor of response to treatment. Int J Gynaecol Obstet. 2003;82:31–40. https://doi.org/10.1016/s0020-7292(03)00079-1.
25. Kamergorodsky G, Ribeiro PA, Galvão MA, Abrão MS, Donadio N, Lemos NL, Aoki T. Histologic classification of specimens from women affected by superficial endometriosis, deeply infiltrating endometriosis, and ovarian endometriomas. Fertil Steril. 2009;92:2074–7. https://doi.org/10.1016/j.fertnstert.2009.05.086.
26. Atkins HM, Lombardini ED, Caudell DL, Appt SE, Dubois A, Cline JM. Decidualization of endometriosis in macaques. Vet Pathol. 2016;53:1252–8. https://doi.org/10.1177/0300985816646433.
27. Fazleabas AT, Brudney A, Chai D, Langoi D, Bulun SE. Steroid receptor and aromatase expression in baboon endometriotic lesions. Fertil Steril. 2003;80:820–7. https://doi.org/10.1016/s0015-0282(03)00982-8.
28. Brosens IA, Brosens JJ. Endometriosis. Eur J Obstet Gynecol Reprod Biol. 2000;90:159–64. https://doi.org/10.1016/s0301-2115(00)00265-7.
29. Giudice LC, Kao LC. Endometriosis. Lancet. 2004;364:1789–99. https://doi.org/10.1016/S0140-6736(04)17403-5.
30. Laganà AS, Garzon S, Götte M, Viganò P, Franchi M, Ghezzi F, Martin DC. The pathogenesis of endometriosis: molecular and cell biology insights. Int J Mol Sci. 2019;20:5615. https://doi.org/10.3390/ijms20225615.
31. Sasson IE, Taylor HS. Stem cells and the pathogenesis of endometriosis. Ann N Y Acad Sci. 2008;1127:106–15. https://doi.org/10.1196/annals.1434.014.
32. Bethune MT, Borda JT, Ribka E, Liu MX, Phillippi- Falkenstein K, Jandacek RJ, Doxiadis GG, Gray GM, Khosla C, Sestak K. A nonhuman primate model for gluten sensitivity. PLoS One. 2008;3:1614.
33. Jabri B, Sollid LM. T cells in celiac disease. J Immunol. 2017;198(8):3005–14.
34. Wagner JD, Jerome CP, Adams MR. Gluten-sensitive enteropathy in a cynomolgus monkey. Lab Anim Sci. 1988;38:592–4.
35. Brewers M. Epidemiology of celiac disease: what are the prevalence, incidence, and progression of celiac disease? Gastroenterology. 2005;128:47–51.
36. Sestak K, Mazumdar K, Midkiff CC, Dufour CC, Dufour J, Borda JT, Alvarez X. Recognition of epidermal transglutaminase by IgA and tissue transglutaminase 2 antibodies in a rare case of rhesus dermatitis. J Vis Exp. 2011;58:3154. https://doi.org/10.3791/3154.

37. Sestak K, Conroy L, Aye PP, Mehra S, Doxiadis GG, Kaushal D. Improved xenobiotic metabolism and reduced susceptibility to cancer in gluten-sensitive macaques upon introduction of a gluten-free diet. PLoS One. 2011;6(4):18648. https://doi.org/10.1371/journal.pone.0018648.
38. Hill ID, Dirks MH, Liptak GS, Colletti RB, Fasano A, Guandalini S, Hoffenberg EJ, Horvath K, Murray JA, Pivor M, Seidman EG. Guideline for the diagnosis and treatment of celiac disease in children: recommendations of the North American Society for Pediatric Gastroenterology, Hepatology and Nutrition. J Pediatr Gastroenterol Nutr. 2005;40:1–19.
39. Villanacci V, Vanoli A, Giovanni A, Salviato T, Bonetti LR, Baronchelli C, Saragoni L, Parente P. Celiac disease: histology-differential diagnosis-complications. A practical approach. Pathologica. 2020;112:186–96.
40. Mazumdar K, Xavier A, Borda JT, Dufour J, Martin E, Bethune MT, Kholsa C, Sestak K. Visualization of transepithelial passage of the immunogenic 33-reside peptide from alpha-2 gliadin in gluten sensitive macaques. PLoS One. 2010;5(4):10228. https://doi.org/10.1371/journal.pone.0010228.
41. Xu H, Feely SL, Wang X, Liu DX, Borda JT, Dufour J, et al. Gluten-sensitive enteropathy coincides with decreased capability of intestinal T cells to secrete IL-17 and IL-22 in a macaque model for celiac disease. Clin Immunol. 2013;147(1):40–9. https://doi.org/10.1016/j.clim.2013.02.012. Epub 2013 Feb 28
42. Laing ST, Merriam D, Shock BC, Mills S, Spinner A, Reader R, Hartigan-O'Conner DJ. Idiopathic colitis in rhesus macaques is associated with dysbiosis, abundant enterochromaffin cells and altered T-cell cytokine expression. Vet Pathol. 2018;55:741–52.
43. Beglinger R, Illgen B, Pfister R, Heider K. The parasite *Trichospirura leptostoma* associated with wasting diseases in a colony of common marmosets, *Callithrix jacchus*. Folia Primatol. 1988;51:45–51.
44. Potkay S. Diseases of the callitrichidae: a review. J Med Primatol. 1992;21:189–236.
45. Chalifoux LV, Broncin RT, Escajadillo A, McKenna S. An analysis of the association of the gastroenteric lesions with chronic wasting syndrome of marmosets. Vet Pathol. 1982;19:141–62.
46. David JM, Dick ED Jr, Hubbard GB. Spontaneous pathology of the common marmoset (*Callithrix jacchus*) and tamarins (*Sanguinis oedipus, Sanguinis mystax*). J Med Primatol. 2009;38:347–59.
47. Baxter VK, Shaw GC, Sotuyo NP, Carlson CS, Olson EJ, Zink MC, Mankowski JL, Adams RJ, Hutchinson EK, Metcalf Pate KA. Serum albumin and body weight as biomarkers for the antemortem identification of bone and gastrointestinal disease in the common marmoset. PLoS One. 2012;8:1–10. https://doi.org/10.1371/journal.pone.0082747.
48. Logan AC, Kahn KNM. Clinical pathologic changes in two marmosets with wasting syndrome. Toxicol Pathol. 1996;24:707–9.
49. Ludlage E, Mansfield K. Clinical care and disease of the common marmoset (*Callithrix jacchus*). Comp Med. 2003;53:369–82.
50. Nakashima E, Okano Y, Nimi K, Takahashi E. Detection of calprotectin and apoptotic activity in the colon of marmosets with chronic diarrhea. J Vet Med Sci. 2013;75:1633–6.
51. Ross C, Davis K, Dobek G, Tardif S. Aging phenotypes of common marmosets (*Callithrix jacchus*). J Aging Res. 2012;2012:1. https://doi.org/10.1155/2012/567143.
52. Richards-Rios P, Wigley P, Lopez J, Wormell D, Barbon A. Changes in the faecal microbiome of pied tamarins (*Sanguines bicolor*) associated with chronic, recurrent diarrhoea and weight loss. Anim Microbiome. 2021;3:1. https://doi.org/10.1186/s42523-020-00062-4.
53. Heffron AS, Lauck M, Somsen ED, Townsend EC, Bailey AL, Sosa M, Eickhoff J, Capuano S III, Newman CM, Kuhn JH, Mejia A, Simmons HA, O'Conner DH. Discovery of a novel pegivirus in common marmosets (*Callithrix jacchus*) with lymphocytic enterocolitis. Microorganisms. 2020;8:1509. https://doi.org/10.3390/microorganisms8101509.
54. Parambeth JC, Ross CN, Miller AD, Austad AD, Lidbury JA, Suchodolski JS, Steiner JM. Serum cobalamin and folate concentrations in common marmosets (*Callithrix jacchus*) with chronic lymphocytic enteritis. Comp Med. 2019;69:135–43.
55. Resnick D, Niwayama G. Radiographic and pathologic features of spinal involvement in diffuse idiopathic skeletal hyperostosis (DISH). Radiology. 1976;119:559–68. https://doi.org/10.1148/119.3.559.
56. Wilson DJ, Kay V, Charig M, Hughes DG, Creasey TS. Skeletal hyperostosis and extraosseous calcification in patients receiving long-term etretinate (Tigason). Br J Dermatol. 1988;119:597–607.
57. Belanger TA, Rowe DE. Diffuse idiopathic skeletal hyperostosis: musculoskeletal manifestations. J Am Acad Orthop Surg. 2001;9:258–67. https://doi.org/10.5435/00124635-200107000-00006.
58. Mader R, Verlaan JJ, Buskila D. Diffuse idiopathic skeletal hyperostosis: clinical features and pathogenic mechanisms. Nat Rev Rheumatol. 2013;9:741–50. https://doi.org/10.1038/nrrheum.2013.165.
59. Nascimento FA, Gatto LA, Lages RO, Neto HM, Demartini Z, Koppe GL. Diffuse idiopathic skeletal hyperostosis: a review. Surg Neurol Int. 2014;5(Suppl 3):122–5. https://doi.org/10.4103/2152-7806.130675.
60. Kandel RA, Renlund RC, Cheng PT, Rapley WA, Mehren KG, Pritzker KP. Calcium pyrophosphate dihydrate crystal deposition disease with concurrent vertebral hyperostosis in a Barbary ape. Arthritis Rheum. 1983;26:682–7. https://doi.org/10.1002/art.1780260518.
61. Livingstone B, Kitchener AC, Hull G, Schwarz T, Vijayanathan S, Allen MJ, et al. Diffuse idiopathic skeletal hyperostosis in captive gorillas (*Gorilla* spp.): appearances and diagnosis. J Zoo Wildl Med. 2020;51:578–90. https://doi.org/10.1638/2019-0180.
62. Lovell NC. Skeletal and dental pathology of free-ranging mountain gorillas. Am J Phys Anthropol. 1990;81:399–412. https://doi.org/10.1002/ajpa.1330810309.
63. Ratliff C, Waller KR, Steinberg H, Clyde VL. Diffuse idiopathic skeletal hyperostosis with secondary dysphagia in a black-handed spider monkey (*Ateles geoffroyi*). J Zoo Wildl Med. 2020;51:455–8. https://doi.org/10.1638/2018-0240.
64. Sokoloff L, Snell KC, Stewart HC. Spinal ankyloses in old rhesus monkeys. Clin Orthop. 1968;61:285–93.
65. Mader R, Verlaan JJ, Eshed I, Bruges-Armas J, Puttini PS, Atzeni F, et al. Diffuse idiopathic skeletal hyperostosis (DISH): where we are now and where to go next. RMD Open. 2017;3:e000472. https://doi.org/10.1136/rmdopen-2017-000472.
66. Ortega M, Gonçalves R, Haley A, Wessmann A, Penderis J. Spondylosis deformans and diffuse idiopathic skeletal hyperostosis (DISH) resulting in adjacent segment disease. Vet Radiol Ultrasound. 2012;53:128–34. https://doi.org/10.1111/j.1740-8261.2011.01891.x.
67. Mader R, Sarzi-Puttini P, Atzeni F, Olivieri I, Pappone N, Verlaan JJ, et al. Extraspinal manifestations of diffuse idiopathic skeletal hyperostosis. Rheumatology (Oxford). 2009;48:1478–81. https://doi.org/10.1093/rheumatology/kep308.
68. Pritzker KPH, Kessler MJ. Arthritis, muscle, adipose tissue, and bone diseases of nonhuman primates. In: Abee CR, Mansfield K, Tardif S, Morris T, editors. Nonhuman primates in biomedical research: diseases. London: Academic Press; 2012. p. 629–97.
69. Fournier DE, Kiser PK, Beach RJ, Dixon SJ, Séguin CA. Dystrophic calcification and heterotopic ossification in fibrocartilaginous tissues of the spine in diffuse idiopathic skeletal hyperostosis (DISH). Bone Res. 2020;8:16. https://doi.org/10.1038/s41413-020-0091-6.
70. Kuperus JS, Westerveld LA, Rutges JP, Alblas J, van Rijen MH, Bleys RL, et al. Histological characteristics of diffuse idiopathic

skeletal hyperostosis. J Orthop Res. 2017;35:140–6. https://doi.org/10.1002/jor.23267.

71. Olivieri I, D'Angelo S, Palazzi C, Padula A, Mader R, Khan MA. Diffuse idiopathic skeletal hyperostosis: differentiation from ankylosing spondylitis. Curr Rheumatol Rep. 2009;11:321–8. https://doi.org/10.1007/s11926-009-0046-9.
72. Beisner BA, Isbell LA. Factors influencing hair loss among female captive rhesus macaques (*Macaca mulatta*). Appl Anim Behav Sci. 2009;119:91–100. https://doi.org/10.1016/j.applanim.2009.03.016.
73. Bellanca RU, Lee GH, Vogel K, Ahrens J, Kroeker R, Thom JP, Worlein JM. A simple alopecia scoring system for use in colony management of laboratory-housed primates. J Med Primatol. 2014;43:153–61. https://doi.org/10.1111/jmp.12107.
74. Kramer J, Fahey M, Santos R, Carville A, Wachtman L, Mansfield K. Alopecia in rhesus macaques correlates with immunophenotypic alterations in dermal inflammatory infiltrates consistent with hypersensitivity etiology. J Med Primatol. 2010;39:112–22. https://doi.org/10.1111/j.1600-0684.2010.00402.x.
75. Lutz CK, Coleman K, Worlein J, Novak MA. Hair loss and hair-pulling in rhesus macaques (*Macaca mulatta*). J Am Assoc Lab Anim Sci. 2013;52:454–7.
76. Steinmetz HW, Kaumanns W, Dix I, Heistermann M, Fox M, Kaup FJ. Coat condition, housing condition and measurement of faecal cortisol metabolites--a non-invasive study about alopecia in captive rhesus macaques (*Macaca mulatta*). J Med Primatol. 2006;35:3–11. https://doi.org/10.1111/j.1600-0684.2005.00141.x.
77. Novak MA, Meyer JS. Alopecia: possible causes and treatments, particularly in captive nonhuman primates. Comp Med. 2009;59:18–26.
78. Steinmetz HW, Kaumanns W, Neimeier KA, Kaup FJ. Dermatologic investigation of alopecia in rhesus macaques (*Macaca mulatta*). J Zoo Wildl Med. 2005;36:229–38. https://doi.org/10.1638/04-054.1.
79. Honess P, Gimpel J, Wolfensohn S, Mason G. Alopecia scoring: the quantitative assessment of hair loss in captive macaques. Altern Lab Anim. 2005;33:193–206. https://doi.org/10.1177/026119290503300308.
80. Kramer JA, Mansfield KG, Simmons JH, Bernstein JA. Psychogenic alopecia in rhesus macaques presenting as focally extensive alopecia of the distal limb. Comp Med. 2011;61:263–8.
81. Luchins KR, Baker KC, Gilbert MH, Blanchard JL, Liu DX, Myers L, Bohm RP. Application of the diagnostic evaluation for alopecia in traditional veterinary species to laboratory rhesus macaques (*Macaca mulatta*). J Am Assoc Lab Anim Sci. 2011;50:926–38.
82. Ardeshir A, Oslunf KL, Ventimiglia F, Yee J, Lerche NW, Hyde DM. Idiopathic microscopic colitis of rhesus macaques: quantitative assessment of colonic mucosa. Anat Rec (Hoboken). 2013;296:1169–79.

Nonhuman Primate Neoplasia

8

Andrew D. Miller and Shannon G. M. Kirejczyk

Abstract

Increased emphasis has been put on the incidence and occurrence of neoplasia in nonhuman primates especially for those naturally occurring cancers that are comparable to the human equivalent. Macaques, mostly rhesus macaques, remain the predominant species at many primate research centers, and much of the literature base is focused on that species grouping. More recently, select tumors of New World primates have received more attention, both from primate research centers and zoologic collections. While many tumors remain uncommon to rare, certain tumors are seen with increased frequency in aging or immunosuppressed animals. This chapter will focus on the diagnostic gross features of select nonhuman primate neoplasms complemented by histology and a basic overview of each neoplastic entity. Since a detailed overview of all nonhuman primate neoplasia is outside of the scope of this chapter, we refer the reader to other resources for more comprehensive overviews (Miller, Nonhuman primates in biomedical research. 2: diseases, 2nd edn. Elsevier, London, 2012).

Keywords

Neoplasia · Carcinoma · Sarcoma

Increased emphasis has been put on the incidence and occurrence of neoplasia in nonhuman primates especially for those naturally occurring cancers that are comparable to the human equivalent. Macaques, mostly rhesus macaques, remain the predominant species at many primate research centers, and much of the literature base is focused on that species grouping. More recently, select tumors of New World primates have received more attention, both from primate research centers and zoologic collections. While many tumors remain uncommon to rare, certain tumors are seen with increased frequency in aging or immunosuppressed animals. This chapter will focus on the diagnostic gross features of select nonhuman primate neoplasms complemented by histology and a basic overview of each neoplastic entity. Since a detailed overview of all nonhuman primate neoplasia is outside of the scope of this chapter, we refer the reader to other resources for more comprehensive overviews [1].

8.1 Musculoskeletal

8.1.1 Osteosarcoma

Tumors arising from the bone are rarely described in nonhuman primates. Osteosarcoma is the most common and detailed most frequently in rhesus macaques, however, is reported in a wide variety of other nonhuman primate species. Osteosarcomas can arise in multiple locations including central, periosteal, and parosteal. Central osteosarcomas predominate in the nonhuman primate case reports. Tumors are expansile, destructive, and widely replace bone and surrounding soft tissues (Figs. 8.1 and 8.2). Metastases are frequently recorded, most commonly affecting the lung.

8.2 The Nervous System

8.2.1 Meningioma

Meningioma is an uncommon tumor in nonhuman primates, with the vast majority of literature published focusing on rhesus macaques. These tumors can occur spontaneously or

A. D. Miller (✉)
Department of Population Medicine and Diagnostic Sciences, Section of Anatomic Pathology, Cornell University College of Veterinary Medicine, Ithaca, NY, USA
e-mail: andrew.miller@cornell.edu

S. G. M. Kirejczyk
Division of Pathology, Emory National Primate Research Center, Atlanta, GA, USA

StageBio, Mount Jackson, VA, USA

I. Kondova - Perseng et al. (eds.), *Atlas of Diagnostic Pathology in Nonhuman Primates*,
https://doi.org/10.1007/978-3-031-41280-6_8

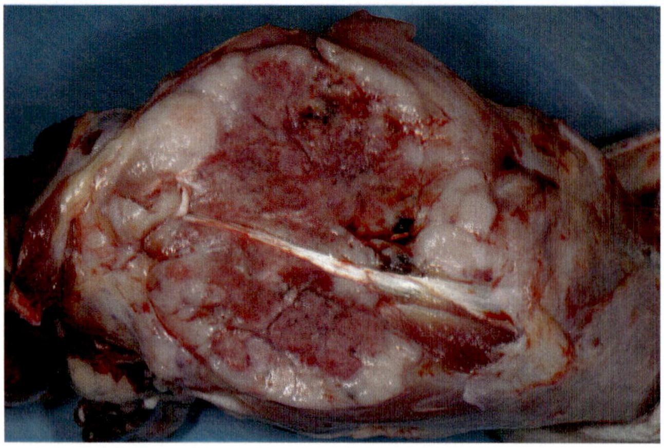

Fig. 8.1 Osteosarcoma, rhesus macaque. A large, expansile mass replaces and destroys the preexisting humeral bone

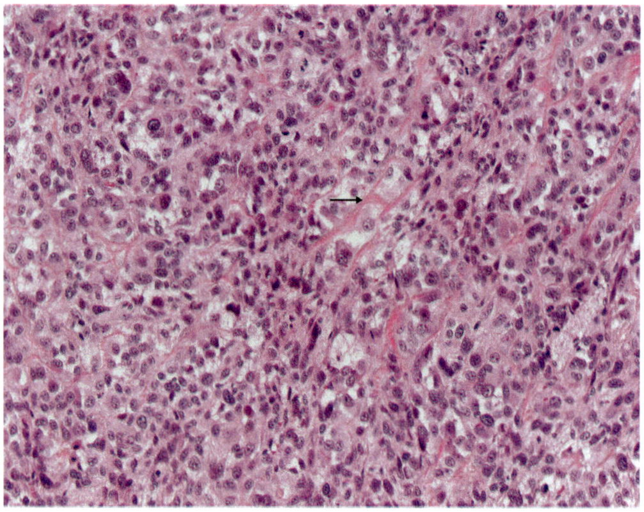

Fig. 8.2 Osteosarcoma, rhesus macaque. Osteosarcomas are composed of a highly variable population of cells. Here, interlacing lobules and vague streams of neoplastic mesenchymal cells surround and produce neoplastic osteoid (arrow). Multinucleated neoplastic cells are frequently seen

arise as part of radiation-induced oncogenesis. They are extra axial, causing a variety of compression and invasion into the underlying nervous tissue (Fig. 8.3). True to their mixed embryologic origin, the histologic pattern of these tumors is incredibly diverse with subtypes ranging from meningothelial, fibrous, microcystic, and papillary among many others. Mineral concretions, either as psammoma bodies or linear deposits, are common in these tumors (Fig. 8.4).

8.2.2 Glioma

Gliomas arise occasionally in nonhuman primates. Reports of astrocytomas predominate over oligodendrogliomas. Astrocytomas may occur spontaneously, following exposure to radiation, or in association with a variety of viruses [2, 3]. Astrocytomas are tan to white, intraaxial tumors that may blend with the surrounding neuroparenchyma or form a bulging mass that causes asymmetry (Fig. 8.5). Histologically, well differentiated (low-grade) astrocytomas have a relatively uniform population of neoplastic cells with small oval nuclei surrounded by a background of astrocytic processes (Fig. 8.6). Secondary structures, including perineuronal satellitosis and perivascular aggregation of neoplastic cells, are features of some tumors (Figs. 8.7 and 8.8). High-grade astrocytomas (glioblastoma) are characterized as such by the presence of necrosis, which often forms serpiginous tracts with nuclear pseudopalisading, increased mitotic activity, microvascular proliferation, nuclear atypia, and multinucleated cells.

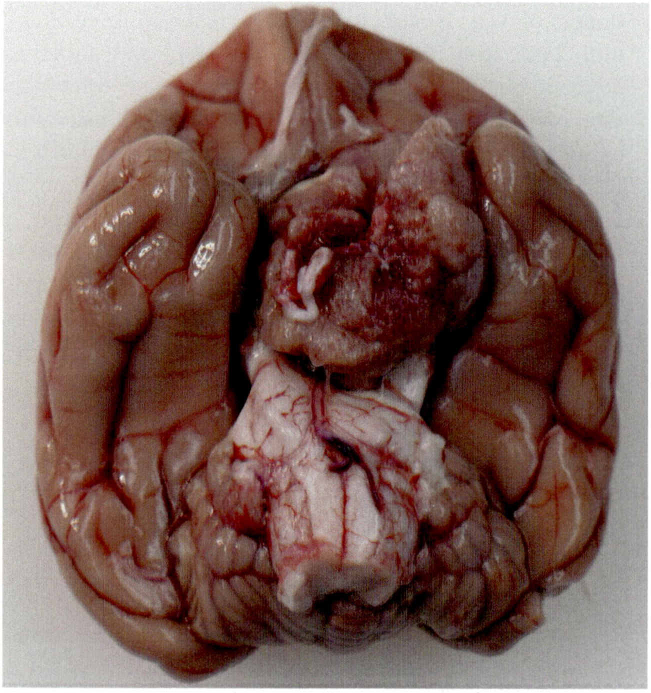

Fig. 8.3 Meningioma, rhesus macaque. A tan, intracranial, extra axial, multilobular tumor is firmly adhered to the ventral aspect of the brain. The tumor compresses the optic chiasm and midbrain. This animal had a history of blindness. By courtesy of Dr. Sanjeev Gumber, Emory National Primate Research Center

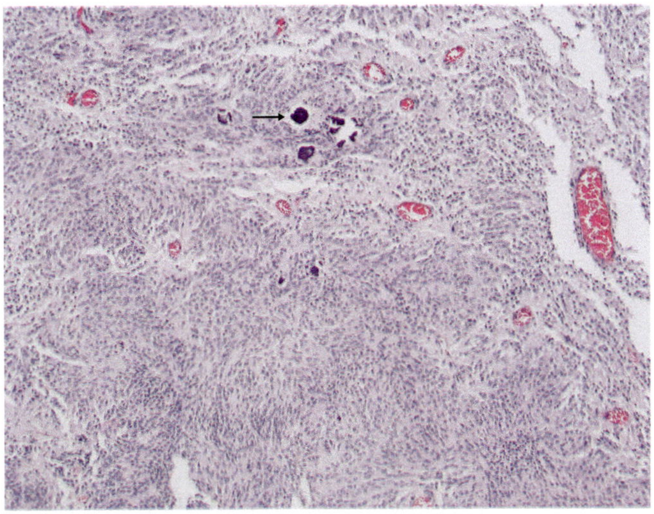

Fig. 8.4 Meningioma, rhesus macaque. An extraaxial mass arising in the parietal lobe is composed of interlacing streams and whorls of neoplastic meningothelial cells that occasionally surround mineral concretions (arrow)

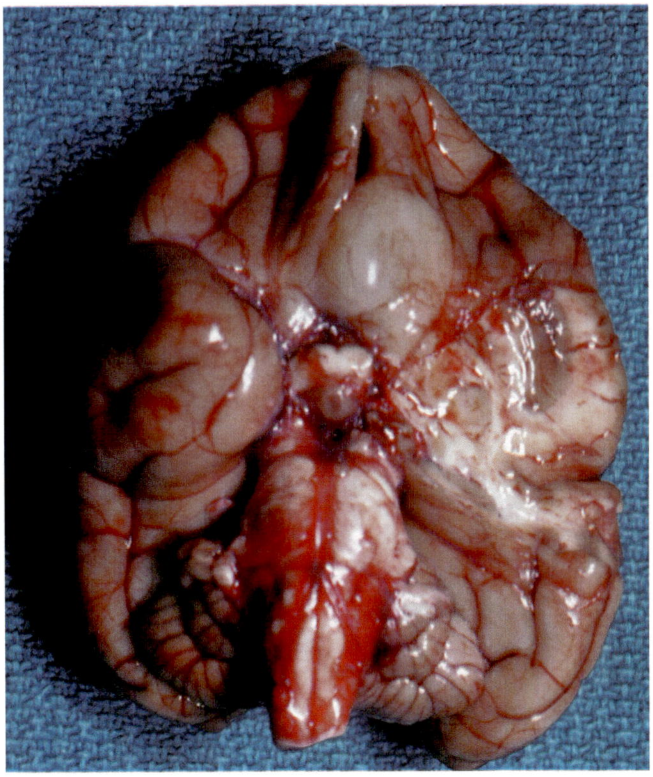

Fig. 8.5 Glioma, rhesus macaque. A pale tan, intraaxial mass expands the ventromedial prefrontal cortex and bulges from the ventral surface of the brain

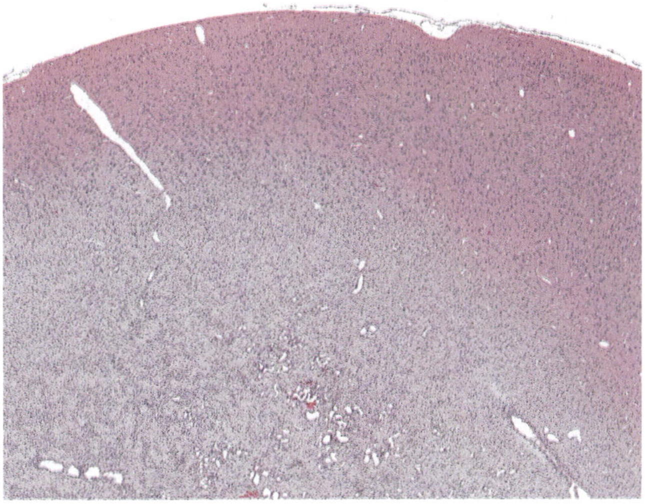

Fig. 8.6 Astrocytoma, rhesus macaque. A moderately cellular, unencapsulated, intraaxial neoplasm focally effaces the internal capsule, obscures the gray and white matter junction, and blends into the overlying cerebral cortex

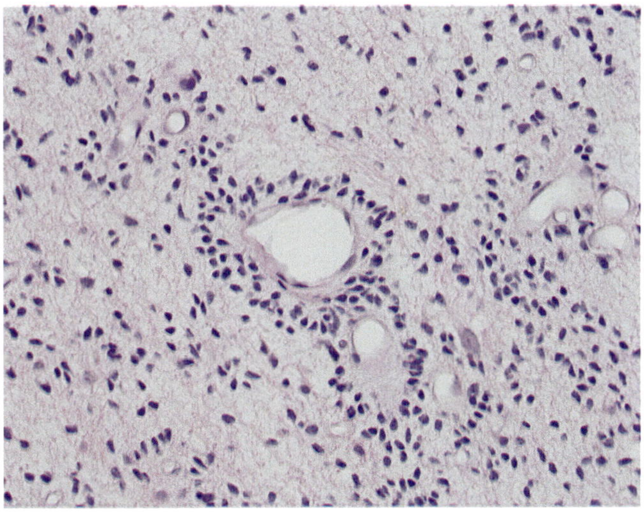

Fig. 8.7 Astrocytoma, rhesus macaque. Higher-magnification image of the same neoplasm shows a uniform population of neoplastic cells with small angular to oval nuclei and a background of astrocytic processes. Secondary structures were observed throughout this tumor. Here, neoplastic cells exhibit perivascular aggregation

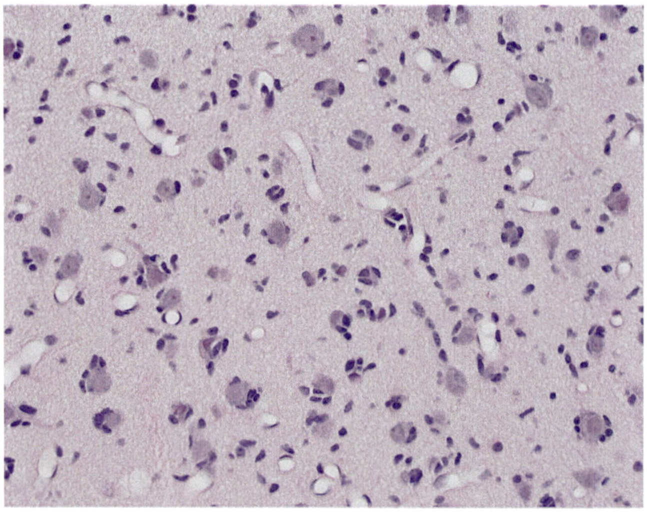

Fig. 8.8 Astrocytoma, rhesus macaque. High-magnification image of the same tumor showing perineuronal satellitosis

8.3 Integument

Benign skin tumors such as lipomas and squamous papillomas are sporadically encountered in nonhuman primates in the diagnostic setting. Malignant tumors of the skin, while rare, are more often reported; these include squamous cell carcinoma (SCC), basal cell carcinoma, melanoma, adnexal gland carcinomas, fibrosarcoma, liposarcoma, and leiomyosarcoma [4, 5]. Squamous cell carcinoma and basal cell carcinoma are further described in the sections that follow.

8.3.1 Squamous Cell Carcinoma

Squamous cell carcinomas (SCCs) occur in a variety of locations in nonhuman primates; reported locations include the digits, face (Fig. 8.9), oral cavity, stomach, cervix, vulva, perineum, and penis. The development of SCCs in NHPs has been associated with a variety of risk factors, including advanced age, viral-induced oncogenesis (papilloma viruses), prolonged sunlight exposure (skin tumors), chronic inflammation, and repetitive trauma [6, 7]. Grossly, cutaneous SCCs begin as plaque-like or nodular to irregular masses, often progressing to ulceration, invasion, and local destruction of tissues (Figs. 8.10 and 8.11). They can present as chronic, nonhealing wounds. Squamous cell carcinomas occasionally metastasize to regional lymph nodes or the lungs. Histologically, well-differentiated SCCs consist of polygonal cells with abundant eosinophilic cytoplasm, intercellular bridges, and variable degrees of keratinization (Figs. 8.12 and 8.13). Poorly differentiated neoplasms may present a diagnostic challenge. Local invasion by tumor cells often incites robust inflammatory cell and scirrhous responses.

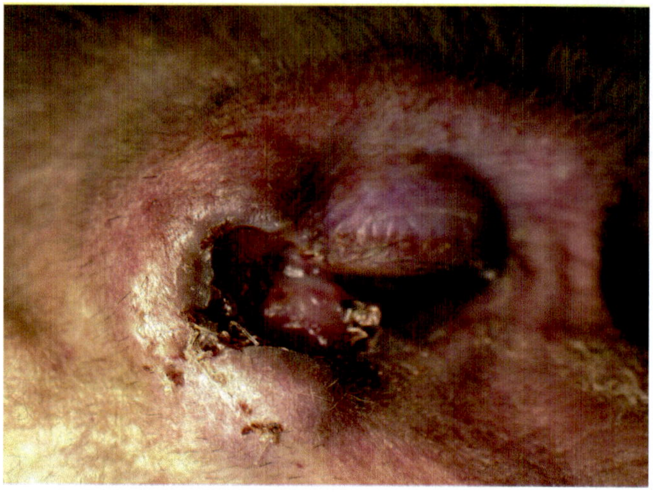

Fig. 8.9 Squamous cell carcinoma, rhesus macaque. There is irregular thickening of the periocular skin and soft tissues overlying the orbital bone and bordering a central focus of deep, crater-like ulceration and cavitation affecting the lateral canthus

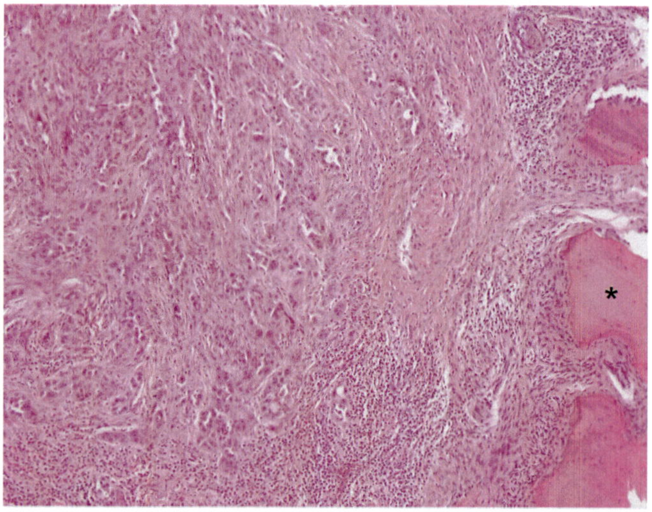

Fig. 8.10 Squamous cell carcinoma, rhesus macaque. Sheets of neoplastic epithelial cells invade and replace the periocular soft tissues to the level of the orbital bone (*). A prominent, mixed inflammatory cell infiltrate surrounds tumor cells multifocally

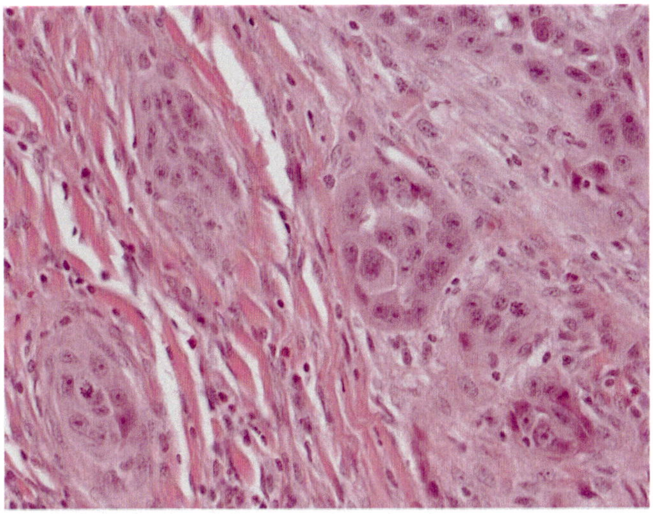

Fig. 8.11 Squamous cell carcinoma, rhesus macaque. Neoplastic cells are polygonal with variably distinct cell borders, abundant eosinophilic cytoplasm, and exhibit moderate anisokaryosis with occasional multinucleation. Occasional mitotic figures are present

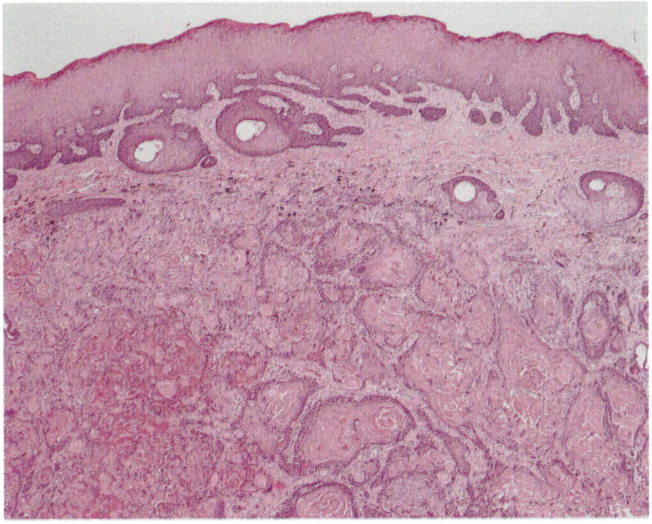

Fig. 8.12 Squamous cell carcinoma, sooty mangabey. A spontaneous, invasive epithelial neoplasm is present within the skin of a forelimb digit. Neoplastic squamous cells replace and expand the dermis

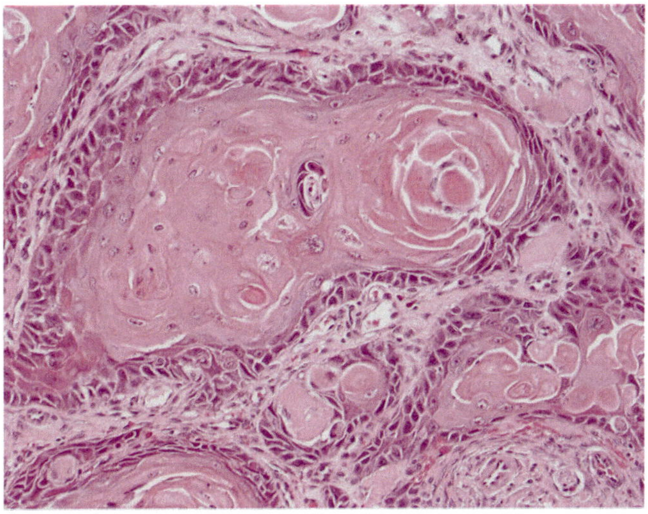

Fig. 8.13 Squamous cell carcinoma, sooty mangabey. A high-magnification image demonstrates polygonal tumor cells with abundant eosinophilic cytoplasm and distinct cell borders. Neoplastic keratinocytes exhibit disorderly maturation. Cells exhibit multifocal dyskeratosis. Anisocytosis and anisokaryosis are moderate to marked

8.3.2 Basal Cell Carcinoma

While basal cell carcinoma (BCC) is a common tumor of humans, this tumor is rarely encountered in nonhuman primates. Histologically, neoplastic basal epithelial cells form invasive nests and trabeculae (Figs. 8.14 and 8.15). Neoplastic cells have higher nucleus-to-cytoplasmic ratios and scant amounts of eosinophilic cytoplasm, which imparts a more basophilic appearance to these tumors at low magnification on the hematoxylin and eosin-stained sections.

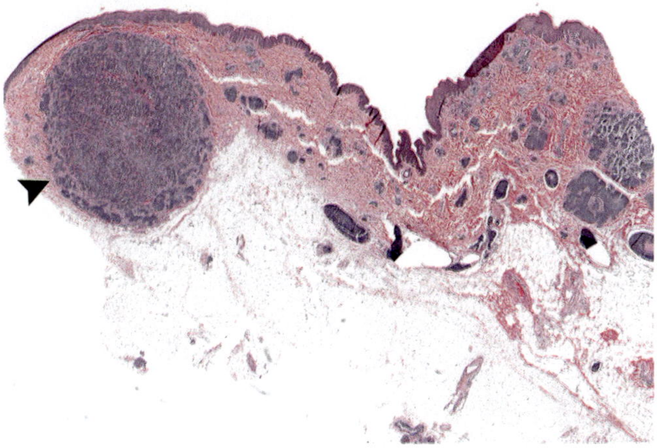

Fig. 8.14 Basal cell carcinoma, rhesus macaque. Neoplastic basal cells form a well demarcated, unencapsulated nodule within the dermis (arrowhead)

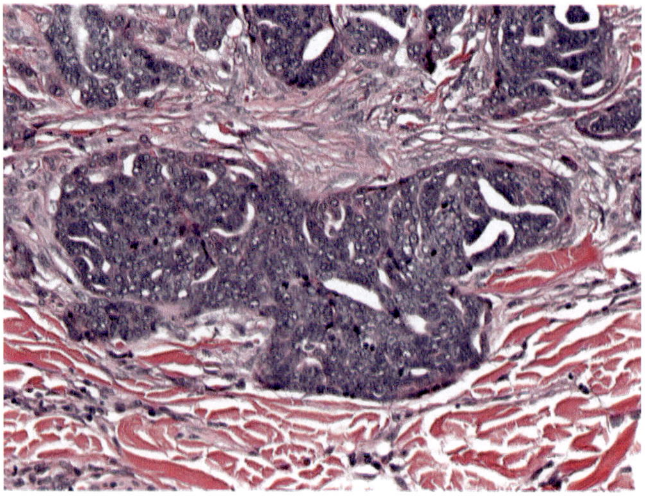

Fig. 8.15 Basal cell carcinoma, rhesus macaque. High-magnification image at the junction of the nodular mass with the dermis showing tumor cell invasion between mature dermal collagen fibers. Islands of neoplastic cells are surrounded by reactive fibroblastic tissue. Neoplastic cells have indistinct cell borders, small amounts of eosinophilic cytoplasm, and densely packed round nuclei

8.4 Alimentary System Including Hepatobiliary and Pancreatic

8.4.1 Large Intestinal Carcinoma of Rhesus Macaques

By far one of the most common spontaneously arising tumors in the aged rhesus macaque with no sex predilection, these tumors predominate in the ileocecal region or colon. They can be associated with melena. Highly invasive, these tumors will commonly extend transmurally through the intestinal wall and are associated with a marked scirrhous response [8]. This scirrhous response often correlates to a napkin ring-like structure at the site of the tumor (Figs. 8.16, 8.17, and 8.18). Prone to spreading widely, these tumors can progress to carcinomatosis and metastatic spread to a number of organs including mesenteric lymph nodes, liver, and lung. Histologically, various growth patterns including tubules, acini, and sheets of neoplastic cells can be seen, often admixed with abundant mucin and mucin-lakes (Fig. 8.18) and embedded in the aforementioned scirrhous response (Fig. 8.19).

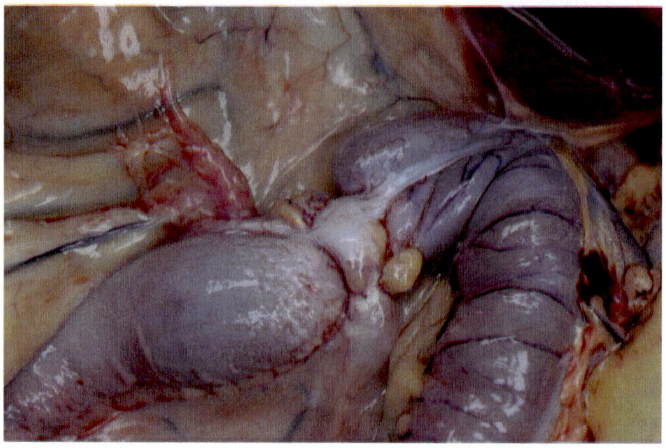

Fig. 8.16 Large intestinal adenocarcinoma, rhesus macaque. The ileocecal junction is thickened by a firm, white, napkin-ring-like mass. The distal ileum is markedly dilated due to stricture at the tumor site and the serosal surface coated by ill-defined, glistening white plaques (carcinomatosis)

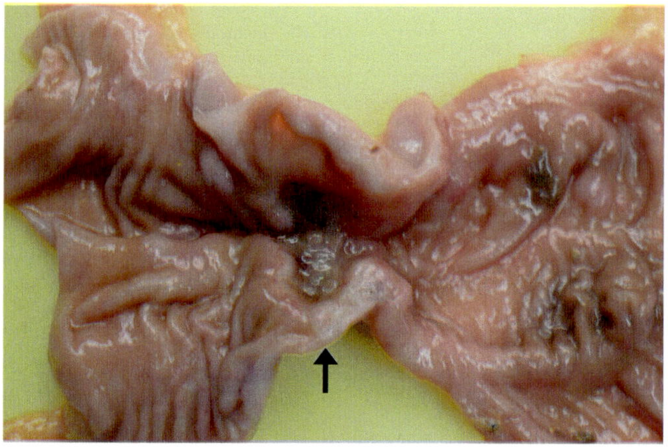

Fig. 8.17 Large intestinal adenocarcinoma, rhesus macaque. The colon is opened longitudinally along a napkin-ring-like stricture to reveal transmural thickening and mucosal ulceration at the tumor site (arrow). By courtesy of Dr. Sanjeev Gumber, Emory National Primate Research Center

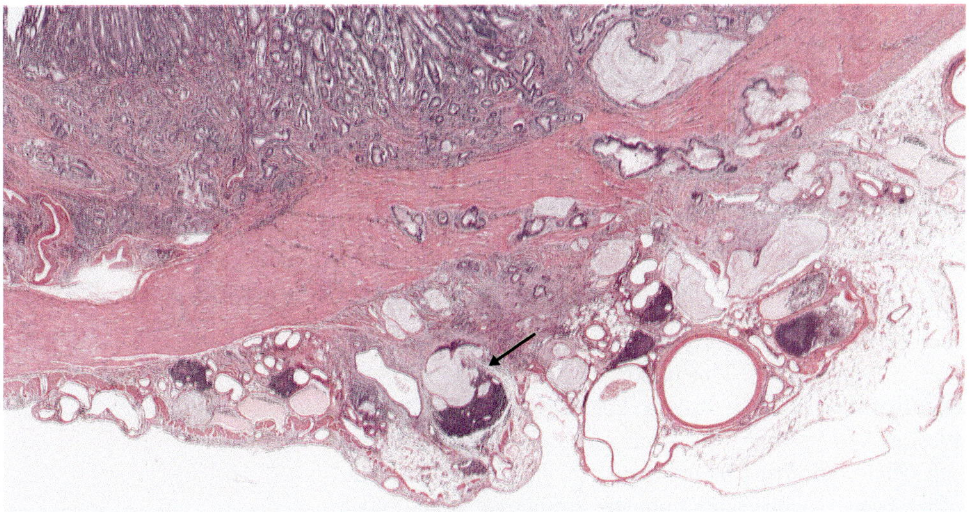

Fig. 8.18 Large intestinal adenocarcinoma, rhesus macaque. Extending transmurally are acini and glandular-like aggregates of neoplastic epithelial cells that often produce and surround large lakes of mucin. Similar lakes fill serosal lymphatics and carpet the serosal layer (arrow; lymphovascular invasion and carcinomatosis)

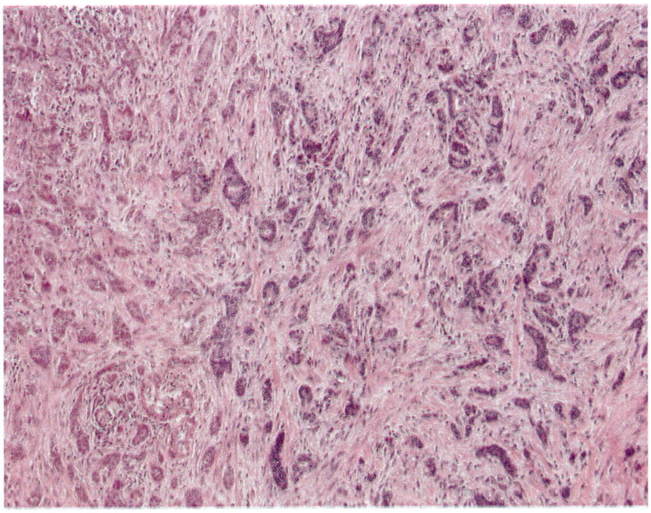

Fig. 8.19 Large intestinal adenocarcinoma, rhesus macaque. Neoplastic cells form small acinar-like structures that are embedded in a dense scirrhous response. This robust fibrous connective tissue matrix helps create the focal stricture that is commonly seen in these tumors

8.4.2 Small Intestinal Adenocarcinoma of Common Marmosets

Contrary to the rhesus macaque, intestinal adenocarcinomas in common marmosets predominate in the proximal small intestine, close to the junction with the duodenum [9]. Most commonly these animals present with a partial to complete obstruction due to marked stricture leading to aboral luminal dilation. These tumors start with regions of dysplasia and carcinoma in situ along the crypt zone before invading transmurally, serosally, and to draining lymph nodes. Older animals are predisposed, and there may be a genetic link as incidences have been higher in some closed colonies. Histologically, there is disordered growth around the crypt region with invasive tubules and acinar-like structures accompanied by a scirrhous response and a variable amount of mucin production (Fig. 8.20). Due to the specific location of this development, care should be taken to process this region in aged common marmosets so that the lesion is not missed.

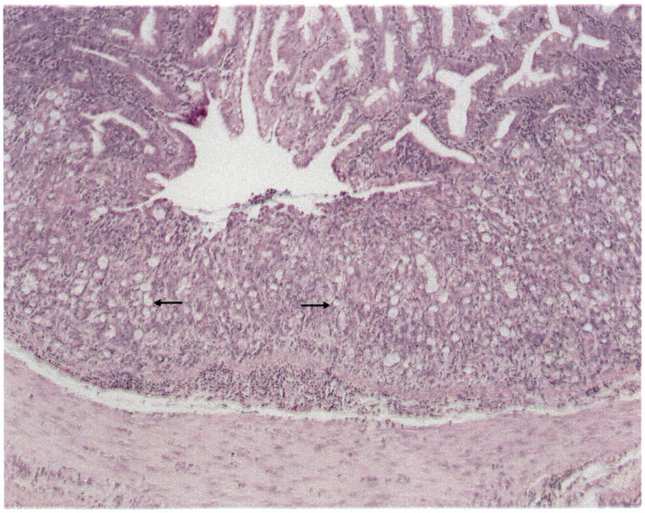

Fig. 8.20 Small intestinal adenocarcinoma, common marmoset. Replacing the crypt region is an expansive neoplasm composed of rudimentary tubules and sheets of neoplastic epithelial cells, many of which have signet ring morphology (arrows)

8.4.3 Colonic Carcinoma of Cotton-Top Tamarins

Cotton-top tamarins were historically used to study spontaneous ulcerative colitis, and up to 40% of animals progressed to develop colonic adenocarcinomas. Current studies on this species are limited to their protected status as an endangered species; however, this lesion remains prevalent in aging populations. Similar to the situation in rhesus macaques and common marmosets, these are invasive neoplasms associated with stricture, sometimes with the presence of melena (Figs. 8.21 and 8.22). This tumor does not progress through adenomatous polypoid changes, but rather is associated with invasion and spread, both locally and distant to liver and lung. Histologic patterns are similar to that in other species with sheets, acini, and tubular structures that invade transmurally and can be associated with variable mucin production (Figs. 8.23 and 8.24).

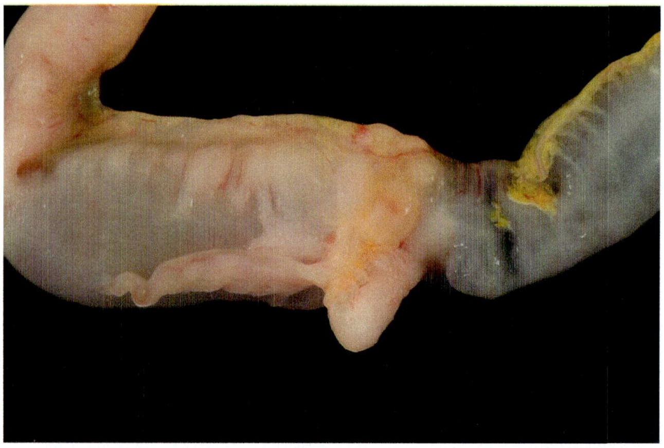

Fig. 8.21 Colonic carcinoma, cotton-top tamarin. A focal stricture (arrow) is present in the mid-colon, and the adjacent colon is markedly dilated and thickened with irregular white serosal plaques

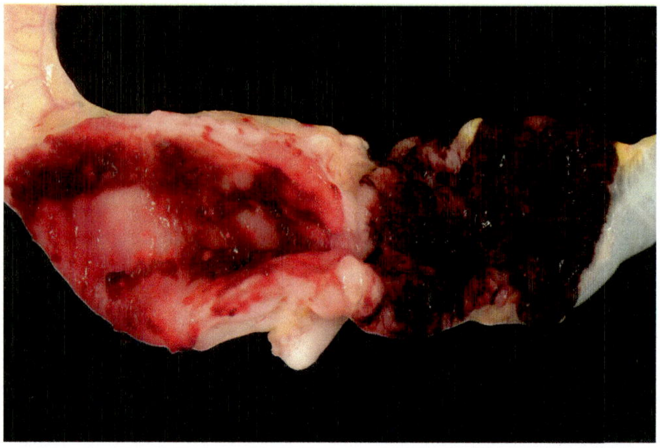

Fig. 8.22 Colonic carcinoma, cotton-top tamarin. The same case as above opened to reveal marked melena and profound thickening of the colon wall

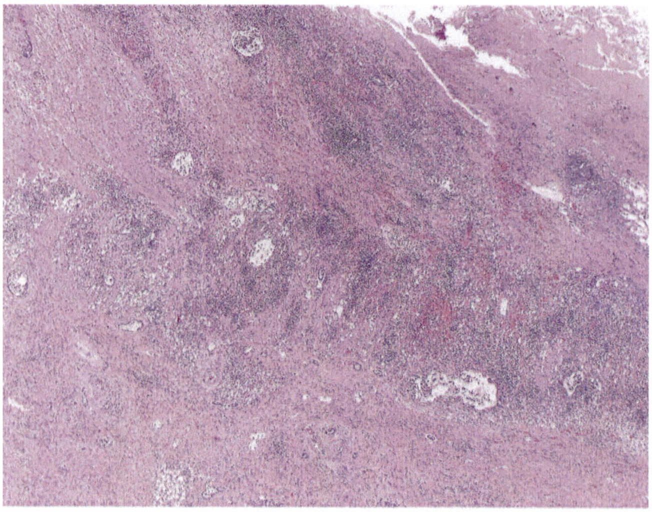

Fig. 8.23 Colonic carcinoma, cotton-top tamarin. The surface mucosa is completely lost with underlying muscle layers invaded by loose clusters, acini, and vague aggregates of neoplastic epithelial cells

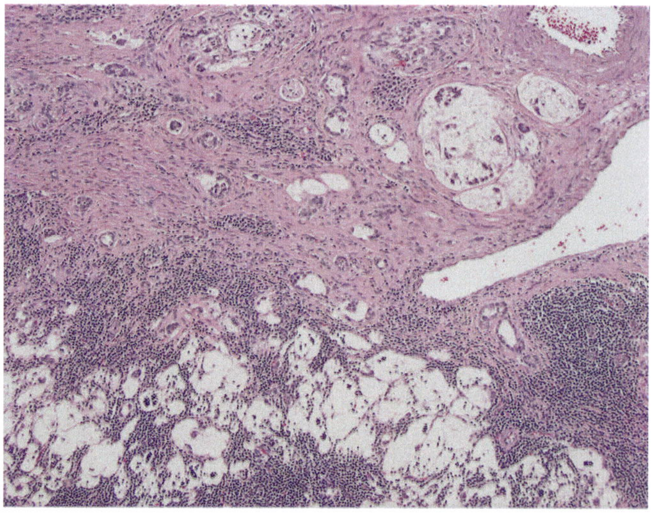

Fig. 8.24 Colonic carcinoma, cotton-top tamarin. Invading a colonic lymph node are numerous epithelial cells admixed with mucin. Similar lakes of mucin and neoplastic epithelial cells are present in nearby lymphatics

8.4.4 Gastrointestinal Stromal Tumor

Gastrointestinal stromal tumors (GISTs) are mesenchymal tumors thought to arise from the intestinal pacemaker cells (interstitial cells of Cajal). Grossly, GISTs vary in appearance from a small intramural nodule to an exophytic, expansile intramural mass that protrudes into the lumen or peritoneal cavity (Figs. 8.25 and 8.26). GISTs may arise anywhere along in the gastrointestinal tract, although tumors of the stomach predominate in the nonhuman primate literature

[10, 11]. Histologically, GISTs are composed of spindle cells with indistinct cell borders, abundant eosinophilic cytoplasm, and elongated nuclei arranged in interlacing bundles and whorls (Fig. 8.27). Strong, diffuse cytoplasmic immunoreactivity for CD117 (c-KIT) is helpful in differentiating GISTs from other spindle cell tumors, particularly leiomyomas, arising in the gastrointestinal tract (Fig. 8.28).

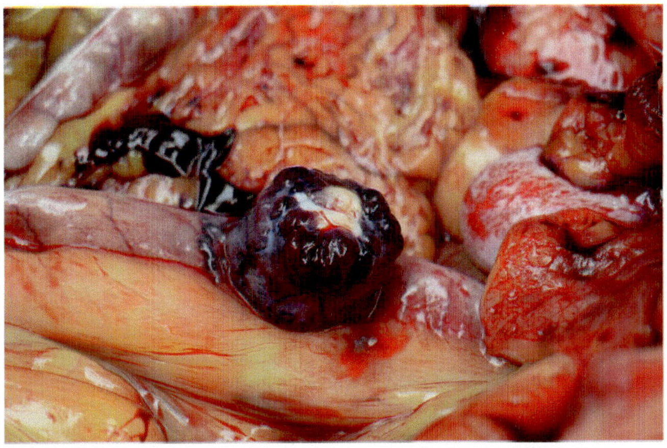

Fig. 8.25 Gastrointestinal stromal tumor (GIST), rhesus macaque. A dark red, broad-based, intramural, exophytic mass projects into the peritoneal cavity from the wall of the distal colon. On the serosal surface, the center of the mass is yellow to white and slightly depressed. This animal presented with hematochezia due to ulceration of the colorectal mucosa overlying the mass

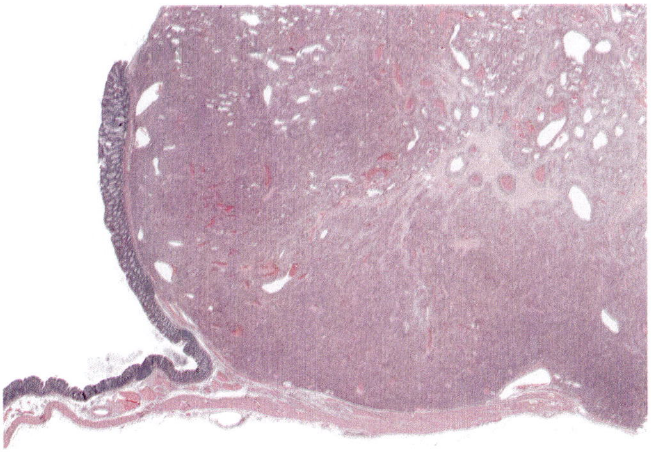

Fig. 8.26 GIST, rhesus macaque. A well-demarcated, unencapsulated, intramural tumor marked expands the colonic wall. The neoplasm is moderately cellular and composed of spindle cells arranged in short, interwoven bundles. The colonic mucosa is focally ulcerated

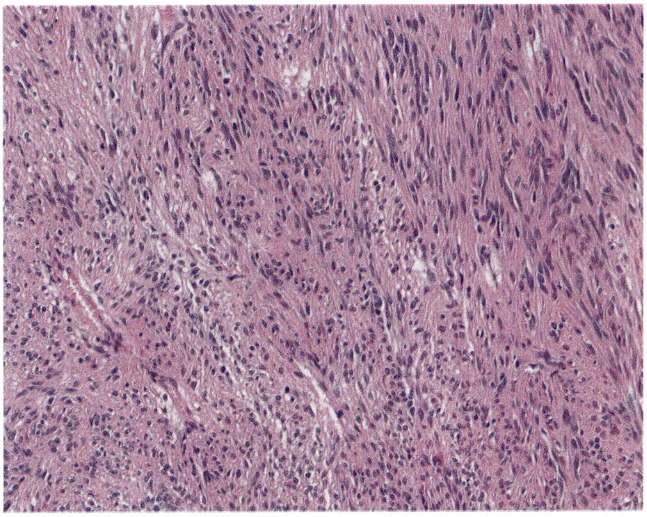

Fig. 8.27 GIST, rhesus macaque. Neoplastic cells are spindle shaped with elongated nuclei, indistinct cell borders and abundant eosinophilic, fibrillar cytoplasm

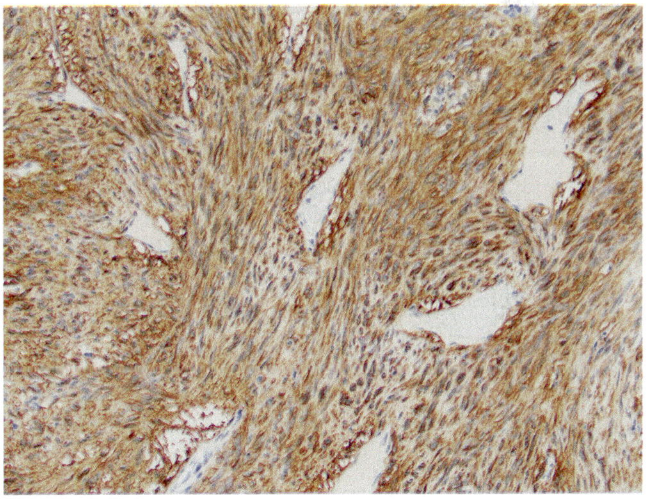

Fig. 8.28 GIST, rhesus macaque. Immunohistochemical stain for CD177 (c-KIT) demonstrates diffuse, strong, cytoplasmic immunopositivity throughout the tumor

8.4.5 Gastric Adenocarcinoma

Primary gastric neoplasia is rare in nonhuman primates. A variety of benign and malignant tumors have been reported, with epithelial-origin predominating over mesenchymal-origin neoplasms [5, 12]. Grossly, epithelial neoplasms may be evident as single or multiple, occasionally polypoid masses (Fig. 8.29), or as regions of gastric wall thickening, often with mucosal ulceration (Fig. 8.30). In macaques, spontaneous gastric carcinomas arise occasionally, and malignant tumors have also been reported following oral administration of a carcinogen [13, 14]. In a closed colony of sooty mangabeys, adenocarcinomas occur sporadically in association with non-*H. pylori* gastric Helicobacteriosis [15, 16]. In mangabeys, these aggressive spontaneous neoplasms occur predominantly in the pyloric region; transmural invasion, mucosal ulceration, and local lymphatic metastasis are common features (Figs. 8.30 and 8.33). Histologically, adenocarcinomas form tubuloacinar structures (Figs. 8.31, 8.32, 8.33, and 8.34). In chemically induced carcinomas of macaques, signet ring cell morphology is a feature of some tumors [13].

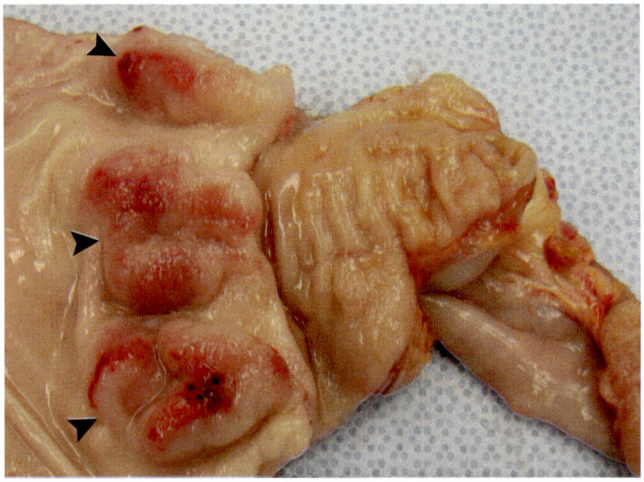

Fig. 8.29 Multiple gastric tumors, rhesus macaque. The gastric pyloric region contains three well-demarcated, tan to red, broad-based, polypoid masses (arrowheads)

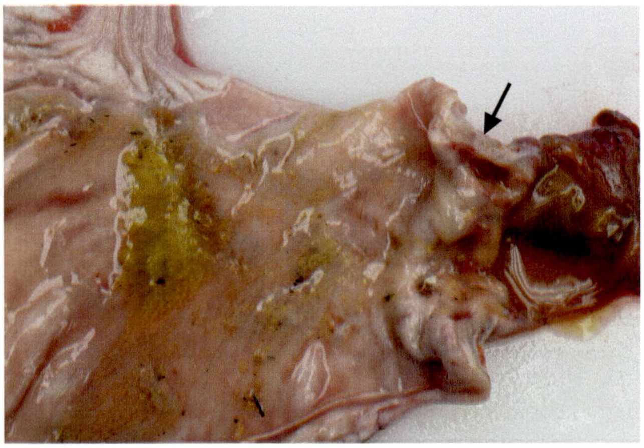

Fig. 8.30 Gastric adenocarcinoma, sooty mangabey. The gastric pylorus is severely thickened (arrow) with a focally extensive area of mucosal ulceration. This animal was euthanized for clinical reasons

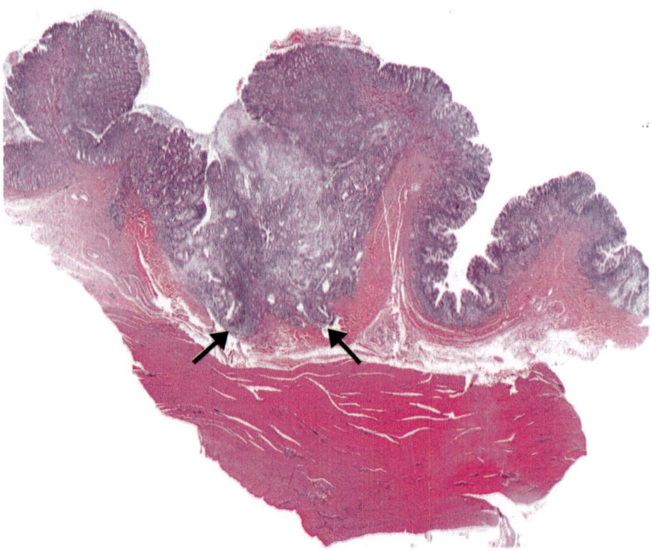

Fig. 8.31 Gastric adenocarcinoma, rhesus macaque. The gastric mucosa is thickened by a proliferation of epithelial cells. Neoplastic cells multifocally invade through the muscularis mucosa into the submucosa (arrows). The mucosal surface is focally necrotic and infiltrated by neutrophils

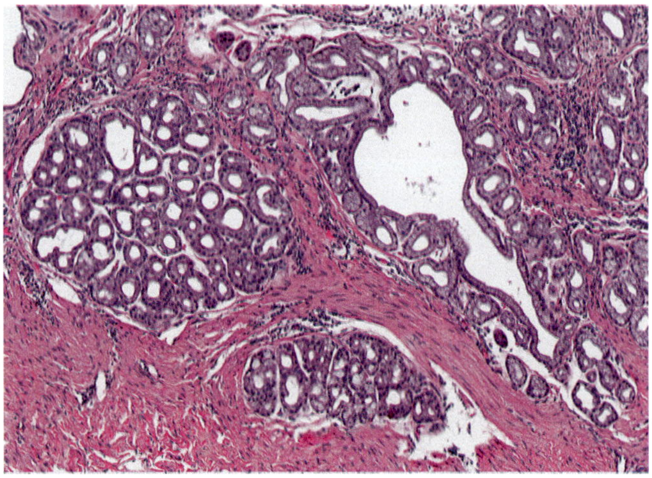

Fig. 8.32 Gastric adenocarcinoma, rhesus macaque. Neoplastic cells form tubules and invade between smooth muscle fibers of the gastric wall

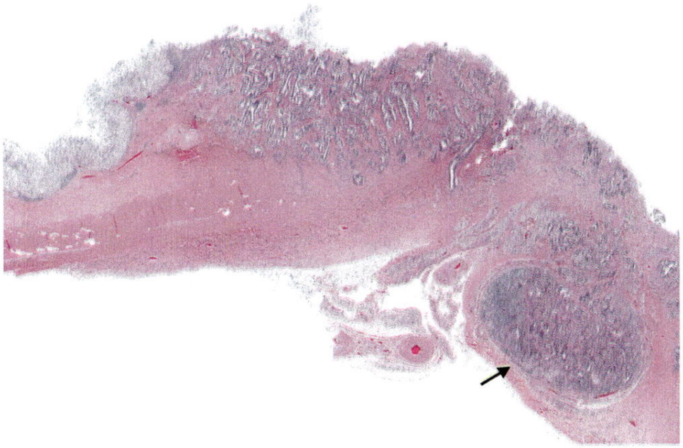

Fig. 8.33 Gastric adenocarcinoma, sooty mangabey. A spontaneous epithelial neoplasm (same tumor as Fig. 8.30) invades the gastric wall in the pyloric region. The mucosa and submucosa are extensively ulcerated. Neoplastic tubules efface much of the local lymph node (arrow). Fibrous connective tissue expands the serosal surface and encapsulates the affected lymph node

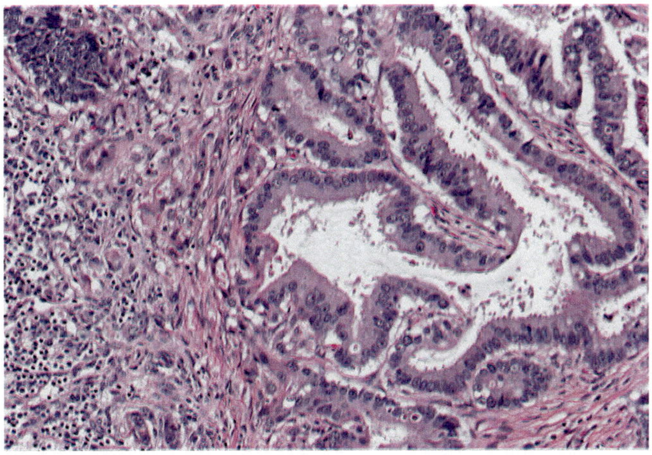

Fig. 8.34 Gastric adenocarcinoma, sooty mangabey. Neoplastic cells form irregular branching tubules and are occasionally individualized or in small clusters

8.4.6 Pancreatic Carcinoma

Tumors of the exocrine pancreatic tissue occur less commonly than those of the neuroendocrine pancreas in nonhuman primates. Epithelial neoplasms may arise from the acinar or duct cells. Histologically, carcinomas form tubules, acini, or sheets; they are unencapsulated and locally invasive (Figs. 8.35 and 8.36).

Immunohistochemistry may be useful in classifying poorly differentiated epithelial neoplasms as well as mixed acinar-neuroendocrine tumors [17]. In humans, cytokeratin 7 is a marker for epithelial cells of ductal origin [18].

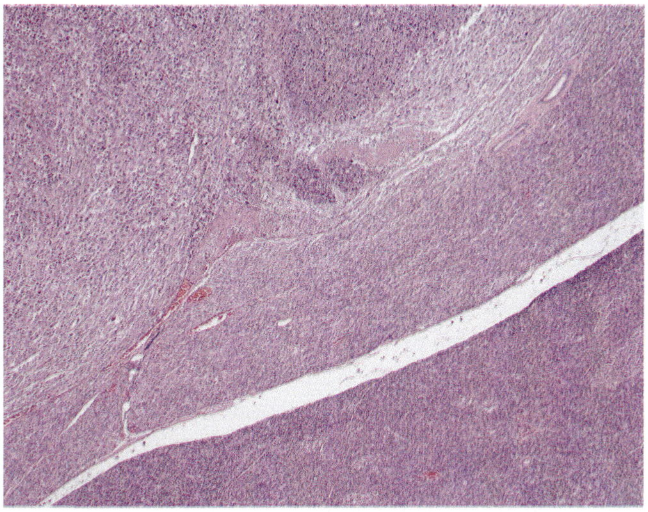

Fig. 8.35 Pancreatic carcinoma, common chimpanzee. An unencapsulated aggressive neoplasm expands a pancreatic lobule, compressing two ducts and preexisting lobular tissue to the periphery. A focus of necrosis is present at the center of the image. This spontaneous tumor was diagnosed in an animal that was euthanized for clinical reasons

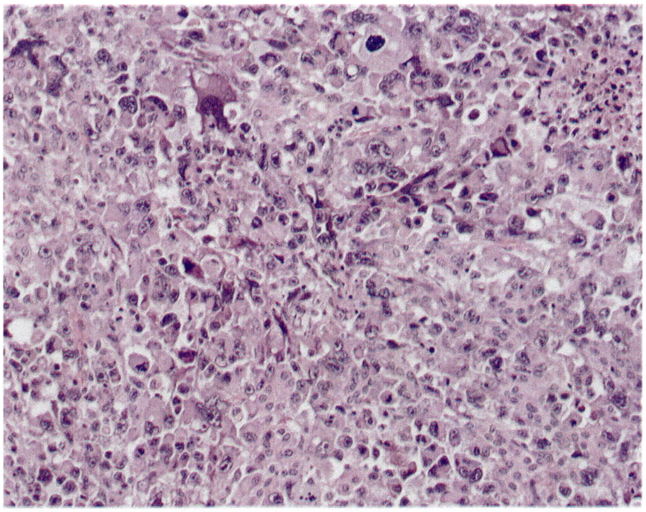

Fig. 8.36 Pancreatic carcinoma, common chimpanzee. Higher magnification of the same tumor. Neoplastic cells are round to polygonal with variably distinct cell borders, abundant eosinophilic cytoplasm, and form sheets. Neoplastic cells exhibit marked anisocytosis and anisokaryosis with multifocal karyomegaly and multinucleation. Numerous mitotic figures are present in this image

8.5 Urogenital

8.5.1 Ovarian Tumors

Ovarian tumors occur occasionally in nonhuman primates. The major types include tumors of the surface epithelium (adenocarcinoma, adenoma), tumors of the germ cells (dysgerminoma, teratoma), and sex cord stromal tumors (granulosa and theca cell tumors). Granulosa cell tumors are the most commonly reported ovarian neoplasms in macaques (Fig. 8.37), followed by teratomas (Fig. 8.38) [19–21]. Contrary to what is n in humans, ovarian adenocarcinoma is rare in nonhuman primates. Benign epithelial ovarian neoplasms (adenomas or cystadenomas) are occasionally diagnosed. Ovarian tumors are usually identified at necropsy as incidental findings and are typically unilateral. Histologically, granulosa cell tumors are solid to cystic tumors composed of sheets of round to polygonal cells with abundant eosinophilic cytoplasm often arranged in follicle-like structures (Fig. 8.39). Neoplastic granulosa cells often surround concretions of extracellular eosinophilic material (Call-Exner bodies) (Fig. 8.40).

Clinicopathologic data are required to determine if these tumors are estrogen-secreting, which can produce secondary lesions such as endometrial hyperplasia, endometrial polyps, and vaginal keratinization [19]. Teratomas are benign tumors composed of tissue deriving from at least two primordial germ layers (ectoderm, mesoderm, or endoderm). The gross and histological features of this tumor vary according to its composition. Tumors may contain haired skin, bone, and teeth (Figs. 8.41 and 8.42).

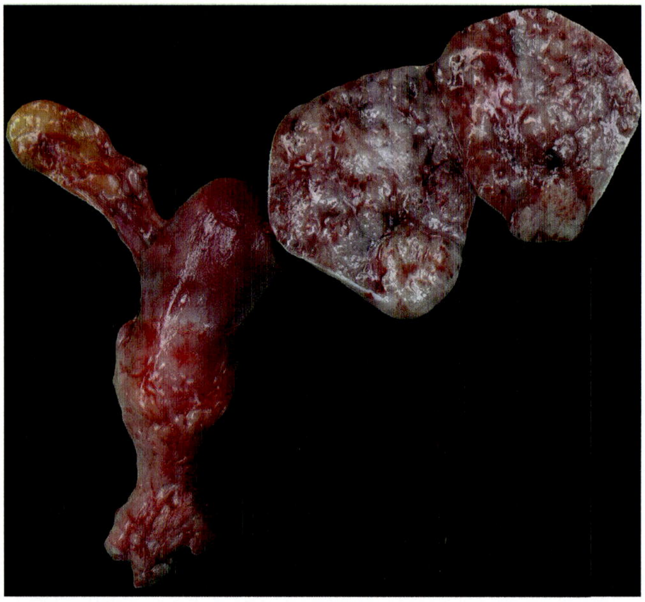

Fig. 8.37 Ovarian granulosa cell tumor, rhesus macaque. The left ovary is replaced by an expansile mass, which on the cut surface is firm, white to tan, and solid with multiple ill-defined lobules and small foci of hemorrhage. The contralateral ovary, oviducts, uterus, cervix, and vagina are grossly normal

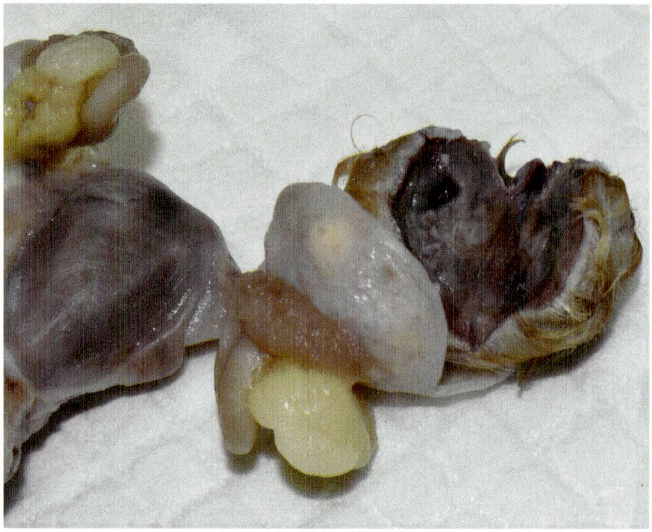

Fig. 8.38 Ovarian teratoma, rhesus macaque. The left ovary is replaced by a round, encapsulated tumor, the surface of which is covered by haired skin. On the cut surface, the mass is composed of white to tan and red soft tissues with few small, centrally located white irregularly shaped foci of firm to hard tissue (bone and cartilage)

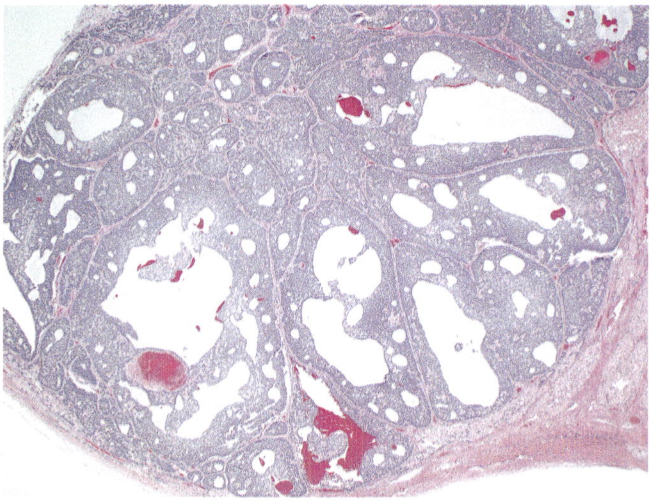

Fig. 8.39 Ovarian granulosa cell tumor, rhesus macaque. Low-magnification image of the same tumor from Fig. 8.37 reveals a neoplasm composed of numerous follicle-like structures separated by thin connective tissue septa. The follicle-like structures are lined by several layers of polygonal cells with scant eosinophilic cytoplasm and frequently surround an irregularly shaped empty space

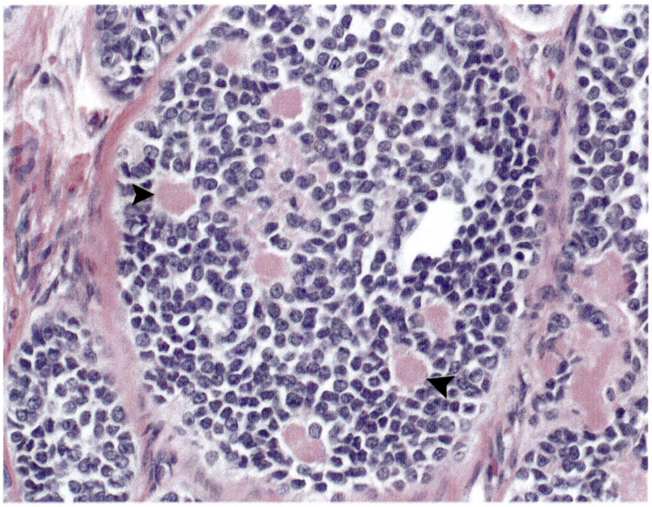

Fig. 8.40 Ovarian granulosa cell tumor, rhesus macaque. High-magnification image of the same tumor demonstrating the follicle-like arrangements of neoplastic granulosa cells separated by fibrous connective tissue septa. Neoplastic cells frequently surround amorphous extracellular accumulations of eosinophilic material (Call-Exner bodies; arrowheads)

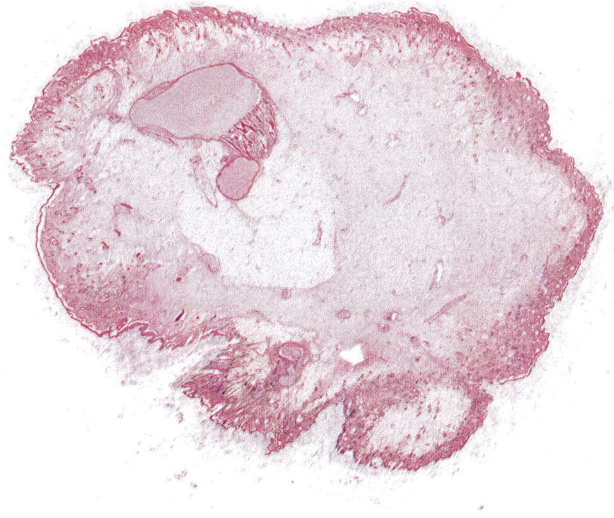

Fig. 8.41 Ovarian teratoma, rhesus macaque. This solid, round tumor is circumferentially covered by haired skin (ectoderm). The solid core of the mass is composed of various mesodermal connective tissue elements including adipose tissue, loose fibrovascular tissue, and a focus of the cartilage and bone

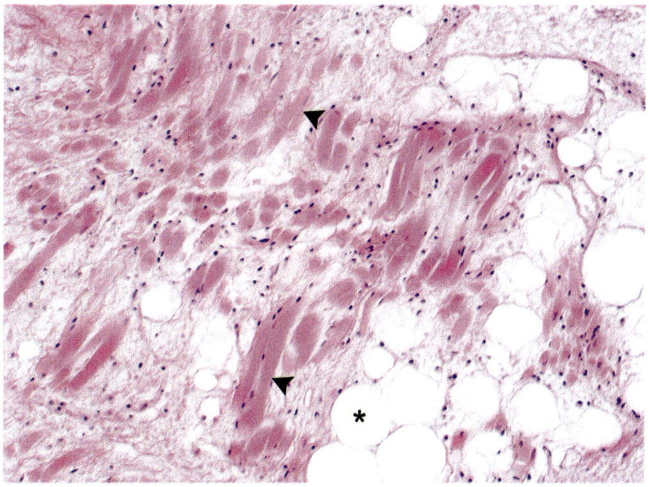

Fig. 8.42 Ovarian teratoma, rhesus macaque. Multiple mesodermal cell lineages are present in this teratoma, including haphazardly arranged striated muscle fibers (arrowheads) and adipocytes (*)

8.5.2 Uterine Leiomyoma/Myosarcoma

Leiomyoma is the most common uterine tumor of nonhuman primates. These benign tumors arise from the myometrium and are pink to tan, firm, nodular, well demarcated, and variably encapsulated. Uterine leiomyomas range from solitary, well-demarcated intramural nodules (Fig. 8.43 and 8.44) to massive, multinodular tumors that enlarge and distort the uterus. Massive tumors can cause clinical signs related to bleeding or compression of adjacent structures. Histologically, leiomyomas are composed of monomorphic spindle cells arranged in bundles and streams (Fig. 8.45). They have a low mitotic rate. Leiomyosarcomas of the female reproductive tract are rarely diagnosed in nonhuman primates (Fig. 8.46). Histologically, they are distinguished from their benign counterparts by higher nuclear-to-cytoplasmic ratios and mitotic rates, local invasion, and necrosis (Figs. 8.47 and 8.48). True to their smooth muscle origin, leiomyosarcomas typically express α-smooth muscle actin (Fig. 8.49).

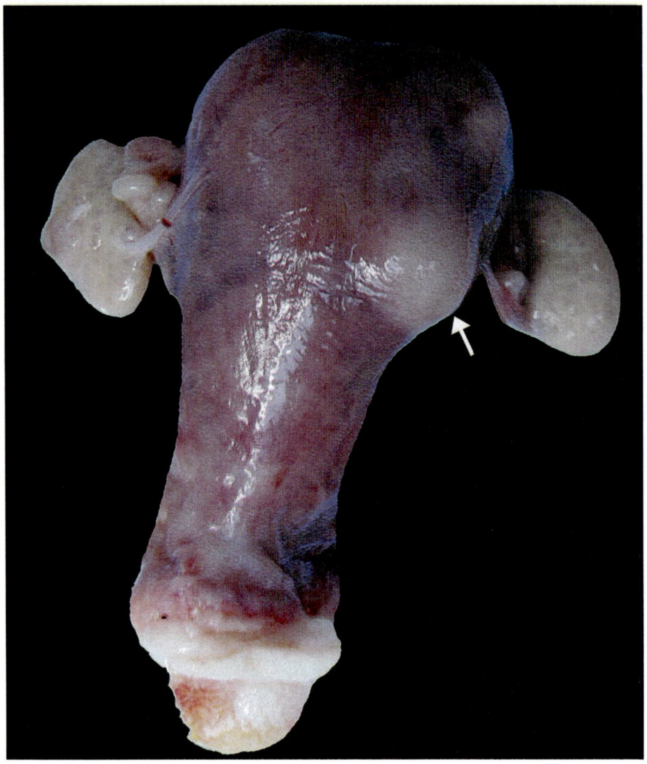

Fig. 8.43 Uterine leiomyoma, sooty mangabey. This image shows the female reproductive tract from an aged mangabey euthanized for unrelated clinical reasons. The uterine body contains a focal, nodular, pale tan, intramural mass (white arrow). By courtesy of Dr. Christopher Pinelli, Emory National Primate Research Center

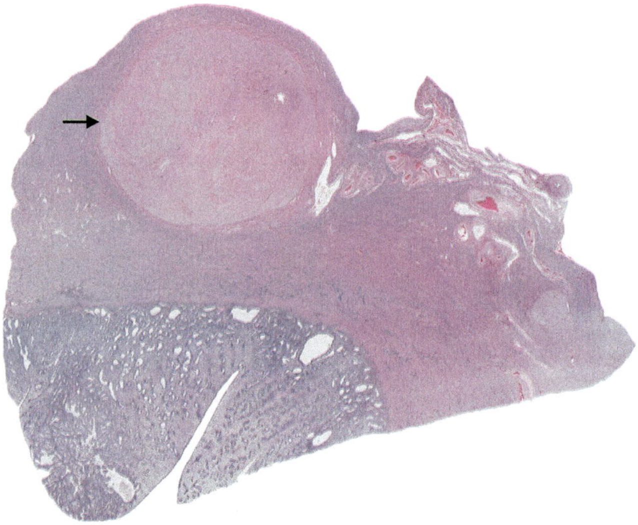

Fig. 8.44 Uterine leiomyoma, sooty mangabey. Photomicrograph of the uterine tumor from the previous image. The uterine wall is focally expanded by an intramural nodule (arrow). The nodule is well demarcated and thinly encapsulated and compresses the adjacent myometrium and elevates the serosal surface. The endometrium is hyperplastic

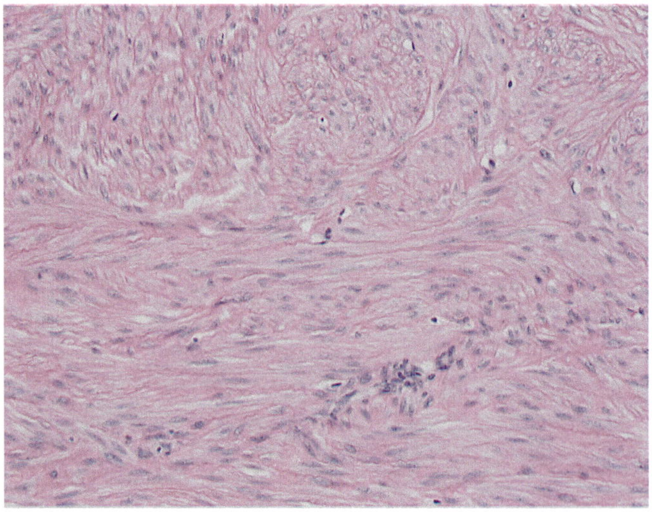

Fig. 8.45 Uterine leiomyoma, sooty mangabey. This high-magnification image of the benign uterine mass demonstrates smooth muscle fibers in longitudinal and cross sections

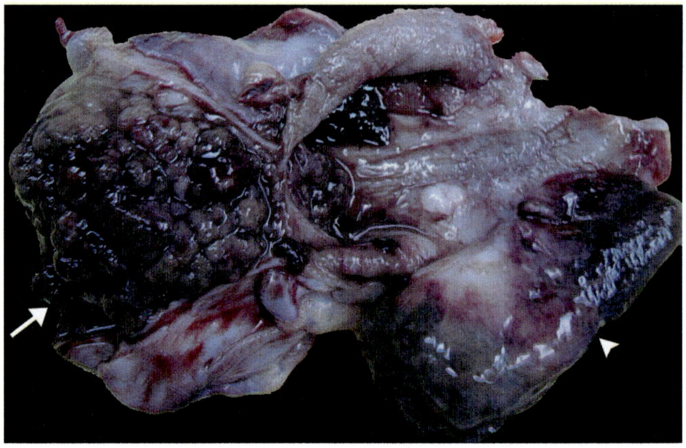

Fig. 8.46 Leiomyosarcoma, rhesus macaque. A malignant neoplasm enlarges and distorts the entire female reproductive tract from the uterus (left of the image) to the vagina (right of the image). The uterine wall is thickened by a red to purple, bosselated neoplasm (arrow). A second broad-based, exophytic mass arises from the vaginal wall (arrowhead). By courtesy of Dr. Christopher Pinelli, Emory National Primate Research Center

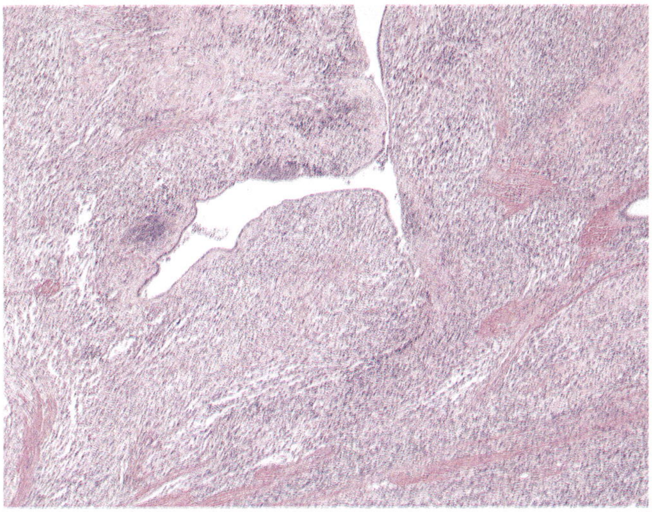

Fig. 8.47 Leiomyosarcoma, rhesus macaque. There is marked expansion and effacement of the uterine wall by an unencapsulated, infiltrative neoplasm composed of densely packed spindle cells

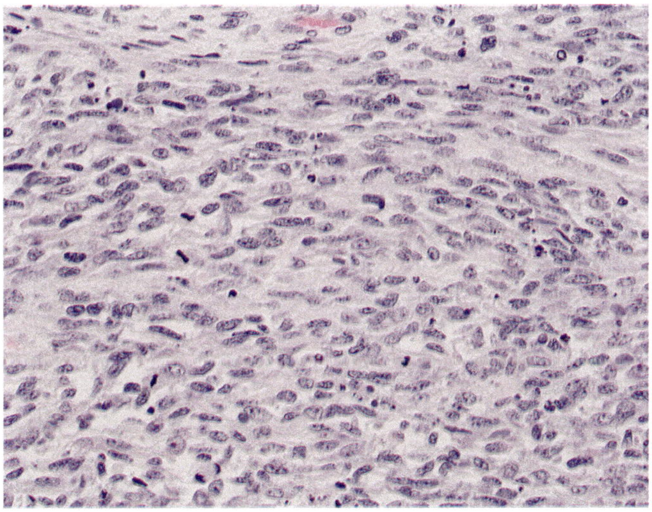

Fig. 8.48 Leiomyosarcoma, rhesus macaque. Neoplastic cells are densely packed with indistinct cell borders, moderate amounts of eosinophilic cytoplasm, irregularly oval to elongate nuclei, and finely stippled chromatin. There are numerous mitotic figures present

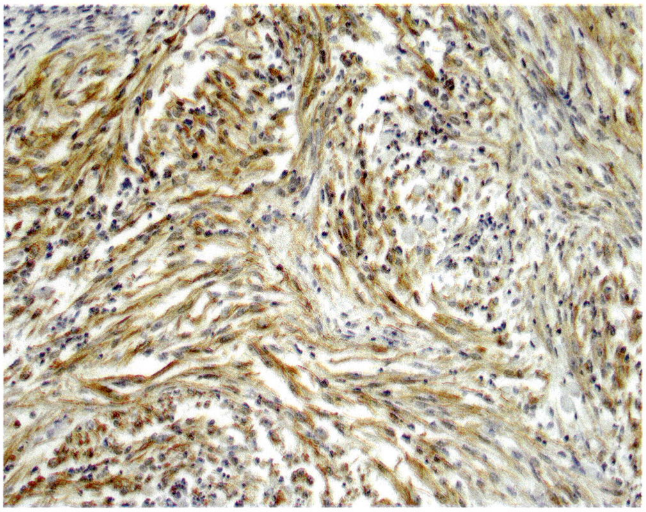

Fig. 8.49 Leiomyosarcoma, rhesus macaque. Neoplastic cells exhibit strong, diffuse cytoplasmic immunoreactivity for alpha smooth muscle actin

8.5.3 Testicular Tumors

Testicular tumors are rare in all nonhuman primate species. The predominant types are broken down into tumors of the germ cells (seminoma), tumors of Sertoli cells (Sertoli cell tumors), and tumors of the interstitial cells (Leydig or interstitial cell tumor). Seminomas are characterized by a bulging, white to tan mass. Sertoli cell tumors are firmer, whiter on cross section, and do not tend to bulge. Leydig cell tumors are tan, may bulge on section, and often contain regions of hemorrhage. Histologically seminomas are composed of sheets of neoplastic germ cells that have enlarged nuclei and a single, prominent nucleolus (Figs. 8.50 and 8.51). Sertoli cell tumors are associated with abundant fibrous bands that dissect the neoplastic cells into rudimentary tubules. Neoplastic Leydig cells have abundant, often vacuolated, eosinophilic cytoplasm (Figs. 8.52 and 8.53).

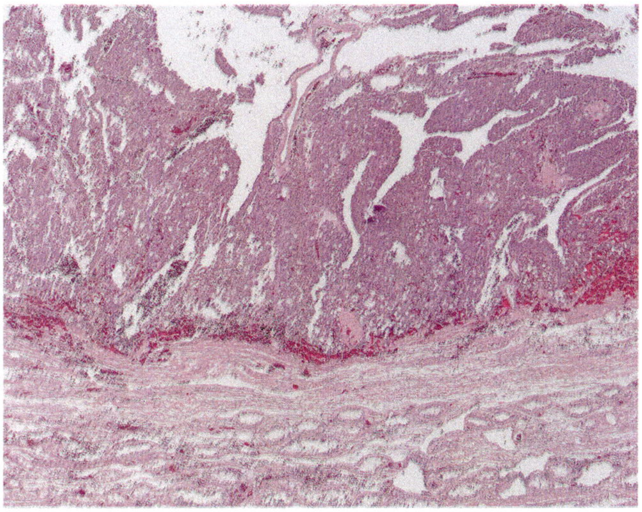

Fig. 8.50 Seminoma, rhesus macaque: The adjacent testicular tissue is compressed and atrophic. The compression is caused by a large, expansile mass composed of sheets of neoplastic germ cells

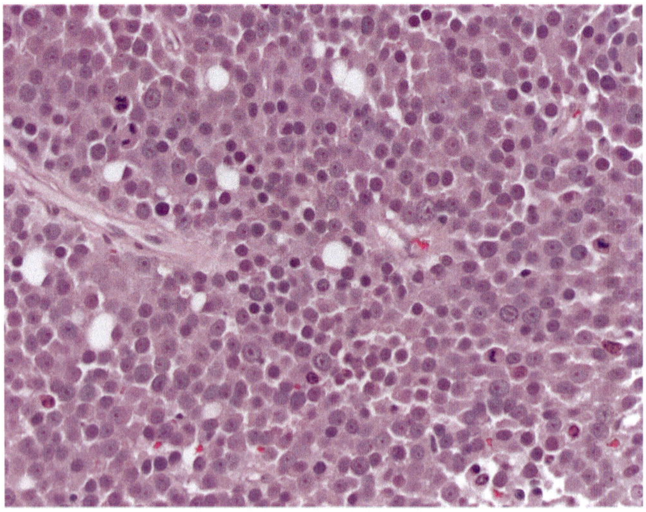

Fig. 8.51 Seminoma, rhesus macaque. Higher magnification of the neoplastic germ cells reveals the stereotypical sheeting pattern with greatly enlarged nuclei, prominent nucleoli, and abundant mitoses

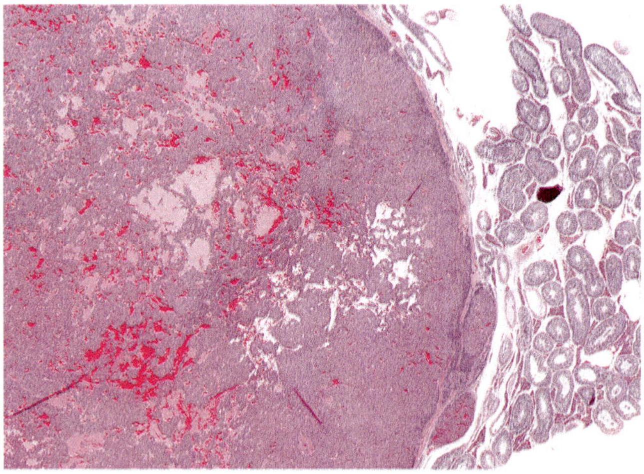

Fig. 8.52 Interstitial cell tumor, squirrel monkey. The testis is focally expanded by an encapsulated, well-demarcated tumor composed of neoplastic interstitial (Leydig) cells that compresses the surrounding seminiferous tubules

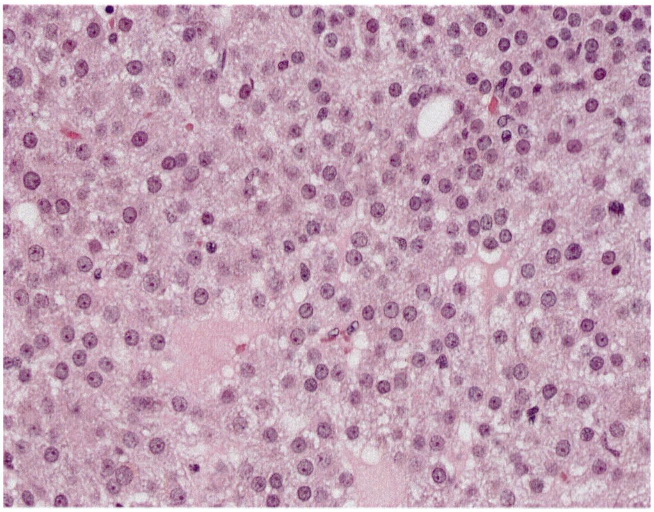

Fig. 8.53 Interstitial cell tumor, squirrel monkey. Neoplastic interstitial cells have distinct cell borders and abundant vacuolated eosinophilic cytoplasm

8.5.4 Renal Tumors

Primary tumors of the kidney occur sporadically in a variety of nonhuman primate species. Tumors of the renal tubular epithelium (adenomas and carcinomas) are more common than nephroblastomas and urothelial neoplasms. Adenomas are typically small, solitary, well demarcated but unencapsulated, tan to white tumors located in the renal cortex (Fig. 8.54); these are often diagnosed incidentally at necropsy. Histologically, adenomas form tubules and papillary proliferations (Fig. 8.55). Renal carcinomas may exhibit a variety of growth pattern(s), including tubular, papillary, solid and cystic, or any combination thereof (Figs. 8.56, 8.57, and 8.58). In humans, carcinomas are classified as such by their size (>1.5 cm), increased pleomorphism, and/or higher mitotic rates compared to adenomas [22]. The rarity of renal carcinomas in nonhuman primates has precluded classification by cell type.

Nephroblastomas are malignant renal tumors of juvenile animals, as in other species. These tumors grow rapidly and replace much of the preexisting renal tissue. Histologically, these embryonal tumors are composed of three elements: epithelium forming tubules and glomerular buds, undifferentiated blastemal cells, and mesenchyme (Figs. 8.59 and 8.60). The composition of individual tumors varies widely; the presence of embryonic epithelium is key to differentiating this tumor from other poorly differentiated neoplasms. In cases where little to no embryonic epithelium is present, immunoreactivity of blastemal cells for the transcription factor Wilms tumor 1 (WT-1) can assist in confirming the diagnosis (Figs. 8.61 and 8.62).

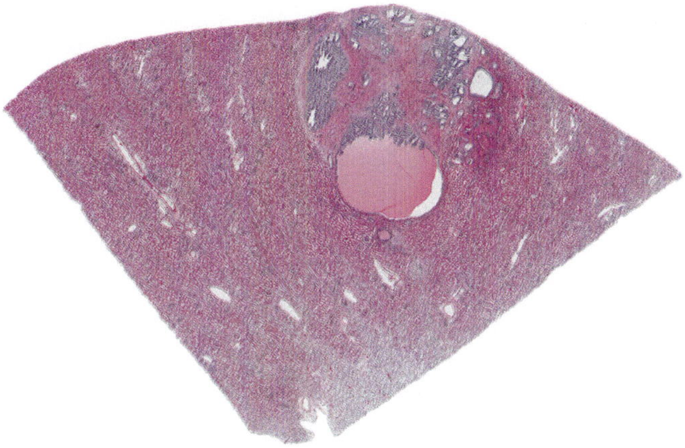

Fig. 8.54 Renal adenoma, southern pig-tailed macaque. The renal cortex contains a well-demarcated, unencapsulated, nodular mass, which slightly elevates the renal capsule and compresses the adjacent cortical parenchyma. This spontaneous tumor was an incidental finding at necropsy

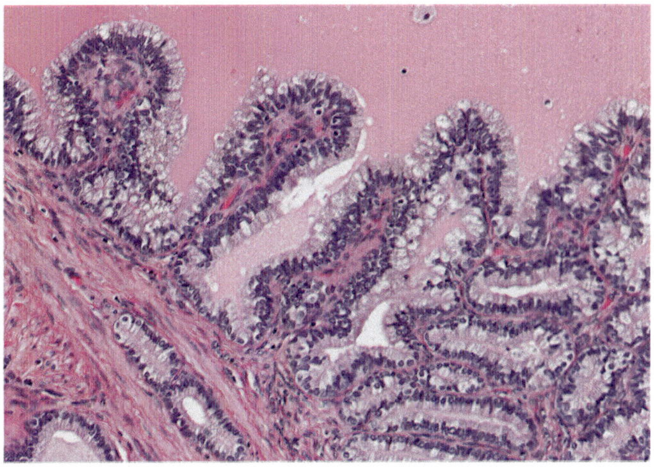

Fig. 8.55 Renal adenoma, southern pig-tailed macaque. Tumor cells exhibit a tubulopapillary growth pattern along the wall of a fluid-filled cyst. Neoplastic cells have abundant, vacuolated to fibrillar eosinophilic cytoplasm, and a small, basally oriented nucleus

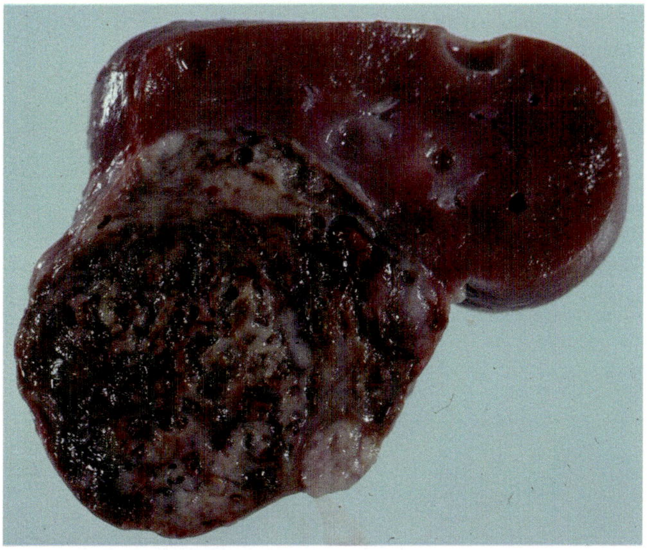

Fig. 8.56 Renal carcinoma, rhesus macaque. A multilobulated, well-demarcated, unencapsulated neoplasm enlarges the kidney and distorts the renal capsule. The tumor is friable, tan to brown and red with numerous small cystic spaces and depressions

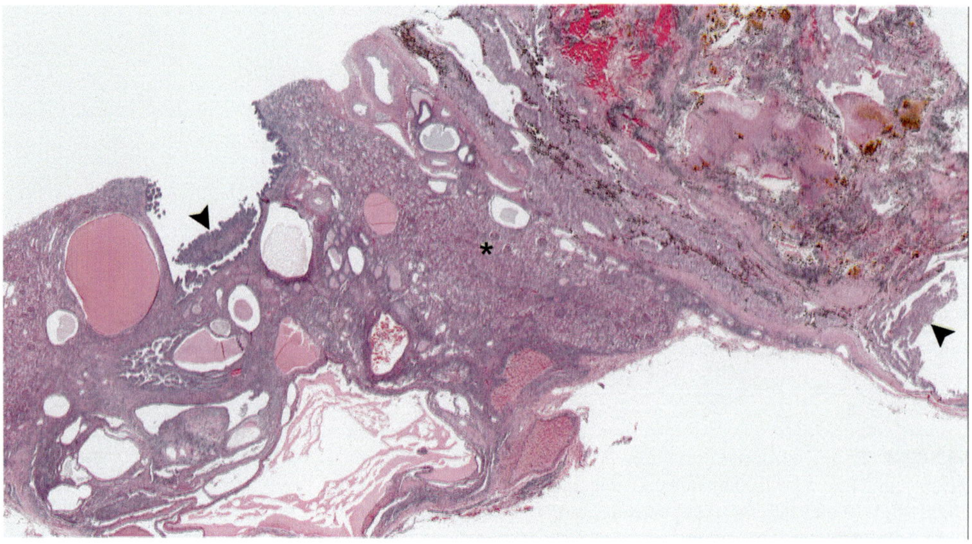

Fig. 8.57 Renal carcinoma, rhesus macaque. A poorly demarcated, malignant renal tubular epithelial neoplasm effaces the majority of the renal cortex and exhibits a variety of growth patterns; neoplastic cells form sheets and ectatic, fluid-filled tubules some of which contain papillary proliferations of tumor cells (arrowheads). There is extensive tumoral hemorrhage and necrosis at the top right of the image

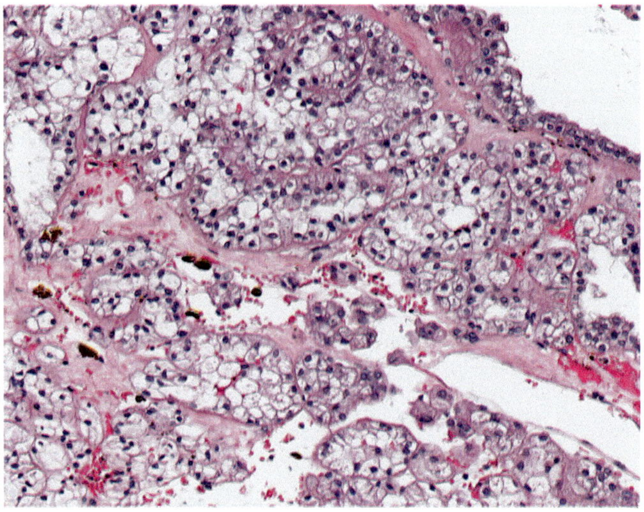

Fig. 8.58 Renal carcinoma, rhesus macaque. Neoplastic renal tubular epithelial cells form nests, sheets, and tubule-like structures, some with small papillary projections. Cells are polygonal with distinct cell borders and abundant cytoplasm that is mostly clear with scant, wispy to granular material. Neoplastic tubules are multifocally ectatic and contain scant fluid, sloughed cells and hemorrhage

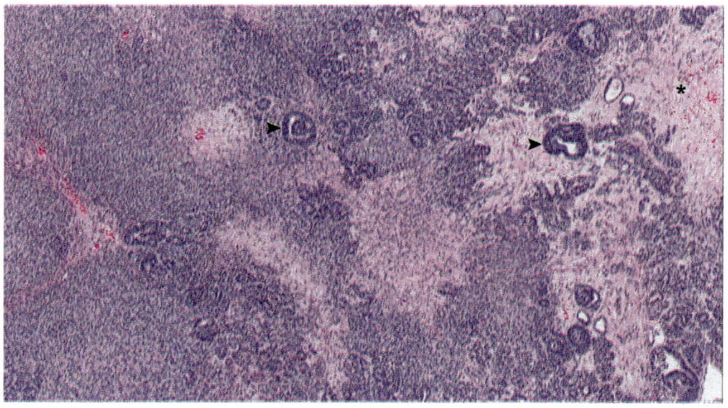

Fig. 8.59 Nephroblastoma, baboon. Low-magnification image of a renal nephroblastoma from a 3-year-old baboon demonstrates the triphasic mixture of embryonic epithelium (arrow heads), undifferentiated blastemal cells, and mesenchyme (*)

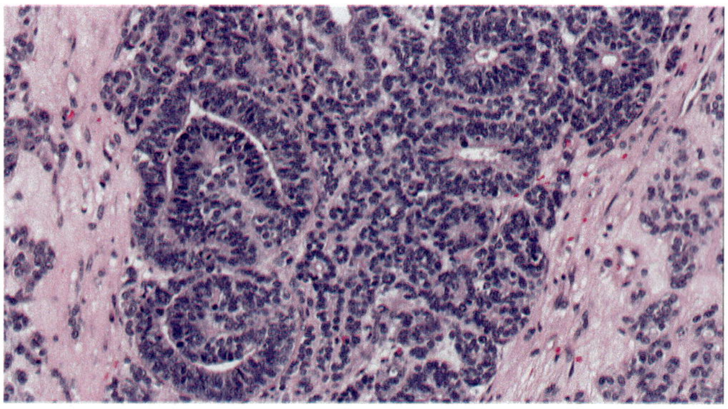

Fig. 8.60 Nephroblastoma, baboon. Embryonic epithelial structures include two glomerular buds on the left and multiple tubules on the right. Blastemal cells occupy the space between the embryonic epithelial structures and mesenchyme is present in the lower left and right corners of the image

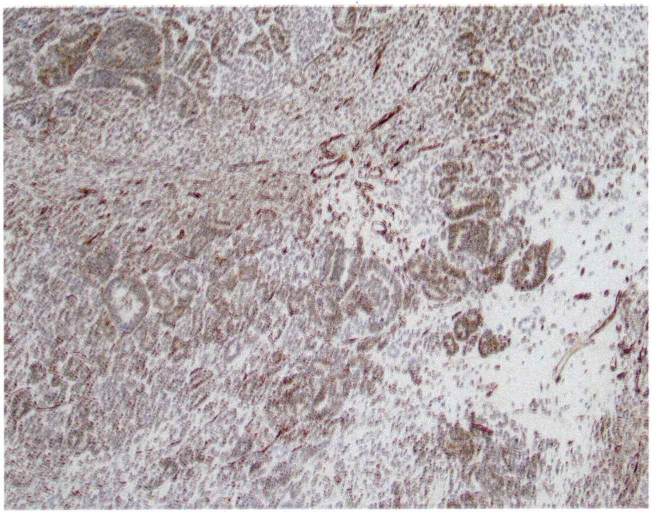

Fig. 8.61 Nephroblastoma, baboon. Immunohistochemical stain targeting the transcription factor Wilms tumor 1 (WT-1) demonstrates strong immunoreactivity within the epithelial and blastemal components and immunonegativity within the mesenchyme

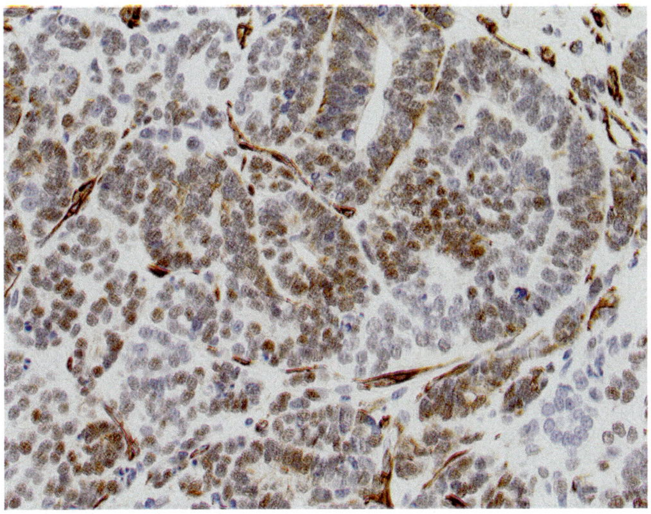

Fig. 8.62 Nephroblastoma, baboon. Higher magnification of the WT-1 immunostain showing mild to strong nuclear immunoreactivity in the majority of epithelial and blastemal cells

8.6 Respiratory

Neoplasia of the respiratory tract is exceedingly uncommon in nonhuman primates. The area of bronchioloalveolar hyperplasia and adenoma formation can be found sporadically in aged nonhuman primates, especially macaques.

8.7 Cardiovascular

Cardiovascular neoplasia is similarly rare and is a sporadic finding in aged nonhuman primates with cardiac sarcomas and vascular neoplasms like hemangioma and hemangiosarcoma rarely reported.

8.8 Hematopoietic

8.8.1 Lymphoma

Lymphoma is one of the most common neoplasms of nonhuman primate, predominately owing to the presence of a number of oncogenic viruses that cause malignant transformation of lymphocytes, especially in animals that are immunosuppressed. This includes lymphocryptovirus (macaques), simian T-lymphotropic virus (baboons), rhadinovirus (macaques), and various oncogenic herpesvirus (wide range of species) [1]. Regardless of the underlying etiology, lymphoma often presents as multicentric disease affecting hematopoietic tissue and a variety of visceral organs including, but not limited to, the kidney, liver, heart, and central nervous system. Lymphoma presents as a space occupying mass that is reddish tan to white, bulges on section, and is relatively soft on palpation (Figs. 8.63 and 8.64). Histology is typified by sheets of neoplastic lymphocytes (round cells) that replace tissue parenchyma (Figs. 8.65 and 8.66).

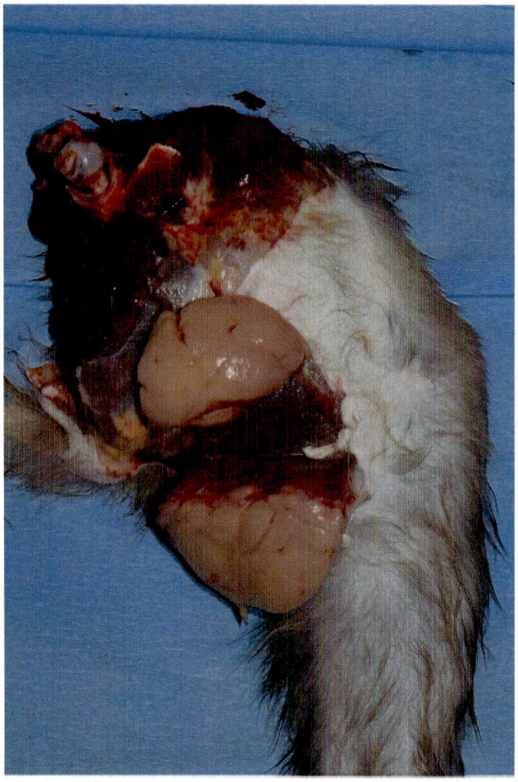

Fig. 8.63 Lymphoma, rhesus macaque. The popliteal lymph node is greatly enlarged and replaced by a tan mass that bulges on cross section. This animal was chronically simian immunodeficiency virus-infected

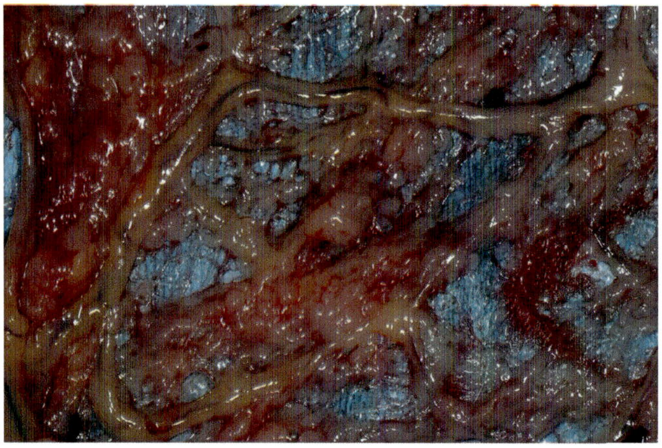

Fig. 8.64 Lymphoma, rhesus macaque. The greater omentum is replaced by hundreds of coalescing tan to white nodules representing widespread omental spread of lymphoma. This animal was chronically simian immunodeficiency virus-infected

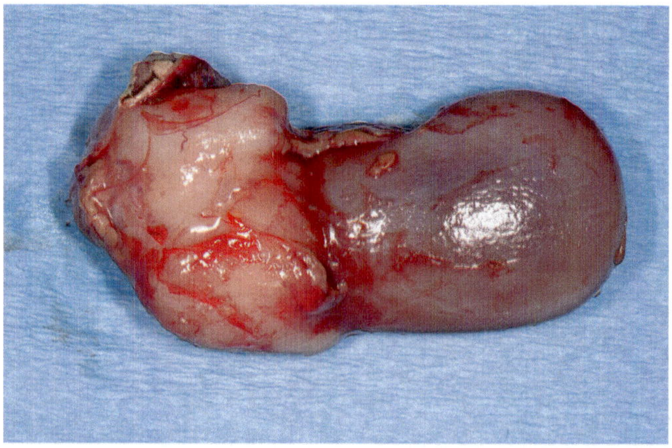

Fig. 8.65 Lymphoma, rhesus macaque. The cranial pole of the kidney is replaced by lymphoma. The adrenal gland is visualized compressed to the upper portion of the mass. This animal was chronically simian immunodeficiency virus-infected

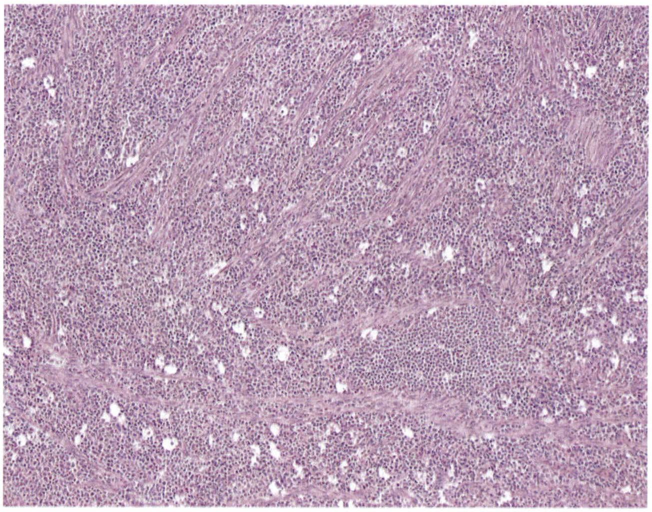

Fig. 8.66 Lymphoma, rhesus macaque. The smooth muscle of the uterus is replaced by sheets of neoplastic lymphocytes

8.9 Endocrine

8.9.1 Pheochromocytoma

Tumors of the enterochromaffin cells of the adrenal medulla are relatively common in a number of nonhuman primate species, namely, cotton-top tamarins and rhesus macaques. These tumors, like their counterparts in other species, cause marked enlargement of the adrenal gland (Fig. 8.67) with abundant intravascular invasion, often into the adjacent caudal vena cava, from where hepatic metastases commonly arise [23]. These tumors form nests, pseudorosettes, and packets of neoplastic cells that have abundant eosinophilic cytoplasm, large round to ovoid nuclei, and a variably apparent nucleolus (Figs. 8.68 and 8.69).

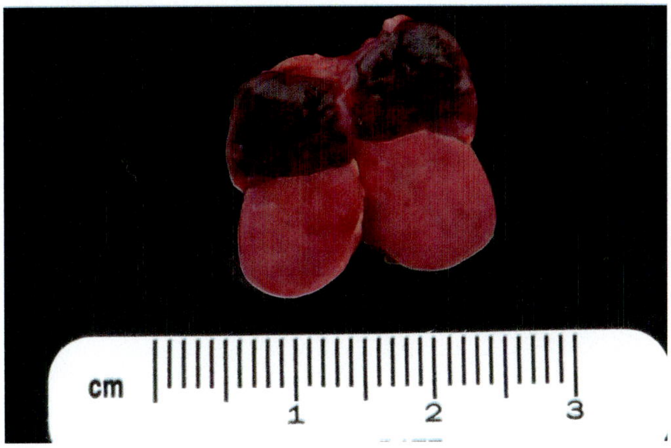

Fig. 8.67 Pheochromocytoma, cotton-top tamarin: The adrenal gland is replaced by a large, hemorrhagic mass that compressed the cranial pole of the kidney

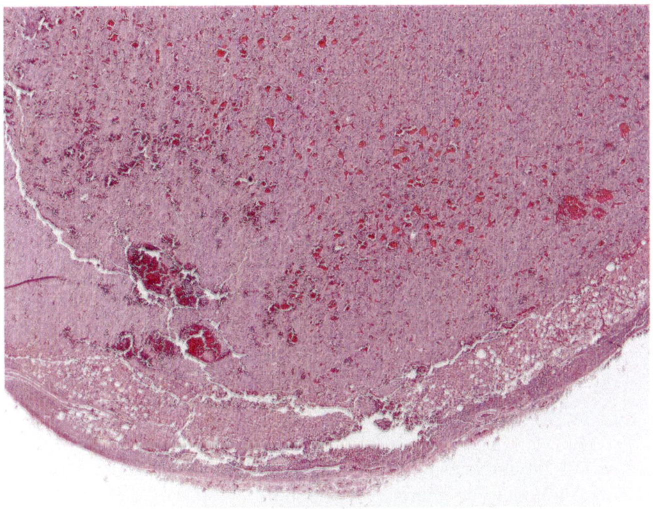

Fig. 8.68 Pheochromocytoma, cotton-top tamarin. The remnant adrenal gland (asterisk) is compressed by a large, unencapsulated neoplasm composed of nests and pseudorosettes of neoplastic enterochromaffin cells

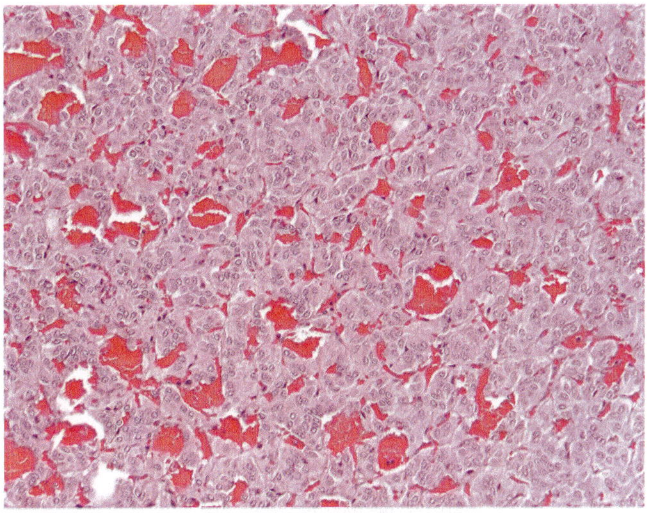

Fig. 8.69 Pheochromocytoma, cotton-top tamarin. Higher magnification of nests and pseudorosettes in a pheochromocytoma

8.9.2 Pituitary Tumors

The majority of pituitary tumors in nonhuman primates are benign tumors (adenomas) of the adenohypophysis (pars distalis). Tumors have been reported spontaneously and following exposure to radiation. Pituitary adenomas are intracranial, extraaxial masses arising in the sellar or suprasellar region (Figs. 8.70 and 8.71). Grossly, these tumors are soft, tan to red, solid to cystic and are typically well demarcated from the surrounding neuroparenchyma although they may exhibit microscopic invasion. Large tumors can cause a mass effect. Histologically, neoplastic cells have variable amounts of granular eosinophilic to basophilic cytoplasm and form nests, packets, or sheets supported by a delicate highly vascularized fibrous connective tissue stroma. Perivascular pseudorosettes and acini formation are features of some tumors (Figs. 8.72, 8.73, 8.74, and 8.75). Pituitary carcinomas, categorized as such by leptomeningeal or systemic spread, are rare in nonhuman primates. Pituitary tumors are subclassified by immunohistochemistry based on their hormonal expression profile using antibodies targeting thyroid-stimulating hormone (TSH) (Fig. 8.76), adrenocorticotropic hormone (ACTH), growth hormone (GH), luteinizing hormone (LH), follicle-stimulating hormone (FSH), and prolactin (PL). A high incidence of lactotroph adenomas has been previously reported in cynomolgus macaques [24]. Hormonally active pituitary adenomas may cause galactorrhea and thyroid or adrenal gland atrophy [25].

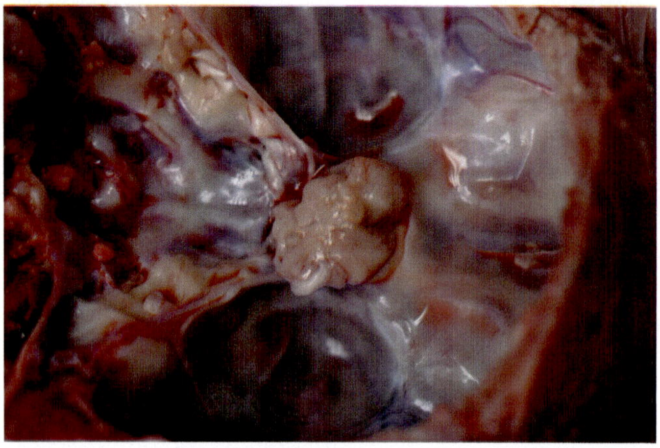

Fig. 8.70 Pituitary adenoma, rhesus macaque. An expansile, tan, intracranial, extraaxial mass effaces the pituitary gland and bulges from the sella turcica. This tumor was an incidental finding at necropsy in a 7-year-old female. It was immunohistochemically determined to be of thyrotroph origin

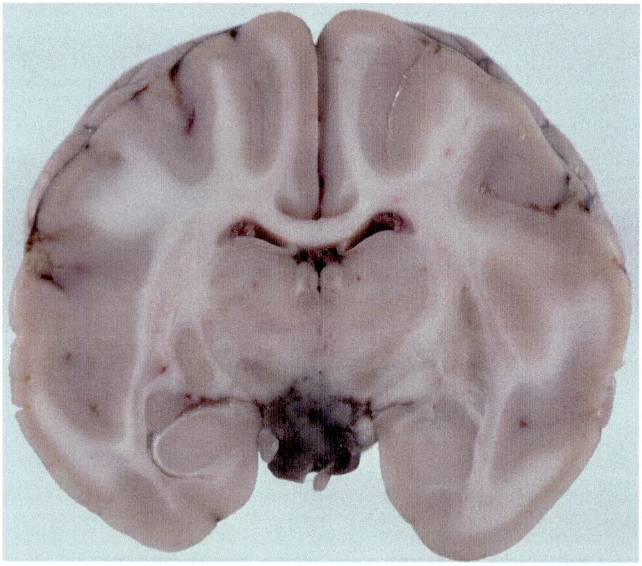

Fig. 8.71 Pituitary adenoma, rhesus macaque. A transverse section of a fixed brain specimen reveals a well-demarcated, dark red tumor, which replaces the pituitary gland and extends dorsally into the hypothalamic region. This tumor was an incidental finding at necropsy in a > 26-year-old animal who was exposed to radiation. It was immunohistochemically determined to be of thyrotroph origin

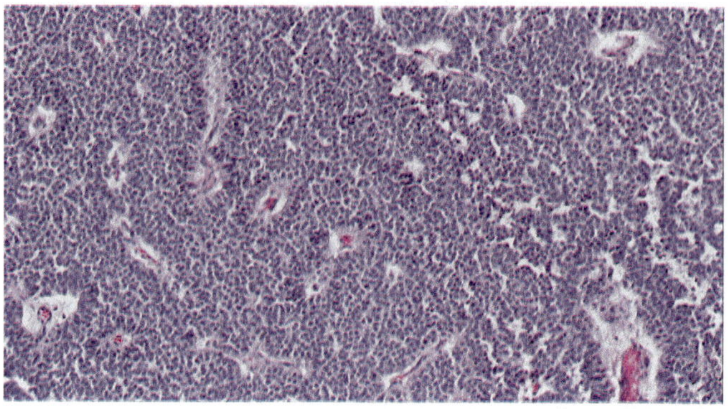

Fig. 8.72 Pituitary adenoma, rhesus macaque. Low-magnification image demonstrates a sheet of closely packed neoplastic cells supported by a delicate, highly vascular stroma

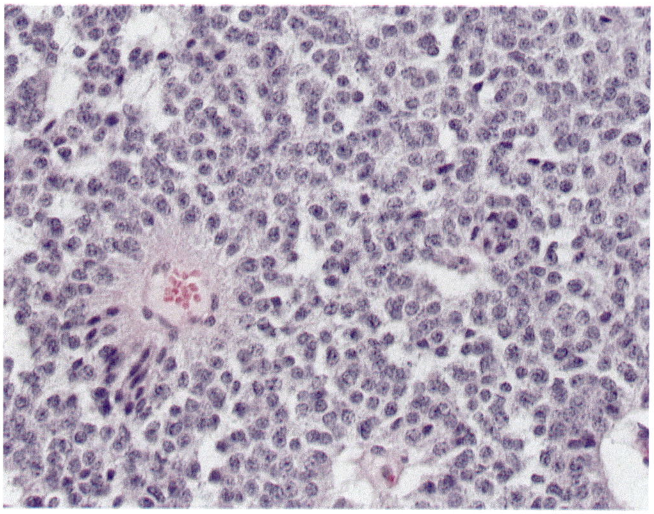

Fig. 8.73 Pituitary adenoma, rhesus macaque. Neoplastic cells have moderate amounts of granular eosinophilic cytoplasm. Cells palisade around a blood vessel forming a pseudorosette

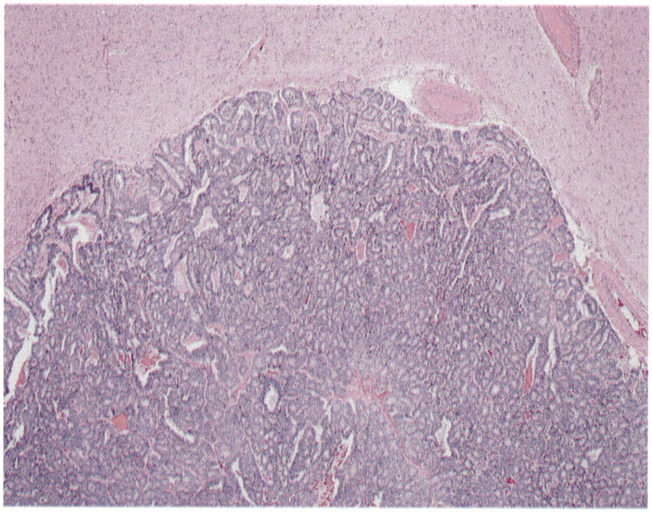

Fig. 8.74 Pituitary adenoma, rhesus macaque. Neoplastic cells form closely packed tubules and acini

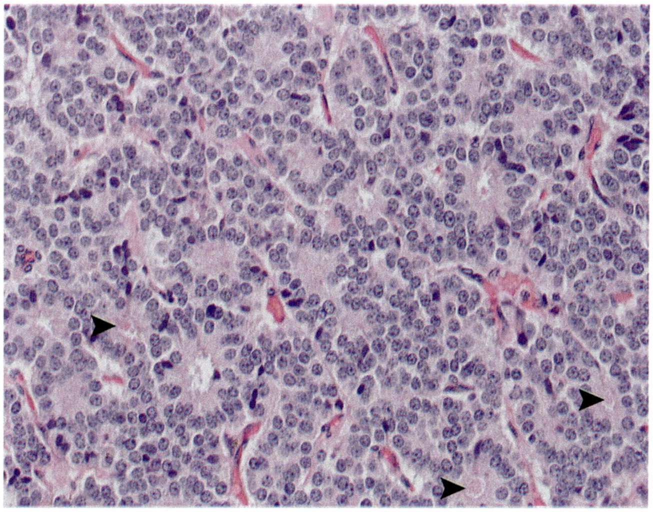

Fig. 8.75 Pituitary adenoma, rhesus macaque. Higher-magnification image of closely packed acini, which occasionally contain eosinophilic secretory material (arrowheads)

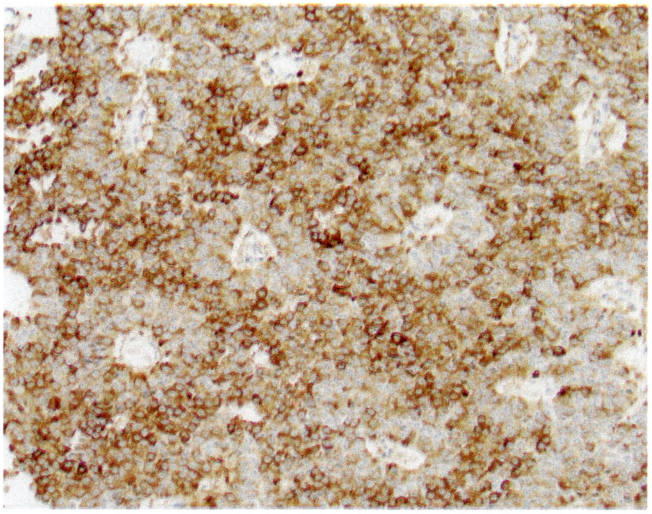

Fig. 8.76 Pituitary adenoma, rhesus macaque. An immunohistochemical stain demonstrates moderate to intense, granular cytoplasmic immunopositivity for thyroid-stimulating hormone (TSH), consistent with a thyrotroph origin

8.9.3 Thyroid Tumors

Thyroid tumors may originate from the follicular cells or C cells, with follicular cell tumors being more common in nonhuman primates. Follicular cell adenomas predominate over malignant tumors and are typically unilateral, solid, nodular, white to tan or brown encapsulated masses that compress the adjacent normal thyroid tissue and may be cystic. Carcinomas have ill-defined margins, often exhibit local tissue invasion and/or metastasis and may have foci of necrosis or hemorrhage. Histologically, follicular cell tumors may exhibit a variety of growth patterns including papillary, follicular and solid (Figs. 8.77, 8.78, 8.79, 8.80, and 8.81).

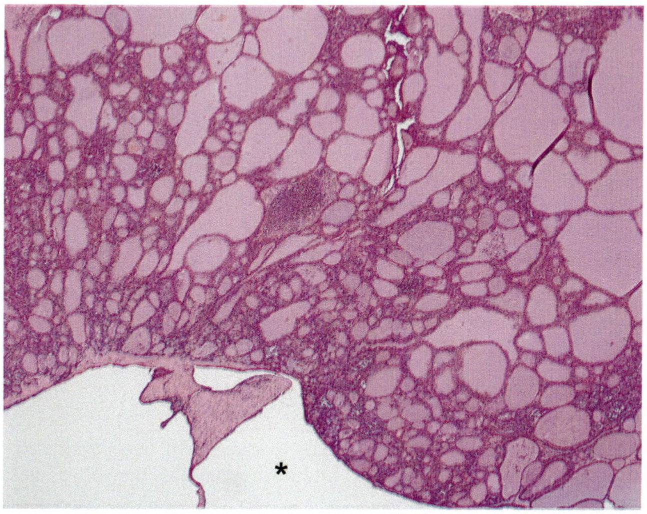

Fig. 8.77 Thyroid cystadenoma, rhesus macaque. A nodular, thinly encapsulated tumor is composed of variably sized follicles filled with colloid. The tumor was contained within a cystic space (*)

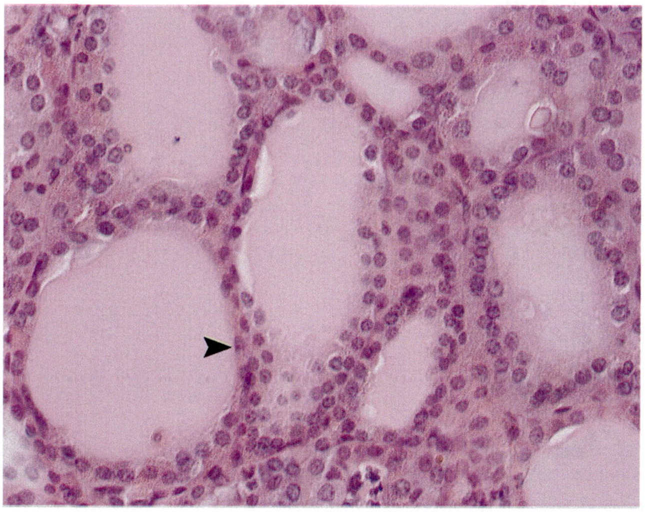

Fig. 8.78 Thyroid cystadenoma, rhesus macaque. High magnification of the benign thyroid tumor showing variably sized colloid-filled follicles lined by a single layer of cuboidal epithelial cells (arrowhead) with abundant, granular to globular cytoplasm

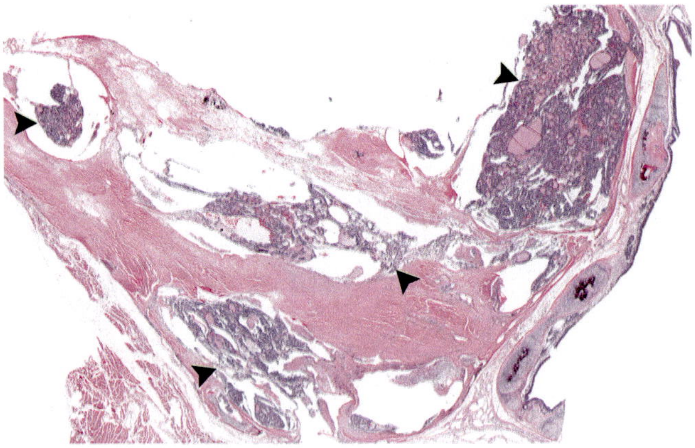

Fig. 8.79 Thyroid carcinoma, rhesus macaque. An unencapsulated, malignant neoplasm (arrowheads) multifocally invades the soft tissues surrounding the trachea

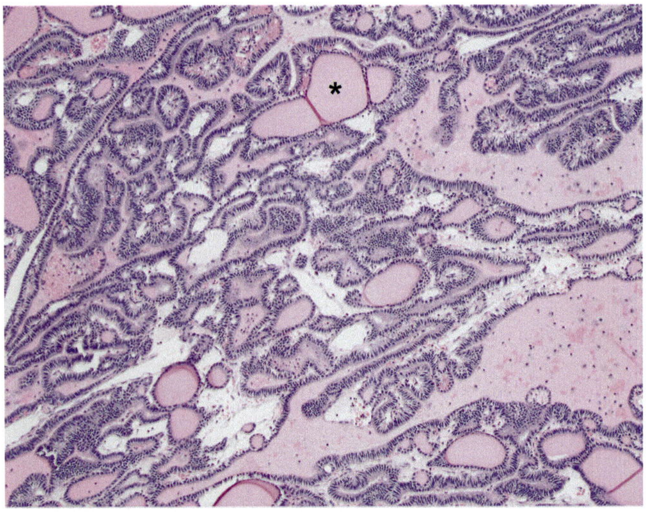

Fig. 8.80 Thyroid carcinoma, rhesus macaque. Neoplastic thyroid follicular cells have a papillary growth pattern with variably sized follicles and colloid production (*)

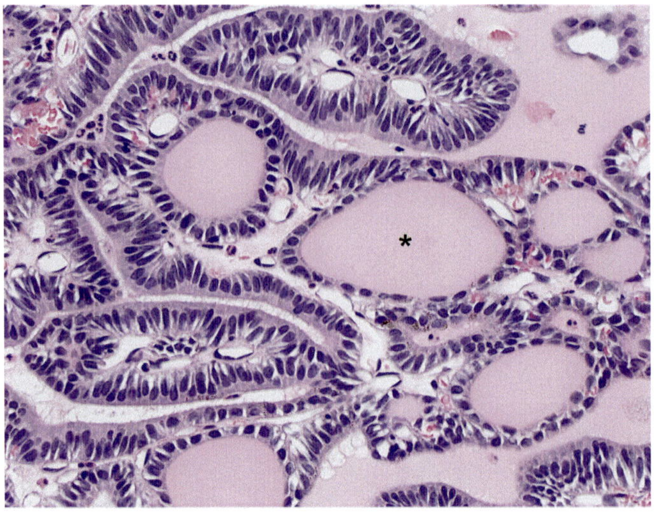

Fig. 8.81 Thyroid carcinoma, rhesus macaque. Papillary projections are lined by a single layer of densely packed cuboidal to columnar follicular cells and supported by a thin fibrovascular core. Numerous follicles contain colloid (*)

8.9.4 Islet Tumors

The majority of pancreatic islet tumors are benign in nonhuman primates. As in other species, many of these tumors arise from the insulin-producing beta cells. Adenomas are tan to red, variably encapsulated, nodular tumors that compress the surrounding tissue (Fig. 8.82). Neoplastic cells form packets, nests, trabeculae, and sheets of monomorphic cells with abundant, granular cytoplasm, and little to no mitotic activity (Fig. 8.83). Malignant islet cell neoplasms, which occur rarely, are characterized by local invasion and/or metastasis. Immunohistochemistry may assist in determining the cell type of origin using antibodies targeting the islet cell hormones insulin (Figs. 8.84 and 8.85), glucagon, somatostatin, gastrin, pancreatic polypeptide, and vasoactive intestinal peptide. Antemortem blood hormone levels are necessary to determine the functionality of these tumors; animals with functional (insulin-secreting) insulinomas are often hypoglycemic.

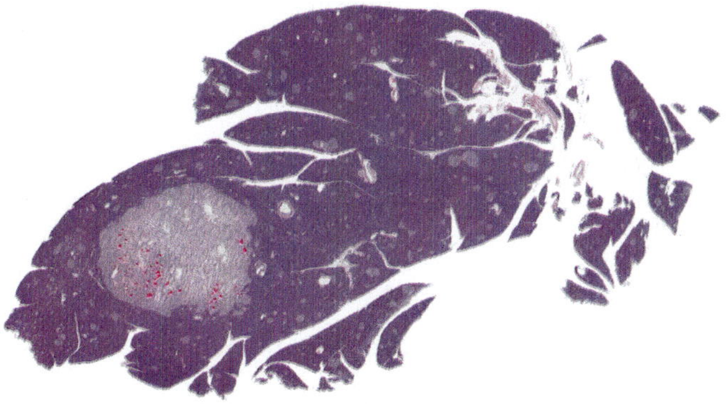

Fig. 8.82 Islet adenoma, rhesus macaque. A nodular, well-demarcated islet cell tumor expands the pancreatic lobule and compresses the surrounding tissue

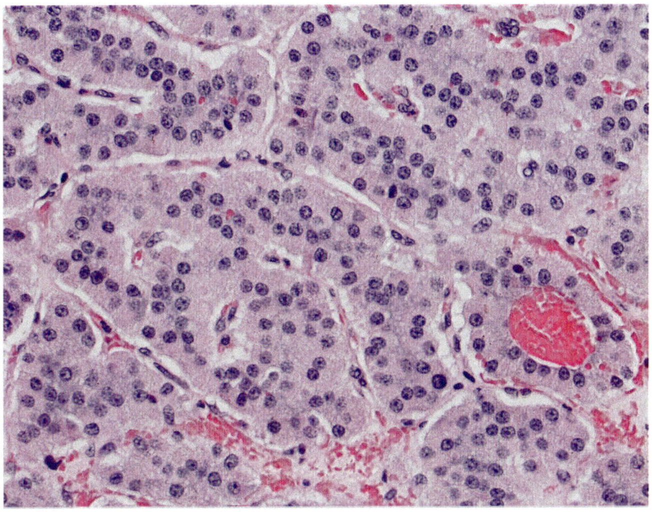

Fig. 8.83 Islet adenoma, rhesus macaque. Higher magnification of islet cell adenoma showing arrangement of neoplastic cells in trabeculae and nests supported by a fibrovascular stroma. Neoplastic cells are monomorphic with abundant eosinophilic, finely granular to vacuolated cytoplasm

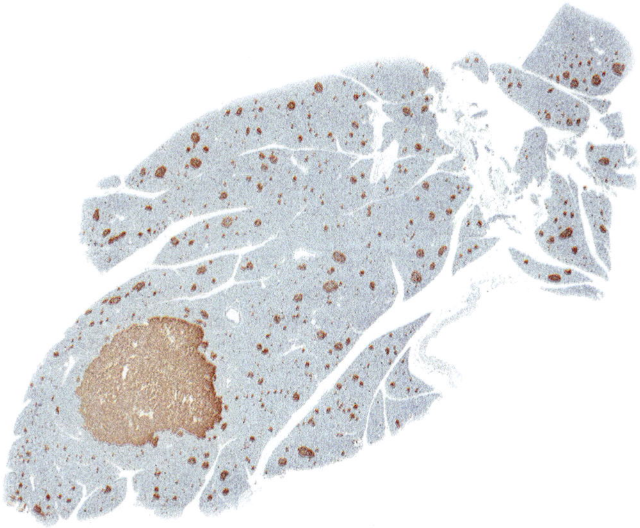

Fig. 8.84 Beta cell adenoma, rhesus macaque. Immunohistochemistry using an antibody directed against insulin demonstrates strong diffuse cytoplasmic immunopositivity for this beta cell marker. The surrounding preexisting islets of Langerhans are also diffusely immunopositive, serving as an internal positive control

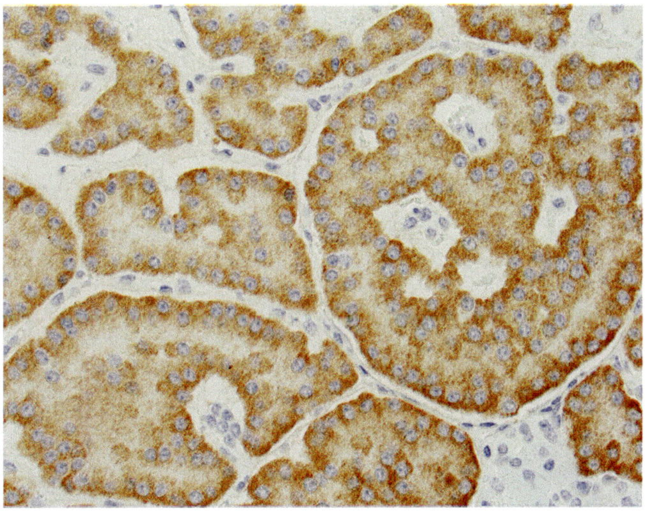

Fig. 8.85 Beta cell adenoma, rhesus macaque. Higher magnification of the insulin immunohistochemical stain demonstrates the strong, diffuse cytoplasmic immunopositivity of neoplastic cells confirms beta cell origin for this tumor

References

1. Miller AD. Neoplasia and proliferative disorders of nonhuman primates. In: Abee CR, Mansfield K, Tardif S, Morris T, editors. Nonhuman primates in biomedical research. 2: diseases. 2nd ed. London: Elsevier; 2012. p. 325–56.
2. Houff SA, London WT, Zu Rhein GM, Padgett BL, Walker DL, Sever JL. New world primates as a model of viral-induced astrocytomas. Prog Clin Biol Res. 1983;105:223–6.
3. Hurley JP, Ilyinskii PO, Horvath CJ, Simon MA. A malignant astrocytoma containing simian virus 40 DNA in a macaque infected with simian immunodeficiency virus. J Med Primatol. 1997;26(3): 172–80.
4. Hubbard GB. Nonhuman primate dermatology. Vet Clin North Am Exot Anim Pract. 2001;4(2):573–83.
5. Simmons HA. Age-associated pathology in rhesus macaques (Macaca mulatta). Vet Pathol. 2016;53(2):399–416.
6. Beck AP, Magden ER, Buchl SJ, Baze WB. Malignant neoplasia of the sex skin in 2 chimpanzees (Pan troglodytes). Comp Med. 2016;66(2):154–61.
7. Haddad JL, Dick EJ Jr, Guardado-Mendoza R, Hubbard GB. Spontaneous squamous cell carcinomas in 13 baboons, a first report in a spider monkey, and a review of the non-human primate literature. J Med Primatol. 2009;38(3):175–86.
8. Harbison CE, Taheri F, Knight H, Miller AD. Immunohistochemical characterization of large intestinal adenocarcinoma in the rhesus macaque (Macaca mulatta). Vet Pathol. 2015;52(4):732–40.
9. Miller AD, Kramer JA, Lin KC, Knight H, Martinot A, Mansfield KG. Small intestinal adenocarcinoma in common marmosets (Callithrix jacchus). Vet Pathol. 2010;47(5):969–76.
10. Bommineni YR, Dick EJ Jr, Hubbard GB. Gastrointestinal stromal tumors in a baboon, a spider monkey, and a chimpanzee and a review of the literature. J Med Primatol. 2009;38(3):199–203.
11. Banerjee M, Lowenstine LJ, Munn RJ. Gastric stromal tumors in two rhesus macaques (Macaca mulatta). Vet Pathol. 1991;28(1): 30–6.
12. DePaoli A, McClure HM. Gastrointestinal neoplasms in nonhuman primates: a review and report of eleven new cases. Vet Pathol Suppl. 1982;19(Suppl 7):104–25.
13. Ohgaki H, Hasegawa H, Kusama K, Morino K, Matsukura N, Sato S, et al. Induction of gastric carcinomas in nonhuman primates by N-ethyl-N′-nitro-N-nitrosoguanidine. J Natl Cancer Inst. 1986;77 (1):179–86.
14. Klumpp SA, McClure HM. Carcinomas, gastrointestinal tract. In: Jones TC, Mohr U, Hunt RD, editors. Nonhuman primates. Berlin: Springer; 1993. p. 53–60.
15. Sharma P, Cohen JK, Paul KS, Courtney CL, Johnson ZP, Anderson DC. Spontaneous gastric carcinomas in sooty mangabeys (Cercocebus atys). Comp Med. 2011;61(6):527–31.
16. Esmail MY, Bacon R, Swennes AG, Feng Y, Shen Z, Garcia A, Sharma P, Cohen J, Fox JG. Helicobacter species identified in captive sooty Mangabeys (Cercocebus atys) with metastatic gastric adenocarcinoma. Helicobacter. 2016;21(3):175–85.
17. Olstad KJ, Kasantikul T, Kiupel M, Reader JR. Mixed malignant pancreatic tumour in a rhesus macaque (Macaca mulatta). J Comp Pathol. 2019;167:46–9.
18. Bouwens L. Cytokeratins and cell differentiation in the pancreas. J Pathol. 1998;184(3):234–9.
19. Cline J, Wood C, Vidal J, Tarara R, Buse E, Weinbauer G, et al. Selected background findings and interpretation of common lesions in the female reproductive system in macaques. Toxicol Pathol. 2008;36(7):142s–63s.
20. Marr-Belvin AK, Bailey CC, Knight HL, Klumpp SA, Westmoreland SV, Miller AD. Ovarian pathology in rhesus macaques: a 12-year retrospective. J Med Primatol. 2010;39(3): 170–6.
21. Kirejczyk S, Pinelli C, Gonzalez O, Kumar S, Dick E Jr, Gumber S. Urogenital lesions in nonhuman primates at 2 National Primate Research Centers. Vet Pathol. 2021;58(1):147–60.
22. Moch H, Cubilla AL, Humphrey PA, Reuter VE, Ulbright TM. The 2016 WHO Classification of Tumours of the urinary system and male genital organs-part a: renal, penile, and testicular tumours. Eur Urol. 2016;70(1):93–105.
23. Miller AD, Masek-Hammerman K, Dalecki K, Mansfield KG, Westmoreland SV. Histologic and immunohistochemical characterization of pheochromocytoma in 6 cotton-top tamarins (Saguinus oedipus). Vet Pathol. 2009;46(6):1221–9.
24. Remick AK, Wood CE, Cann JA, Gee MK, Feiste EA, Kock ND, et al. Histologic and immunohistochemical characterization of spontaneous pituitary adenomas in fourteen cynomolgus macaques (Macaca fascicularis). Vet Pathol. 2006;43(4):484–93.
25. Daviau JS, Trupkiewicz JG. Pituitary adenoma with galactorrhea in an adult male cynomolgus macaque (Macaca fascicularis). Contemp Top Lab Anim Sci. 2001;40(5):57–9.

Index

I. Kondova - Perseng et al. (eds.), *Atlas of Diagnostic Pathology in Nonhuman Primates*,
https://doi.org/10.1007/978-3-031-41280-6